LABORATORY SAFETY
FOR CHEMISTRY STUDENTS

LABORATORY SAFETY FOR CHEMISTRY STUDENTS

ROBERT H. HILL, JR.
DAVID C. FINSTER

A JOHN WILEY & SONS, INC., PUBLICATION

Published by John Wiley & Sons, Inc., Hoboken, New Jersey
Published simultaneously in Canada

Limit of Liability/Disclaimer of Warranty: While the publisher and author have used their best efforts in preparing this book, they make no representations or warranties with respect to the accuracy or completeness of the contents of this book and specifically disclaim any implied warranties of merchantability or fitness for a particular purpose. No warranty may be created or extended by sales representatives or written sales materials. The advice and strategies contained herein may not be suitable for your situation. You should consult with a professional where appropriate. Neither the publisher nor author shall be liable for any loss of profit or any other commercial damages, including but not limited to special, incidental, consequential, or other damages.

For general information on our other products and services or for technical support, please contact our Customer Care Department within the United States at (800) 762-2974, outside the United States at (317) 572-3993 or fax (317) 572-4002.

Wiley also publishes its books in a variety of electronic formats. Some content that appears in print may not be available in electronic formats. For more information about Wiley products, visit our web site at www.wiley.com.

Library of Congress Cataloging-in-Publication Data:

Hill, Robert H., 1945-
  Laboratory safety for chemistry students / Robert H. Hill, Jr., David C. Finster.
    p. cm.
  Includes index.
  ISBN 978-0-470-34428-6 (pbk.)
  1. Chemical laboratories–Safety measures. I. Finster, David C., 1953- II. Title.
  QD51.H55 2010
  542.028′9–dc22
                        2009052126 h

Printed in the United States of America

10 9 8 7 6

To those who have suffered personal loss, injuries, and even death in laboratory incidents that were preventable. May we use the knowledge from these incidents to teach the next generation of scientists about laboratory and chemical safety.

CONTENTS

CONTENTS

CONTENTS

PREFACE: TO THE STUDENTS

THERE IS probably no single course in "laboratory safety or chemical safety" at your college or university. Why not? Chemistry curricula have developed over many decades with a focus on the main topics of chemistry: organic, inorganic, physical chemistry, analytical chemistry, and (more recently) biochemistry. For decades, the topic of chemical safety was included at the margins of lab courses, mostly taught in a small way as a footnote to various lab experiments and procedures. Some chemists and chemistry teachers were aware of the importance of safety, while many were not. In the late twentieth century, and now even more in the twenty-first century, for a variety of legal, ethical, and educational reasons, the topic of chemical safety has been taught much more, but it is still not considered by most as "mainstream content area" of chemistry. The absence of good resources (a void we hope this book fills) contributed to this stature. In summary, many chemistry faculty simply don't consider instruction in laboratory and chemical safety to be very important—or at least important enough to devote a whole course to the topic.

While this textbook could easily be used as a primary textbook for a course in chemical safety, the authors actually strongly prefer that it be used instead throughout the curriculum. We believe that safety instruction is so important that it should be included in *all chemistry laboratory courses*. Additionally, the small "bites" of lab safety included among the 70 sections used separately over an extended four-year period provide constant reinforcement of the importance of safety that nurtures a strong safety ethic. This book has been written with that use in mind.

How so? As you will see, the eight chapters in the book are "layered" in three tiers, with a variety to topics suited to introductory, intermediate, and advanced courses. Each section presents information on a "need to know" basis. For example, there's actually a lot to know about wearing gloves in labs, but you don't need to know everything right away. The first section about gloves is written for introductory courses; a later section is written for organic and advanced students. The same is true for eye protection and for chemical hoods. In this regard, the book is structured unlike any other college textbook you've ever seen. It really is a book that will last for four years (and beyond).

We expect that most of the sections in this book will be tied to various experiments that you are conducting in labs. Again, let's learn what we need to learn on a "need to know" basis. Working with flammable chemicals? Read about solvents and fires. Working with a strong acid or oxidizing agent? Read about corrosives. Worried about lab emergencies or lab incidents? Read about emergency response. This may be the most practical textbook you use in college!

Why should you learn about safety? Well, to stay safe, of course, in the laboratory. This reason alone is enough, but there are additional advantages to knowing about safety. First, it's cheap. Accidents always cost more money than whatever is spent on safety equipment and materials that help prevent these incidents. Second, being safe prevents injuries, damage to health, perhaps even death, and these outcomes have costs that obviously go beyond money. Third, it's environmentally responsible. Knowing how to use chemicals and dispose of wastes legally and appropriately is being environmentally conscious (in a way, frankly, that the chemical industry was not for many decades in the twentieth century). Fourth, you develop habits that will make you a valuable employee someday. Chemical companies now understand, better than many colleges and universities, that being safe is the soundest financial practice a company can adopt. And as more laws and regulations have been developed over the past several decades, employers and employees really have no choice about many aspects of laboratory safety. Your understanding of this situation, upon graduation, will make you an attractive candidate for a job.

While much of this book is very practical and "informational" in nature, some early sections discuss the issue of one's mental attitude about safety, which may seem more philosophical in nature at first. But, in reality, adopting a positive attitude about safety is *the* most important, practical step you can take to be safe. With this mindset, all other actions in a laboratory are performed only after stopping to think about hazards and risks and the means by which you can stay safe in the lab.

We hope you find this book valuable as part of your chemical education. As chemists, the authors have the same passion for chemistry as do your teachers. Understanding nature through the "filter of chemistry" provides great insight and intrinsic joy to most chemists, in addition to the tremendous power of chemistry to improve the quality of the human condition. We are passionate about safety, too, and hope that your time in the lab is both intellectually rewarding and safe! There is much to learn, as the size of this book indicates, and the book offers not much more than an introduction to most topics. We hope that you continue your "safety education" long after you graduate from college.

Finally, you will notice that each section begins with an "Incident". Stories are powerful, and often memorable, ways to learn a principle or to reveal a danger. We hope you find these incidents useful and we encourage you to share your story about safety with us! Hopefully, the story is a happy one about what "almost happened" (although you will see that most of our incidents are not "near misses"). If you have a story that will help some future student learn from your experience, please contact us at dfinster@wittenberg.edu or roberth_hill@mindspring.com. Maybe your story will be in the next edition of the book! We'd also like to hear how you like the book or have suggestions for improvement. Stay safe!

ROBERT H. HILL, JR.
DAVID C. FINSTER

Atlanta, Georgia
Springfield, Ohio
March 2010

TO THE INSTRUCTOR

Purpose

THE PRINCIPAL purpose of this textbook is to provide a resource that can be used to help teach undergraduate chemistry students the basics of laboratory and chemical safety. This textbook is not designed for a single course but rather its concept is to use short sections in laboratory sessions (or perhaps some lecture sessions) over the four years of undergraduate study. It can be used as a companion text for each laboratory chemistry course throughout the curriculum, including research, using specific sections that fit the topics and hazards of the laboratory experiments.

It is the vision and hope of the authors that if the chemistry academic community has a textbook about laboratory and chemical safety that they will use parts or all of it in the laboratory or classroom curriculum. This book was written from the heart as a result of a passion for laboratory and chemical safety. The authors recognize, as do many others, that there is a need to improve the level of knowledge and education about laboratory and chemical safety among new and upcoming chemists and other laboratory scientists who work in laboratories and handle chemicals and other hazardous materials in their operations.

We believe that laboratory and chemical safety should be integral parts of the entire chemistry educational process, touching virtually all fields of chemistry, since we see laboratory and chemical safety as subdisciplines of the field of chemistry that cross-cuts virtually all areas of chemistry. Thus, teaching safety is a long-term effort that requires attention as each area of chemistry is introduced and advances so that a strong knowledge and positive attitude toward laboratory and chemical safety can be developed. Our approach is to teach laboratory and chemical safety in small sections throughout the chemical education process. This iterative process is practical from a learning point of view and sends the message to students: safety is always important.

Audience

This textbook is written primarily for undergraduate chemistry students, but we believe other laboratory science students, scientists, technicians, and investigators will also find it useful. Many graduate and working chemists will find this book useful since it is likely that they are unfamiliar with the level of laboratory and chemical safety education found in this book. Those working in industrial, government, and other independent laboratory situations will also find this book useful. Although designed as a teaching tool and not a resource text, it can serve in the latter capacity and contains many references to other resources.

Scope

This book is broad in scope since it introduces most areas of laboratory and chemical safety. This book is not a comprehensive treatise on laboratory and chemical safety and it does not go into great detail with specific procedures or methods. It presents various topics on a "need to know" basis, targeting different levels of instruction throughout a chemistry curriculum. This book will help chemists and other scientists use four simple principles of laboratory and chemical safety to:

1. Recognize hazards;
2. Assess the risks of those hazards;
3. Minimize, manage, or control those hazards; and
4. Prepare to respond to emergencies.

We use the acronym RAMP to remind the student of these principles—RAMP up for safety.

Unique Approach and Organization

This is a unique textbook designed to be used throughout the four years of undergraduate study. Topics are targeted toward each level (year) of study by the students over their undergraduate experience. Topically, it is divided into eight chapters, and further into 70 sections for introductory (year 1) intermediate (year 2) and advanced topics (years 3 and 4).

- Chapter 1 Principles, Ethics, and Practices
- Chapter 2 Emergency Response
- Chapter 3 Understanding and Communicating About Laboratory Hazards
- Chapter 4 Recognizing Laboratory Hazards: Toxic Substances and Biological Agents
- Chapter 5 Recognizing Laboratory Hazards: Physical Hazards
- Chapter 6 Risk Assessment
- Chapter 7 Minimizing, Controlling, and Managing Hazards
- Chapter 8 Chemical Management: Inspections, Storage, Wastes, and Security

Each section begins with a preview, a quote, and a laboratory incident that asks "What lessons can be learned from this incident?" This is followed by the text that is relevant to the topic and incident with references that often contain links to the Internet. Dispersed through out the book are *Chemical Connections* that seek to demonstrate how safety uses chemical principles and *Special Topics* that seek to explain relevant topics of interest to a particular section. Each section also concludes with a series of multiple choice questions about the topic.

Safety, like other disciplines, is principle driven. The student must be encouraged to use critical thinking in applying safety principles and practices to conduct chemical work safely and to identify the need for additional information about the safety in operations handling chemicals or other hazardous agents.

How This Book Can Be Used

We anticipate several ways in which the book may be used. It may be used directly by the student and taught by an instructor. However, the authors are well aware of the difficulty of adding more to the curriculum and believe that each section can be used as a prelaboratory assignment session. The student can be directed to go to a web site to take an electronic quiz for each section with results going to the laboratory instructor to ensure that each student has been successful in understanding the basic topics presented in a section before the laboratory session.

More specifically, we anticipate two models for using the sections as prelab assignments:

1. An instructor can assign a reading and electronic quiz, and do little more. This practice alone may represent an improvement in safety instruction, requires virtually no additional work on the part of the instructor and no allocation of class/lab time, and provides some form of assessment of student learning.

2. An instructor can assign a reading and the electronic quiz, and follow this up in a prelab session with discussion of the topic, probably making specific reference to the experiment of the day,

which is likely to be related to the safety topic. The degree to which the instructor elaborates on the topic can be considerable. Discussion questions and "what if" scenarios are easy to develop. The value of the book is that precious lab time is not spent on "covering the basics" and "information transfer." Students will come to the lab with some background knowledge, which allows for a more productive, and likely more sophisticated, discussion of a particular safety topic.

Ultimately, our goal in providing this resource is to minimize, if not eliminate, the activation energy barrier that prevents many faculty from discussing safety more in their classes and labs. The excuse that "there's not enough time" is eliminated when no class or lab time, in the first model above, is used. The excuse that "I'm not trained in safety" is eliminated since the book provides the expertise and thoughtful presentation of the safety topics. The American Chemical Society Committee on Professional Training requires (as stated in the Guidelines and Evaluation Procedures for Bachelor's Degree Programs) the "approved programs should promote a safety-conscious culture in which students understand the concepts of safe laboratory practices and how to apply them." Use of this book meets that learning goal.

Ideally, this book would be purchased in the first year for chemistry majors and used as a supplementary text throughout the entire undergraduate chemistry curriculum. However, the authors recognize that many students in introductory courses are not chemistry majors and will not continue in the chemistry curriculum. Using the Wiley Custom Select option, there is also the opportunity to make single sections of the book available for clustering in faculty—designed packets that are individually suited to particular teachers, courses, and/or campuses. This will be at an attractive price that makes use of the packets reasonably as a supplementary purchase for students. The strategy can be pursued throughout the curriculum, although at some point the purchase of the entire book, particularly for chemistry majors, would seem prudent.

ACKNOWLEDGMENTS

WE THANK all our friends and colleagues for their support and encouragement during the writing of this text; the value of this support is indeterminable. We also recognize those who took a more active role in helping us review the many sections of the book and who provided some stories of incidents and some of our figures. The following list highlights those who have helped us in one or more ways. Their input and unbiased criticism has been invaluable and has helped us create a much better text. We thank:

Janice Ashby, Centers for Disease Control and Prevention

David Ausdemore, Centers for Disease Control and Prevention

Emily Bain, Kenyon College

Kathy Benedict, University of Illinois, Urbana-Champaign

George Bennett, Millikin University

Mark Cesa, INEOS USA, LLC

Debbie Decker, University of California, Davis

Larry Doemeny, ACS Committee on Chemical Safety

Amina El-Ashmawy, Collin County Community College

Harry Elston, Midwest Chemical Safety LLC

Barbara Foster, University of West Virginia

Cheryl Frech, University of Central Oklahoma

Ken Fivizzani, ACS Committee on Chemical Safety

Jean Gaunce, Centers for Disease Control and Prevention

Pete Hanson, Wittenberg University

Dennis Hendershot, Center for Chemical Process Safety

Mary Hill, Memorial University Medical Center

Bill Howard, Centers for Disease Control and Prevention

David Katz, Pima Community College

Neal Langerman, Advanced Chemical Safety

Mark Lassiter, Montreat College

Gary Miessler, St. Olaf College

Larry Needham, Centers for Disease Control and Prevention

Rick Niemeier, National Institute for Occupational Safety and Health

Alice Ottoboni, Retired, California State Department of Public Health

Les Pesterfield, Western Kentucky University

Russ Phifer, WC Environmental, LLC

Gordon Purser, University of Tulsa

Jonathan Richmond, Jonathan Richmond & Associates, Inc.

ACKNOWLEDGMENTS

Joyce Rodriguez, Centers for Disease Control and Prevention

Eileen Segal, Retired

Linda Stroud, Science & Safety Consulting Services

Ralph Stuart, University of Vermont

Erik Talley, Weill Cornell Medical College

Paul Voytas, Wittenberg University

Doug Walters, Environmental and Chemical Safety Educational Institute

Stefan Wawzyniecki, University of Connecticut

Mark Wilson, Centers for Disease Control and Prevention

Frankie Wood-Black, Trihydro

Tim Zauche, University of Wisconsin—Platteville

These people share our enthusiasm and passion for improving laboratory and chemical safety among our new and upcoming scientists.

We also thank the students enrolled in the following courses at Wittenberg University in Spring 2009 who provided feedback for early versions of many sections of the book: Chemistry 162, Chemistry 281, Chemistry 302, Chemistry 321, and Chemistry 372.

And finally, we thank our families without whose support and understanding this work would have not been possible.

R. H. H., Jr.
D. C. F.

ACRONYMS

ACGIH	American Conference of Governmental Industrial Hygienist	EPA	U.S. Environmental Protection Agency
ACS	American Chemical Society	FAS	Fetal alcohol syndrome
AIDS	Acquired immunodeficiency syndrome	FIFRA	Federal Insecticide, Fungicide, and Rodenticide Act
ALARA	As low as reasonable achievable	GFCI	Ground fault circuit interrupters
APHIS	Animal and Plant Health Inspection Service	GHS	Globally Harmonized System of Classification and Labeling of Chemicals
ANSI	American National Standards Institute	HAZWOPER	Hazardous Waste Operations and Emergency Response
ASSE	American Society of Safety Engineers	HBV	Hepatitis B virus
BBP	Blood-borne pathogens	HEPA	High efficiency particulate air
BEI	Biological exposure index	HHS	Department of Health and Human Services
BLEVE	Boiling liquid expanding vapor explosion	HVAC	Heating, vacuum, and air conditioning
BMBL	Biosafety in Microbiological and Biomedical Laboratories	IARC	International Agency for Research on Cancer
BSL	Biosafety Levels	IDLH	Immediately dangerous to life and health
CAS	Chemical Abstracts Service		
CCW	Counterclockwise	IEC	International Electrotechnical Commission
CDC	Centers for Disease Control and Prevention	IR	Infrared
CFR	Code of Federal Regulations	LCSSs	Laboratory Chemical Safety Summaries
CGA	Compressed Gas Association		
CHO	Chemical Hygiene Officer	MRI	Magnetic resonance imaging
CHP	Chemical Hygiene Plan	MSDS	Material Safety Data Sheet
CLIPS	Chemical Laboratory Information Profiles	NFPA	National Fire Protection Association
CNS	Central nervous system	NIH	National Institutes of Health
CPR	Cardiopulmonary resuscitation	NIOSH	National Institute for Occupational Safety and Health
CSB	U.S. Chemical Safety and Hazard Investigation Board		
CW	Clockwise	NMR	Nuclear magnetic resonance
DOL	Department of Labor	NRC	Nuclear Regulatory Commission
DHS	Department of Homeland Security	NSF	National Sanitation Foundation (can also be National Science Foundation)
DOT	Department of Transportation		
ELF	Extremely low frequency		
EMF	Electromagnetic frequency	NTP	National Toxicology Program
EMT	Emergency medical technician	OEL	Occupational exposure limit
EPCRA	Emergency Planning and Community Right-to-Know Act	OJT	On the job training

OSHA	Occupational Safety and Health Administration	STEL	Short-term exposure limit
PEL	Permissible Exposure Limit	SWDA	Safe Water Drinking Act
PHA	Process hazard analysis	TLV	Threshold Limit Value
PSM	Process Safety Management of Highly Hazardous Materials	TSCA	Toxic Substances Control Act
		TWA	Time-weighted average
RCA	Root Cause Analysis	UL	Underwriter's Laboratory
RCRA	Resource Conservation and Recovery Act	USDA	U.S. Department of Agriculture
		UV	Ultraviolet
RAMP	Recognize, Assess, Minimize, Prepare	VLF	Very low frequency
		WHO	World Health Organization
REL	Recommended exposure limit	RSO	Radiation Safety Officer
RF	Radio frequency	RSP	Radiation Safety Program
SDS	Safety Data Sheet	RSC	Radiation Safety Committee
SI	Système International	SCBA	Self-contained breathing apparatus

CHAPTER 1
PRINCIPLES, ETHICS, AND PRACTICES

THERE ARE hundreds of "things to learn about safety." But, without some framework into which to place all of this information, it just becomes rules and guidelines that are too easy to forget over time. Understanding the reasons *why* safety is important is crucial to understanding the reasons that stand behind the facts, rules, and guidelines. This chapter introduces the *safety ethic* and the *four principles of safety* that reappear in most sections of the book. We also present three sections about green chemistry, which implies a strong connection between this new way to think about environmentally responsible chemistry and chemical safety. Finally, the Level 3 topics discuss various aspects of safety that may seem less relevant to you, a student, but become very important in your life as a chemist after college when you move into the working world of chemistry.

INTRODUCTORY

1.1.1   The Four Principles of Safety   Discusses the importance of safety in the laboratory and introduces the *four principles of safety* and *the student safety ethic*.

1.1.2   What Is Green Chemistry?   Introduces the fundamental tenets of green chemistry and discusses how these principles can used in the college chemistry curriculum.

INTERMEDIATE

1.2.1   Rethinking Safety: Learning from Laboratory Incidents   Describes how you can learn lessons from a critical analysis of an incident using the "five whys."

1.2.2   Green Chemistry in the Organic Curriculum   Discusses applications of green chemistry in organic chemistry laboratories.

ADVANCED

1.3.1   Fostering a Safety Culture   Discusses approaches to establishing and promoting a strong safety culture among those people who are working under your supervision or leadership.

1.3.2   Employers' Expectations of Safety Skills for New Chemists   Provides suggestions about what employers might expect in safety skills for new employees (new graduates) who begin working in their facilities.

1.3.3 Laws and Regulations Pertaining to Safety   Provides a brief overview of the laws and regulations pertaining to safety and chemicals in the laboratory.

1.3.4 Green Chemistry—The Big Picture   Reviews selected principles of green chemistry that apply most directly to laboratory safety.

1.1.1

THE FOUR PRINCIPLES OF SAFETY

Preview   This section discusses the importance of safety in the laboratory and introduces the *four principles of safety* and *the student safety ethic*.

> *The greatest discovery of my generation is that a human being can alter his life by altering his attitudes of mind.*

William James, American physician, philosopher, and psychologist (1842–1910)[1]

INCIDENT 1.1.1.1   MIXING ACID AND WATER[2]

John was doing an experiment that required the use of dilute sulfuric acid. The instructor said that students should mix 1 part of concentrated sulfuric acid with 4 parts of water and that everyone should always *add acid to water* and not water to acid. John was not paying attention when the instructions were given and he added water to acid. There was a violent popping noise, the beaker became hot, and a mist formed over the solution, and some solution splattered out onto his skin and his partner's skin.

What lessons can be learned from this incident?

Hazards and Risks

This book was written to teach you how to work safely in the laboratory. To be safe in the laboratory or elsewhere, you need to do only four things:

- *R*ecognize hazards.
- *A*ssess the risks of hazards.
- *M*inimize the risks of hazards.
- *P*repare for emergencies.

We will use the acronym RAMP as a mnemonic guide to remember: *recognize, assess, minimize, prepare*. This will be a recurring theme in this book. Let's see how these four steps make sense by considering various aspects of safety.

 Safety is freedom from danger, injury, or damage. Being safe requires actions by *you* and by others. When you decide to adopt safety as an integral part of your college laboratory experiences it means that you always seek to do those things that prevent incidents that might cause injury and harm. This sounds easy, but when doing chemistry experiments in laboratories, it is often easier to get "caught up" in the procedures and trying to understand what you are doing and to forget about making sure you're doing it safely. Throughout this book, each section will start with a real incident that describes a dangerous episode in a laboratory. All of the people in these incidents are "just like you." They had some safety education (just as you will experience) but were not thinking about safety when something bad happened.

 Let's explore some basic ideas about risks and hazards and how to prevent incidents in laboratories.

 A *hazard* is a potential source of danger or harm. The word *potential* means something that is capable of being dangerous or harmful. Many chemicals may have inherent hazardous properties and

Laboratory Safety for Chemistry Students, by Robert H. Hill, Jr. and David C. Finster
Copyright © 2010 John Wiley & Sons, Inc.

these hazardous properties never change. The practice of safety is really about minimizing, managing, or controlling these hazards. You'll learn a great deal in this book and in your chemistry courses about various categories of chemicals and the hazards they pose.

Risk is the probability of suffering harm from being exposed to a hazard or unsafe situation. The level of risk depends on many things beyond the inherent hazard of a chemical. For example, the amount of the chemical, the form it is in (gas, liquid, or solid), and how you handle the chemical all affect the level of risk.

Exposure means coming in direct contact with a hazard or chemical in a fashion that causes injury or harm. A dose is the amount that you might ingest, breathe, or spill on your skin. An exposure might also arise from a fire or explosion. The dose, the length of exposure, and the path of exposure play significant roles in the extent of harm. The important thing to remember is that we want to minimize or eliminate exposure to hazards.

Hazards Are a Part of Our World!

Many things present hazards or have hazardous properties, but we have learned to use them safely every day. Often, the very properties that make chemicals useful are also the same properties that make them hazardous (see Figure 1.1.1.1). Just because something is hazardous does not mean that we would want to stop using it. In fact, our lives and our comfort often depend on the use of chemicals and equipment with hazardous properties. Let's look at an example of one of the most common chemical hazards that most of us encounter frequently.

Gasoline is extremely flammable and under the right conditions it can easily catch fire or even explode. Yet we use gasoline in our cars everyday without experiencing its potential adverse effects because we have *recognized* its hazardous properties, *assessed* how we could be exposed, developed methods to effectively *minimize* or control this hazard, and learned how to *prepare* for and handle emergencies with gasoline. We go to gas stations that store thousands of gallons of gasoline in tanks under our feet, we pump gallons of gasoline into our cars, and we routinely drive around with gallons of gasoline in the car's gas tank near us. The very properties that make gasoline hazardous are the same properties that make it useful. We have learned how to use this hazardous substance safely.

Can you think of materials or things that have hazardous properties that are very useful to us but can present a hazard to us if we are careless with them or mishandle or misuse them? Do you think that most 21st century citizens would approve of pumping a high-pressure, explosive chemical that can deprive you of oxygen in high concentrations into their homes? Probably not! But many of us do this when we use natural gas. What about electricity? What is the hazard? What is the overall risk? Common hazards are generally viewed as being less risky than uncommon hazards. Most hazards in the laboratory, even though they are "uncommon" at first, are no more dangerous than these common hazards encountered in our daily lives.

FIGURE 1.1.1.1 A Gallon of Gasoline. There are many hazardous chemicals in and around our homes, although we might treat them as more hazardous in a lab than in the home since we are so used to them at home.

Safety from the Experts

Three factors have been identified by Geller that contribute to safety: (1) *environmental factors* including facilities, location, equipment, procedures, and standards; (2) *person factors* including attitude, beliefs, personality, knowledge, skills, and abilities; and (3) *behavior factors* including safe and risky practices.[3] These factors are interconnected so that each factor influences others. Suppose, for example, that you see someone fall over something that was out of place, an environmental factor. You think to yourself (a personal factor), "I need to be more careful about things on the floor that I might trip over." When you see a wastepaper basket out of place in an aisle, you move it to its place out of the way (a behavior factor).

Being safe requires attention to all three factors since they have significant influence on your safety. If your college laboratory is safe and you observe other students and teachers there working safely by observing safety rules and safe practices, then you will begin to adopt an attitude that "safety is important in the laboratories and I want to be safe, too."

How Do We Learn Safety?

Safety is an empirical discipline. This means that we often learned how to be safe from past mistakes and incidents. Experience can teach us a lot, but if you learn safety by making a lot of your own mistakes, you may not survive for long! Most of us do not want to have to personally experience fires, explosions, toxic exposures, or other potentially dangerous incidents. Instead, we should be learning safety guidelines that have developed from the adverse experiences of others.

We can reduce our exposure to hazards through continuous and diligent efforts to include considerations about safety in our daily decisions. This simply involves thinking about safety and taking steps to prevent incidents, especially in the laboratory. Unfortunately, there is rarely positive feedback for being safe, but there is often negative feedback for *not* being safe. The consequence of exposure to unnecessary risk is likely to be injury or harm to you or others.

It's All About Minimizing Risk!

To maintain a safe lab environment, it is critical that everyone minimizes and/or eliminates the risk of exposure to hazards. We do this everyday with precautions that we have learned to prevent or reduce injury and harm. Table 1.1.1.1 shows common risk factors associated with these risks and proven risk reduction actions.[4-6] For example, if we wear seat belts, we are less likely to sustain life-threatening injuries in a collision. The phrase "less likely" is important because wearing a seat belt does not guarantee that you will be injury-free if you are in a collision, but it does mean that your chances of

TABLE 1.1.1.1   Common Risks, Risk Factors, and Risk Reduction Actions[4-6]

Common risks	Risk factors	Risk reduction actions
Cardiovascular disease, death from—heart disease and stroke, 1st and 3rd leading causes of death	High blood pressure, high blood cholesterol	A 12–13 point reduction in blood pressure decreases cardiovascular diseases by 25%; a 10% decrease in total blood cholesterol decreases coronary disease up to 30%
Lung cancer, death from	Cigarette smoking—early death: 22 times higher for male smokers; 12 times higher for female smokers	Stop cigarette smoking
Vehicle accidents, death from	No seat belt use; alcohol-impaired driving	Seat belt use laws; seat belt enforcement laws; sobriety checkpoints; reducing blood alcohol concentration to 0.08%; minimum legal drinking age laws

survival are better if you wear a seat belt. And they will be even better if you observe other public safety rules, such as driving within the speed limit and using defensive driving techniques.

Similarly, you will be required to wear splash goggles when working in the laboratory in order to reduce the risk of injury to your eyes from chemical splashes. The goggles don't guarantee that you will not be injured but experience has shown that wearing splash goggles significantly reduces the chance of injury to your eyes. And, when combined with other safety measures like using a hood or face shield or wearing gloves, risks are decreased even more.

Taking Unnecessary Risks: The Cause of Most Incidents

An important cause of many injuries and incidents is taking unnecessary risks; this is sometimes called practicing at-risk behavior.[7] Unnecessary risks are actions that violate safety principles, safety rules, and safe practices. If you can avoid taking unnecessary risks, you will likely prevent or reduce chances of an incident.

Why would someone take unnecessary risks? Someone may make a willful decision to violate rules, may unconsciously act based on past experiences, or may be unaware of a risk. Examples of unnecessary risk behavior are speeding, not wearing a seat belt, smoking cigarettes, overeating, not exercising, using illicit drugs, or skipping safety steps, such as deciding not to wear protective equipment needed for the job. In the laboratory such behavior includes not wearing safety goggles, eating or drinking in the laboratory, wearing inappropriate clothing, or taking unnecessary chances with hazards.

Avoiding unnecessary risks is not always easy. Often it means resisting human nature that causes us to do things that are convenient, comfortable, or expedient. Some people have learned through experience that if they cut corners they can save time and resources, and they sometimes can do this without negative consequence. That is, they get away with violating safety principles and they begin to think that it is okay to do something that is inherently unsafe or dangerous. This becomes a bad habit. As they get more careless and violate more rules and principles, they increase their chances of having an incident.

Types of Laboratories: Teaching, Research, and Industry

In the first few years of college, most laboratories in science courses are "cookbook" laboratories. By this we mean that students read and follow a set of laboratory instructions in order to perform an experiment to collect data. There are many variations on this theme but beginning students mostly "just follow instructions." Some experiments might involve elements of experimental design but this will be fairly limited in your early science laboratory courses. Because new students don't have to design experiments, they also don't have to think about designing good safety procedures into laboratory experiments because that has already been done by the author of the procedure. Unfortunately, in a sense, students get trained *not* to think as much about safety because they assume, with justification, that laboratory experiments that they are doing are already "safe."

In some upper-level courses and in undergraduate research laboratories, students start to participate in the design of laboratory experiments, usually under the watchful eye of an instructor or research mentor. For these experiences, it is very important to consider the safety of a procedure in the process of designing a new experiment.

Chemists with undergraduate degrees who enter a graduate program or work in an industrial laboratory setting become far more independent with regard to the design and implementation of laboratory experiments, so the burden for designing safe experiments rests heavily with the chemist in the laboratory. Many incidents occur in laboratories with relatively new chemists or graduate students running new experiments under situations where a thorough assessment of the hazards and risks was not undertaken.

It is the goal of this book to help educate chemistry students to work and function safely in laboratories. We will not focus much on the design of experiments in the early sections of each chapter since newer students don't design experiment and we believe that it is best to carefully focus on the hazards for the experiments that are to be encountered. In later sections in each chapter we will address more "advanced" topics, including considering safety in the design of an experiment and more in-depth discussions of hazards and other safety topics.

The Four Principles of Safety

We introduced RAMP (*recognize, assess, minimize, prepare*) above. Let's look at each of these steps using the language of hazards and risks. (See Figure 1.1.1.2.)

The first principle, to *recognize the hazards of chemicals, equipment, and procedures*, requires that you know and recognize the hazards of the chemicals that you are using. Sounds simple, right? Well, depending on your knowledge, it can be a very challenging process. There are millions of chemicals, of course, and knowing the hazards of all of them is not possible. But, we are helped in two ways.

First, most chemicals will fall into one (or two) of a handful of categories that have generally known hazards. To understand these hazards you must first understand the terminology and information that describes these various chemical properties. What does "flammable" mean? What is "toxic" or "corrosive"? And, how will you know if a chemical has any of these properties?

Second, and more specifically, "getting to know your chemical" requires that you review and understand available information about its hazards, such as container labels, Material Safety Data Sheets (MSDSs, see Section 3.1.3), reference books, online hazard information, and talking with experienced people.

Chapters 3–5 address these issues about hazards of chemicals in laboratories.

The second principle, *assess risks of hazards associated with exposures and procedures*, is perhaps the most important of all the principles.[4] This requires that you consider what kind of exposure to various chemicals could or will occur during a procedure or reaction as well as the risk associated with the use of equipment. Is this reaction exothermic (releasing energy) in a way that might lead to a fire or explosion? Are there any flammable chemicals involved that might pose a fire hazard? What is the chance of some exposure to a toxic chemical?

It is important *not to underestimate risks*, particularly in "familiar" situations.[2] This book and the chemistry courses that you take will help you learn to make good judgment about risks. Chapter 6 discusses this topic in more detail.

The third principle, *minimize risks*, requires careful attention to both the design and execution of an experiment. This requires that you take whatever reasonable steps are necessary to minimize, manage, or eliminate your exposure to a hazard by using good laboratory safety practices. This can only be done after a careful consideration of risk. The key steps in minimizing risk are designing and performing experiments with safety in mind, using personal protective equipment (such as splash goggles) and other safety equipment (such as chemical hoods), and applying good housekeeping practices. Many accidents are caused by sloppy and cluttered work areas. Chapter 7 discusses how to manage and minimize risks in laboratories.

Finally, despite efforts to prevent incidents (accidents) and exposure in the laboratory, it is prudent to prepare for them. Thus, we present the fourth principle: *prepare for emergencies*. What kinds of emergencies can happen in a laboratory? Fires, explosions, exposures to chemicals, personal injuries—all the sorts of hazards that have already been considered! Preparing for emergencies involves knowing what safety equipment is readily available and how to operate it (see Figure 1.1.1.3). You also need to

---

The Four Principles of Safety

Recognize Hazards
Assess Risks
Minimize Hazards
Prepare for Emergencies

FIGURE 1.1.1.2 The Four Principles of Safety. These four principles appear in nearly every section of this book. Memorizing, and using, these ideas whenever you think about safety issues will lead to "incident-free" laboratories.

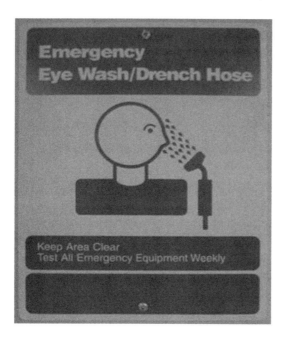

FIGURE 1.1.1.3 Eyewash Safety Sign. There are usually several signs indicating the location of safety equipment in laboratories. These often have pictograms that indicate the use of the safety equipment.

know when and how to exit a laboratory safely and what first aid equipment is available. Since knowing how to handle laboratory emergencies is one of the "first things" that you need to learn about when working in chemistry laboratories, this is the content of Chapter 2.

The Student Safety Ethic

We conclude this introduction to safety with a key element to guide you in the undergraduate laboratory—*the student safety ethic*.[8] In Section 1.3.2 we will introduce the *the safety ethic*, which expands the *student safety ethic* to include a full range of topics for practicing chemists and researchers.

Ethics are principles of right or good conduct. When you work in the laboratory, it will be helpful to have a set of guiding principles relating to safety. Early on, we learn safety ethics from family members and teachers. In college, your instructors will teach about safety and encourage a safety ethic. Later in life, you will learn about safety from your employer, fellow employees, managers, and safety professionals in the workplace. Many employers have strong safety programs to protect their employees as well as their own financial interests and property. And since some employers may not emphasize safety, it is in these environments where your own safety ethic may be the most important or even challenged.

Why do you need strong safety ethics? People with weak safety ethics and a poor education in safety frequently put themselves at higher risks and are more likely to injure themselves or others. And beyond the workplace someday, you will want strong safety ethics to take care of your family and friends.

We encourage you to adopt the student safety ethic: *I work safely, avoid unnecessary risk, and accept responsibility for safety*.[8] This is a simple statement but it has broad implications and defines broad actions to implement (see Figure 1.1.1.4).

- To *work safely* means that you are educated in safety, you continue to learn about safety, you learn to recognize and evaluate hazards, you practice safe procedures, and you maintain a high level of safety awareness.

- To *avoid unnecessary risk* means that you learn to recognize risks and minimize and manage those risks when working in the laboratory.

- To *accept responsibility for safety* as an act of caring for others means that you are responsible and accountable for your safety and for the safety of others. This requires a constant awareness in

The Student Safety Ethic

I work safely,
avoid unnecessary risk, and
accept responsibility for safety

FIGURE 1.1.1.4   The Student Safety Ethic. This safety ethic reflects a mindset of attitude and responsibility that keeps you, and others, safe in the laboratory.

the laboratory of what you are doing and what other students are doing. Instructors and teaching assistants share the responsibility of maintaining safety in undergraduate teaching laboratories but this does not release you from sharing in this responsibility.

Adopting the *student safety ethic* will require a significant effort by you largely because it is easy to assume that whoever designed an experiment did this with safety already in mind and it is very easy to get so focused on trying to understand the experiment itself in a learning environment that you forget to think about safety. And, quite simply, you probably have not yet been taught to constantly think about safety.

This book is designed to progressively teach you many topics about safety over the course of several years. But, it all begins with the right attitude, which is embedded in the *student safety ethic*.

Safety Rules!

We finish this section with a list of basic safety rules for laboratory work. Most of these rules will become better understood as you study various other chapters in this book. However, this is a good place to start so that you can get into a chemistry laboratory and function safely.

TABLE 1.1.1.2   Basic Safety Rules for Laboratories Handling Chemicals

1. Follow instructor and laboratory instruction directions carefully.
2. Wear proper eye protection for use around chemicals at all times in the laboratory.
3. Wear clothing that protects against exposure and provides protection from spills. Wear chemically resistant gloves when prudent to do so.
4. Do not eat, drink, smoke or use smokeless tobacco products, chew gum, apply cosmetics, or take medications in the laboratory.
5. Use the chemical hood when working with volatile chemicals, flammable liquids or gases, or odorous chemicals, or when there is a possibility of the release of toxic chemical vapors, powders, or dusts.
6. There should be no boisterous conduct, excessive noise (radios, DVD players, iPods), or practical jokes in the laboratory.
7. Never taste any laboratory chemical. When smelling a chemical (if you are instructed to do so), gently waft the vapors toward your nose. Do not directly inhale the vapors.
8. If any chemical spills on your skin or in your eyes, immediately flush the affected area with water and notify the instructor.
9. Do not work alone in the laboratory.
10. Notify the instructor immediately of all accidents, incidents, injuries, spills, or hazardous situations.
11. Dispose of waste chemicals in the containers provided.
12. Do not heat flammable liquids with a Bunsen burner or other open flame.
13. Label all containers with chemicals or solutions of any kind.

References

1. WILLIAM JAMES. The Quotations Page; available at http://www. quotationspage.com/quote/1971.html.
2. ROBERT HILL. Personal account of incident.
3. E. SCOTT GELLER. *The Psychology of Safety Handbook*, Lewis Publishers, Boca Raton, FL, 2001, p. 163.
4. Centers for Disease Control and Prevention. Heart Disease and Stroke: The Nation's Leading Killers; available at http://www.gov.gov/nccdphp/publications/aag/cvh.htm (accessed November 15, 2006).
5. Centers for Disease Control and Prevention. Health Effects of Cigarette Smoking: Fact Sheet; available at http://www.cdc.gov/tobacco/factsheets/HealthEffectsofCigaretteSmoking_Factsheet.htm (accessed November 15, 2006).
6. Centers for Disease Control and Prevention. Community-Based Interventions to Reduce Motor Vehicle-Related Injuries: Evidence of Effectiveness from Systemic Reviews; available at http://www.cdc.gov/ncipc/duip/mvsafety.htm (accessed November 15, 2006).
7. H. W. HEINRICH, D. PETERSEN, and N. ROOS. *Industrial Accident Prevention: A Safety Management Approach*, 5th edition, McGraw-Hill, New York, 1980.
8. R. H. HILL. The Safety Ethic: Where can you get one? *Journal of Chemical Health and Safety* **10**(*3*):8–11 (2003).

QUESTIONS

1. In RAMP, the R stands for
   (a) Risk
   (b) Repair
   (c) Report
   (d) Recognize

2. In RAMP, the A stands for
   (a) Avoid
   (b) Assign
   (c) Assess
   (d) Accident

3. In RAMP, the M stands for
   (a) Maintain
   (b) Materials
   (c) Monitor
   (d) Minimize

4. In RAMP, the P stands for
   (a) Prevent
   (b) Prepare
   (c) Plan
   (d) Protect

5. Which definition is *incorrect*?
   (a) A *hazard* is a known source of danger or harm.
   (b) *Risk* is the possibility or probability of suffering harm from being exposed to a hazard or unsafe situation.
   (c) *Exposure* means coming in direct contact with a hazard or chemical in a fashion that causes injury or harm.
   (d) *Safety* is freedom from danger, injury, or damage.

6. Which statement is true?
   (a) All chemicals are hazardous.
   (b) Only chemicals that are toxic are considered hazardous.
   (c) Only chemicals in open beakers are hazardous.
   (d) Only liquid and gaseous chemicals are hazardous.

7. Which statement is true?
   (a) Usually chemicals used in general chemistry labs are nonhazardous.
   (b) The more useful a chemical is in a lab, the more hazardous it is.
   (c) The more useful a chemical is in a lab, the less hazardous it is.
   (d) Most chemicals that we encounter in labs and elsewhere are hazardous.

8. Learning to be safe in chemistry laboratories is best accomplished by
   (a) Performing experiments to see what's safe and what isn't
   (b) Memorizing safety rules
   (c) Thinking about the hazards and risks associated with experiments
   (d) Learning from one's own mistakes

9. Facilities, location, equipment, procedures, and standards are examples of
   (a) Environmental factors
   (b) Person factors
   (c) Behavior factors
   (d) None of the above

10. Attitude, beliefs, personality, knowledge, skills, and abilities, are examples of
    (a) Environmental factors
    (b) Person factors
    (c) Behavior factors
    (d) None of the above

11. Safe and risky practices are examples of
    (a) Environmental factors
    (b) Person factors
    (c) Behavior factors
    (d) None of the above

12. Most accidents in labs occur when
    (a) Experienced lab workers try a new experiment
    (b) Inexperienced lab workers use chemicals
    (c) Safety rules and safe practices are not followed
    (d) "Accidents happen" beyond the control of the lab worker

13. Why do some lab workers take unnecessary risks?
    (a) They willfully violate rules that they know about.
    (b) They are unaware of the risks.
    (c) They repeat experiments with high risk that have not failed before.
    (d) All of the above.

14. Safety incidents generally occur in
    (a) Introductory labs
    (b) Advanced labs
    (c) Research labs
    (d) Labs at all levels of the curriculum

15. Why is it sometimes hard to recognize the hazards of chemicals?
    (a) For the most part, we do not know what the hazards can be.
    (b) The terms used to describe hazards are not well defined.
    (c) While hazards are generally known, it is not easy to access them in print or on the Web.
    (d) Recognizing hazards requires an understanding or the language of safety *and* knowing where to access safety information.

16. What document may contain useful information about the hazards of a chemical?
    (a) Chemical Hygiene Plan
    (b) Chemical Risk Document
    (c) Course syllabus
    (d) Material Safety Data Sheet

17. Assessing risks in labs can be challenging because
    (a) Only experts can really assess risk accurately
    (b) It requires the anticipation of what kinds of exposures are possible for a given procedure
    (c) Chemical reactions are generally not very reproducible and therefore unpredictable
    (d) Predicting the hazards of chemicals is almost impossible

18. Many accidents in labs are caused by
    (a) Poor housekeeping

    (b) Selecting chemicals incorrectly
    (c) Running reactions at temperatures that are too high
    (d) Using personal protective equipment inappropriately

19. Which is *not* a component of the student safety ethic?
    (a) Work safely.
    (b) Avoid unnecessary risk.
    (c) Accept responsibility for safety.
    (d) Rely on your teacher to keep you safe.

20. Eating, drinking, and smoking in a lab
    (a) Are not allowed in labs
    (b) Are allowed whenever the lab instructor says it is OK to do so
    (c) Are allowed in labs whenever there are no open beakers of chemicals in use
    (d) Are allowed in labs as long as there is no reasonable chance of ingesting lab chemicals

21. Tasting chemicals
    (a) Is allowed for solutions that are very dilute.
    (b) Is never allowed in chemistry labs
    (c) Is allowed for lab chemicals that we also know are "common chemicals" (such as vitamin C)
    (d) Is allowed for chemicals that are known to be nontoxic

22. If there is an accident (incident) or chemical spill in the lab, you should
    (a) Notify your instructor immediately
    (b) Notify your instructor only if emergency equipment had to be used
    (c) Alert other nearby students, but tell the instructor only if the event is "serious"
    (d) Notify your instructor before the end of the lab period

1.1.2

WHAT IS GREEN CHEMISTRY?

Preview   This section introduces the fundamental tenets of green chemistry and discusses how these principles can be used in the college chemistry curriculum.

Man shapes himself through decisions that shape his environment.

Rene Dubos, French Scientist[1]

INCIDENT 1.1.2.1   WASTED CaO REAGENT[2]

Approximately 25 g CaO (lime) was used in a synthesis on an occasional basis. To save money, CaO was purchased in bottles containing 500 g. A researcher failed to tighten the lid adequately after removing 25 g, water vapor entered the bottle, reacted exothermically with the CaO, and the bottle broke. Some of the reacted material was hard as a brick and much of the unreacted CaO scattered widely as a fine powder. Several hundred grams of CaO was unusable and was a mess to clean up amidst broken glass.

What lessons can be learned from this incident?

What Is Green Chemistry?

"Green chemistry is the design and use of methods that eliminate health and environmental hazards in the manufacture and use of chemicals."[3] The main features of green chemistry are described below. In addition to the health and environmental advantages of green chemistry, many of these methods can potentially be less expensive for industrial processes. The latter advantage can work as an important driving force in adopting green chemistry procedures in the chemical industry.

In 1998 *Green Chemistry: Theory and Practice*[4] was published, introducing the *Twelve Principles of Green Chemistry* that characterize the main features of green chemistry (see Section 1.3.4 for more information about these principles). Application of these twelve principles leads to the following desirable objectives in chemical reactions and processes:

- Select the least hazardous chemicals possible with regard to the environment and human health.
- When synthesizing new compounds, minimize hazardous features and optimize the ability of the compound to degrade to innocuous products in the environment.
- Design reaction procedures that minimize energy consumption and/or run at ambient temperature and pressure, preferably using catalysts.
- Use chemicals that are renewable as starting reagents (feedstocks). (*Renewable* refers to resources that can be replaced by natural processes at a rate comparable to or faster than they can be consumed by humans. *Feedstocks* are the raw materials used by industry to make products.)
- Design reactions with high yields to minimize waste and inefficiency.
- Design procedures that make recycling reagents and solvents easy.
- Design procedures that eliminate wastes or produces wastes that can be recycled.

How can we use these ideas in college chemistry laboratory experiments?

Laboratory Safety for Chemistry Students, by Robert H. Hill, Jr. and David C. Finster
Copyright © 2010 John Wiley & Sons, Inc.

"Going Green" in the Chemical Industry

Throughout most of the 19th and 20th centuries, little attention was paid to the effects of chemicals on the environment (and even on chemists, in some instances). Standard procedure in the chemical industry, at least until the last third of the 20th century, was to use the chemicals that performed the necessary reactions without much regard to the hazards of the chemicals or the environmental fate of products and wastes. Simply put, for a very long time the accepted waste disposal procedure was to simply dump them into the ground or into a river. Many small and large environmental disasters occurred due to these practices until the early 1970s when federal regulations such as the Clean Air Act (1963, with amendments in 1970, 1977, and 1990), the Clean Water Act (1974), the Toxic Substances Control Act (TSCA, 1976), and the Resource Conservation and Recovery Act (RCRA, 1976) began limiting the discharge of hazardous substances. These regulations limit the pollution of the environment to "acceptable" levels but one goal of green chemistry is to reduce hazardous discharges to minimal levels.

With the publication of *Green Chemistry: Theory and Practice* in 1998 the chemical industry and the Environmental Protection Agency[5] began to partner to promote the agenda of green chemistry. The American Chemical Society has established the *Green Chemistry Institute®* (GCI)[6] and the GCI *Pharmaceutical Round Table.*[7] In the United Kingdom, the Royal Society of Chemistry publishes the journal *Green Chemistry*,[8] and launched the Green Chemistry Network[9] in 1998. While some green chemistry initiatives have been adopted by the chemical industry, this is not yet widespread. The degree to which this happens in the future will depend on the economic viability of green chemistry and the commitment of future chemists and managers to the principles of green chemistry.

"Going Green" in the Lab

The main goal of the green chemistry initiative is to change how the chemical industry operates since the chemical industry generates significant volumes of wastes. The volumes and quantities of chemical wastes from academic labs are miniscule compared to that generated by the chemical industry. However, what students learn in chemistry courses in colleges will affect how they later function in industrial environments. In addition to *learning* about the principles of green chemistry, it is important to *practice* green chemistry in academic laboratories when possible as training for the future. In some instances this may be somewhat symbolic with regard to the actual impact on the environment, but learning the green techniques and principles is best done experientially in the lab. Another advantage is that laboratory safety is likely to be improved since less hazardous material will mean less risk to students.

Experiments that students perform when taking chemistry have traditionally been designed with the goal of teaching various techniques and demonstrating important principles of chemistry. Since about 2003, some colleges and universities have reviewed their course-related laboratory experiments asking: "How can we teach our laboratory courses using the principles of green chemistry?" The goal in answering this question is not always to become "perfectly green" but at least to consider how to make experiments "greener."

What changes are possible in college chemistry labs?

- Use safer solvents. When possible consider using solvents that are less toxic and less flammable, or even run "solvent-free" reactions.
- Reduce volumes and amounts. Consider reducing the scale of an experiment, for example, by running a reaction with 3 grams instead of 10 grams of a starting compound. This saves the costs of starting materials, uses less solvent, uses less energy (if heated), and produces less waste. Fifty years ago, before instrumental analysis, products could only be analyzed with subsequent chemical tests. Today, modern instrumentation often requires only a very small sample size for analysis. This allows some reactions to be "scaled down" considerably. Many chemistry departments have adopted "microscale" techniques in various parts of the curriculum that use smaller-scale glassware, particularly in organic chemistry labs.

- Minimize hazardous by-products and wastes. For example, nitric acid is an excellent oxidizing agent but produces noxious gases. Chromium(VI) is an excellent oxidizing agent but it is carcinogenic when inhaled. Using hydrogen peroxide as an oxidizing agent has the advantage of producing water as a by-product.

- Use less toxic reagents. For example, if one needs to precipitate a particular cation, using anions that are less toxic, such as oxalate or carbonate, is preferred over chromate or sulfide.

- Minimize waste. Few reactions occur in 100% yield so there are almost always leftover reagents and/or waste by-products. Integrating the reuse and recycling of these materials into the experiment itself eliminates or minimizes the need for waste disposal.

Your Role as a Student

Students generally do not design laboratory experiments and it is in the experiment design process where it is best to implement green chemistry. Apart from asking "how green" your experiments are, you can also simply follow procedures carefully and, most importantly, work safely and dispose of all wastes in appropriate containers.

Starting to "think green in the lab" will be an important component of your education. The market for chemists who have been educated with chemical sustainability in mind will almost certainly expand in the years to come.

References

1. Rene Dubos. Brainy Quotes; available at http://www.brainyquote.com/quotes/authors/r/rene_dubos.html (accessed June 12, 2009).
2. Lester Pesterfield. Personal anecdote.
3. P. ANASTAS and J. WARNER. *Green Chemistry: Theory and Practice*, Oxford University Press, New York, 1998, pp. 8–9.
4. P. ANASTAS and J. WARNER. *Green Chemistry: Theory and Practice*, Oxford University Press, New York, 1998.
5. Environmental Protection Agency. Green Chemistry; available at http://www.epa.gov/greenchemistry/ (accessed September 17, 2009).
6. American Chemical Society. ACS Green Chemistry Institute; available at http://portal.acs.org/portal/acs/corg/content?_nfpb=true&_pageLabel=PP_TRANSITIONMAIN&node_id=830&use_sec=false&sec_url_var=region1&__uuid=62f1c35e-db51-45b4-8b73-ba437b12a682 (accessed September 17, 2009).
7. American Chemical Society. ACS GCI Pharmaceutical Round Table; available at http://portal.acs.org/portal/acs/corg/content?_nfpb=true&_pageLabel=PP_TRANSITIONMAIN&node_id=1422&use_sec=false&sec_url_var=region1&__uuid=10148fad-5c1e-40e7-9615-b1aaad66cc95 (accessed September 17, 2009).
8. Royal Society of Chemistry. RSC Publishing. Green Chemistry; available at http://www.rsc.org/Publishing/Journals/GC/index.asp (accessed September 17, 2009).
9. Royal Society of Chemistry. Green Chemistry Network; available at http://www.rsc.org/chemsoc/gcn/index.htm (accessed September 17, 2009).

QUESTIONS

1. Which is not a goal of green chemistry?

   (a) Select the least hazardous chemicals possible.
   (b) Optimize the ability of a compound to degrade to innocuous products in the environment.
   (c) Design reactions that minimize energy consumption.
   (d) Use feedstock chemicals that are not renewable.

2. Which of the following is (are) necessary for green chemistry to be successful in the chemical industry?

   I. Economic viability
   II. Commitment of chemists and managers
   III. Passage of the Green Chemistry Act by Congress

   (a) I
   (b) I and II
   (c) I and III
   (d) II and III

3. When comparing the three oxidizing agents, nitric acid, Cr(VI), and hydrogen peroxide,

   (a) Nitric acid is the "greenest" because it has the highest oxidizing potential
   (b) Cr(VI) is the "greenest" because it is a single-atom ion
   (c) Hydrogen peroxide is the "greenest" because it produces innocuous by-products
   (d) Nitric acid is the "greenest" because it is the cheapest reagent on a mole basis

4. Which of the following features of a solvent make it a "greener" solvent?

   I. Less flammable
   II. Less toxic
   III. Lower molar mass
   IV. Easily recycled

   (a) I and III

(b) II and IV

(c) II and III

(d) I, II, and IV

5. What is the main role for undergraduate chemistry students in the lab with regard to being "green"?

(a) Follow all lab instructions, particularly with regard to chemical waste disposal.

(b) Use smaller amounts of expensive chemicals.

(c) Use water as a solvent instead of organic solvents.

(d) Use less toxic substances.

1.2.1

RETHINKING SAFETY: LEARNING FROM LAB INCIDENTS

Preview This section describes how you can learn lessons from a critical analysis of an incident using the "five whys."

I have vivid memories of several "accidents" that threatened or caused serious injury. I would say that I experienced some lessons about safety through these accidents.

Robert H. Hill, Jr.[1]

INCIDENT 1.2.1.1   HAIR ON FIRE[2]

Armand had long hair. The instructor told all students to be sure that any long hair was tied up so they could safely work at their laboratory bench. The girls in the class followed the instructions. But some of the boys with long hair did not follow these instructions and the male instructor did not strongly enforce the rule for boys. While using a Bunsen burner, Armand leaned too close to the burner and his hair caught fire. He was able to quickly extinguish the flames by beating them out with his hand. A fellow student also helped by pouring water on Armand's hair. Armand received slight burns to his hand, face, and ear.

What lessons can be learned from this incident?

Prelude

We assume that you may be reading this section at the beginning of your second year of college chemistry classes, or at least after you have taken some chemistry courses with labs. Now that you have been "in the lab" for awhile, you probably have a better appreciation of what can or might go wrong during experiments.

As you think back on your own experiences, you can review any incidents through the filter of RAMP. With an incident in mind, or reviewing Incident 1.2.1.1, think about the following questions:

- In what way did someone not recognize a hazard?
- In what way did someone not assess the risk posed by that hazard?
- In what way did someone not manage the risk properly, which allowed something bad to happen?
- Was everyone prepared for the emergency when it happened?

You can use the four RAMP concepts in analyzing "what went wrong" for any incident. This is one way to think about the incident that can easily reveal mistakes. Often, there is more than one mistake that leads to an incident. In this section we will introduce another method of incident analysis that can reveal more about causes of incidents and how to prevent incidents.

That Was No Accident!

Incidents are unplanned, unexpected, and undesirable events that have adverse impacts (injury, death, damage) and consequences on health, property, materials, or the environment.[3] *Accidents* are the same

---

as an incident but over time many people have understood or implied that they were chance happenings, being unavoidable and without specific preventable causes. For that reason many safety professionals don't like to use the term accident because every incident invariably has one or more preventable causes—some of which are obvious while other preventable causes are subtle and often unrecognized without an in-depth or careful review of the facts of the incident. We will use the term incident, instead of accident.

Near misses are unplanned events (also sometimes called "close calls" or "near hits"), which did not have severe adverse impacts on health or the environment, but just narrowly missed causing severe injury or damage. Often, these near miss incidents are very scary when reviewed in retrospect but we can learn much from them. Recognizing near misses is important because they could be precursors to future serious incidents. Using the information from an analysis of the near miss can help develop "lessons learned," a term that describes how we can derive actions to prevent future incidents.

This section is about learning lessons from incidents using critical thinking. It does not describe details for conducting a formal incident investigation. There are established methods to do these investigations that are beyond the scope of this text. Nevertheless, as previously pointed out, safety is an empirical discipline based on our experiences—that is, our past mistakes, errors, incidents, accidents, or near misses. You should understand on a basic level that incidents have causes and by using some critical thinking about an incident you may be able to arrive at the cause or causes, which in turn could lead you to identify recommendations or steps that can prevent future incidents similar to this one. This is an important skill, especially for those working in laboratories as a profession, since it is likely that some incidents that will happen to you will not be subject to investigation. You can also apply these same skills to incidents that happen to you outside the laboratory in your everyday life.

It's About Learning Lessons for Prevention, Not Blaming Someone

In examining incidents, there are a few important considerations to keep in mind. It is important NOT to fix "the blame" on any individual for an incident, but rather to focus on determining factors that caused the incident and how these factors can be avoided to prevent future incidents. Many thorough investigations have shown that the causes of incidents in a place of work are frequently related to a lack of proper management, and rarely relate to intentional, irresponsible, reckless, or blatantly dangerous individual acts.[4] Incidents are often the result of at-risk behavior, but the at-risk behaviors may not be recognized as such and may often be subtle. These are often a combination of several small actions that when examined individually might not have caused an incident but when taken together under the circumstances of time and place resulted in an unexpected and adverse event.

The extent of any investigation depends on the seriousness and impact of a given incident. For a minor incident there is not likely to be formal investigation but rather a "lessons learned" scenario as described below. In cases of serious injury or major laboratory or equipment damage, a more formal investigation by safety professionals will likely be conducted. In any case we can usually learn a lot from examination of incidents, and it is not unusual to find that a similar incident has occurred in some laboratory somewhere. Collections of incidents reveal that we continue to make the same mistakes over and over again! There are compilations of laboratory and chemical incidents worthy of examination, both in print[5,6] and online.[7-11] The American Industrial Hygiene Association's Laboratory Health and Safety Committee has an outstanding Internet site describing laboratory incidents with lessons learned and recommendations that are especially useful.[7]

In each incident you should ask yourself: (1) What happened? (2) How did it happen? (3) Why did it happen? This series of questions is one form of *root cause analysis* (RCA),[12] a standard procedure in the investigation of incidents in business and industry. The last question is the most important one and you will find you may have to ask and answer this question several times in sequence before you come close to the "real" or "root" causes of an incident. RCA generally recommends asking "Why?" five times to get to a fundamental, rather than superficial, cause. Root causes are the basic causes of an incident that can be reasonably identified, that can be controlled, and for which recommendations or lessons learned can be derived. Many times root causes are not immediately obvious, but can be identified from careful inquiry. Once you know the "root" causes you should be able to develop recommendations or steps to prevent this from happening again.

Learning How to Learn Lessons

So let's demonstrate how this might work, using an incident described earlier. Keep in mind that this is a simple exercise, and because you have not conducted a thorough investigation to collect all of the facts, it is possible your conclusions could be different from those shown in the example below. This is likely because each of us makes different assumptions that may or may not be correct. Remember that you can use this process to examine any incident that has involved you to learn more about the root causes and derive lessons that might prevent those from happening again.

INCIDENT 1.2.1.2   HOT GLASS[12]

During a first-year laboratory session, students were asked to learn to bend a glass tubing to form a 90° bend using a Bunsen burner. A student performed the operation and took the glass tubing to the instructor. He handed the tubing to his instructor, who promptly dropped the glass tubing after burning her hand because the tubing was still hot from being held in the burner. The tubing broke when it hit the floor.

What lessons can be learned from this incident?

Let's pretend for this example that we know the "facts." Let's ask the three questions: *What happened? How did it happen? Why did it happen?* When you ask the latter question, let's continue to ask why for at least five times to arrive near a root cause. Note that this will identify *a* root cause. It is likely there is more than one root cause. Most incidents occur because a series of missteps or mistakes were made that ultimately came together to result in the adverse event.

> *What happened?* A hot piece of glass tubing was placed in the hand of the instructor, causing a burn and causing the glass tubing to be damaged when it dropped to the floor. There were two adverse outcomes—a burn and a broken glass tube.

> *How did it happen?* The tubing was heated to red hot, bent, and then removed from the flame. Once the glass was removed from the flame, it began to cool; however, as most cooks know, glass retains heat well and the glass tubing was still very hot. The student must have quickly walked to the instructor with glass tubing, holding it at the ends. Not being aware that the glass was still hot, it was handed to her to examine to determine if it was a good bend. Thus, the student did not recognize the hazard. Furthermore, the instructor did not recognize (the hazard) that the glass was hot or could have been hot, and grabbed the hot glass in her hand.

We present below two series of "five questions" that analyze this incident.

> *Why did it happen? [#1]* The student did not recognize the hazard. *Why? [#2]* Probably because the instructor did not alert the student to the potential hazard (hot glass) that resulted from the heating. *Why? [#3]* The instructor failed to recognize the potential hazard of hot glass, as indicated by her acceptance of the hot glass. *Why? [#4]* The instructor had not taken time to consider (assess) the potential hazards of the experiment prior to the start of the laboratory. *Why? [#5]* The instructor had not been taught about the hazards of this experiment.

> *Why did it happen? [#1]* The instructor did not recognize the hazard. *Why? [#2]* Hot glass looks like cold glass and the instructor did not think a student would hand a hot object to another person. *Why? [#3]* The instructor did not have enough safety education or experience in working with students. *Why? [#4]* The safety education the instructor received did not address this topic. *Why? [#5]* The organization (college) failed to adequately educate instructors about the specific and general risks in chemistry labs.

While each series starts with a different answer to the first "Why?" they both end up at the same root cause: inadequate instructor education in safety.

You could at this point continue to ask "Why?" Also, other potential reasons could be offered, depending on the "facts"—these may have entered your thoughts as we went through this example. For instance, the instructor had not been taught how to conduct risk assessments of experiments, or the instructor failed to conduct the needed assessment of the hazards. If you were able to question the student and instructor you might be able to more precisely determine the root causes.

How Might We Prevent This Incident From Happening Again?

We have now answered the question "Why?" five times, and we are near a "root" cause for the incident. So what would be the recommendations or steps to prevention (lessons learned) for this incident?

1. Instructors should be educated about assessing, recognizing, and managing the hazards of experiments.

2. Instructors need to know or be able to determine the specific hazards of the specific experiments that are being carried out under their direction.

3. Instructors should communicate the hazards to the students before the experiments begin.

4. For this specific experiment, instructors should not hold out their hands to receive tubing that may be hot; instructors should keep their hands in their pockets or behind their backs.

5. The instructor should pass on this lesson learned to other instructors so they are also aware of this hazard and will not make the same mistake.

This analysis of a simple laboratory incident demonstrates the principle of "critical" thinking that is an essential part of your job in keeping yourself and others safe. For other incidents it may not be as easy to arrive at conclusions because it may not be so easy to determine exactly what happened. Nevertheless, this method is useful in many instances to examine incidents that might occur during your time in the laboratory. Perhaps the most important thing for you to learn during an incident analysis is that you are in control of your decisions and actions. Incidents can be prevented if you can learn lessons from these incidents and then through actions prevent them from happening again! Now you may understand why many of us choose not to use the term "accident," which implies this was a chance happening, when in reality there are almost always causes that contributed to an incident.

Formal Incident Investigations

There are more formal methods of analysis to identify "root" causes of incidents that use maps to assist investigators.[13] In these procedures the source of the incident is divided into equipment and personnel difficulties, and each of these is divided into potential categories of problems. Each category then leads to major root cause categories, such as Management Systems, Procedures, Training, Communications, and so on. In Incident 1.2.1.2 above, we identified a lack of education for the instructor as a cause. However, if we learned during our incident investigation that the instructor had been educated in the hazards of the experiment, then there would have been a different answer to the last "Why?" It may have been one of communication in which the instructor failed to provide adequate safety instructions to the student for some reason, or perhaps the instructor gave the proper instructions but the student failed to hear or understand or forgot the instructions because there were too many things to remember. The point here is that the causes are dependent on the incident and the facts of the incident as best and as honestly as they can be determined. If you are involved in an incident or near miss, it is important that you consider carefully what happened and attempt to determine why this happened so you can develop steps to prevent it from happening again.

Incident Analysis in Academic Labs: RAMP

In Section 1.1.1, we introduced the RAMP acronym. We will use this frequently throughout the book to illustrate how this protocol can help prevent incidents in chemistry labs.

- *Recognize* hazards.
- *Assess* the risks of hazards.

- *Minimize t*he risks of hazards.
- *Prepare* for emergencies.

Performing root cause analysis on incidents can shed light on all four parts of RAMP and such analysis usually demonstrates several mistakes, not just one, that ultimately can be identified for most incidents. Furthermore, incidents can sometimes be prevented by eliminating *just one* of the several mistakes.

References

1. ROBERT H. HILL. Changing the way chemists think about safety. *Journal of Chemical Health and Safety* **11**(*3*):5–8 (2004).
2. FARIBA MOJTABAI and JAMES A. KAUFMAN (editors). Learning by Accident, Vol. 1, Laboratory Safety Institute, Natick, MA, 1997; modeled after incident #28, p. 5.
3. PETER C. ASHBROOK. Evaluation of EHS incident response actions at a large research university. *Journal of Chemical Health and Safety* **11**(*2*):20–23 (2004).
4. LARRY RUSSELL. Incident investigation: Fix the problem—not the blame. *Journal of Chemical Health and Safety* **6**(*1*):32–34 (1999).
5. FARIBA MOJTABAI and JAMES A. KAUFMAN (editors). Learning by Accident, Vol. 1, Laboratory Safety Institute, Natick, MA 1997; FARIBA MOJTABAI and JAMES A. KAUFMAN (editors). Learning by Accident, Vol. 2, Laboratory Safety Institute, Natick, MA, 2000; TERESA ROBERTSON and JAMES A. KAUFMAN (editors). Learning by Accident, Vol. 3, Laboratory Safety Institute, Natick, MA, 2003.
6. NORMAN V. STEERE (editor). Safety in the Chemical Laboratory, Vol. 1, ACS Division of Chemical Education, reprinted from *Journal of Chemical Education*, Easton, PA, 1967, pp. 116–121; NORMAN V. STEERE (editor). Safety in the Chemical Laboratory, Vol. 2, Division of Chemical Education, reprinted from *Journal of Chemical Education*, Easton, PA, 1971, pp. 121–122; NORMAN V. STEERE (editor).

Safety in the Chemical Laboratory, Vol. 3, Division of Chemical Education, reprinted from *Journal of Chemical Education*, Easton, PA, 1974, pp. 152–153.
7. American Industrial Hygiene Association. Laboratory Health and Safety Committee. Laboratory Safety Incidents; available at http://www2.umdnj.edu/eohssweb/aiha/accidents/index.htm (accessed September 17, 2009).
8. Princeton University. Environmental Health and Safety. Section 11. Anecdotes; available at http://web.princeton.edu/sites/ehs/labsafety manual/sec11.htm (accessed September 17, 2009).
9. University of Arizona. Risk Management and Safety. Chemical Safety Bulletins; available at http://risk.arizona.edu/healthandsafety/chemicalsafetybulletins/ (accessed September 17, 2009).
10. U.K. Chemical Reaction Hazard Forum. Incidents; available at http://www.crhf.org.uk/index.html (accessed September 17, 2009).
11. Chemical Information Network. Chemical Accidents and Incidents; available at http://www.cheminfonet.org/accid.htm (accessed September 17, 2009).
12. R. HILL. Personal account of an incident.
13. EQE International, Inc. Root Cause Analysis Handbook: A Guide to Effective Incident Investigation, Government Institutes, Rockville, MD, 1999.

QUESTIONS

1. Laboratory incidents most commonly
   - (a) Have one or more preventable causes
   - (b) Are unavoidable even with extensive safety training
   - (c) Are called "accidents" because they mostly occur for random reasons
   - (d) Do not have the adverse effects of injury or damage

2. What kinds of episodes can be informative with regard to learning about safety?
   - (a) Incidents that caused harm
   - (b) Incidents that caused lab damage, but no physical injuries to personnel
   - (c) "Near miss" episodes that that did not cause harm
   - (d) All of the above

3. What is the *least* important question to answer when investigating an incident?
   - (a) How did the incident occur?
   - (b) What happened?
   - (c) Who caused the incident?
   - (d) Why did the incident occur?

4. Which incidents should be reported as "lab accidents"?
   - (a) Incidents that did not cause harm or damage, but easily could have

   - (b) Incidents that caused damage but no harm
   - (c) Incidents that caused harm and damage
   - (d) All of the above

5. In doing an incident analysis, which of the following step(s) in RAMP are commonly not considered carefully enough?
   - I. *Recognize* hazards.
   - II. *Assess* the risks of hazards.
   - III. *Minimize* the risks of hazards.
   - IV. *Prepare* for emergencies.
     - (a) I and II
     - (b) I and III
     - (c) I, II, and IV
     - (d) I, II, III, and IV

6. In reviewing incidents it is usually discovered that
   - (a) There is one main cause of the incident
   - (b) There are many causes for an incident and prevention of any single cause might have prevented the incident
   - (c) There are many causes for an incident and it would have been necessary to eliminate all of the causes in order to have prevented the incident
   - (d) Inexperienced workers caused the incident

1.2.2

GREEN CHEMISTRY IN THE ORGANIC CURRICULUM

Preview   This section discusses applications of green chemistry in organic chemistry laboratories.

At 25 percent market penetration in 2010, this technology could save 9.25 trillion British thermal units per year (Btu/yr) and eliminate over 360,000 tons of CO_2 emissions per year.

EPA Green Presidential Chemistry Challenge report[1]

INCIDENT 1.2.2.1   ORGANIC SOLVENT FIRE[2]

While holding a flask containing 200 mL of 10% ethylacetate/90% petroleum ether, a researcher decided that she needed some glassware in an oven. As she was removing the glassware with one hand, the flask in the other hand came within inches of the hot oven and the solvent vapors caught fire. The cotton glove on her hand holding flaming solvent caught fire and startled the researcher. She dropped the flask, it broke, and the flaming solvent spread on the floor. She put out the fire on her glove, found a near-by dry chemical extinguisher, and put out the fire on the floor. Her only injury was a third degree burn to one of her fingers.

What lessons can be learned from this incident? How might the application of green chemistry principles have reduced the chance of this incident occurring?

Why Organic?

The goals of green chemistry are to make the enterprise of chemistry a sustainable venture by eliminating waste and using and generating fewer hazardous compounds and less hazardous compounds. A survey[3] of articles in the *Journal of Chemical Education* from 2000 to 2008 that describe "new green experiments" shows that 74% are organic experiments. Why does organic chemistry emerge as a primary application area for green chemistry?

Most experiments in introductory chemistry courses use hydrophilic solutes with water as the solvent. While there are many chemical reasons for this (such as the solubility of common acids and bases in water), water is also a convenient solvent since it is inexpensive, it is nonflammable, and, if it doesn't have nasty things dissolved in it, it is easy to dispose of by just pouring it down the sink. The transition to organic chemistry traditionally involves a change from *hydrophilic* chemistry to *hydrophobic* chemistry in the laboratory.

The most common solvents in the organic laboratory are carbon-based such as toluene, chloroform, acetone, acetonitrile, ethyl acetate, and various ethers and alcohols. None of these chemicals should be poured down a sink because they are mostly water-insoluble, toxic, and flammable, and in addition this is unlawful. Many of the solutes that will be dissolved in these solvents will also have varying degrees of toxicity and flammability that make them inappropriate for easy disposal. All of these features make organic chemistry a prime target for the application of green chemistry.

It is easy to see the importance of organic chemistry with regard to green initiatives when considering the high volume usage of organic chemicals and solvents in the petrochemical industry and the pharmaceutical industry. Greening these industries will have an enormous effect on the global production and disposal of hazardous wastes and on the safety of these industries as illustrated in the opening quote to this section.

Laboratory Safety for Chemistry Students, by Robert H. Hill, Jr. and David C. Finster
Copyright © 2010 John Wiley & Sons, Inc.

Green Initiatives in Organic Chemistry

Solvents are widely used in organic reactions. These liquids are often flammable and have varying degrees of toxicity. As illustrated in Incident 1.2.2.1 there is always the possibility of fires when handling flammable solvents, especially if there are unrecognized ignition sources in the vicinity. While we really don't know specifically what the researcher was doing or why the particular 10% ethyl acetate/90% petroleum ether solvent was chosen for her work, one can speculate that perhaps a less hazardous substitute or a smaller scaled experiment could have been considered for her experiments.

- In some instances, it is possible to eliminate the solvent entirely.[4] Some organic reactions can be run under solvent-free conditions, particularly, when using a microwave oven as the energy source.[5] (This is also likely to be an energy-saving measure, meeting another goal of green chemistry.)

- Often a safer solvent can be substituted for a more hazardous solvent. "Safer" can be either, or both, of the goals of reducing flammability or toxicity. Two publications have described "green solvents."[6,7] Supercritical carbon dioxide has proved to be an excellent substitute for some organic solvents, particularly if newly designed polymers are added, which can increase solute solubility.[8]

- Running reactions on a smaller scale (at least in academic laboratories) will reduce solvent (and solute) use, reduce waste generation, and likely also reduce energy consumption.

Other strategies include the following:

- Use catalysts to reduce energy consumption and to change reaction pathways to increase yield and reduce wastes.

- Eliminate steps in multistep syntheses and/or eliminate the need to isolate intermediates.

- Design reactions with higher yields and better atom economy. See *Chemical Connection 1.2.2.1* "Yield" versus "Atom Economy."

CHEMICAL CONNECTION *1.2.2.1*

"YIELD" VERSUS "ATOM ECONOMY"

For over 100 years, one important metric by which the "success" of a chemical reaction has been measured is the concept of "yield," or "percent yield." As you know, percent yields can range from less than 1% to nearly 100%. The percent yield of a reaction can be limited by equilibrium conditions but is more often limited by the presence of side reactions that divert the reactants to undesired products. Sometimes a chemist performing a synthesis will be very pleased with a 30% yield and for other reactions the desired yield may approach 100%. In industry, high yields are important since "more product means more profit" and the side reactions are likely generating waste, which requires costly disposal.

In 1991, Barry Trost[12] introduced the concept of "atom economy" (for which he received a Presidential Green Chemistry Challenge Award[13]). The essence of this concept is the notion that another useful way to measure the "success" of a chemical synthesis is to consider the fate of *all* of the reactant atoms (apart from solvent). For example, if a catalyst is used to simply facilitate a rearrangement of molecules with 100% conversion (and the catalyst can be recycled with 100% recovery), then all of the atoms in the reactant molecule will end up on the product molecule. This is 100% atom economy. Alternately, an elimination reaction will necessarily have less than 100% atom economy since the atoms eliminated from the reactant represent "waste."

The formula developed to measure atom economy is based on the total masses of reactants and the total masses of the atom utilized in the product (MM = molar mass):

$$\%\text{Atom Economy} = (\text{MM of atoms utilized}/\text{MM of all reactants}) \cdot 100\%$$

Let's use the Wittig[14] reaction (see Figure 1.2.2.1) to demonstrate how percent yield and atom economy can differ.

This reaction might typically have a percent yield of 80–90%, which is reasonably good. However, the calculation of the atom economy (even at 100% yield) is less comforting in the context of green chemistry:

$$\text{Atom Economy} = (96/374) \cdot 100\% = 26\%$$

In terms of the masses of reactants and products, the bulk of the mass of the reactants ends up in the by-product, $C_{18}H_{15}PO$. If we multiply the atom economy by the percent yield ($0.85 \cdot 0.26 = 0.22$), we end up with an even poorer industrial reaction.

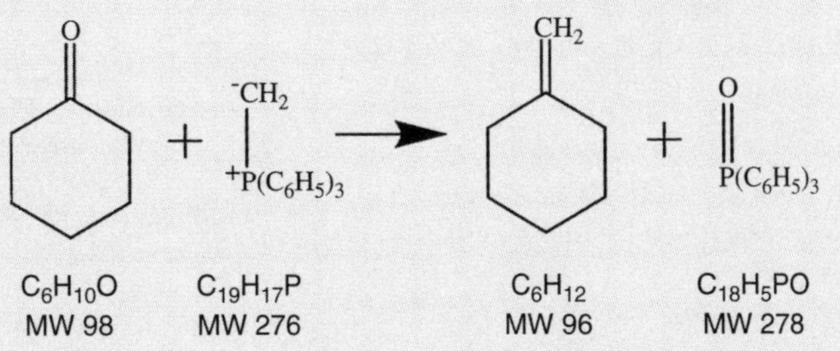

$C_6H_{10}O$ MW 98 $C_{19}H_{17}P$ MW 276 C_6H_{12} MW 96 $C_{18}H_5PO$ MW 278

FIGURE 1.2.2.1   Wittig Reaction on Cyclohexanone. Even at 100% yield, this reaction has a poor atom economy of only 26%.

Thus, the goal of green chemistry is to consider *all* of the atoms used in a synthesis. Atom economy, more so than percent yield, is the best way to do this.

Since 1996 the EPA has conducted the Presidential Green Chemistry Challenge Awards Program,[9] which recognizes approximately five individuals and organizations each year. One award is typically presented in each of the following categories:

Small Business: A small business for a green chemistry technology

Academic: An academic investigator for a green technology

Focus Area 1: An industry sponsor for a technology in the use of greener synthetic pathways

Focus Area 2: An industry sponsor for a technology in the use of greener reaction conditions

Focus Area 3: An industry sponsor for a technology in the design of greener chemicals

Some recent award winners demonstrated the following:

- The use of soy-based (renewable feedstock) chemicals to prepare toner for laser printers and copiers[1]
- A newly synthesized pesticide that replaces (toxic) organophosphate pesticides, with the synthesis using catalysts that saved energy and recycling solvents and reagents[10]
- A new method for synthesizing the precursors for the "Suzuki coupling reaction," designed to avoid harsh conditions and to generate far less waste[11]

Many examples of green chemistry awards can be found at http://www.epa.gov/greenchemistry/pubs/pgcc/past.html.

Being Green, and Safe, in Organic Chemistry Laboratories: RAMP

- *Recognize* the hazards of solvents, reactants, catalysts, and wastes present in organic chemistry experiments.
- *Assess* the level of risks from these hazards.
- *Minimize*, or eliminate, the risks from these hazards by reducing scale, substituting safer (greener) chemicals used as solvents and reagents, and designing processes that are inherently safer and eliminate, or produce biodegradable, wastes.
- *Prepare* for emergencies by understanding the chemical and health hazards of experiments.

References

1. Environmental Protection Agency. Green Chemistry. 2008 Greener Synthetic Pathways Award; available at http://www.epa.gov/greenchemistry/pubs/pgcc/winners/gspa08.html (accessed September 17, 2009).
2. Risk Management & Safety Department, University of Arizona. Incident: Solvent Fire; available at http://risk.arizona.edu/healthandsafety/chemicalsafetybulletins/solventfire.shtml (accessed August 27, 2009).
3. D. C. Finster. Survey of *Journal of Chemical Education*, 2000–2008 using "green chemistry" in the search engine produced 89 articles. Of these, 65 were new laboratory experiments and 47 of those were organic experiments.
4. L. Tanaka. *Solvent-Free Organic Synthesis*, Wiley–VCH Verlag, Weinheim, Germany, 2002.
5. A. Loupy. *Microwaves in Organic Synthesis*, Wiley–VCH Verlag, Weinheim, Germany, 2002.
6. W. M. Nelson. *Green Solvents*, Oxford University Press, New York, 2003.
7. M. A. Abraham and L. Moens (editors). *Clean Solvents*, American Chemical Society, Washington, DC, 2002.
8. S. K. Ritter. Green challenge, *Chemical & Engineering News* **80**(*26*):26–30 (2002).
9. Environmental Protection Agency. Green Chemistry. Presidential Green Chemistry Challenge; available at http://www.epa.gov/greenchemistry/pubs/pgcc/presgcc.html (accessed September 17, 2009).
10. Environmental Protection Agency. Green Chemistry. 2008 Designing Greener Chemical Award; available at http://www.epa.gov/greenchemistry/pubs/pgcc/winners/dgca08.html (accessed September 17, 2009).
11. Environmental Protection Agency. Green Chemistry. 2008 Academic Award; available at http://www.epa.gov/greenchemistry/pubs/pgcc/winners/aa08.html (accessed September 17, 2009).
12. B. M. Trost. The atom economy—A search for synthetic efficiency, *Science* **254**:1471–77 (1991).
13. Environmental Protection Agency. Green Chemistry. 1998 Academic Award; available at http://www.epa.gov/greenchemistry/pubs/pgcc/winners/aa98a.html (accessed September 17, 2009).
14. Royal Society of Chemistry. Green Chemistry—The Atom Economy; available at http://www.rsc.org/images/PDF1_tcm18-40521.pdf (accessed September 17, 2009).

QUESTIONS

1. What are the main reasons that organic chemistry is a primary target for "green chemistry"?

   I. Organic solvents are often flammable.
   II. Organic solvents are often toxic.
   III. Organic solutes are often flammable.
   IV. Organic solutes are often toxic.

   (a) I and II
   (b) I, II and III
   (c) III and IV
   (d) I, II, III and IV

2. Which is *not* a "green" approach to replacing organic solvents?

   (a) Solventless reactions
   (b) Supercritical carbon dioxide
   (c) Using water
   (d) Using the least-expensive organic solvent

3. Which strategy is not a viable "green" option in the chemical industry?

   (a) Using catalysts to reduce energy consumption
   (b) Reducing the number of steps in a multistep synthesis
   (c) Maximizing yield
   (d) Running the reaction on a smaller scale

4. Access the web site that describes the use of soy-based chemicals to prepare toner for laser printers and copiers. What was the main "green chemistry" success in this process?

   (a) Use of a catalyst
   (b) Use of a renewable feedstock chemical
   (c) Replacing an organic solvent with water
   (d) Reducing the number of steps in a multistep synthesis

5. Access the web site that describes the synthesis of a new pesticide that replaces (toxic) organophosphate pesticides. What was the main "green chemistry" success in this process?

   (a) Reducing the number of steps in a multistep synthesis
   (b) Use of a renewable feedstock chemical
   (c) Replacing an organic solvent with water
   (d) Design and synthesis of a less toxic product

6. Access the web site that describes a new method for synthesizing the precursors for the "Suzuki coupling reaction." What was the main "green chemistry" success in this process?

   (a) Use of a catalyst
   (b) Use of a renewable feedstock chemical
   (c) Replacing an organic solvent with water
   (d) Design and synthesis of a less toxic product

1.3.1

FOSTERING A SAFETY CULTURE

Preview This section discusses approaches to establishing and promoting a strong safety culture among those people who are working under your supervision or leadership.

Example is not the main thing influencing others, it is the only thing.

Albert Schweitzer[1]

INCIDENT 1.3.1.1 METHANOL FIRE[2]

A high school teacher, carrying out a chemistry demonstration, was using methanol with some chemical salts when a sudden explosion occurred that burned several students in the front row. Three students received serious burns to their faces, necks, arms, hands, and legs. The other students in the classroom ran from the room. Media reported that there was a lack of safety oversight that is common in many schools and inspections are rare.

What lessons can be learned from this incident?

Safety Follows the Leader—What If You Are the Leader?

"Safety is everyone's responsibility." You have probably heard this before. For some of you junior or senior chemistry students, it is possible that you could find yourself playing a significant role assisting in or overseeing a laboratory session. This may require that you set the tone and standard for safety. In fact, if you are a paid laboratory assistant or instructor (an employee), then you do have responsibilities for safety.

When you graduate and become employed, you may find the safety of a laboratory and the people working in it are now your responsibility. That is, you will be officially charged with ensuring the safety of the laboratories that you will manage. You may be in charge if you become a chemistry teacher in a local school system, and suddenly you are the "resident expert" who is responsible for establishing and maintaining safety for students using the laboratories in the school where you work. This encompasses safety for all parts of programs and facilities, including conducting experiments safely, maintaining the facilities themselves (laboratory, stockroom) and safety equipment, preparing reagents, managing chemical waste, teaching safety, and conducting safe demonstrations. This may also be the case if you go to work in an industrial operation, where you must not only do your own work but are assigned responsibility for safety of laboratory operations and those people working in those laboratories.

The most important thing that you must do is to establish and promote a strong, enthusiastic, vibrant safety culture. This chapter provides some information about how you might build a safety culture.

Using Albert Schweitzer's example, leadership in safety is the key to success—and that means not only telling people how to do things safely, but you must follow the same guidance that you gave to those under your charge. Much has been written on leadership, and learning about leadership and leaders may become a critical part of continuing education as you move into new roles in the future. This was probably not part of your education as a chemistry student major. Finding books and courses about leadership and organizational behavior is easy, though, and you may find yourself consulting such books someday.

Laboratory Safety for Chemistry Students, by Robert H. Hill, Jr. and David C. Finster
Copyright © 2010 John Wiley & Sons, Inc.

Understanding What Motivates People

Scott Geller, a psychologist who specializes in safety, has also written about safety leadership in *The Psychology of Safety Handbook*.[3] In his book Geller points out that leadership is different from management. Leaders inspire people to follow and take actions by seeking to instill the need for responsibility, while managers work to ensure that people meet goals and outcomes and that they are held accountable for achieving these.

There are many traits that are common in leaders. Passion for their vision is one of those, and this must be one of the traits that you exhibit as a leader in safety. You must be honest and trustworthy to be a safety leader and you must be able to motivate people to do their jobs in a safe manner. You must lead by example so that you always follow safety procedures every time you enter the laboratory. Always wear appropriate safety gear, and always consider safety in all that you do. Overall, you should be one who has adopted *the safety ethic* (see Figure 1.3.1.1) so that you not only believe in the value of safety but you strive to ensure that safety becomes an integral part of your everyday life.

Leaders learn to communicate effectively so that they can get their message across to those who need to hear it. Their words should be inspiring and positive to emphasize what is possible. A safety leader should continually strive to learn more about safety and how to relate the importance of safety to other people. Leaders seek to educate in safety rather than relying only on safety training. Safety education imparts basic knowledge so that people themselves understand safety principles and will be able themselves to select appropriate safety procedures and approaches.

Leaders seek to provide an understanding about why safety is important, offering examples about learning lessons in safety. Leaders are good at listening to others to better understand their perspective, which in turn allows leaders to offer sound advice, guidance, support, or leadership. To get others to accept the importance of safety, to adopt a positive attitude about safety, and to conduct safety in a safe manner requires the leaders to lead. Those in leadership roles must have positive attitudes toward safety, vibrant passions for safety, and strong personal commitments to safety exemplified by their examples. Some traits of a good safety leader[4] are found in Table 1.3.1.1.

A Battle Between Safety and Human Nature

Geller provides 50 principles for establishing a strong safety culture.[5] Some of these principles are: safety should be an internally driven value, people should understand safety theory, people should teach safety, safety leaders can be developed, and the focus of safety should be safety processes not safety outcomes. These principles lead you to continually work on how to do things safely rather than just emphasizing a "zero" incident rate or time away from work for an incident. Geller points out that safety continually conflicts with human nature, which seeks to do those things that are convenient, comfortable, and expedient since safety can sometimes lead to discomfort, inconvenience, and inefficiency of time.

Geller[5] notes the importance of observation and feedback in establishing strong safety behaviors, and people tend to see their failures as caused by external factors not under their control rather than their internal failures. Geller places strong emphasis on active caring, a proactive process that is critical to building a strong safety culture; empowering people to make good safety decisions can build self-confidence, optimistic outlook, and a sense of personal control of safety. For more details about his valued approaches to safety, read *The Psychology of Safety Handbook*.

The Safety Ethic

I work safely,
value safety,
prevent at-risk behavior
promote safety, and
accept responsibility for safety

FIGURE 1.3.1.1   The Safety Ethic. In the workplace, this safety ethic reflects a mindset of attitude and responsibility that keeps you, and others, safe in the laboratory.

TABLE 1.3.1.1   Traits of Safety Leaders

- Inspire people so that they want to be safe—so they are passionate about safety.

- Seek open and transparent communication to build an honest and trustworthy relationship.

- Lead by example, supporting safety in their actions and in their investments in time and resources.

- Hold people accountable for safety processes and activities, rather than specific outcomes.

- Educate others in safety with examples and rationale, using opportunities for instructing, coaching, mentoring, and delegating.

- Listen first. Try to understand the other person's perspective before offering advice, support, or direction.

- Promote ownership of the safety process by seeking others' involvement in the safety process and allowing them opportunities to achieve desired outcomes.

- Provide expectations rather than mandates—giving people opportunities to make their own decisions.

- Express some "uncertainty" as to how to reach safety goals—giving great latitude to others to figure out how to accomplish these goals.

- Understand that some things cannot be measured, but it is important to increase self-esteem, personal control, optimism, and a strong sense of being a part of the safety culture.

- Perceive that people have skills and attributes over a continuum, so that individuals have particular abilities to accomplish certain things well and others not as well.

Source: Adapted from Geller.[4]

Building Positive Attitudes Toward Safety

To instill a strong positive attitude toward safety requires continual reinforcement of safety's importance in every experiment that is conducted. To have a strong understanding of safety requires that a person recognize hazards, be able to assess how exposures to these hazards might occur, and know how to manage and control hazards so that exposure and risk are minimized. Allowing people with whom you work to learn about the hazards, to identify the routes of exposure, and to learn how best to manage these hazards is the key to bringing an understanding of safety. Also, learning about emergency procedures, how emergency equipment operates, and how to make decisions about emergencies can bring a better understanding for the need for safety and at the same time prepare the person to work in the laboratory where incidents may occur.

Becoming a good leader requires that you have good people skills, that you are able to listen and communicate well with people, and that people come to trust your judgment. These are not easy skills to learn, but it is possible for people to learn to be good leaders.[6] If you are interested in learning more about leadership, there are many books on leadership and you should make an effort to continue to build your skills by reading these kinds of books.

References

1. A. SCHWEITZER. BrainyQuote; available at http://www.brainyquote.com/quotes/authors/a/albert_schweitzer.html (accessed December 20, 2008).

2. T. WEBBER. School chemistry accidents expose lack of safety oversight. Daily Chronicle, July 6, 2002; available at http://www.daily-chronicle.com/articles/2002/07/06/news/export9289.prt (accessed December 22, 2008).

3. E. SCOTT GELLER. *The Psychology of Safety Handbook*, Lewis Publishers, Boca Raton, FL, 2004.

4. E. SCOTT GELLER. *The Psychology of Safety Handbook*, Lewis Publishers, Boca Raton, FL, 2004, pp. 453–475.

5. E. SCOTT GELLER, *The Psychology of Safety Handbook*, Lewis Publishers, Boca Raton, FL, 2004, pp. 478–499.

6. J. C. MAXWELL. *The 21 Irrefutable Laws of Leadership: Follow Them and People Will Follow You*. Nelson Business, Nashville, TN, 1998.

QUESTIONS

1. What is one of the most important features of managing safety as a leader?

   (a) Establishing punitive responses for errors in safety

   (b) Being familiar with laws and regulations

   (c) Setting a good example

   (d) Setting a "chain of command" hierarchy

2. What is the most effective motivation for lab personnel with regard to maintaining a safe environment?

   (a) The desire to maintain a zero accident (incident) rate
   (b) An internally driven value for safety
   (c) Bonuses in pay
   (d) Making safety procedures convenient and expedient

3. According to Geller, people tend to see failures as caused by

   (a) Their own mistakes
   (b) Unclear rules

   (c) Factors beyond anybody's control
   (d) External factors

4. In Incident 1.3.1.1, what factor led to the injury of the students?

   (a) Failure of the students to take adequate steps to protect themselves
   (b) Failure of the teacher to anticipate problems
   (c) A faulty piece of equipment
   (d) Inadequate preparation for a common emergency

1.3.2

EMPLOYERS' EXPECTATIONS OF SAFETY SKILLS FOR NEW CHEMISTS

Preview   This section provides suggestions about what employers might expect in safety skills for new employees (new graduates) who begin working at their facilities.

Each one of us has a vision of how we would like our co-workers to embrace the value of working safely as standard practice, the way things should be done.

Kenneth Fivizzani[1]

INCIDENT 1.3.2.1   ALLERGIC REACTION TO FORMALIN[2]

A new employee was working in a laboratory that required preparation of a formalin solution. She prepared a small carboy with this solution in a deep sink in the laboratory, but spilled some of the solution. She did not clean up the spill, but left a few minutes later. Shortly thereafter another employee who had worked in the laboratory for many years arrived and began doing her work. After a short time there she began to develop a cough and as she continued to work it became more severe and soon she realized that she was having trouble breathing. She exited the laboratory and called for help. Her colleagues took her to the workplace clinic. She was put on oxygen and transported to a hospital. She later recovered. Investigators learned of the formaldehyde (formalin) exposure. The older employee had become sensitized to formaldehyde and was no longer able to do her work in this laboratory.

What lessons can be learned from this incident?

How Employers View Safety

Many of you will leave academia to pursue a chemistry-based career with industry, business, government, or nonprofit organizations. The chemical industry has learned the value of safety through many hard lessons.[3] In the early part of the 20th century, industry had high accident rates that produced big liabilities and lawsuits. Over the years since then, the vast majority of industry has learned that a strong safety program is good for business, and although it requires an investment of resources, in the end it is better than ignoring safety, which can jeopardize the very existence of a business.

Nevertheless, incidents do occur in all organizations, institutions, and companies. You will likely never hear about most of these incidents as most are not publicly reported. Good organizations learn lessons from incidents and take actions to strengthen their safety processes and programs and foster a strong safety culture. However, you read in the news about chemical incidents occurring because an organization failed to value safety, failed to establish a strong safety process and program, and allowed unsafe conditions to escalate into a serious event. Most major players in the chemical industry not only have strong safety programs but they are leaders in safety and some actually have organizations that teach safety management to others.

Industry has learned the importance of making its employees partners in safety. Safety is accomplished when a company promotes safety and supports safety programs with funding, training, and time devoted to carrying out jobs safely. Employees are regarded as safety partners, helping keep the overall company safe. That means that they not only expect employees to be skilled in safety, but they expect that employees will help develop, implement, and follow safety policies, practices, and procedures. Furthermore, most will encourage employees to actively help strengthen safety programs through reporting unsafe conditions and helping develop ways to remedy those shortcomings. So when you join a company, your job is to help the company produce products and an important part of that job is safety. This may be different from your experience in academia, and often new employees are "shocked" by the emphasis on safety in industry and business. But this really should not be too surprising since this affects the bottom line—profits.

The Working World

Employers have a business to run and they want to protect their interests. They will expect you to adopt their interest and become a viable and strong part of the team—especially the safety team. Lapses in safety cost money, loss of time, possible injuries, property damage, and often bad publicity in the local or even national media. This is bad for the bottom line and companies will want you to understand that they expect you to do your job safely and keep their workplace safe.

Companies need to make money or they go out of business and their employees will be without jobs. They will want to do all they can to make the workplace safe, but there will be limits on what they can or will want to do in the safety arena. There is always a balance between risk and safety. This means that they do not want you to blindly do your job without considering the consequences of bypassing (or ignoring) safety procedures; they want you to do your job safely. The name of the game is making good judgments about risk assessment and risk management, which involves taking steps to minimize risk in environments where some level of risk is inevitable.

The company's safety record and the management's attitude toward safety are likely to be something that you consider when you select an employer. Do not hesitate to ask a prospective employer about safety. You may learn something about their attitude and approach to safety that may help you make important judgments. It is important that you take a positive attitude about safety to your new job, and you will want to work for an employer who has a similar interest in safety.

Attitude Is Everything

A positive and strong attitude toward safety is critical to maintain a safe workplace. If you have been using this book, you will have learned safety in little bites over the past few years. Maintaining a good attitude toward safety requires a continuous reminder that it is important, and by learning safety in small doses over a long period of time, you have established the kind of pattern needed in your new workplace. It will be up to you to maintain this attitude through continued learning and through continued mentoring of others in your sphere of influence. Consider the truth in the famous quote by Charles Swindoll[4]:

> The longer I live, the more I realize the impact of attitude on life. Attitude, to me, is more important than facts. It is more important than the past, the education, the money, than circumstances, than failure, than successes, than what other people think or say or do. It is more important than appearance, giftedness or skill. It will make or break a company... a church... a home. The remarkable thing is we have a choice everyday regarding the attitude we will embrace for that day. We cannot change our past... we cannot change the fact that people will act in a certain way. We cannot change the inevitable. The only thing we can do is play on the one string we have, and that is our attitude. I am convinced that life is 10% what happens to me and 90% of how I react to it. And so it is with you... we are in charge of our Attitudes.

While you may or may not agree with this quote, we believe that you will find that your attitude toward safety plays a key role in how you do or do not incorporate safety into your life. And it also affects the others around you—a positive attitude can be contagious and is often welcomed by all. It does not take long to learn who merely gives "lip service" to safety and who actually believes and practices safety.

Safety Skills Are Essential

In support of a positive attitude toward safety, you must have the skills necessary for the tasks ahead. Remember (How could you forget?) our continuing mantra—RAMP. You must be able to recognize hazards, assess the risks of those hazards, manage and minimize those hazards, and prepare for emergencies. These safety skills will serve you well in your new job. We have given you a lot of information about hazards that you might encounter in the laboratory. We expect that much of this will be a valuable background for your new job, but we are also aware that every job has its own particular set of hazards and it will be up to you to learn about these. Similarly, once you learn to assess how you can be exposed and determine the relative risk of the hazards you will be using, you will be able to do this in your new workplace.

Managing and minimizing the hazards will be dependent on the facilities, equipment, and other measures provided to you by your employer. It is likely that you will have what is needed to do your job in that area. Your employer has hired you because he/she believes that you can contribute to the success of the company's operations, and your expertise in safety can help your employer operate safely. Lastly, you must be prepared for handling emergencies. The longer you work in any given place, the more likely that you will encounter some sort of emergency. Be as prepared as you can be, and learn as much as you can, about what might go wrong, so that you can deal with emergencies calmly and effectively. Remember, simulation and practice are essential in emergency preparation.

Your Expectations as an Employee

We have discussed what your employer expects of you—but what should you expect of your employer? While every employer is different, it is not unreasonable to expect a safe and healthy workplace to do your job. Most of us also know that sometimes employers may not provide the best possible working conditions, but as a new employee you should not expect to be put in harm's way. This again does not mean that you will be working in a risk-free environment—there is risk in everything we do, including our jobs. You have been hired in part because you have some expertise in how to safely handle hazards, and your employer will expect that you will have these skills and be able to use them effectively. Employers will expect you to accept some risk and do your job to the very best of your ability by minimizing those risks. Some employers actually welcome ideas for improvements, particularly if they don't cost much in time and money. Some employers recognize that upfront investment for safety, in the long term, saves them money and greatly reduces their liabilities. Other employers may not have this perspective. It will take a little time to learn about your employer's stance on safety. Your job will be to help your employer recognize when safety needs to be addressed.

Leading, Supervising, and Managing Employees—The Path to Safety

Some of you may be called upon to become supervisors and managers over one or more employees. This requires another whole set of skills that you will have to learn on the job and through more education and training. However, when it comes to safety, you set the tone and the attitude for your team. People will look to you for safety guidance and will trust that you will not allow them to be put in dangerous situations. It is important that you earn their trust early. It is quite possible that you will have to teach them how to do their jobs safely. This is where you can pass on the RAMP message. Some chemists' careers migrate into the area of health and safety. See *Special Topic 1.3.2.1* Chemical Health and Safety as a Career or Collateral Duty.

SPECIAL TOPIC *1.3.2.1*

CHEMICAL HEALTH AND SAFETY AS A CAREER OR COLLATERAL DUTY

There are many fields of endeavor in chemistry. As you are deciding what you want to do and where you want to do it, you may want to consider chemical health and safety as a career. Alternatively, you may have an opportunity to take on some responsibilities for chemical health and safety as a collateral (additional) duty. For example, your employer might designate you as their Chemical Hygiene Officer (CHO, see Section 3.3.1). In many smaller companies without a formal Environmental Health and Safety division, the person designated as the CHO (a mandatory position required by OSHA) isn't necessarily someone with any particular, specialized training in this area, but rather the "best qualified" person on site to take on the task.

A strong understanding of chemistry is an invaluable tool in chemical health and safety. Laboratory work is also a valued experience for one considering chemical health and safety since many organizations involved in chemistry have laboratories. It is easier to understand those working in laboratories if you have direct experience and knowledge of the kind of work that they are doing. Most of you will find that chemical health and safety touches you in virtually any job in the chemical enterprise—be it industry, government, or academia.

What kinds of things do you do as a chemical health and safety officer or specialist? Your primary job would be to make sure that the health and safety of people in your area of responsibility are protected and measures to prevent incidents are developed, implemented, and maintained. As you might guess from looking at the scope of this book by reviewing the table of contents, health and safety covers all areas of working with chemicals. This includes preparing for emergencies, working with solvents or reactive materials, developing plans to minimize exposures to radiation or biological materials, and managing hazardous wastes or storing chemicals. Many health and safety people focus on a few areas and work with others as a team to cover all areas.

In these jobs you might be asked to help:

- Examine procedures and assist in finding ways to best minimize exposures
- Recommend the best personal protective equipment or safety equipment for an operation
- Establish emergency procedures and liaisons between employees and emergency responders
- Carry out air sampling of an environment to assess potential exposure
- Analyze those samples to determine exposure
- Learn about laws and regulations and ensure your organization follows requirements
- Conduct inspections to find hazards and areas that need improvement

You will probably need some additional education in these areas. Some people actually move into a field known as industrial hygiene to learn about worker health and safety. Many people also learn what is necessary to know when they have a need to learn it—known by many as OJT (on job training).

Perhaps the most important aspect of your job will be working with people, gaining their trust that you know your field and that you are there as a colleague to help, not hinder, them to do their jobs safely without incidents or injuries. Working in chemical health and safety can be a rewarding career because your work will be about incident prevention and saving people from injury or worse. Your work will be keeping your organization safe in order to prevent legal and financial trouble, and your work will be helping your fellow employees and colleagues and your organization accomplish its goals or mission in a safe and healthful way. This combination of technical and "people" skills is well suited to some chemists.

It is common for new employees to have little or no education in safety, so it will be up to you to give them direction and guidance. Teaching them about the hazards of their workplace is a good start. Explaining the principles of assessing the risks of hazards may be something new to people, although they probably do this everyday but don't realize that it is assessment. Helping your team members to minimize the risks of the hazards in their workplace is one of the most critical skills they can learn. Finally, teaching them to be prepared to handle emergencies will help them have confidence in themselves and in you.

Protecting your employees and your employer requires some knowledge of legal requirements. There are many of these and while you will not need to be an expert, you will need to know and understand the basics of these. There are laws and regulations to protect workers on the job—including laboratory workers. These come from the Occupational Safety and Health Administration (OSHA) in the United States. Similarly, the Environmental Protection Agency (EPA) regulates hazardous waste, and mismanagement of hazardous waste can become enormously expensive through clean-up costs and fines. The Department of Transportation (DOT) oversees the safe shipping of hazardous materials. The Nuclear Regulatory Commission (NRC) regulates the use of radioisotopes and the Centers for Disease Control and Prevention (CDC) with the U.S. Department of Agriculture (USDA) together regulate Select Agents that you may encounter in the laboratory. You will learn more about these in Section 1.3.3.

The Safety Ethic

In Section 1.1.1 we introduced the student safety ethic: *I work safely, avoid unnecessary risk, and accept responsibility for safety*. When making the transition from the academic to a workplace environment, it is appropriate to modify this ethic to reflect the new responsibilities and conditions associated with being an employee.

Why do you need strong safety ethics? Your first thought is probably self-preservation and long life—and for good reason. People with weak safety ethics and a poor education in safety frequently put themselves at higher risks and are likely not to survive long without injury to themselves or others.

We encourage you to memorize the safety ethic: *I value safety, work safely, prevent at-risk behavior, promote safety, and accept responsibility for safety*.[5] This is a simple statement but it has broad implications and defines broad actions to implement. So let's look at each element and provide some interpretative guidelines to give us a common understanding about what is meant by each part.

- To *value safety* means that you make safety a positive, integral part of your everyday activities; safety is a value not to be compromised, safety is an inseparable part of your daily activities requiring prudent behavior, and safety prevents and protects you, your family, your co-workers, and others from harm or suffering.

- To *work safely* means that you are educated in safety to minimize risks of injury or illness, you continually strive to learn about safety, you learn to recognize and evaluate hazards, you identify and practice safe procedures, you seek ways to minimize, reduce, and control risks from hazards, and you maintain a high level of awareness of safety at home, on the job, and at leisure.

- To *prevent at-risk behavior* means that you learn to recognize at-risk behavior, you do not practice at-risk behavior, you seek to prevent at-risk behavior by others, and you maintain awareness of at-risk behavior.

- To *promote safety* means that you promote safety to others through your daily actions, you act as an example and a leader for others in safety, you act as a mentor educating the unknowing and inexperienced in safety, you recognize others for their safe acts, and you pass *the safety ethic* to others.

- To *accept responsibility for safety* means that you are responsible for your safety, you are responsible and accountable for the safety of your family, co-workers, and employees, ensuring that they know and understand about its importance, you actively pursue the safe way to do things, you ensure your employees' safety with training, proper safety equipment, and safe facilities, and you will not compromise the safety of your family, co-workers, employees, or others.

Adopting *the safety ethic* will require a significant effort by you. For example, to work safely requires that you learn to recognize and evaluate hazards—this is one of the principal objectives of this entire text and it requires a base of knowledge that you must build on continuously throughout your lifetime, not just in this course of study.

RAMP

- A new job requires that you *recognize* the importance of a positive and strong attitude toward safety, and a need to bring your knowledge of safety to this new workplace.

- You will need to *assess* the risks of this new workplace and determine the best way to minimize those risks within the limitations of the facilities, equipment, and measures provided by your employer.

- You will need to assist your employer by *minimizing* the risks of your new workplace, and if you are supervising people you will need to also help your employees learn about safety and its importance, so that they will be able to minimize the risks of its hazards.

- You will need to be *prepared* to deal with emergencies that might occur in a new workplace, and similarly help *prepare* your employees with simulations and practice of emergency procedures.

References

1. K. P. Fivizzani. Transforming employees into safety partners. *Journal of Chemical Health and Safety* **11**(*3*):9–11 (2004).
2. R. H. Hill. Personal account of an incident.
3. R. H. Hill. The emergency of laboratory safety. *Journal of Chemical Health and Safety* **14**(*3*):14–19 (2007).
4. ThinkExist.com Quotations. Charles R. Swindoll quotes; available at http://thinkexist.com/quotes/charles_r._swindoll/ (accessed October 14, 2009).
5. R. H. Hill. The Safety Ethic: Where can you get one? *Journal of Chemical Health and Safety* **10**(*3*):8–11 (2003).

QUESTIONS

1. New employees in most chemical industry positions find that

    (a) They have been adequately trained with regard to safety
    (b) Safety regulations in industry are generally less stringent than in academia
    (c) Learning about safety and functioning safely is part of "the job" in industry
    (d) Most companies find that safety programs cost too much money to implement

2. In industry, lapses in safety cause

    (a) Bad publicity
    (b) Property damage
    (c) Injuries and loss of work time
    (d) All of the above

3. When interviewing for a job in industry, it is best

    (a) Not to mention anything about safety since employers will not want to hire you
    (b) Ask questions about safety practices so that you can impress the interviewer
    (c) Ask if the company has ever been cited for OSHA violations
    (d) Ask questions about safety to display your own interest in safety and to learn if the company values safety

4. It is reasonable for a lab employee to expect

    (a) To work in a risk-free environment
    (b) That all employers will value safety
    (c) To work in a environment where safety is valued
    (d) That all recommendations about safety improvement will be immediately implemented

5. After having worked at a company for a while, you can expect new employees to

    (a) Be familiar with all state and federal safety regulations
    (b) Value safety on a daily basis
    (c) Display a wide range of skills and attitudes about safety in the workplace
    (d) Handle all emergencies efficiently and confidently

6. What is a good reason to adopt the safety ethic?

    (a) It will help you to function effectively and safely as an employee.
    (b) It will likely lead to larger pay raises.
    (c) It is required by OSHA.
    (d) It makes your job easier.

7. Which activity is *not* part of the safety ethic?

    (a) Valuing safety
    (b) Preventing at-risk behavior
    (c) Reporting incidents immediately
    (d) Accepting responsibility for safety

1.3.3

LAWS AND REGULATIONS PERTAINING TO SAFETY

Preview   This section provides a brief overview of the laws and regulations pertaining to safety and chemicals in the laboratory.

> *Abandon hope, all ye who enter here.*
>
> *The Divine Comedy*, by Dante[1]

INCIDENT 1.3.3.1   LITHIUM ALUMINUM HYDRIDE EXPLOSION[2]

A student working in a university laboratory needed lithium aluminum hydride (LAH) for part of an experiment. When he opened the can, LAH came in contact with moist air and it exploded causing burns to the face and hands of the researcher. He was transported to a local hospital for treatment. The local fire department was finally able to contain the fire, but there were long delays in returning to the building as the HazMat team worked hours to ensure that it was safe for all to return. It was decided by the authorities that the building had to remain closed until environmental testing could verify that it was safe to reenter the building. There were additional concerns because there were experimental animals in the building and their welfare was in question.

What lessons can be learned from this incident?

Laws and Regulations

Laws are a set of rules established by an authority—our government (federal, state, local). These are established for the common good of our people and are usually broad documents that express the desired outcome. *Regulations* are the specific set of requirements established to carry out the details of an established law.

The Occupational Safety and Health Act of 1970 was the law established by Congress to provide protection for workers in American workplaces.[3] Its purpose was to provide a safe and healthful workplace for every man and woman in our nation. This law was assigned to the Department of Labor (DOL) and the Occupational Safety and Health Administration (OSHA) was established to carry out the law. OSHA set up regulations under the Code of Federal Regulations (CFR) to administer the law. In the CFR the DOL is assigned to 29 CFR and the occupational safety and health regulations are found under Part 1910.

SPECIAL TOPIC *1.3.3.1*

WHAT IS THE CODE OF FEDERAL REGULATIONS?

When Congress passes a law it becomes incorporated into the U.S. Code (USC), and this code is divided into 50 titles (sections) that cover all of the various areas of our society requiring and having laws. Various federal agencies are charged with carrying out those laws and they do so by developing and implementing (called promulgation)

---

regulations that are the detailed steps to implement the spirit of the law. When federal agencies formulate, advertise, and finally adopt regulations, they initially publish these in a publication called the *Federal Register*—a daily governmental publication. The regulations published here are then encoded into the Code of Federal Regulations, also known as the CFR.

When reading about laws about health and safety you will often encounter seemingly arcane references such as "29 CFR 1910.1200." What is this?

There are 50 "titles" in the CFR and to get a sense of the range of the CFR it is best to simply examine the 50 titles at http://www.access.gpo.gov/nara/cfr/cfr-table-search.html#page1. Title 29 deals with "Labor" (see Figure 1.3.3.1). There are nine "volumes" in 29 CFR and Volume 5 contains "Part 1910," which is where the laws and regulations associated with the Occupational Safety and Health Administration are listed. Within 29 CFR 1910, there are 26 "subparts" labeled with the letters "A–Z." Subpart Z has the title "Toxic and Hazardous Substances" and is the location of the "Hazard Communication" standard (29 CFR 1910.1200) and the "Occupational Exposure to Hazardous Chemicals in Laboratories" standard (29 CFR 1910.1450—commonly called the "Lab Standard"). (See Figure 1.3.3.1.)

Title 29:

Volume 5 (Parts 1900-1910)

1910 – Occupational safety and health standards

1910 Subpart Z – Toxic and Hazardous Substances

1910.1200 Hazard Communication

1910.1450 Occupational Exposure to Hazardous Chemicals in Laboratories

FIGURE 1.3.3.1 Location of the Hazard Communication Standard and "Lab Standard" in the Code of Federal Regulations. The CFR is millions of pages of federal legislation. Finding a particular regulation requires knowledge of the organization of the CFR.

In addition to the general regulations about chemicals in the Lab Standard, there are also 29 specific regulations in Subpart Z about particular chemicals and these supersede the rules in the Lab Standard. For example, 29 CFR 1910.1048 is about formaldehyde.

To establish these specific regulations, OHSA must go through a "rule-making" process requiring them to:

- Publish a notice of rule-making
- Solicit input and information from organizations or people (accept documents, hold hearings)
- Arrive at a proposed standard after its own internal deliberations considering all input
- Publish the proposed standard for final rule-making that solicits further review and input
- Publish the final rule

As you can understand, this is a difficult, time-consuming, onerous process and often evokes strong disagreements and not infrequently results in legal challenges. For example, in 1989 OSHA sought to update its 1971 PEL list (found in 1910.1000, Subpart Z, referred to commonly as the Z-tables). In this update OSHA lowered PELs for 212 substances, raised the PEL for 1 substance, and added 164 new PELs for previously unregulated substances. However, a number of legal challenges by industry and labor groups resulted in an appeals court issuing a decision to vacate (throw out) the new standard based on the need for more information about health effects and feasibility analyses. As a result, OSHA reverted back to using the 1971 PEL list for compliance, and this is where it stands as 2010 starts. This may answer any questions you have as to why OSHA has not updated its "old" PELs with the newer TLVs.

There are many laws and regulations but we will focus on those that affect laboratories. At times when reading through regulations it seems that you have entered *The Inferno* of Dante's *Divine Comedy* due to the complexity of the regulations. Furthermore, regulations are not always driven "only" by science since lobbyists, companies, unions, and other political entities are part of the process of constructing them. Nonetheless, ignorance of the law is no excuse for failing to adhere to these regulations and one must strive to learn as much as possible about laws that affect the operations of chemistry laboratories.

Occupational Safety and Health Regulations — Legal Requirements

OSHA established regulations to clearly set out the minimum legal requirements, called standards, for a safe and healthful workplace.[4] While there are many regulations, you probably only need to know about a select few that pertain to laboratories[5] and we will focus on those. Keep in mind that OSHA regulations are mandatory, legal requirements—your employer is required to comply with all aspects of these regulations and is subject to inspections and potential citations for violations by OSHA. As a university student you are not covered by OSHA regulations, but employees working for the university and in the laboratories are covered. This includes lab assistants, lab managers, teaching assistants, instructors, faculty, and staff. Examples of occupational standards are provided below.

> *29 CFR 1910 Subpart H (1910.120), Hazardous Waste Operations and Emergency Response* (known as HAZWOPER). This regulation requires that anyone who responds to an emergency involving hazardous waste receive training and demonstrate competence in emergency response operations.

> *29 CFR 1910 Subpart I Personal Protective Equipment*. This regulation describes general standards for head, foot, face, and eye protection, skin protection, and respiratory protection.

> *29 CFR 1910 Subpart Z Toxic and Hazardous Substances*. These are standards for specific toxic and hazardous substances. Certain parts of these standards apply to laboratories and other parts do not. Most specific airborne standards for chemicals are applicable to laboratories, but other requirements do not apply. Two standards within Subpart Z are:

>> *29 CFR 1910.1200 Hazard Communication*. This regulation, known as HazComm, established the requirements for notifying workers of hazards in the workplace. It is the regulation that requires Material Safety Data Sheets (MSDSs) and the need for training workers to recognize hazards. OSHA issued a proposed rule-making notice in October 2009 to change HazComm to implement the Globally Harmonized System of Classification and Labeling of Chemicals (GHS). This will change the requirements of HazComm to require all suppliers of chemicals to conform with the GHS hazard rating and labeling system (see Sections 3.2.1 and 6.2.1 to learn more about GHS).

>> *29 CFR 1910.1450 Occupational Exposure to Hazardous Chemicals in Laboratories*. This regulation, known as the Lab Standard, established the broad requirements for handling hazardous chemicals in a laboratory. This is a performance standard, which means that OSHA did not specifically identify methods, procedures, or practices to accomplish the purpose—to protect laboratory workers from chemicals. Rather, the management of a laboratory can decide how to protect workers as long as they meet certain minimal requirements, such as having a Chemical Hygiene Plan and a Chemical Hygiene Officer or Committee. OSHA recognized that handling chemicals in a laboratory is very different from the general chemical industry. Thus, the Lab Standard supersedes many of the requirements for the general industry.

Other parts of the OSHA standards may regulate your specific workplace, but you will have to learn more about these on your own.

Hazardous Waste Regulations — Legal Requirements

All hazardous waste generated in the United States is subject to regulations established by the U.S. Environmental Protection Agency (EPA) under the Resource Conservation and Recovery Act (RCRA, 1976). RCRA (pronounced "reck-rah") established the requirement for a "cradle-to-grave" system for hazardous waste—requiring it be tracked from the time it was initially generated until the time that it is finally disposed of. These regulations are complicated and generally most companies rely on in-house experts designated within its management to handle these issues. They frequently hire contractors to dispose of their hazardous wastes and track the disposal with documentation that can be presented in legal reviews. The regulations for hazardous waste management begin at 40 CFR Part 260.[6] You should know the following about hazardous waste:

- There are specific requirements depending on whether an organization is a small scale generator or a large scale generator; there are more stringent requirements for large scale. In November 2008 the EPA amended its regulations to allow academic hazardous waste generators an alternative approach for handling hazardous laboratory waste under the Academic Labs Rule (40 CFR 262).[7]
- There are very stringent requirements for records and documentation.
- There are specific time lines for disposing of hazardous material once it has been generated and declared as hazardous waste—an institution only has so many days to take action to dispose of this waste.
- Organizations treating, storing, and disposing of hazardous waste must obtain a permit from the EPA or the local state agency assigned that responsibility. Typically, those working in laboratories do not treat their hazardous waste, but rather use a contractor to dispose of their hazardous waste.
- Hazardous waste processors must have permits to treat and dispose of waste.
- The generator of hazardous waste is responsible for the ultimate disposal of its hazardous waste. This means that a hazardous waste processor must be chosen carefully because if the processor does not dispose of waste properly, the generator is still responsible—even if that means the generator has to pay for it again if it is uncovered at a later time. This is a potentially large liability.
- Pouring hazardous waste down the drain and into the water waste stream is prohibited by law.
- There are potentially huge fines for improper treatment of hazardous waste. These can be hundreds or thousands of dollars per day going back to the time of generation. This is not an area that you want to ignore.

Additional information about handling wastes is described in Section 8.3.4.

Radioactive Materials Regulations — Legal Requirements

It is quite possible that, sometime in your laboratory operations, you will have a need to use or come in close proximity to radioactive materials. These may be radioactive sources that are contained in "closed" systems, such as analytical instrumentation (X-ray diffraction, analytical detectors, etc.), or that are not contained but rather are "open" sources of radioactive materials, such as ^{14}C-labeled compounds or ^{3}H-labeled compounds.

If you do encounter these kinds of radiation sources, your facility will be regulated by a state nuclear regulatory agency or by the Nuclear Regulatory Commission (NRC). These agencies were established to protect the public from radioactive hazards. The NRC closely monitors the safety of large scale sources of radiation, such as nuclear reactors and nuclear power plants, but it also has a strong program to monitor the use of radioisotopes in laboratories. To use radioactive materials your facility must be licensed by the NRC and must have a documented Radiation Safety Program.

NRC regulations governing the use of radioisotopes in facilities including laboratories is found in 10 CFR Part 20.[8] Briefly, you should know that NRC requires:

- Laboratory workers keep exposure "as low as reasonably achievable" (ALARA)
- Compliance with maximum dose limits per year for radiation exposure
- Licensed facilities designate a Radiation Safety Officer or a Radiation Safety Committee
- Monitoring, comprehensive inventory, and security of sources
- Designated locations for work with radiation and signs identifying these areas
- Extensive training and record keeping

The NRC or authorized state agencies conduct unannounced inspections of all licensed facilities. These inspections should go well if your facility has a strong radiation safety program. The NRC issues citations and fines for failures to protect workers and to follow requirements.

Select Agent Regulations — Legal Requirements

In 1996 Congress passed the Antiterrorism and Effective Death Penalty Act that established criminal prosecution and punishment for anyone who used chemical or biological agents for terrorist purposes. A group of more than 80 of these agents were called "Select Agents" and were composed of a large number of biological agents and biological toxins—these agents were selected for their extremely hazardous properties and their potential to cause significant harm to the public.[9] The regulations have established requirements on the use of these agents in laboratories with an emphasis on security.

Registration is required for use of these agents. Different agents require registration with particular government agencies:

- The Centers for Disease Control and Prevention (CDC) for human pathogens and toxins [part of the Department of Health and Human Services (HHS)]
- The Animal and Plant Health Inspection Service (APHIS) for animal pathogens [part of the U.S. Department of Agriculture (USDA)]

Violations of these Select Agent regulations may be deemed criminal and subject to investigation by the Federal Bureau of Investigation and prosecution by the U.S. Department of Justice.

Because there is much overlap between chemistry and biology, especially in molecular biology, it is possible that you may become involved in research with these agents or toxins. Regulations governing these Select Agents are found in 42 CFR Part 73, 7 CFR Part 331, and 9 CFR Part 121. Briefly, you should know that the Select Agent Program (SAP) requires that:

- A strong security plan be in place to limit access of these agents
- Criminal background checks be done for all persons having access to these agents
- There are designated locations and areas for Select Agent work
- Laboratories using these materials follow the safety guidelines established by the CDC and the National Institutes of Health, which are found in *Biosafety in Microbiological and Biomedical Laboratories*, 5th edition (or latest edition)
- Laboratory work be limited to the agents and activities listed in the lab's permit—you can't start using another agent on the list unless you have included that in your application
- Transfers of agents can only be to other laboratories that are registered and approved to receive these agents—you can't just bottle up a sample of a Select Agent and send it to a colleague, special transfer procedures are required

Because there can be funding for work with these Select Agents, there are many laboratories involved in research with these materials and thus many institutions have become registered in SAP.

The Toxic Substances Control Act

The Toxic Substances Control Act, known as TSCA ("tos ka"), was established in 1976 to give authority and responsibility to the U.S. Environmental Protection Agency (EPA) to track and screen chemicals produced in large quantities that go into commerce.[10-12] TSCA's premise was to protect the public and the environment against unreasonable risks from toxic substances, before chemical products enter public commerce. While the idea of TSCA may sound like a reasonable and significant step in protecting the public, TSCA has largely been ineffective as it has been hamstrung with restrictions and limitations imposed to protect innovations from the chemical industry that might find productive and commercial uses for chemicals. TSCA requires the EPA to use a formal rule-making process in its implementation and the result of this stringent requirement has been that only five chemicals have been regulated under TSCA, even though thousands are suspected or known to have significant toxic properties. Although the

EPA has more than 80,000 chemicals in its TSCA inventory, it does not require chemical producers to generate or report information on health and environmental safety of these chemicals.[13] While a useful strategy of TSCA might be improvements in green chemistry efforts, there is no funding mechanism under TSCA to support this kind of activity. A report by the University of California concluded that the inadequacies of TSCA resulted in gaps in available significant toxicity data and slowed the development of new greener technology.[13] The Obama administration is considering options that might strengthen TSCA, better protect the public, and foster innovation.[14]

Emergency Planning and Community Right-to-Know Act (EPCRA)

The Emergency Planning and Community Right-to-Know Act[15] (EPCRA) sets requirements for federal, state, and local governments, Indian tribes, and industry regarding emergency planning and "Community Right-to-Know" reporting on hazardous and toxic chemicals. EPCRA falls under the jurisdiction of the Environmental Protection Agency. The Community Right-to-Know requirements were designed to help increase the public's knowledge and access to information on chemicals at individual facilities, their uses, and releases into the environment. EPCRA was passed due to concerns regarding the environmental and safety hazards posed by the storage and handling of toxic chemicals. A disaster in Bhopal, India, that killed and injured thousands of people from the release of methyl isocyanate played a large role in the passage of EPCRA in the United States (see *Special Topic 7.1.1.2 A Case Study in Risk Management—The Tragedy at Bhopal, India*).

While EPCRA has several requirements regarding emergency planning and response, the part that impacts laboratories pertains to emergency response planning. All facilities (including laboratories) involved in manufacturing, processing, or storing designated hazardous chemicals must make Material Safety Data Sheets (MSDSs) describing the properties and health effects of these chemicals available to state and local officials and local fire departments. Furthermore, these facilities must also report, to state and local officials and local fire departments, inventories of all on-site chemicals for which MSDSs exist. Thus, information about chemical inventories at facilities and MSDSs must be available to the public under the Community Right-to-Know portion of EPCRA. So when you are required by your management to provide a chemical inventory for your laboratory, it is in part due to the requirements of EPCRA and is your part in being a good citizen to the community.

Chemicals That Could Be Used As Potential Terrorist Agents—Legal Requirements

In 2007 the Department of Homeland Security (DHS) established "Chemical Facility Anti-terrorism Standards" that seek to minimize the risk of chemicals being stolen and used for terrorist purposes. These regulations cover the use of 325 chemicals in chemical facilities that present high levels of security risk, including laboratories. Many of these chemicals are found in laboratories.

Because the DHS is most concerned about large stockpiles of chemicals that could be exploited by terrorists, its standards have threshold quantities for reporting—that is, if the quantities of those listed chemicals are below the threshold, then they are not reportable. While laboratories use relatively small quantities of any particular chemical, the rule applies to the quantities totaled together in all laboratories of an institution—so a university would have to take an inventory of all of the DHS listed chemicals throughout the campus and determine if the total exceeds the threshold. Many universities have found some of the listed chemicals on their campuses, although not in the large quantities that concerned the DHS. This report highlights the likelihood that some of these chemicals are in your laboratory building or your specific laboratory. DHS regulations may be found in 6 CFR Part 27 and the listing of the chemicals and reportable quantities are found in Appendix A of that regulation.[16]

RAMP

- *Recognize* that there are regulations covering the use of chemicals in laboratories.
- *Assess* the risks of using chemicals and take into consideration that the use of certain chemicals may have risks that are subject to regulations.

- *Minimize* the risks of exposure to chemicals in your laboratory and, where applicable, adhere to requirements of regulations.
- *Prepare* for your work in the laboratory in such a way that it has taken regulations into account.

References

1. Dante Alighieri. *The Divine Comedy*.
2. J. Montanaro. FAMU Researcher Burned in Experiment. WCTV Eyewitness News. Posted 12:35 PM, April 10, 2008; available at http://www.wctv.tv/news/headlines/17461924.html# (accessed October 14, 2009).
3. The Occupational Safety and Health Act of 1970; available at http://www.osha.gov/pls/oshaweb/owasrch.search_form?p_doc_type= OSHACT (accessed November 15, 2008).
4. Occupational Safety and Health Administration. Occupational Safety and Health Standards. 29 CFR 1910; available at http://www.osha. gov/pls/oshaweb/owasrch.search_form?p_doc_type=STANDARDS& p_toc_level=1&p_keyvalue=1910 (accessed August 28, 2009).
5. Occupational Safety and Health Administration. Safety and Health Topics: Laboratories; available at http://www.osha.gov/SLTC/ laboratories/index.html (accessed August 28, 2009).
6. EPA regulations; available at http://www.epa.gov/epawaste/laws-regs/ regs-haz.htm (accessed December 15, 2008).
7. EPA regulations. Academic Labs Rule. 40 CFR 262; available at http://ecfr.gpoaccess.gov/cgi/t/text/text-idx?c=ecfr&sid=5f9a97d1803 bb6e85e8695fb69e2e233&rgn=div5&view=text&node=40:25.0.1.1.3 &idno=40#40:25.0.1.1.3.11 (accessed August 28, 2009). See also http://www.epa.gov/epawaste/hazard/generation/labwaste/rule.htm (accessed August 28, 2009).
8. NRC regulations; available at http://www.nrc.gov/reading-rm/doc-collections/cfr/part020/ (accessed December 15, 2008).
9. National Select Agent Registry; available at http://www.cdc.gov/ od/sap/ (accessed August 28, 2009).
10. Environmental Protection Agency. The Toxic Substances Control Act; available at http://www.epa.gov/lawsregs/laws/tsca.html (accessed October 29, 2009).
11. F. Hoerger and R. Hagerman. Understanding the Toxic Substances Control Act: Compliance and Reporting Requirements. In: Handbook of Chemical Health and Safety (R. J. Alaimo, ed.), American Chemical Society, Washington, DC Oxford University Press, New York, 2001, Chapter 10, pp. 53–60.
12. L. Schierow. Congressional Research Service Report RL 30022, Toxic Substances Control Act; available at http://www.cnie.org/NLE/ CRSreports/BriefingBooks/Laws/k.cfm (accessed October 29, 2009).
13. M. P. Wilson, D. A. Chia, and B. C. Ehlers. *Green Chemistry in California: A Framework for Leadership in Chemicals Policy and Innovation*, California Policy Research Center, University of California, Berkeley, 2006. Prepared from the California Senate Environmental Quality Committee and the California Assembly on Environmental Safety and Toxic Materials; available at http://coeh.berkeley.edu/FINALgreenchemistryrpt.pdf.
14. T. Avril. Obama plan would tighten rules on toxic chemicals. *The Philadelphia Enquirer*, September 30, 2009; available at http://www. philly.com/inquirer/business/20090930_Obama_plan_would_tighten_ rules_on_toxic_chemicals.html (accessed October 29, 2009).
15. Environmental Protection Agency, Emergency Planning and Community Right-to-Know Act; available at http://www.epa.gov/oecaagct/ lcra.html (accessed October 29, 2009).
16. DHS 6 CFR Part 27, Appendix to Chemical Facility Anti-Terrorism Standards; Final Rule; available at http://www.dhs.gov/xlibrary/ assets/chemsec_appendixafinalrule.pdf (accessed August 15, 2009).

QUESTIONS

1. Which statement is *false?*
    - (a) Laws are a set of rules established by an authority.
    - (b) Regulations are a specific set of requirements within a law.
    - (c) Laws can be established by local, state, or federal governments.
    - (d) When working in a lab, it is important to distinguish between laws and regulations.

2. What groups of people influence how science-related laws and regulations are written?
    - (a) Lawmakers
    - (b) Scientists
    - (c) Lobbyists and unions
    - (d) All of the above

3. Who is under the jurisdiction of OSHA regulations?
    - I. Only nonmanagerial employees
    - II. Students
    - III. All employees
    - IV. Faculty (at colleges and universities)
        - (a) I and IV
        - (b) III and IV
        - (c) III

4. "1910.1200" is also known as
    - (a) The Lab Standard
    - (b) HazComm
    - (c) Subpart Z
    - (d) The Chemical Hygiene Plan

5. "1910.1450" is also known as
    - (a) The Lab Standard
    - (b) HazComm
    - (c) Subpart Z
    - (d) The Chemical Hygiene Plan

6. Standards for specific chemicals and substances are in
    - (a) 29 CFR 1910.1200
    - (b) CFR 1910.1450
    - (c) CFR 1910 Subpart Z
    - (d) CFR 1910 Subpart I

7. Standards for personal protective equipment are in
    - (a) 29 CFR 1910.1200
    - (b) 29 CFR 1910.1450
    - (c) CFR 1910 Subpart Z
    - (d) CFR 1910 Subpart I

8. RCRA

   (a) Is managed under the U.S. EPA
   (b) Is no longer in effect
   (c) Is not part of the CFR since it is a state-based program
   (d) Deals with hazardous waste only from industry and not from academic labs

9. RCRA requirements

   (a) Vary depending on the quantity of waste that the facility generates
   (b) Establish time lines for disposing of hazardous materials
   (c) Hold waste generators responsible for the ultimate disposal of waste even when contractors handle the waste
   (d) All of the above

10. The NRC governs the use of

   (a) Devices that produce X rays and gamma rays, but not radioactive materials
   (b) Radioactive materials, but not devices that produce X rays and gamma rays
   (c) Radioactive materials and devices that produce X rays and gamma rays
   (d) Nonradioactive chemicals that can decay into radioactive chemicals

11. "Select Agents" are

   (a) Substances with extremely hazardous properties and potential to cause significant harm to the public
   (b) Defined in the Antiterrorism and Effective Death Penalty Act of 1996
   (c) Sometimes encountered in molecular biology labs
   (d) All of the above

12. TSCA was implemented in what year?

   (a) 1976
   (b) 1979
   (c) 1986
   (d) 2007

13. TSCA was designed to

   (a) Measure the toxicity of commercially used chemicals
   (b) Protect the public and the environment from unreasonable risks to toxic substances, before chemical products enter public commerce
   (c) Protect the public and the environment from unreasonable risks to toxic substances, after chemical products enter public commerce
   (d) Protect the public and the environment from reasonable risks to toxic substances, before chemical products enter public commerce

14. What percentage of the chemicals in the TSCA inventory have been regulated under TSCA.

   (a) 0.006%
   (b) About 1%
   (c) About 37%
   (d) About 93%

15. The Chemical Facility Anti-terrorism Standards

   (a) Were established in 2002 by the Department of Homeland Security
   (b) Involve a relatively small list of chemicals that could be used for terrorist purposes
   (c) Require that colleges and universities register dangerous chemicals with the DHS regardless of quantity
   (d) Are not part of the CFR since the DHS operates outside of normal federal regulations

1.3.4

GREEN CHEMISTRY—THE BIG PICTURE

Preview   This section reviews selected *principles of green chemistry* that apply most directly to laboratory safety.

> *Green chemistry does not stand for the old model of regulation that industry fears. Regulation is a means of trying to fix something that is already broken. Green chemistry and engineering is about making sure we don't create problems from the beginning. We are attempting to make regulation obsolete.*

> Dr. Robert Peoples, Director of the American Chemical Society's Green Chemistry Institute[1]

INCIDENT 1.3.4.1   BROKEN THERMOMETERS[2]

In the latter part of the 20th century, mercury thermometers were commonly used in laboratories and also in the medical community to measure temperature. In the 1960s, a lab instructor was leading a laboratory session composed of student nurses. After the laboratory session had started, a student came to tell him that she had broken her thermometer. Moments later two more students came to tell him the same thing. He was puzzled by the rash of thermometers being broken and he walked about the laboratory. He saw one student begin "shaking" the thermometer down and, sure enough, the long thermometer hit the lab bench and was shattered. He asked why she was doing this, and she informed him that she was following instructions received in nursing classes to "shake down" thermometers. The lab instructor immediately called all the students together to explain that the thermometers being used in the laboratory did not need shaking, and the epidemic of broken thermometers ceased.

(This incident shows why we have moved away from using devices with mercury. Learn more the hazards of mercury spills in Incident 6.2.2.1.)

What lessons can be learned from this incident?

The Twelve Principles of Green Chemistry

In 1998 *Green Chemistry: Theory and Practice*[3] was written by Paul Anastas and John Warner. This book introduced the *Twelve Principles of Green Chemistry* presented in Table 1.3.4.1. These twelve principles are the explicit recipe for the overall goal of green chemistry as stated in Section 1.1.2. "Green chemistry is the design and use of methods that eliminate health and environmental hazards in the manufacture and use of chemicals."

This book is about laboratory safety, and the discussions in Sections 1.1.2 and 1.2.2 focused on how green chemistry can be applied in introductory and organic courses. We will continue our discussion of green chemistry and how it relates to improving laboratory safety but as a matter of your general education in chemistry we encourage you to consider all twelve principles as you progress through your career since the general goal of green chemistry and laboratory safety are similar in many respects.

Since the properties of many chemicals that make them useful as reagents are the very properties that make them hazardous (such as acid/base or redox properties), it is a practical necessity that chemistry cannot be "perfectly green." We take many of the principles in Table 1.3.4.1 as important signposts about the *direction* in which the enterprise of chemistry should change rather than statements of absolute goals. Furthermore, since it seems inevitable that some chemical reactions will always have hazardous

TABLE 1.3.4.1   Twelve Principles of Green Chemistry

1. Prevent waste. Design chemical syntheses to prevent waste, leaving no waste to treat or clean up.

2. Design safer chemicals and products. Design chemical products to be fully effective, yet have little or no toxicity.

3. Design less hazardous chemical syntheses. Design syntheses to use and generate substances with little or no toxicity to humans and the environment.

4. Use renewable feedstocks. Use raw materials and feedstocks that are renewable rather than depleting. Renewable feedstocks are often made from agricultural products or are the wastes of other processes; depleting feedstocks are made from fossil fuels (petroleum, natural gas, or coal) or are mined.

5. Use catalysts, not stoichiometric reagents. Minimize waste by using catalytic reactions. Catalysts are used in small amounts and can carry out a single reaction many times. They are preferable to stoichiometric reagents, which are used in excess and work only once.

6. Avoid chemical derivatives. Avoid using blocking or protecting groups or any temporary modifications if possible. Derivatives use additional reagents and generate waste.

7. Maximize atom economy. Design syntheses so that the final product contains the maximum proportion of the starting materials. There should be few, if any, wasted atoms.

8. Use safer solvents and reaction conditions. Avoid using solvents, separation agents, or other auxiliary chemicals. If these chemicals are necessary, use innocuous chemicals.

9. Increase energy efficiency. Run chemical reactions at ambient temperature and pressure whenever possible.

10. Design chemicals and products to degrade after use. Design chemical products to break down to innocuous substances after use so that they do not accumulate in the environment.

11. Analyze in real time to prevent pollution. Include in-process real-time monitoring and control during syntheses to minimize or eliminate the formation of by-products.

12. Minimize the potential for accidents. Design chemicals and their forms (solid, liquid, or gas) to minimize the potential for chemical accidents including explosions, fires, and releases to the environment.

Source: Anastas and Warner.[3]

features, it is important for students to learn to work in a hands-on fashion with hazardous substances under the safest possible conditions.

Being Green, and Safe

Let's review which of the twelve principles speak most directly to the issue of lab safety. As we discuss these you should consider that someday *you* are likely to be part of a team that will develop new lab products and processes so you will have the opportunity to make chemistry greener and safer. Consider, too, that someday your "lab" may be an industrial plant where you are working with chemical engineers and other scientists to run chemical reactions on much larger scales than anything experienced in a chemistry curriculum.

Principle #2: *Design safer chemicals and products*. The stated goal of "little or no toxicity" is certainly desirable, but also quite ambitious. A significant fraction of this book is devoted to the discussion of the toxicity of chemicals because many chemicals are toxic! And it is not uncommon to find after a chemical has been used for some time that it has previously unrecognized toxic consequences, such as cancer, reproductive effects, or other health effects. So it seems likely that we will not eliminate toxic chemicals, but the green approach will focus on using "less toxic" reagents and products. However, there is no doubt that "less toxic" is inherently safer with regard to the production, transportation, use, and disposal of chemicals. The Toxic Substances Control Act (1976) currently regulates the 80,000 chemicals in industrial use[4] but, perhaps surprisingly, does not require that manufacturers test chemicals with regard to toxicity before they are used in commerce. Thus, the design, production, and use of chemicals in the United States are largely driven by function, performance, and price with relatively little regard to toxicity. For both chemists and consumers, the burden of answering the question "Is this safe?" lies almost entirely with the user. The goal of "designing safer chemicals and products" requires an understanding of the toxicity of a particular chemical and considering a less toxic chemical.

Including "less toxic" along with "function, performance, and price" as a design parameter will be a new paradigm for the chemical industry.

A few examples of Principle #2 precede the era of green chemistry. Lead poisoning, particularly in children, is known to cause neurological damage. Many years ago it was discovered that small children sometimes ingested paint chips containing lead-based paint pigments. Removing the lead from paint was a relatively easy chemical change that made a product less toxic. Another lead-based example is the use of tetraethyl lead as an antiknock agent in gasoline for several decades. The combustion of "leaded" gasoline produced significant concentrations of lead in the atmosphere that was unhealthy to breathe. As you know, we now use "unleaded" gasoline, and this practice started in the 1970s. While this is safer from a toxicological perspective, the reason for replacing tetraethyllead with other antiknock agents was not to reduce lead levels in the atmosphere but because it "poisoned" catalytic convertors that were introduced in almost all cars in the mid-1970s.

Principle #3: *Design less hazardous chemical syntheses*. While Principle #2 addresses the toxic properties of the products, Principle #3 addresses the design of the chemical process to make the products. Ideally, in reactions with 100% yield and no by-products, the toxicity of reagents might seem irrelevant since they are consumed in the process. However, the use of toxic reagents requires their production, transportation, and use under circumstances where accidents can happen and lead to exposures. Furthermore, since few reactions actually result in 100% yield, unreacted hazardous materials and by-products present risks and disposal problems. The desire to modify reaction schemes to use less hazardous chemicals throughout a process is both greener and safer. For example, twenty years ago benzene, chloroform, and carbon tetrachloride were solvents regularly used in academic and industrial labs. As the health hazards of these solvents became apparent, they were replaced by less toxic solvents and are now rarely used.

Principle #5: *Use catalysts, not stoichiometric reagents*. Catalysts are widely used in chemistry laboratories and in naturally occurring biochemical systems. Acids and bases are used as catalysts in many syntheses. Hydrogenation reactions also use catalysts, often at high temperature and pressure. Enzymes are used to catalyze biochemical reactions, such as using trypsin for the hydrolysis of peptides. Finally, many industries utilize biochemical systems with microbiological organisms to produce desired biosynthetic products. Since biological systems make extensive use of catalytic systems, these kinds of approaches may become more prevalent in laboratories to make reactions and processes greener. Chemists who develop new processes using new reagents are often the first to encounter previously unrecognized hazards, including those of catalysts. It is important that laboratory processes be developed to use these reagents safely.

An example of Principle #5 is the use of immobilized enzymes instead of strong acids in the synthesis of esters used in the cosmetic industry.[5] This not only eliminated a potentially hazardous solvent but also saved considerable energy in the process. Many other such examples of the use of catalysts are listed in reference[6].

Principle #8: *Use safer solvents and reaction conditions*. This is an extension of Principle #3 that specifically considers the use of solvents and other auxiliary chemicals. The use of solvents is so commonplace in chemical reactions that not using a solvent may seen impossible or unreasonable, but in fact may industrial syntheses are solvent-free, including the syntheses of benzene, methanol, MTBE, phenol, and polypropylene.[7] Also, some syntheses are conducted in the gas phase using heterogeneous catalysts. As a matter of scale, research labs can use solvents with less concern about hazards since the scale of reactions is relatively small. However, transferring the same reaction to industrial production scales can greatly increase the hazards and risks. Thus, eliminating a solvent in the "scale-up" process is both greener and safer. Finally, the use of safer solvents with regard to toxicity and other hazards is obviously desirable. Water and supercritical carbon dioxide both have many advantages with regard to reduced hazards and toxicity and have been widely employed in many synthetic and extraction processes.

Principle #9: *Increase energy efficiency*. Since energy consumption always has some concomitant and negative side effect for the local and/or global environment, the reduction of energy is always a desired component of a "green" approach to any process. Beyond that, running reactions at milder or ambient temperatures and low or lower pressure is inherently safer since the likelihood and severity of accidents and fires will be reduced. The use of catalysts (which is also favored in Principle #5) often allows reactions to be run at milder conditions.

Principle #12: *Minimize the potential for accidents*. As noted above, many of the steps already recommended will reduce the chances and severity of accidents in both lab and industrial settings. Often, however, taking the time to specifically consider the hazards and risks associated with a particular chemical process, and ways to reduce them, can lead to redesign of a process to make it safer. Although not explicitly stated in Principle #12, it is also wise to consider what responses are appropriate for various kinds of accidents that might occur. A quick and appropriate response can often greatly lessen the hazards and damages associated with accidents.

We have not discussed all of the twelve principles, but only those most directly having an impact on safety. However, the other principles *can* impact safety. For example:

- If the wastes are hazardous, reducing wastes will be safer.
- If the derivatives are hazardous, avoiding chemical derivatives will be safer.
- Real-time monitoring of chemical processes might not only "minimize or eliminate the formation of by-products" but may also provide alerts about unsafe reaction conditions.

Industrial Examples

The following examples are taken from the Environmental Protection Agency (EPA) sponsored Presidential Green Chemistry Challenge Awards Program.[6]

Making Alkali Metal Reactions Safer [8] Alkali metals, such as sodium and potassium, are excellent reducing agents but their high reactivity makes them dangerous to handle and store. Because of this, some commercial reactions have been developed to avoid using these metals, but the alternate synthetic pathways are often inefficient and wasteful. A method has been developed to incorporate reactive metals in porous metal oxides to form sand-like powders so that the metals can be used with greater control of their reactivity. Furthermore, these materials are safer to store and reduce waste disposal.

Microbial Production of 1,3-Propanediol [9] 1,3-Propanediol is a useful monomer for the production of various polyesters, but its use has been hampered by the fact that its production uses an expensive process and starts with a petroleum feedstock. Using genetic engineering, a microbe was developed that biocatalytically generates 1,3-propanediol from glucose that is derived from cornstarch, a renewable feedstock.

Enhancing the Solubility Properties of Supercritical Carbon Dioxide[10] Supercritical liquid carbon dioxide ($scCO_2$) has been recognized as an environmentally benign and nonflammable solvent, but its "solvent power" is less than that of many non-environmentally benign and highly flammable alkanes that it could potentially replace. Some fluorinated "CO_2-philic" materials were discovered that enhanced the solvent power of $scCO_2$, but these were quite expensive and also had high levels of persistence in the environment. Professor Eric J. Beckman and his group at the University of Pittsburgh developed a set of three criteria that would provide guidance for the development of alternates CO_2-philes. This led to the synthesis of poly(ether-carbonates) that were found to be very miscible with $scCO_2$, biodegradable, and far less expensive than fluorinated compounds.

As these examples show, the drive toward green chemistry and its application is likely to generate entirely new areas of pursuit in tomorrow's chemistry enterprise. Finding ways to apply the principles of green chemistry will surely lead to new discoveries and more effective and safer ways to produce needed chemical products. This will be increasingly important as the principal source of chemical products, petroleum, grows scarcer. These challenges will present new opportunities for chemists and other scientists to work together to make a safer world.

Sustainability—The Key Role for Green Chemistry

A popular term heard today is sustainability. But just exactly what is sustainability and how do we achieve it? Perhaps the clearest definition is found at the EPA sustainability web site. *Sustainability* means "meeting the needs of the present without compromising the ability of future generations to meet their own needs."[11] That would mean that our generation (including you) needs to live and work in such a way that we get what we need to sustain our health and quality of life while at the same time we ensure that future generations have a habitable planet that can sustain them with clean air, clean water, a clean environment, adequate food supply, and adequate renewable resources. We are not doing this now, and it will take work to change our behavior as individuals and a society function in a sustainable fashion.

All of this translates into a key and critical role for green chemistry in transforming the chemical enterprise. Green chemistry will not just be the trendy thing to do, but it will become an essential approach for our chemical enterprise in the future. In time, green chemistry must become the normal way we do things. Working together, the American Chemical Society, the American Institute of Chemical Engineers, and the National Research Council outlined eight challenges to be addressed to develop a sustainable chemical enterprise, including green and sustainable chemistry and engineering.[12,13]

To produce our high-technology products, chemistry uses large amounts of our energy resources, large quantities of water, and large amounts of petroleum-based materials and generates large amounts of waste. In the near future we will be unable to sustain our quality of life following this mode of operation. Chemists (including you) will need to find innovative ways to do chemistry and discover ways to conduct chemistry that sustains our world and our survival. With the challenges of global warming, global competition for business and natural resources, and increasing world populations,[14] we will need leadership from the chemical enterprise to turn green chemistry into the normal way of doing chemistry—using less energy and less water, and making no significant impact on our environment.

RAMP

- *Recognize* hazards that are associated with reactants, solvents, products, and by-products and the procedures in which they are used.

- *Assess* the risks of hazards by considering the probability of various exposures.

- *Minimize* the risks of hazards by eliminating or reducing quantities of hazardous chemicals, substituting more hazardous chemicals with less hazardous chemicals, and eliminating or changing solvents.

- *Prepare* for emergencies, since preparation can reduce the resulting damage if accidents occur.

References

1. *Chemical & Engineering News* **86**(*17*):54 (April 28, 2008).
2. R. H. HILL. Personal account of incident.
3. P. ANASTAS and J. WARNER. *Green Chemistry: Theory and Practice*, Oxford University Press, New York, 1998. Environmental Protection Agency. Green Chemistry, Twelve Principles of Green Chemistry (adapted from Anastas and Warner); available at http://www.epa.gov/greenchemistry/pubs/principles.html (accessed October 27, 2009).
4. Green Chemistry, Cornerstone to a Sustainable California; available at http://coeh.berkeley.edu/docs/news/green_chem_brief.pdf (accessed August 15, 2009).
5. Greener Synthetic Pathways Award; available at http://www.epa.gov/greenchemistry/pubs/pgcc/winners/gspa09.html (accessed August 15, 2009).
6. Environmental Protection Agency. Green Chemistry, Award Winner; available at http://www.epa.gov/greenchemistry/pubs/pgcc/past.html (accessed September 29, 2009).
7. M. LANCASTER. *Green Chemistry: An Introductory Text*, Royal Society of Chemistry, London, 2002, p. 132.
8. Environmental Protection Agency. 2008 Small Business Award; available at http://www.epa.gov/greenchemistry/pubs/pgcc/winners/sba08.html (accessed August 15, 2009).
9. Environmental Protection Agency. 2003 Greener Reaction Conditions Award; available at http://www.epa.gov/greenchemistry/pubs/pgcc/winners/grca03.html (accessed August 15, 2009).
10. Environmental Protection Agency. 2002 Academic Award; available at http://www.epa.gov/greenchemistry/pubs/pgcc/winners/aa02.html (accessed August 15, 2009).
11. Environmental Protection Agency. Sustainability. What Is Sustainability? Available at http://www.epa.gov/Sustainability/basicinfo.htm#sustainability (accessed September 29, 2009).
12. National Research Council. *Sustainability in the Chemical Industry*, National Academy Press, Washington, DC, 2005.
13. American Chemical Society. Sustainability and the chemical enterprise, Washington, DC, 2008; available at http://portal.acs.org/portal/acs/corg/content?_nfpb=true&_pageLabel=PP_SUPERARTICLE&node_id=1906&use_sec=false&sec_url_var=region1&__uuid=fe03360a-14e5-4e00-8ab4-60d773bce982 (accessed September 29, 2009).
14. T. L. FRIEDMAN. *Hot, Flat, and Crowded*, MacMillan, New York, 2008.

QUESTIONS

1. The green chemistry goal of designing chemicals with "little or no toxicity" is a challenge because

   (a) All chemicals are toxic at a high enough dose

   (b) TCSA does not require chemical companies to assess toxicity, therefore we know little about the toxicity of many chemicals

   (c) The "least toxic" chemicals are already being used and produced by chemical manufacturers

   (d) Both (a) and (b)

2. What are the green features of a chemical synthesis?

   (a) High yield, approaching 100%

   (b) Use of less-toxic reagents

   (c) Production of less-toxic by-products

   (d) All of the above

3. Using catalysts in reactions is a green step because

   (a) Catalysts are inherently nontoxic

   (b) Catalysts are generally used in safer solvents

   (c) Catalysts allow reactions to be run under less energy-intensive conditions

   (d) Most catalysts are biological in nature and therefore less toxic

4. Which solvent is the greenest?

   (a) Hydrocarbon without halogens

   (b) Perhalogenated hydrocarbon

   (c) Supercritical carbon dioxide

   (d) Solvent with very high boiling point

5. Sustainability means meeting the needs of the present

   (a) And ensuring a high standard of living for U.S. citizens

   (b) And allowing the future generation to take care of themselves

   (c) Without compromising the ability of future generations to meet their own needs

   (d) Without interference from government regulations and laws

CHAPTER 2
EMERGENCY RESPONSE

"IN AN emergency, what do you do when...?" The answer to this question can save your life and the lives of others, or simply turn a potentially harmful situation into a "minor incident." One of the four principles of safety is *preparing for emergencies*. Of course, planning in advance about how to respond to emergencies is necessary since there isn't time to learn during the emergency. This chapter reviews common emergencies and situations in labs where something has gone wrong and quick action is necessary to minimize the damage. Some laboratory emergencies share many common features with nonlaboratory emergencies, but others do not. However, much of what you learn in this chapter can easily be transferred to many other settings, so becoming safer in the lab makes your safer elsewhere, too. And, once you learn more about the nature of laboratory emergencies, you'll also better understand the relatively easy steps to avoid them in the first place!

INTRODUCTORY

2.1.1 Responding to Laboratory Emergencies Provides basic information about what you should do when there is an emergency in the laboratory.

2.1.2 Fire Emergencies in Introductory Courses Describes the categories of fires, what kinds of fires are most likely in introductory laboratory courses, and how to extinguish these common fires.

2.1.3 Chemical Spills: On You and in the Laboratory Describes actions that should be taken in the event of a chemical spill on you or in your laboratory.

2.1.4 First Aid in Chemistry Laboratories Provides basic information about what you should do if there is some kind of emergency in the laboratory.

INTERMEDIATE

2.2.1 Fire Emergencies in Organic and Advanced Courses Describes the likely nature of fires in intermediate and advanced labs and how to extinguish them.

2.2.2 Chemical Spills: Containment and Cleanup Discusses how you should respond to a chemical spill and how common chemical spills can be cleaned up.

2.1.1

RESPONDING TO LABORATORY EMERGENCIES

Preview   This section provides basic information about what you should do when there is an emergency in the laboratory.

> *Be Prepared.*
>
> Scout motto[1]

INCIDENT 2.1.1.1   SULFURIC ACID SPILL[2]

A student was working in a laboratory handling an Erlenmeyer flask containing sulfuric acid. Someone was not careful and knocked the flask off the bench and it hit the floor, spattering acid onto the student's shirt and jeans. He went to the emergency shower across the room and on the way took off his shirt that was already in shreds from the acid—he dropped it into the sink. Using the safety shower he was able to wash off the acid quickly so that it only left temporary red marks on his skin. His jeans were also shredded by the acid. His rapid response prevented any serious burns.

What lessons can be learned from this incident?

Prelude

This may be one of the first sections that your instructor has you read in this book. Even before knowing much about chemicals and the wide variety of hazards that they can pose, you may be working in a chemistry lab during the first week of your college chemistry course. Since lab accidents can happen at any time, it is best to learn first some basic emergency response actions. This section of the book gives the first set of basic answers to the question, "What do I do if. . .?" Most of the information in this section applies to other emergency situations at home or elsewhere, too.

Emergencies Requiring Evacuation—Being Prepared to Act Immediately!

The most likely laboratory emergencies, although rare, are fires, chemical spills, or common injuries such as minor burns and cuts. Section 2.1.4 will deal with the first aid responses to minor injuries, and what to do for more serious injuries. This section is about responding to fires and chemical spills.

In introductory laboratory classes, major fire and chemical spill emergencies are unlikely since relatively small amount of chemicals are used. Nonetheless, it is important to know what to do if something unexpected happens. Sections 2.1.2 and 2.1.3 will present more information about both of these topics that follow these introductory comments.

The first and most natural response to some emergency is a moment of panic, particularly if you fear that your own safety is in jeopardy. So, the first, best response is simply to recognize that moment of panic, take a deep breath to calm down (a bit), and then decide how to respond.

Laboratory Safety for Chemistry Students, by Robert H. Hill, Jr. and David C. Finster
Copyright © 2010 John Wiley & Sons, Inc.

"Fight or Flight?" When You Need to Leave the Lab

If you hear a fire alarm, you should immediately leave the laboratory. If you are using a gas burner or something electrical and you have time to turn this off without putting yourself at risk, then turn it off.

What if the emergency is closer to you? First, you should loudly call or shout to everyone in the vicinity that there is an emergency. The next choice that you have to make in any emergency is to either try to deal with the emergency or simply to leave the area or building (while being sure that someone in charge knows about the problem). Don't put yourself at unnecessary risk. Professional emergency responders, such as firefighters, have a clear priority of objectives when faced with hazardous situations:

1. Life safety.
2. Minimize property loss.

Firefighters make careful judgments about their own safety as they try to rescue others and/or minimize property loss. The 21st century motto in the United States fire service is: "Everybody goes home." No one should ever die trying to save a *building*; this is not good risk-reward analysis.

The firefighters' priorities should be your priorities, too. If you have any concerns about your own safety and/or the safety of others, the first and best response is to leave the area. In doing so, you should alert others in the same lab, or nearby labs, of the emergency. *As a student, you should also alert a teaching assistant or instructor about the emergency. These individuals may have more safety training and ultimately are more responsible for the lab and lab incidents than you are (as a student).*

Depending on the size of the emergency, you may also wish to consider pulling a fire alarm since this will effect an evacuation of the building. Fire alarm stations are almost always located near exterior exit doors or stairwell doors. Pulling a fire alarm to evacuate a building in the event of a significant chemical spill that presents a health hazard to building occupants is a reasonable action, even if there is not a fire.

In order to leave a building you must know where the exits are! Fire codes require that exit signs be clearly marked so finding an exit should be easy. Exit signs are typically located well above eye level near exit doors and stairwells, but it is possible, under heavy fire conditions, that smoke will obscure these signs. This discussion assumes that you are making a choice to leave the building far sooner than when "heavy smoke conditions" develop.

Knowing where exits are located is a good personal safe practice, not only for the laboratory, but for other places you go, such as your dormitory, the theater, shopping malls, a hotel, a restaurant, or a place of business. Take notice of these when you enter a building. When staying in a hotel with a long corridor of rooms, count the number of doors between your room and the exit so you can find the exit when crawling in heavy smoke!

Perhaps surprisingly, humans tend to leave a building or room by the same door that they entered, even if that is not the closest exit! Modern laboratories are required to have at least two exits; when you first enter a lab you should take note of the location of the "other" exit since it may be the preferred route in an emergency.

For a significant event, someone needs to call 911 and/or campus security. Teaching assistants or other instructors will likely do this since they are in more responsible positions, but if you have any doubt about whether this was done, it is safest to call 911. Do not assume that pulling a fire alarm notifies the fire department; sometimes it does and sometimes it doesn't depending on how the system is designed.

In many business and industrial facilities, employees are instructed to gather at a predetermined assembly location outside a structure so that all persons can be accounted for during an evacuation. Ideally, this should happen at colleges and universities too, but the transient nature of the population of a building at any point in time makes 100% accountability impossible from a practical point of view. However, your instructor may request that your class or lab section gather at a particular location so that some level of accountability can be determined. If you know that someone has been left behind, you should notify an emergency responder.

You should not reenter a building after an evacuation until specifically instructed to do so by emergency personnel.

Mitigating the Emergency

There are a limited number of situations where you, as a student, may be the best person to handle an emergency. Often a quick response to a small fire can prevent it from becoming a big fire. Some chemical spills can be handled quickly and easily. Both of these situations are discussed in greater depth in Sections 2.1.2 and 2.1.3.

As noted above, your primary responsibility is your own safety and you should not attempt to fight fires that are too big or that you have not been trained to fight. Nor should you try to clean up chemical spills unless you can so this safely and appropriately. Your instructor will help you determine what is a reasonable, and an unreasonable, response.

Some emergency procedures that require immediate evacuation present ethical dilemmas if injured persons who cannot remove themselves might be left behind. There is no simple answer to this dilemma. Most people would put themselves at some risk to help save another, but there are circumstances when a victim's injuries may be fatal or it may simply be the case that a small rescuer simply cannot successfully remove a very large victim/patient. In situations where someone is "left behind," it is critical to immediately inform emergency responders who are better trained and equipped to rescue people. A "worst case" scenario is for one person to try to help another unsuccessfully and become yet another victim.

Nonlaboratory-Related Emergencies

There may be emergencies that require you to seek shelter rather than evacuating. These emergencies could be storms, tornadoes, or police emergencies. Usually there are designated places known as "shelter-in-place" locations within buildings and these are the safest locations for people to gather during these emergencies. These locations may be interior hallways, stairwells, closets, or other interior rooms without windows but with adequate architectural support for shelter. Always avoid windows or glass doors that may be shattered and become flying shrapnel that could seriously injury you or others.

It is best to learn about shelter-in-place locations before emergencies take place. Ideally, your instructor should be "in charge" in these situations and help you determine what to do, but it wise to know what to do without such guidance, too.

Shelter-in-place locations are important for persons with physical disabilities who may not be able to easily exit the building. If there are any persons with physical disabilities in your area, you should make an effort to know about the shelter-in-place locations. If an emergency occurs and you can assist a person with a disability without putting yourself in danger, then consider helping that person to the shelter-in-place location. If you cannot assist them, let them know that you will notify the fire department of their location.

In the case of a police emergency, you will need to follow the instructions of law enforcement officials. Many campuses now have elaborate protocols in place to respond to terrorist acts or other similar threats and situations.

As in the situation with fires or spills, it is best to turn off equipment before leaving a lab if it is possible to do so safely. If electrical power is lost in a building, laboratory fume hoods will not work. Any materials in the hoods should be "capped," if it is possible to do so safely before leaving.

Summary

In an emergency:

- Stay calm. ℧
- Mitigate the emergency, if you are trained to do so and can do so safely.
- Leave the area if instructed to do so, or if you think it is unsafe to remain.
- Pull fire alarms to evacuate buildings, if necessary.
- Call 911 and provide specific details about the emergency.
- Know where the exits are located.
- Help injured persons, if it is safe to do so.

- Know where "sheltered locations" are in your building.
- Have someone with specific knowledge of the emergency meet emergency responders to provide them with information they will need about the emergency.

References

1. Boy Scouts of America. To the Speaker of the House of Representatives. Boy Scouts of America, 2001 Annual Report; available at http://www.scouting.org/Media/AnnualReports/2001/speaker.aspx?print=1 (accessed May 15, 2008).

2. F. MAJTABAI and J. KAUFMAN (editors). *Learning by Accident*, Vol. 1, The Laboratory Safety Institute, Natick, MA, 1997, #258.

QUESTIONS

1. In introductory chemistry labs, most emergencies involving fires or spills are not likely to be major events since
   (a) Most of the chemicals are not flammable
   (b) Most of the chemicals are relatively safe to use
   (c) Only small quantities of chemicals are used
   (d) Lab instructors can be relied upon to quickly respond to any emergencies

2. If you hear a fire alarm, you should
   (a) Finish the particular procedure that you are involved with before leaving the lab
   (b) Ask your instructor if it is safe to leave the lab
   (c) Leave the lab immediately, taking a moment to shut off electrical equipment and gas burners, if it is safe to do so
   (d) Make sure that you see something on fire before over-reacting to the situation

3. If a fire starts near you in the lab, you should
   (a) Extinguish the fire with a fire extinguisher if you are trained to do so and can do this safely
   (b) Immediately leave the lab and building
   (c) Tell your instructor only if the fire seems like it may become larger
   (d) Ask another student what to do

4. When leaving a lab during an emergency, you should
   (a) Always leave through the same door that you entered
   (b) Leave through the nearest exit
   (c) Follow the crowd
   (d) Leave the lab only after given permission to do so by your instructor

5. When pulling a fire alarm
   (a) You can safely assume that this also notifies the fire department
   (b) You should also call 911 after exiting the building
   (c) You should stay near the pull station so that you can turn off the alarm when the fire is out
   (d) You can safely assume that this will notify campus security

6. You can reenter a building after an emergency when
   (a) You no longer believe that any emergency exists
   (b) The fire alarm is silenced
   (c) You see other people reentering the building
   (d) Emergency personnel allow you to do so

7. If, during an emergency, you cannot find your lab partner outside, you should
   (a) Reenter the building to search for your lab partner
   (b) Tell an emergency responder that you believe someone may still be in the building
   (c) Call 911
   (d) Reenter the building to search for your lab partner if you think that there really isn't an emergency

8. In the event of a terrorist attack in a science building, you should
   (a) Leave the building immediately, but not pull the fire alarm
   (b) Leave the building immediately, and pull the fire alarm
   (c) Hide in a safe location or follow the instructions of emergency personnel
   (d) Make sure all lab experiments are shut down before taking any other action

9. Fire alarms should be pulled
   (a) Only if you are absolutely sure that there is a fire
   (b) Only if you think that there is a fire
   (c) Any time that it is prudent to evacuate a building, whether there is a fire or not
   (d) Only if your instructor tells you to do so

10. During an emergency you should help an injured person only if
   (a) The injuries are serious
   (b) You can do so without putting yourself at risk
   (c) You have specific training to help the person
   (d) You are certain that you know how they were injured

2.1.2

FIRE EMERGENCIES IN INTRODUCTORY COURSES

Preview This section describes the categories of fires, what kinds of fires are most likely in introductory lab courses, and how to extinguish these common fires.

Great emergencies and crises show us how much greater our vital resources are than we had supposed.

William James[1]

INCIDENT 2.1.2.1   FIRE FROM FRAYED ELECTRICAL WIRING[2]

A beaker of acetone, a very flammable organic solvent, was placed near a hot plate. The acetone fumes, heavier than air, crept along the top of the bench and at some point the frayed electrical wiring of the hot plate generated a spark and ignited the fumes. An instructor's clothing caught fire. One alert student safely extinguished the fire and another wrapped the instructor in a fire blanket. There were no serious injuries or damage.

What lessons can be learned from this incident?

Prelude

Fires in laboratories can be incredibly dangerous. Besides the danger of receiving burns, burning chemicals can produce toxic fumes and the risk of explosions. Fortunately, in introductory lab courses the nature and amounts of flammable substances are quite limited so that "worst case" scenarios and explosions are not likely. This section discusses the most common situations that might occur in introductory laboratory courses and explains the necessary background information to help you understand the risks of fires in laboratories. Section 2.2.1 presents more about fires in advanced chemistry and research labs where the hazards are likely to be more significant.

It may be that your instructor is most interested in having you learn when and how to use a fire extinguisher in a laboratory. If so, it is possible to read only the section below on "Using Fire Extinguishers." While that minimally prepares you to use an extinguisher we encourage you to learn more about the various classes of fires since we believe that the more you understand the nature of fire and kinds of fire that can occur the more likely you will be successful in extinguishing a small fire (or *preventing* a fire). If possible you should participate in any hands-on training opportunity to practice using a fire extinguisher since this experience is more realistic.

Classes of Fires

Fires that you might expect to encounter in introductory laboratories typically will involve "ordinary combustibles materials" or "flammable liquids." These phrases introduce us to the four general categories of fire, shown in Table 2.1.2.1.

Let's look at an overview of the four classes of fire and how the most common extinguishing agent, water, interacts with them.

Most fires that people (and firefighters) encounter in *nonlaboratory* situations are Class A fires. Some examples are burning houses and their contents, clothes and wood, burning cars, a trash dumpster on fire, and forest fires. As we will see below, these Class A fires are readily extinguished using water.

Laboratory Safety for Chemistry Students, by Robert H. Hill, Jr. and David C. Finster
Copyright © 2010 John Wiley & Sons, Inc.

TABLE 2.1.2.1   Classes of Fires

Class	Description	Examples
A	Fires involving ordinary combustible materials	Paper, wood, clothing, furniture, plastics
B	Fires involving flammable liquids	Ether, hexanes, gasoline, oil
C	Fires involving energized electricity	Hot plates, spectrometers, computers
D	Fires involving reactive metals	Sodium, lithium, metal hydrides

Most laboratories do not use water (extinguishers) because fires with chemicals and electrical equipment should be extinguished with other agents.

Class B fires are burning organic liquids. Gasoline that is burning inside an automobile engine is a controlled Class B fire. Using "lighter fluid" on a charcoal grill is a Class B fire. Many organic solvents used in chemistry labs generate Class B fires. Trying to extinguish a fire involving organic solvents with water is counterproductive since most organic solvents do not mix with water and will float on top of water. A stream of water is more likely to spread the solvent and fire. For this reason, water should be not used to extinguish solvent fires. A different type of extinguishing agent is needed for Class B fires (see below).

Class C fires are any fires (Class A or Class B) that *also* involve energized electricity. A burning computer is a Class C fire. Toast burning in a toaster (that is "on") is a Class C fire. Since water conducts electricity, using water on these fires allows for the possibility of spreading the electrical charge back to the person using water to extinguisher the fire. Since this raises the possibility of fatal electrical shock, *water should never be used on a Class C fire*.

Class D fires are pretty rare outside of specific chemical laboratories or workplaces. Some industrial processes that involve very hot metals such as aluminum or magnesium can lead to a fire if water contacts these metals. In labs, we often use elemental sodium or lithium and these are very reactive with water. Compounds called "hydrides" are also very reactive and will catch fire upon contact with water. Class D fires are called "active metal" fires.

Class C and D fires are discussed more in Section 2.2.1.

But, what are we most likely to find on fire in a chemistry lab? While it's possible that some Class A materials might be on fire, there is a very good chance the some Class B organic liquid will also be on fire. And, since there is often equipment involving electricity in labs, some fire can easily be Class C as well. So, these are not fires that you should try to extinguish with water, and you'll probably have a hard time locating a pressurized water fire extinguisher in a lab. You will learn more about other types of extinguishers below.

To understand what extinguisher to use, and why, we should next learn about the fire triangle and the fire tetrahedron.

The Fire Triangle and the Fire Tetrahedron

For many years, fire scientists considered fire to consist of three components: oxygen, fuel, and heat. These three features comprised the fire triangle (Figure 2.1.2.1). We can use the fire triangle to think about how to prevent a fire from starting by not allowing all three components to meet. Keeping any one of them away from the other two will prevent a fire from starting. We use the fire triangle to understand how to *prevent* fires. We now better understand the details of the chemistry of fire and this triangle has been replaced by the fire tetrahedron because there is a fourth element that is needed to explain how fires are extinguished. (Figure 2.1.2.2). The fourth component is the chain reaction. By *removing* any one of these four components we can *extinguish* a fire. We use the fire tetrahedron to understand how to *extinguish* fires.

Let's look at each part of the tetrahedron.

Fuel is an obvious component of a fire. The most common fuels are Class A materials like paper, wood, cloth, or plastic or Class B liquids such as gasoline or some common organic solvents used in labs. Removing the fuel is usually not easy to do although sometimes containing the fuel or moving it

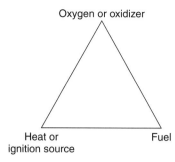

FIGURE 2.1.2.1 The Fire Triangle. The fire triangle helps explain how fires work and how to prevent fires.

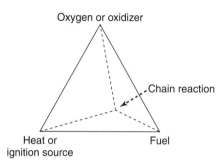

FIGURE 2.1.2.2 The Fire Tetrahedron. The fire tetrahedron helps explain how fires can be extinguished.

(such as taking a burning trash can outside) may stop the spread of the fire by limiting the availability of additional fuel.

All fires involve *oxidizing agents* (that oxidize or burn the fuel) and this is almost always atmospheric oxygen, O_2. If we can remove air from the fire, or vice versa, the fire will extinguish. This is sometimes easy. A small beaker of flammable liquid on fire will quickly extinguish if a watch glass (or other noncombustible "lid") is placed on the top of the open beaker. The remaining air in the beaker will be quickly consumed and the fire will self-extinguish. Similarly, a piece of paper on fire will extinguish if a book is placed on top of it. The book excludes air from the site of combustion. Or, the traditional "stop, drop, and roll" technique extinguishes burning clothing when one "rolls" on the fire and momentarily excludes oxygen. Finally, if we can "coat" a burning liquid with a powder or other substance that excludes the atmospheric oxygen from the liquid, the fire will extinguish.

A *source of energy*, usually heat or an electric spark, is required to start and sustain a fire. Removing the heat is done with water since water is such a great cooling agent. And since fuel and atmospheric oxygen are commonly available, most fire prevention measures are designed to prevent the initial source of heat from starting a fire.

The *chain reaction* is the least obvious component of the fire tetrahedron since this is occurring at the molecular level and we don't observe it directly. The exact mechanism by which chemicals burn is complicated but we know it involves a catalytic chain reaction. Some extinguishers work not by removing fuel, oxygen, or heat but by interfering with the chemical reaction in a fashion that stops the fire.

The value of understanding the components of the fire tetrahedron is that *removing any one* of the components stops the fire.

How Fires Burn

It may surprise you to learn that liquids and solids don't burn! This statement seems at odds with the common experience of paper, wood, or gasoline burning, but in fact, at the molecular level, the solid or liquid has to be vaporized before it will burn. When a match ignites paper or some flammable liquid, the heat from the match first pyrolyzes (decomposes) the material so that some vapor is produced. (See *Special Topic 2.1.2.1* Pyrolysis and Fires.) The heat also starts the chemical reaction of the flammable vapor combining with oxygen gas, which becomes the chain reaction: the exothermic (heat-releasing) reaction provides more heat to pyrolyze more of the flammable liquid or solid and the fire keeps burning.

SPECIAL TOPIC *2.1.2.1*

PYROLYSIS AND FIRES

Pyrolysis is the decomposition of solid or condensed material by heating.[3] The term pyrolysis come from the Greek term for fire, *pyro*, and the Greek term meaning losing or breaking down, *lysys*. When sufficient heat is applied to a material, both intermolecular and intramolecular bonds break, and smaller components or compounds are generated that are volatile and often flammable. At the elevated temperatures these breakdown products can burn in air. When you see a solid material burning, it is really not the solid that is aflame, but the gases and volatiles that are released as the pyrolysis process continues releasing more gases.

In some homes there is inadequate clearance between chimneys or metal exhaust systems from fireplaces or furnaces and surrounding structural wood such that the radiant heat can pyrolyze the wood and, over time, eventually generate enough volatile flammable materials to ignite. This is the cause of some house fires. In fact, all burning wood and vegetation is a result of pyrolysis.

Pyrolysis is a major method of processing in the chemical industry. Petroleum is subjected to pyrolysis processes, known as cracking and catalytic cracking, to produce alkanes or alkenes that can be further used to produce fuels, such as gasoline. Pyrolysis has long been used to make charcoal from wood and other similar products such as shells. Pyrolysis occurs in our foods when we cook them—so when you take a bite of that golden brown apple pie remember it is produced by pyrolysis chemistry.

Why is this important? Just momentarily cooling a burning liquid can stop the fire, which eliminates the rapid vaporization of the liquid. This is a common way to extinguish Class B fires, as we will see below. The same is sometimes true of solids, although sometimes a burning solid will be hot enough (retaining enough residual heat through its own heat capacity) that a momentary cooling does not help. The very hot solid might be momentarily cooled, but the fire can restart spontaneously once air comes in contact with the fuel again.

Classes of Fires and Types of Fire Extinguishers

Fire extinguishers are categorized by the class of fires (see Table 2.1.2.1 above) that they extinguish. Wouldn't it be great if there was just one extinguisher that worked on all four classes of fires that was also cheap to manufacture and purchase? Well, we don't quite have that happy a situation, but the options are still pretty good. Table 2.1.2.2 shows the types of fire extinguishers. Below we'll discuss carbon dioxide extinguishers and dry chemical extinguishers since these are the most likely fire extinguishers to be used in a lab. In Section 2.2.1 we'll discuss Class D fires and the appropriate fire extinguishers. We'll also make brief mention of water as an extinguishing agent, although it is usually not the best agent for lab fires.

TABLE 2.1.2.2   Types of Fire Extinguishers

Type of extinguisher	A	B	C	D	Comments
Pressurized water	X				Usually in a silver container; uncommon in labs
Carbon dioxide	X	X			Usually red; common in labs; large cone; no pressure gauge
Dry chemical (BC)		X	X		Usually red; common in labs; small cone; pressure gauge
Dry chemical (ABC)	X	X	X		Usually red; common in labs; small cone; pressure gauge
Class D powder				D	Commercial versions uncommon (but available); a bucket of sand is more common

Using Water to Extinguish a Fire

Since most lab fires are Class B and/or Class C fires, water is usually *not* a good first choice to extinguish a fire in a lab. The only exception to this would be, for example, a situation where some papers caught on fire in a lab. This is clearly a Class A fire and can be doused easily with water or by, if possible, moving the papers into a nearby sink and turning on the water.

Water extinguishes a Class A fire primarily by cooling the burning fuel. Type A extinguishers are not likely to be found in laboratories. This removes the "heat" from the fire tetrahedron and the fire stops burning. (See *Chemical Connection 2.1.2.1* Why Firefighters Love Water.)

CHEMICAL CONNECTION *2.1.2.1*

WHY FIREFIGHTERS LOVE WATER

Water has several advantages as an extinguishing agent for Class A fires. Some of these advantages may seem "obvious" or even unimportant, but when compared to other possible extinguishing agents, this list becomes very important.

1. It is cheap.

2. It is abundant.

3. It is nontoxic.

4. It doesn't react in a fire, and therefore produces no toxic by-products.

5. It has a very high heat capacity. This means that, on a per gram basis, it is able to absorb heat very well without increasing its own temperature dramatically. Since the main mode of "action" of water on a fire is to reduce heat (by absorbing heat), having a high heat capacity is great.

6. It is possible to dissolve other chemicals in water so that it becomes an even more effective extinguishing agent. Firefighters sometimes add a foaming agent to water in a system embedded in the pump in a fire engine. "Class A foam" is sometimes used on burning houses to allow the water to better soak into the wood, which provides additional protection against a "rekindle." Class A foam is a surfactant, a kind of detergent molecule that you will learn about in an introductory chemistry course. "Class B foam" is an agent that creates a nonflammable foam that will cover the surface of a Class B liquid fire, thus preventing oxygen from contact with the vapor of the liquid and stopping the fire.

The only minor disadvantage to water for firefighters is that in an enclosed room with a very hot fire, some of the water used to cool the fire may turn into steam. This steam can penetrate through openings in a firefighter's protective gear and cause serious burns. Firefighters typically have no exposed skin while fighting an interior fire, but the steam can circumvent this imperfect protection.

Carbon Dioxide Extinguishers

A common type of fire extinguisher to be found in a lab is a heavy metal container with several pounds of liquid carbon dioxide inside and a large wide-mouth black nozzle (see Figure 2.1.2.3). The CO_2 extinguisher is for Class B or Class C fires. The vapor pressure of liquid CO_2 at $20\,^{\circ}C$ is about 58 atm. When this extinguisher is discharged, gas immediately exits the extinguisher through a large nozzle and is reduced to local atmospheric pressure. The method of extinguishing the fire is to reduce the concentration of atmospheric oxygen to a very low level by creating a "blanket" of CO_2 gas at the site of the fire. This momentarily extinguishes the fire, and in the case of a burning liquid, the rate of vaporization of the liquid is greatly reduced. For just a moment, no more fuel or heat is available to allow the fire to continue to burn. After the CO_2 quickly dissipates and oxygen returns, the heat from the fire is gone and the fire can't restart. Thus, even though the fuel and oxygen are present, interrupting the fire eliminates the heat needed to continue chemical reaction. Most liquids have lower heat capacities so the liquid itself never gets very warm.

FIGURE 2.1.2.3 CO_2 Fire Extinguisher. This extinguisher should be used only on Class B and Class C fires.

What happens if a CO_2 extinguisher is used on a Class A fire such as a burning piece of wood? The fire is momentarily extinguished due to the lack of oxygen, but when the CO_2 dissipates the very hot wood can easily reignite when oxygen returns. Thus, CO_2 extinguishers should not be used on Class A fires.

CO_2 extinguishers are good for Class B fires (organic liquids) but can also be used successfully on some Class C fires. An electrical fire might be extinguished by a CO_2 extinguisher. However, unless the equipment is deenergized by shutting off the electricity supply, the fire may restart. Similarly, solvent fires might reignite if there is sufficient source of heat (besides the heat of the fire itself, which will be eliminated once the fire is extinguished).

Dry chemical extinguishers are preferred for solvent fires, as described below.

Dry Chemical Extinguishers

There are two types of dry chemical extinguishers: BC and ABC, referring to the classes of fires that they extinguish.

BC dry chemical extinguishers contain powder that is designed to coat the surface of a flammable liquid and to eliminate the vaporization of the liquid. This stops fire since no more fuel is available.

ABC dry chemical extinguishers work like BC, except that the powder used is also selected so that it forms a sticky solid layer on solid (Class A) materials. This layer prevents oxygen from attacking the fuel, even though it may still be hot enough to burn (see Figure 2.1.2.4).

Which Extinguisher Should I Use?

With the information provided above, you should be able to carefully select the correct extinguisher. And, the labels on fire extinguishers (A, BC, ABC) help you select the correct extinguisher, too. But, happily and with prior preparation, in the moment of panic that often accompanies a fire, you probably won't have to think about classes of fire and types of extinguishers. Why not?

Fire codes require that buildings have the kinds of fire extinguishers that are appropriate for the most likely kinds of fires that occur in those buildings. So, in an emergency, you can always rely on the fact that the fire extinguishers on the wall or in the hallway will be effective for the fire that you encounter. Since chemistry and other science labs might have a wide range of flammable and

FIGURE 2.1.2.4   ABC Fire Extinguisher. This extinguisher can be used on Class A, B, and C fires. It is the most common type of extinguisher found in chemistry laboratories. It leaves a powdery mess after use, but this is better to clean up than the destruction a fire may cause.

combustible materials in them, it's likely that the available extinguisher will be an ABC dry chemical extinguisher. Look at the extinguisher; it is labeled with large letters and pictograms showing what fires it can, and cannot, extinguish.

It is possible that some labs may have BC extinguishers—either CO_2 or dry chemical. Don't use these on a Class A fire since they may not be as effective as a type A fire extinguisher.

If you use dry chemical extinguishers on electronic equipment, such as computers or laboratory instrumentation, the electronics will likely be seriously damaged. A CO_2 extinguisher is a preferable "first choice" extinguisher on electronic equipment, if you have the choice of what type of extinguisher to use. It is more likely that you will use whatever extinguisher is available. Most electronic equipment in the introductory lab is relatively inexpensive, in the range of thousands of dollars, particularly in comparison to advanced instruments, which can cost hundreds of thousands or even millions of dollars. Extinguishing the fire at the cost of destroying an instrument, to prevent a *larger* fire with *more* damage, is a top priority.

The most common extinguishing agent is still water and it works great on Class A fires. If a piece of paper is on fire in a lab, water will put out the fire. If a person's clothing is on fire, the safety shower will put out the fire. But, don't assume that the first large beaker of clear liquid you see is water! It may be a flammable organic solvent and throwing this on a Class A fires generates another fire, now Class B!

Using Fire Extinguishers

The best way to learn how to use an extinguisher is to practice in a fire class with a firefighter. Many fire departments give free hands-on training to the public and educational institutions. Don't pass up this opportunity to handle an extinguisher. The instructions below are appropriate to read but there is no substitute for hands-on training.

All portable fire extinguishers work the same way. This commonality of design makes them easy to use. An easy way to remember how to use the extinguisher is the PASS technique: pull, aim, squeeze, and sweep.

- *Pull* the pin. Fire extinguishers have a safety pin near the handle to prevent accidental discharge. This pin should be secured in place with a plastic tie that can be broken with a good tug on it. If, even with adrenaline flowing, you cannot break the tie by pulling, try to twist it to break it or place the extinguisher flat (sideways) on the ground, kneel on it and pull the pin to break the plastic tie.

- *Aim* the extinguisher at the base of the fire. Sometimes there is a detachable nozzle connected to the extinguisher by a flexible hose. In other designs the nozzle may simply "swing up" from the side.
- *Squeeze* the handle to begin the discharge. Dry chemical extinguishers make a little noise and CO_2 extinguishers make a loud noise!
- *Sweep* the discharge back-and-forth horizontally across the base of the fire.

If the handle if released, the extinguisher will stop discharging the agent. But, it is usually best to discharge the *entire* extinguisher. A flammable liquid fire that is 99% extinguished will quickly reignite and thwart whatever initial attempt you made to extinguish it. Your first effort will have been wasted and used part of the extinguisher. Dry chemical extinguishers produce a real mess when discharged, but cleaning up a powdery mess is better than watching a building burn to the ground or risking lives. Extinguishers don't last long, 30 seconds to a few minutes, depending on the size. If you discharge a fire extinguisher, report it immediately to your lab supervisor or instructor. This will ensure that the empty extinguisher is refilled and put back in place.

Of course, you can only use a fire extinguisher if you know where it is located. Fire code requires that portable extinguishers be located in conspicuous locations, along normal paths of travel and no more than 5 feet above the floor. Most labs will have fire extinguishers inside the lab. In labs that have Class B hazards (i.e., contain flammable liquids) the fire extinguisher must be no more than 50 feet from the door in a hallway. You should always visually scan any lab that you work in to see the location(s) of the fire extinguisher(s) and then also determine what type they are.

What If *You* Are on Fire?

If an incident occurs that sets your clothing on fire, there are a few ways to respond to this. Burning clothing, or skin, is a Class A fire. Since responding very quickly is important, the fastest effective response is the best one when several choices are available.

Modern laboratories all have safety showers that are designed primarily to quickly rinse the whole body in the event of some chemical contamination from a spill or explosion. These safety showers are also an extremely effective and rapid way to extinguish burning clothing. If you do need to use one, be prepared for the shock from the cold water—these showers don't use warm water. The only disadvantage to safety showers is the distance that someone may need to run to get to one.

Some labs may have faucets with long flexible attachments (commonly called drench hoses) on them, perhaps even with eyewash capabilities. For a small fire, this may be an effective option and may allow a victim to be lowered to the ground near a sink so that the water stream can quickly be applied to the fire.

Burning clothing can sometimes be extinguished by the "stop, drop, and roll" method, where the victim rolls on the floor. The rolling action will extinguish some of the fire; it is helpful if others also "pat" the area of the fire starting at the head and moving down the body. "Patting" with a towel or a jacket will help protect the hands of those who are helping. Do not "pat" a burning person with your bare hands or while wearing gloves that could melt.

Some labs may have fire blankets available. These nonflammable blankets can be used to wrap around a victim to smother the fire. If you are on fire, using a safety shower is the best option; but if this is not quickly available you can "stop, drop, and roll." If a fire blanket is available, it is best used by someone assisting you to help smother the fire. You should be aware that if you are on fire and standing, wrapping in a fire blanket could produce a "chimney" effect that not only may not extinguish the fire but could promote continued burning and injuries from the fire. Fire blankets can also be used to warm someone who has stepped into a safety shower or held up to provide privacy for someone who needs to remove burnt or contaminated clothing.

Should a portable fire extinguisher be used on a person? Ideally, no, since a CO_2 extinguisher (BC) is not designed to extinguish a Class A fire and using either a BC or ABC dry chemical extinguisher exposes the victim to possibly inhaling some of the powder. However, any of these extinguishers is better than no extinguisher. These should be used only when water or fire blankets are not available, and the "stop, drop, and roll" method has not been effective.

Should You Fight the Fire?

A small fire can usually be extinguished by a trained person with the proper extinguisher. But, life preservation trumps property loss. Only try to fight a fire if:

- It is a "small fire." There is no clear definition of small, but "flames from floor to ceiling" or "an entire lab bench" is not a small fire. If it *seems* too big, it *is* too big.
- You have the correct extinguisher and can retrieve is rapidly.
- You know how to use the extinguisher.
- You always keep an exit available away from the fire. Never allow a fire to get between you and your only exit.

What Else to Do

In an introductory lab, there should always be some instructor available and there will likely be several students around if a fire occurs. If you choose to attempt to quickly put out a fire, you should:

- Yell to others that there is a fire and you are attempting to put it out.
- Have someone call 911.
- Make sure that someone else tells the instructor what is happening.
- Make sure that someone else is starting to evacuate the lab.

A quickly extinguished fire represents a very brief and very exciting and alarming moment during a lab experiment. It can start and be over in 30 seconds. Or, the fire can grow and the science building can be destroyed. In many situations it is perfectly appropriate to pull the fire alarm and call 911, even when only minutes later the situation is under control and no further hazard exists. Firefighters *do not like false alarms that are pranks* for a host of good reasons, but firefighters do not at all mind responding to a genuine emergency that was ultimately taken care of before they arrived. It is always safe to assume that the worst can happen when something starts to go wrong and take the necessary steps to keep yourself and others safe.

There's More

This has been a fairly brief introduction to fires, written as an appropriate introduction for the first year of chemistry labs at most colleges and universities. There is more to learn about fires and various fire hazards that can present themselves in advanced and research labs. Section 2.2.1 continues this discussion.

The RAMP paradigm can be used to think about fire hazards and how to prepare for fire emergencies.

- *Recognize* the flammable and combustible materials, particularly organic solvents, in the laboratory. *Recognize* any electrical equipment that may pose a fire hazard or the presence of any active metals.
- *Assess* the risk of the fire hazard by considering the quantities of materials and possible ignition sources.
- *Minimize* the risk by using and storing flammable liquids and active metals appropriately and by checking electrical equipment.
- *Prepare* for fire emergencies by knowing where extinguishers are located, how to use them, and where exits and fire alarms are located.

References

1. WILLIAM JAMES. From a letter to W. LUTOSLAWSKI in 1906. In: *Bartlett's Familiar Quotations*, 17th edition (J. KAPLIN, ed.), Little, Brown, and Co., Boston, 2002, p. 581, line 3.

2. F. MAJTABAI and J. KAUFMAN (editors). *Learning by Accident*, Vol. 1, The Laboratory Safety Institute, Natick, MA, 1997, #46.

3. *The American Heritage® Dictionary of the English Language*, 4th edition; available at Dictionary.com web site, http://dictionary.reference.com/browse/pyrolysis (accessed April 10, 2009).

QUESTIONS

1. The best way to learn how to use a fire extinguisher is to

   (a) Use one during a real fire emergency
   (b) Read the label on the extinguisher
   (c) Practice using a fire extinguisher during a training course
   (d) Practice using a fire extinguisher anytime you are not busy during a lab experiment

2. Class A fires involve

   (a) Energized electricity
   (b) Flammable liquids
   (c) Ordinary flammables
   (d) Reactive metals

3. Class B fires involve

   (a) Ordinary flammables
   (b) Reactive metals
   (c) Energized electricity
   (d) Flammable liquids

4. Class C fires involve

   (a) Ordinary flammables
   (b) Energized electricity
   (c) Flammable liquids
   (d) Reactive metals

5. Class D fires involve

   (a) Ordinary flammables
   (b) Flammable liquids
   (c) Energized electricity
   (d) Reactive metals

6. A burning computer that is being used is what class of fire?

   (a) A
   (b) B
   (c) C
   (d) D

7. A burning lab notebook is what class of fire?

   (a) A
   (b) B
   (c) C
   (d) D

8. If some organic solvent has spilled onto an operating hot plate and caught fire, what class of fire is this?

   (a) B only
   (b) C only
   (c) B and C
   (d) A and C

9. A fire involving sodium hydride is what class of fire?

   (a) A
   (b) B
   (c) C
   (d) D

10. What part of the fire tetrahedron was not originally included in the fire triangle?

    (a) Fuel
    (b) Oxidizing agent
    (c) Ignition source
    (d) Chain reaction

11. Most fire prevention methods involve the elimination of what part of the fire tetrahedron?

    (a) Fuel
    (b) Oxidizing agent
    (c) Ignition source
    (d) Chain reaction

12. What phase(s) of matter do not burn?

    I. Solid
    II. Liquid
    III. Gas

    (a) I and II
    (b) I and III
    (c) Only I
    (d) II and III

13. Water is a useful extinguishing agent on what class(es) of fire?

    (a) Only A
    (b) A and B
    (c) A and C
    (d) Only D

14. Why are pressurized water extinguishers not found in chemistry laboratories?

    (a) They are too expensive.
    (b) They are too heavy for some people to operate.
    (c) Most lab fires are Class B or Class C fires.
    (d) Most lab fires are Class D fires.

15. Why is a carbon dioxide extinguisher not always effective against a Class A fire?

    (a) The CO_2 will further "feed" the Class A fire.
    (b) The fire might momentarily be extinguished but the hot fuel can reignite when air hits the fire.
    (c) Carbon dioxide cannot cool the flames enough.
    (d) They typically last only 10–15 seconds, which is not long enough to extinguish the fire.

16. BC fire extinguishers work by

    (a) Cooling the fire
    (b) Stopping the chain reaction
    (c) Preventing vaporization of a flammable liquid
    (d) "Deactivating" the fuel

17. ABC fire extinguishers work by

    (a) Cooling the fire
    (b) Chemically neutralizing the fuel
    (c) Preventing vaporization of a flammable liquid
    (d) Forming a sticky layer that prevents oxygen (in the air) from reacting with the fuel

18. Why is the most handy fire extinguisher in a chemistry lab almost always the appropriate fire extinguisher?

    (a) It is the cheapest to use, and therefore easily replaced.
    (b) It is probably an ABCD extinguisher that works on all classes of fires.
    (c) Fire code requires that available fire extinguishers be the appropriate type for the most likely type of fire.
    (d) It is the type of extinguisher that most folks know how to use.

19. What is the correct sequence of actions in order to use any fire extinguisher properly?

    (a) Aim, pull, sweep, and squeeze
    (b) Aim, pull, squeeze, and sweep
    (c) Pull, aim, squeeze, and sweep
    (d) Pull, aim, sweep, and squeeze

20. Portable fire extinguishers typically discharge for about

    (a) 5–10 seconds
    (b) 10–30 seconds
    (c) 30 seconds to a few minutes
    (d) 5–10 minutes

21. When using a fire extinguisher it is best to

    (a) Use it only in "short bursts" until the fire is out
    (b) Use it continuously until the fire is out
    (c) Discharge the entire extinguisher to optimize the use of extinguishing agent
    (d) Discharge it until it starts to make too much of a mess in the lab

22. The best extinguishing agent for a person on fire is

    (a) Type BC
    (b) Type ABC
    (c) Water
    (d) Type ABCD

23. A student should attempt to use a fire extinguisher

    (a) Only if an instructor says it is OK
    (b) Always, before sounding an alarm or alerting anyone else
    (c) Only if the fire is small enough and the student can confidently use the available extinguisher
    (d) On all fires, no matter how small or large because the fire will certainly get larger and cause considerable damage

2.1.3

CHEMICAL SPILLS: ON YOU AND IN THE LABORATORY

Preview   This section describes actions that should be taken in the event of a chemical spill on you or in your laboratory.

It is by presence of mind in untried emergencies that the native metal of man is tested.

James Russell Lowell, American poet[1]

INCIDENT 2.1.3.1   PHENOL CHEMICAL BURN[2]

A student was using a phenol solution in his laboratory when he spilled the solution and it splashed on his pants. Although there was a safety shower nearby the spill area, he went past the shower, to the men's restroom. There he removed the pants, washed the chemical off his leg, and rinsed the pants. After this, however, he put on the contaminated pants again and then proceeded to the university clinic. As a result of continued wearing of contaminated pants, he received second degree chemical burns from the phenol.

What lessons can be learned from this incident?

Prelude

Spills of chemicals in laboratories come in various sizes, and present various levels of hazard. Knocking over a 100-mL graduated cylinder with 5 mL of 0.1 M NaCl in it is a fairly trivial event. Dropping a large glass container that has 4 L of concentrated nitric acid or 4 L of (flammable) acetone is a serious event. What to do in the event of some chemical spill depends on the nature of the spill.

Except in spills of small amounts of innocuous solids or liquids, it is not likely that you, as a student, should attempt to clean up the spill. (Your instructor should supervise any efforts you undertake with regard to cleaning up a spill.) In most cases, spill cleanup should be handled by someone with appropriate training and who knows what to do with the chemical once is it collected. This does not mean that you should "do nothing," however, if a spill occurs, especially if some chemical spills on you or your clothing.

Chemical spills can sometimes lead to injuries to students in the form of inhaling toxic fumes or skin exposure. In all cases, dealing with injuries and effecting a quick evacuation of the area supersedes any other action regarding the spill. Information about how to treat persons exposed to chemicals is presented in Section 2.1.4, "First Aid in Chemistry Laboratories." In some situations, calling 911 for emergency medical help is appropriate.

Spills that Don't Contaminate People

Most solids don't present high hazards when spilled since they do not disperse easily. If some solid chemical is spilled at your lab bench, a dispensing area, or in a balance room, you should have someone "protect the area" so that other students are aware of the spill and can avoid any contamination and then notify your instructor about the spill. Your instructor may tell you how to clean up the spill or someone else may take care of it. When cleaning up a solid spill, it is usually best to use a small brush and dustpan to gather most of the solid. This should not be returned to the original bottle, since it is

Laboratory Safety for Chemistry Students, by Robert H. Hill, Jr. and David C. Finster
Copyright © 2010 John Wiley & Sons, Inc.

almost certainly contaminated with trace amounts of dirt or dust. Your instructor will tell you what to do with the chemical. In most cases it will need to be treated as hazardous waste, which should never be placed in a trash can. Only rarely would it be appropriate to place a chemical in a wastepaper basket. Wiping down the area with a wet sponge is probably a good, final step in the cleanup.

Liquid spills can be innocuous or quite dangerous. Liquids spread much more easily and some may generate toxic or flammable vapors. Flammable vapors can find an ignition source and turn a spill into a fire or explosion. Toxic vapors can overcome someone who, with good intentions, attempts to clean up the spill.

The first and best response to any liquid spill is to immediately clear the area of all people and quickly notify an instructor. In academic labs, it is very likely that identifying what was spilled will be fairly easy. Since liquids spread easily, if possible to do so safely, it is best to build a small "dike" around the spill area using sand, an absorbent material, or a "spill pillow" that both absorbs the liquid and prevents further spread. Containing the spill is best, but you should not put yourself at risk to do this. If you use a spill kit, you should tell someone in authority about the incident and particularly note that the spill kit was used. Returning an "empty" spill kit to a shelf is unsafe since it will not be ready for a subsequent emergency!

It is not likely that you will be responsible for cleaning up a liquid spill, except in cases of very innocuous solutions or water. Your instructor will make the decisions about cleaning up liquid spills.

Splashes in Your Eyes

Probably the most harmful thing that could happen to you is to have a chemical splashed into your eyes. This is why you are required to wear chemical splash goggles in the laboratory. But what should you do if you do get a chemical splash into your eyes? Each laboratory will be equipped with eyewash stations that are made especially for rinsing your eyes in the event of a splash. Figure 2.1.3.1a shows a typical eyewash; your eyewash may be a different design than this eyewash but the purpose is the same. Figure 2.1.3.1b shows the eyewash when it has been pulled down and activated.

Most eyewash units are designed to operate hands-free once activated by pushing a plate or pulling the unit down to activate water flow. The valve stays open and water continues as long as it is in the open position. Here are several important things that you should know about eyewashes.

- Know where your eyewash is located.
- Make sure access to the eyewash is kept clear and unobstructed—not blocked by equipment or objects in front of it or around it.
- Learn how to operate it.
- Flush it at least once a week to remove any debris or bacterial growth that might have accumulated in the eyewash over time. Allow the water to run for about 3 minutes. Some eyewashes have

(a)

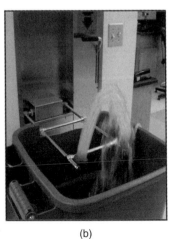
(b)

FIGURE 2.1.3.1   A Laboratory Eyewash. (a) Pushed up in off position and (b) pulled down in on position.

covers over the heads to provide protection against dust accumulation. These should be easily removed.

- Make sure water flow comes evenly from each eyewash head. Water streams should cross.
- Tell your instructor if you find the eyewash is not operating correctly.

Using Emergency Eyewashes

In the event of splash in your eyes, you should remember to do the following:

1. Flush your eyes immediately at the eyewash for a minimum of 15 minutes—this is a long time but you want to be sure that the chemical is removed as much as possible when you seek medical assistance.

2. If possible, seek assistance to find the eyewash and to help you in washing your eyes. Hold your eyelids open or get someone else to hold your eyelids open. It is a natural reflex to want to close your eyes in the event of a splash but this only retains the chemical and it is critical that you remove as much of the chemical as possible from your eye. If you have contacts, remove them if this is easy to do—but if it is difficult just continue to rinse. Move your eyeball around as the water continues to wash to make sure that you have covered all areas of the splash.

3. Seek medical assistance immediately after washing to ensure the chemical has been removed, to assess any eye damage, and to determine if further treatment is needed. Get information about the chemical, including its name and properties, preferably from a Materials Safety Data Sheet so the physician or medical person can evaluate the need for treatment. However, do not delay getting the person to medical care while someone searches for this information.

Spills that Contaminate People

Spills on your clothing or skin can also be fairly innocuous or quite hazardous. In introductory labs, strong acids and strong bases are likely to be the most hazardous substances in this regard. Explosions are rare in introductory chemistry classes since the circumstances that could lead to explosions are generally avoided. However, an explosion or splash can deposit chemicals on your skin and even a spill on a lab bench might lead to contamination of your torso, legs, or feet. Wearing appropriate clothing and personal protective equipment (such as gloves, and certainly splash goggles, apron, or lab coat) will minimize exposures.

Most solids do not react rapidly with skin, but there are some exceptions, so it is always a good idea to remove the solid as quickly as possible. Washing with water may seem appropriate but some chemicals react with water so the best first action is to scrape the solid off the skin using a credit card or similar object. Scraping the solid into a beaker or onto a piece of paper is best, but don't delay scraping unnecessarily. Scraping onto an open lab bench or into a sink is less ideal, but speed and safety may trump "ideality" in these situations. After most of the solid is removed, flushing the skin with water is OK and appropriate. Report such incidents to your instructor, but do not delay treatment in order to do this.

Washing Splashed Chemicals from Your Body—Using Emergency Showers

Each laboratory should have an emergency shower. Sometimes these showers are located in the hall just adjacent to the laboratory so they serve several laboratories at once. These showers are to be used if you spill large quantities of chemicals on your body and clothing. Some safety showers are called deluge showers because they release a large volume of water very rapidly—usually >75 liters per minute and do not shut off until all of a specified volume of water has been released. Other safety showers have a handle that pulls down to the on position and when the handle is released it shuts off. Be prepared, this water will be cold! All areas where there are corrosives or other chemicals that could injure the body should have emergency showers. A typical emergency shower is shown in Figure 2.1.3.2. Below are listed important points about using emergency showers.

FIGURE 2.1.3.2   A Laboratory Safety Shower. It is very important to know the locations of the nearest safety showers in laboratories or in nearby hallways. During an emergency, time is critical.

- Emergency showers should be used in the event of a chemical splash or spill on your body and clothes. They can also be used to extinguish fire if your clothing is burning.
- Stand under the shower and pull the handle or chain to activate the valve that releases the water flow.
- Clothing soaked in chemicals can do serious injury to the body so you need to remove all contaminated clothing, including shoes, socks, and jewelry that might have chemicals trapped on them. Do not attempt to wipe off the chemical from the clothing.
- Remove the contaminated clothing carefully so that the chemical does not get splashed in your eyes or on some other part of the body.
- Residence time of the chemical on your body is critical so removing contaminated clothing and washing the area under a shower should be done as quickly as possible.
- Modesty can be a hindrance that can result in more serious injury. Assist modest people, reassuring them of the importance of removing contaminated clothing. If possible try to provide screening of a sort to assist in preserving modesty.
- Try to get assistance to help remove clothing. This is particularly important if the person is dazed and does not respond quickly.
- Clothing may need to be cut off—do this with care to avoid further injury to the person and to you if you are assisting in removing clothing.
- A clean laboratory coat can be used for warmth and modesty after the shower.
- Clothing that is contaminated needs to be handled as hazardous waste.
- After the shower has removed the chemical, seek medical assistance to evaluate potential injuries from the chemical. If possible bring a Material Safety Data Sheet to the physician; however, do not delay seeking medical treatment to look for this information.

Summary

The key steps in responding to a chemical spill are:

- *Evacuate*. Depending on the size of the spill and the hazards posed by the chemical, it may be necessary to evacuate the immediate area or the entire building.

- *Communicate*. You should alert someone, preferably your laboratory instructor or assistant, that there has been a spill. If you know the nature of the spill, you should relay this information so that appropriate action can be taken. You should alert others of the possible hazard and advise them to avoid this area.

- *Isolate*. If possible, try to limit the extent of the spill (for liquids) by containing the spill with sand or other appropriate material.

- *Mitigate*. With your instructor's permission and guidance you may participate in cleaning up the spill. However, you should not attempt to clean up any but the most trivial spill without help from someone with more training.

References

1. J. R. LOWELL. BrainyQuote; available at http://www.brainyquote.com/quotes/quotes/j/jamesrusse137191.html (accessed March 15, 2009).
2. Princeton University, Environmental, Health and Safety Department, Anecdotes. Failure to Remove Contaminated Clothing Exacerbates Chemical Burns; available at http://web.princeton.edu/sites/ehs/labsafetymanual/sec11.htm (accessed September 8, 2009).

QUESTIONS

1. Most spills in academic lab settings are fairly small,

   (a) But still need to be dealt with immediately and properly
   (b) And don't need attention immediately since they present little risk of exposure
   (c) But often the chemicals are so toxic that evacuation by everyone is recommended
   (d) But almost always are flammable, which requires immediate evacuation

2. Which of the following is appropriate for chemical spills in academic labs?

   I. Should be cleaned up only by the instructor
   II. Should be cleaned up only by certified HazMat technicians
   III. Should be cleaned up only by appropriately trained persons
   IV. Should be cleaned up by the student under the appropriate supervision

   (a) Always only I
   (b) II
   (c) III, and either I or II
   (d) Always only IV

3. Liquid spills tend to be the most dangerous since

   (a) Solids are rarely flammable or toxic
   (b) Liquids spread more easily than solids
   (c) Almost all liquids are flammable
   (d) Solids can be present in a lab for days or weeks without any significant exposure hazard

4. Spill kits

   (a) Can easily be used by students
   (b) Should be returned after use immediately to their storage area so that their use can be tracked by monthly inspections
   (c) Should be used by instructors only
   (d) Can be used by students or instructors as long as they are used properly and not returned to storage without notifying someone in charge

5. Good procedures related to using eyewash stations include

   (a) Knowing where they are located
   (b) Having had them flushed for several minutes on a regular basis
   (c) Flushing the eyes for at least 15 minutes
   (d) All of the above

6. It is almost always safe to wash off a chemical on the skin with lots of water unless

   (a) It is a strong acid or a strong base
   (b) It is a solid that reacts with water
   (c) Using a shower causes public embarrassment
   (d) Using the shower creates a mess in the lab

7. When using an emergency shower it is important to

   (a) Almost always remove clothing
   (b) Assess the nature of the harm after sufficient washing
   (c) Have a lab coat or other extra clothing available
   (d) All of the above

2.1.4

FIRST AID IN CHEMISTRY LABORATORIES

Preview   This section provides basic information about what you should do if there is some kind of medical emergency in the laboratory.

A half doctor near is better than a whole one far away.

German Proverb[1]

INCIDENT 2.1.4.1   CPR REVIVES RESEARCHER AFTER ELECTRICAL SHOCK[2]

A lab worker attempted to wipe some moisture from a high-voltage power supply and received a severe shock and burns. After screaming, he stumbled into a hallway and was met by a co-worker. He then collapsed, stopped breathing, and had no pulse. A public safety officer was nearby and started CPR. An ambulance crew subsequently restored his heartbeat using defibrillation, and he survived. (He later acknowledged that he knew the power was on but didn't think that he was contacting high voltage. Safety interlocks had been overridden and safety guards had been removed.)

What lessons can be learned from this incident.

Prelude

Every experienced lab worker can relate stories of injuries sustained in labs. Many are minor, some are serious, and virtually all of the injuries could have easily been avoided by following fairly simple guidelines discussed throughout this book. Whether injuries *should* happen, and what causes them, is a very worthwhile discussion to have with peers and mentors; whether injuries will happen seems statistically inevitable. Acknowledging this, it is prudent to have reasonable safety equipment handy (such as fire extinguishers and safety showers) and also to be prepared to render help to anyone who is injured. How to respond to lab emergencies in general is discussed in Section 2.1.1. This section discusses specific "first aid" procedures.

Sometimes, knowing even just a little first aid (along with what *not* to do) can change the outcome of a serious injury. Incident 2.1.4.1 illustrates how CPR (cardiopulmonary resuscitation) almost certainly saved a life since "no pulse, no breathing" is not a condition that can wait for an ambulance to arrive. Although not actually contained in the Hippocratic oath, the phrase "*do no harm*" is also a good starting point for rendering aid. "Good Samaritan" laws generally protect persons with no formal medical training from prosecution if mistakes are made while well-intended aid is administered.

Being prepared to administer first aid to an injured person doesn't require much training or education. If possible, it is smart to take a course in CPR and basic first aid techniques—this could save the life of a friend, loved one, or family member. This section of the book outlines basic procedures but cannot replace hands-on practice and training.

Types of Lab-Related Medical Problems

Many kinds of medical emergencies can occur. We will focus on lab-related emergencies and avoid the discussion of situations (such as a diabetic emergency) that require medical response but are likely unrelated to the circumstances of working in a chemistry lab.

Laboratory Safety for Chemistry Students, by Robert H. Hill, Jr. and David C. Finster
Copyright © 2010 John Wiley & Sons, Inc.

We will categorize lab-related emergencies as shown in the list below. Beside each category are other sections of the book that discuss in more detail the effects and prevention of these injuries.

- Inhalation of gases and vapors (Sections 4.1.1, 4.1.2, 4.2.1, 7.1.4, 7.2.3)
- Skin and eye exposure to chemicals (Sections 2.1.3, 2.2.2, 5.1.1, 5.2.1, 7.1.1, 7.1.2, 7.2.1, 7.2.2)
- Burns (Sections 2.1.2, 5.1.2)
- Electric shock (Section 5.3.5)
- Exposure to extreme cold (Section 5.3.9)
- Cuts or open wounds (Section 7.1.3)
- Traumatic injuries, such a bruises and broken bones (Sections 5.3.3, 5.3.4)
- Exposure to biological agents (Sections 4.3.2, 7.3.4)
- Radiation exposure (Sections 5.3.7, 7.3.2, 7.3.3)

If any of these incidents cause serious injuries or loss of consciousness, you should call for help and ask that 911 (or the emergency number for your campus) be called. It is always better to err in calling for emergency response personnel rather than waiting when minutes might make a difference in the outcome. Try to ensure that someone is assigned to bring the emergency response personnel to the location of the victim.

Inhalation of Gases and Vapors

Inhaled gases and vapors can cause asphyxiation (due to lack of oxygen) and/or react with the lungs to cause other breathing difficulties. In either case, it is necessary to remove the patient to prevent further inhalation of damaging substances and to get the person to an area of fresh air. This action must be taken judiciously, however, so that you, as a rescuer, do not become the next victim. *Before* you attempt rescue, you must assess the safety of the atmosphere in the area of the patient. *Before* attempting any rescue, call for additional help. For a nonbreathing patient, administer CPR (but be sure to avoid inhaling the exhalation of the patient). In some circumstances high-flow oxygen may need to be delivered, which requires an ambulance crew.

Skin and Eye Exposure to Chemicals

In almost all cases, rinsing the skin and eyes with copious amounts of water for at least 15 minutes is advised. The only exceptions to this are skin exposure to chemicals that react with water, which are usually solids. (A common example of this is lime, CaO.) In these cases, using a credit card or similar tool to first scrape most of the chemical off the skin is required. For eye exposures and significant skin exposure, immediate transport to an emergency room or a physician is advised.

Burns

We discuss here only "thermal burns" due to exposure or contact with a hot object. Chemical burns, radiation burns, and electrical burns are discussed in other sections. Thermal burns are categorized as first, second, and third degree, as shown in Table 2.1.4.1.

All but the most modest first degree burns must be treated with care, and quickly. First, of course, the source of heat must be removed. Second, remove clothing and jewelry in the affected area if it easy to do so. If clothes are burned or melted against skin, do not pull if attached to skin. For burns with no blisters or closed blisters, tap water can be used to continue to cool the affected area briefly—do not cool torso or head for an extended period as this could cause hypothermia. Do not use nonsterile water on open blisters since this can cause infections. Never apply any ointment, cream, or salve to a burn. Never apply ice to a burn.

Many burn patients should be transported to an emergency room or burn center.

TABLE 2.1.4.1   Burn Severity

Result of burn	First degree	Second degree	Third degree
Outer layer of skin damaged	Yes	Yes	Yes
Second layer of skin damaged	No	Yes	Yes
Tissue below second layer damaged	No	No	Yes
Color of burn	Red	Dark red or red with blisters	Charred black or white, or brown/dark-red wax coated appearance
Pain level	Low–moderate	High	High, or absent (there is no pain with sufficient nerve damage)
Presence of blisters	No	Yes	Yes or no

Electric Shock

First, make absolutely sure that the source of the electricity is off or that there is *no chance* that the patient is still in contact with energized electricity. Touching a patient still in contact with the electricity will make you the next victim.

Assess the patient for breathing and heartbeat. Administer CPR if necessary, *after* calling for additional help and having someone call 911.

Exposure to Extreme Cold

The use of liquid nitrogen and dry ice in labs can lead to "burns" from exposure to the very cold agents. Most likely, the exposure time will have been relatively brief and not have caused severe damage. Removing the cooling agent and warming the affected area with warm (not hot) water (assuming no open wound) or even just warming a cold hand inside one's armpit will be an effective treatment. For more serious cases of frostbite where the skin is white and hard (but perhaps not frozen solid), transport to an emergency room is advised. *Do not rub* affected areas as a means to warm the area.

Cuts or Open Wounds

The size of the cut dictates the response. Obviously, a small cut can be cleaned and a dressing can be applied. A large wound, with significant bleeding, presents a considerable risk. To stop copious bleeding, apply direct pressure to the wound, preferably with a sterile dressing but immediate action overrides a delayed search for a sterile dressing. Have someone call 911.

Do not remove an impaled object since removing the object will likely increase the rate of bleeding. Apply a dressing around the object to stabilize in place and minimize bleeding.

Excessive blood loss can lead to shock (which is the lack of oxygen perfusion at the cellular level). Keep the patient warm and elevate the feet until emergency personnel arrive.

Traumatic Injuries

An explosion or fall can lead to trauma to the body and/or broken bones. Call 911 immediately. Do not move the patient unless the local environment is life-threatening. For any fall, even from as little as a few feet, assume that there may be a spinal injury and do not move the patient unless absolutely necessary. Stabilize the head and neck area. Cut away clothing as needed to make certain that any hidden wounds are identified.

Unless you are specifically trained to do so, do not try to "reset" a broken bone. Broken bones in limbs can sometimes puncture the skin (called an "open" wound). These can be disturbing to view but are rarely life-threatening. The ambulance crew will decide how to treat the patient in this regard.

Exposure to Biological Agents

While less likely in most chemistry labs, any exposure to a pathogen (virus or bacteria) requires transport to an emergency room for assessment. As in other circumstances, make sure that you, as a rescuer, do not become the "next victim" by becoming exposed to the same pathogen. Beyond separating the patient and the pathogen, there is no "first aid" procedure. Have as much information about the exposure (agent and duration and nature of exposure) available for ambulance personnel. In some instances, the receiving hospital may elect to treat the patient as "contaminated," which requires special actions on their part.

Radiation Exposure

As with biological pathogens, there is little "first aid" possible after exposure to alpha, beta, X-ray, or gamma radiation. With due regard for your own safety, shut down or remove the source of the radiation or otherwise separate the patient from the radiation source. Call 911. Obtain as much information as possible about the exposure to give to the ambulance personnel.

Other General Issues Regarding Emergency Response

As noted in Section 2.1.1, the first task of any rescuer is to protect themselves and not become the "next victim." Always take a moment to stop and think about the situation before rushing to help someone.

Ambulance and hospital personnel will want as much information as possible about the nature of the injury or exposure and the circumstances in which it occurred. For any patient who is not conscious or is otherwise unable to respond, you can best help by getting the patient's name, age, and, if possible, any medical history and any medications that are being taken. For chemical exposures, try to have an MSDS available for transport to the hospital.

You will notice that ambulance personnel are *always* wearing gloves. This universal procedure is to better protect against bloodborne pathogens. Fortunately, such gloves should be readily available in all chemistry labs and you, too, should wear gloves as a matter of routine when involved in any first aid procedure.

Finally, always alert others when involved in any kind of rescue or first aid. Notifying lab assistants, professors, and campus security should be standard procedure. It is always better to err on the side of calling emergency response personnel where there are injuries. In most lab settings, there are plenty of people available to help. Assign tasks, if necessary, to make sure all relevant people are informed of the situation.

RAMP

- *Recognize* the presence of any immediate hazards before you attempt to assist or rescue a victim.
- *Assess* the type of injury.
- *Minimize* further damage by calling for help, ensuring that you and the victim are not in danger, and administering first aid.
- *Prepare* for emergencies that require first aid by taking a first aid course.

References

1. The Quote Garden. Quotations on Medical Subjects; available at http://www.quotegarden.com/medical.html (accessed September 18, 2009).

2. American Industrial Hygiene Association Laboratory Safety and Health Committee. Laboratory Safety Incidents; available at http://www2.umdnj.edu/eohssweb/aiha/accidents/electrical.htm.

QUESTIONS

1. If you see an apparently unconscious person on the floor in a lab what is the first thing to do?

   (a) Call for extra help before entering the room.

   (b) Determine the nature of the injury.

   (c) Determine if there are any hazardous conditions in the room.

   (d) Identify who the person is.

2. It is generally a good idea to rinse any solid or liquid chemical from the skin unless

   (a) The chemical is a strong oxidizing agent

   (b) The chemical is a strong acid or base

   (c) It is a chemical that reacts with water

   (d) The victim indicates that the chemical isn't causing any pain

3. Which category of burns generally requires medical attention?

   (a) First degree only

   (b) Third degree only

   (c) First and second degree

   (d) Second and third degree

4. Which substance can safely be applied to a burn?

   (a) Ice

   (b) Burn cream

   (c) Tap water, if there are no open blisters

   (d) Zinc oxide cream

5. If you encounter an unconscious person who you believe may have been electrocuted, you should

   (a) Start CPR immediately

   (b) Check for burn marks

   (c) Make absolutely sure that the patient is not still in contact with live electricity before rendering aid

   (d) Shut off all electrical breakers to the room before entering the room

6. If you encounter a person with a large piece of glass impaled into the body from an apparent explosion, you should

   (a) Immediately remove the object since it is probably contaminated with chemicals

   (b) Immediately remove the object and then apply pressure to the wound

   (c) Remove the object only if the patient complains of severe pain

   (d) Leave the impaled object intact and stabilize it with a dressing

7. If you encounter a person that has fallen from a 5-foot platform and is unconscious, you should

   (a) Start CPR immediately

   (b) Not move the person unless the environment is hazardous

   (c) Rearrange the person's position so that they are lying flat on the back and all arms and legs are in a "neutral," straight position

   (d) Search for identification

8. It is important to tell an ambulance crew that a patient might have been exposed to a pathogen since

   (a) The crew and receiving hospital will need to prepare to handle a patient who might be "contaminated"

   (b) The crew will need to wear special protective equipment

   (c) The crew will need to notify the Department of Homeland Security

   (d) All of the above

9. When dealing with any medical emergency in a lab, you should always

   (a) Notify others about the situation

   (b) Wear gloves

   (c) Try to determine if any condition in the lab is related to the medical emergency

   (d) All of the above

2.2.1

FIRE EMERGENCIES IN ORGANIC AND ADVANCED COURSES

Preview This section describes the likely nature of fires in intermediate and advanced labs and how to extinguish them.

I've done made a deal with the devil. He said he's going to give me an air-conditioned place when I go down there, if I go there, so I won't put all the fires out.

Red Adair, renowned oil-well firefighter[1]

INCIDENT 2.2.1.1   LITHIUM ALUMINUM HYDRIDE FIRE[2]

While pouring lithium aluminum hydride (LAH) from a plastic bag into a flask containing THF, some of the LAH spilled and burst into flame on the surface of the fume hood. The researcher dropped the bag and then removed his lab coat in an attempt to smother the fire. The coat and LAH were then pulled out of the hood onto the floor. The LAH burned itself out and a dry chemical extinguisher was used to put out the lab coat.

INCIDENT 2.2.1.2   SOLVENT FIRE[3]

A defective shelf collapsed spilling 12 containers of hexane and leading to an explosion and fire. About 50 additional gallons of flammable liquids were in the storage area. The damage was estimated to be $200,000 to $300,000.

What lessons can be learned from these incidents?

Classes of Fires

Class A (ordinary combustibles) and Class B (organic liquids) fires can occur in all labs and are the most common types of fires (see Table 2.1.2.1 in Section 2.1.2). Organic labs and other advanced labs are likely to have more exotic chemical reagents. Some labs might also have "homemade" equipment that may be more prone to electrical failure. Class C fires involve energized electricity and Class D fires involve "active metals." We discuss these in more detail in this section.

We assume that you are also familiar with the *fire tetrahedron*, as described in Section 2.1.2. Understanding the fire tetrahedron helps one to understand how various fires burn and how to extinguish them.

Class B Fires: Organic Liquids

Although briefly discussed in Section 8.1.1, the quantity of flammable organic liquids and solvents used in organic labs and other advanced labs often greatly exceeds the amounts used in introductory courses. Thus, the likelihood of a Class B fire is greater in organic labs. (See *Chemical Connection 2.2.1.1* Flammability of Halogenated Solvents and Halons.)

Laboratory Safety for Chemistry Students, by Robert H. Hill, Jr. and David C. Finster
Copyright © 2010 John Wiley & Sons, Inc.

CHEMICAL CONNECTION *2.2.1.1*

FLAMMABILITY OF HALOGENATED SOLVENTS AND HALONS

The NFPA fire rating is an indication of the flammability of a chemical. Table 2.2.1.1 shows the NFPA fire ratings for two series of chlorinated hydrocarbons.

TABLE 2.2.1.1   NFPA Fire Ratings for Selected Chlorinated Hydrocarbons[4]

Molecular formula	CAMEO NFPA rating[4]	Carbon oxidation state
CH_4	4	-4
CH_3Cl	4	-2
CH_2Cl_2	1	0
$CHCl_3$	0	$+2$
CCl_4	0	$+4$
C_2H_6	4	-3
ClC_2H_5	4	$-3, -1$
$1,1-Cl_2C_2H_4$	3	$-1, +1$
$1,1,1-Cl_3C_2H_3$	1	$-3, +3$
$1,1,2,2-Cl_4C_2H_2$	Noncombustible[5]	$+1$
C_2Cl_6	Noncombustible[5]	$+3$

Unsubstituted hydrocarbons are very flammable and fully substituted halogenated hydrocarbons are non-flammable. Methane (CH_4) and carbon tetrachloride (CCl_4) are good examples of each category, respectively. Partially substituted hydrocarbons, such as dichloromethane, have limited flammability—dichloromethane (CH_2Cl_2), for example. The notion of "degree of flammability" or "degree of oxidation" is also illustrated by examination of the oxidation state of the carbon atoms.

For many years the very popular classes of compounds, known as Halons, were used as extinguishing agents—these were methane and ethane derivatives being highly substituted with halogens. The halogens used were various mixtures of fluorine, chlorine, and bromine. These gases were used primarily as agents for electrical fires, where the extinguishing action was the displacement of oxygen in the room. Their proposed mechanism of reaction was the consumption of hydroxyl radicals as part of the chain reaction of combustion—thus removing one of the four sides of the "fire tetrahedron." There were several advantages to using Halons over CO_2 including better effectiveness on a per gram basis and less damaging physiological effects on humans at the required concentrations. In 1989 the Montreal Protocol on Substances that Deplete the Ozone Layer called for the halt to the production and use of substances implicated in the destruction of the ozone layer. Although the use of Halons as firefighting agents was only a tiny fraction of the worldwide use of ozone-destroying molecules, new extinguishing systems no longer use Halons (although existing systems are grandfathered). Current substitutes for Halons are perfluorinated molecules and only partially halogenated molecules.[6]

First, and most importantly, water is not a useful extinguishing agent for Class B fires. Most organic solvents have a density less than that of water and are usually immiscible with water. The water neither "covers" the fire nor does it "dilute" the organic liquid to a nonflammable condition. Thus, the force of the stream of water from either a pressurized water (PW, Class A) portable extinguisher or from a fire hose serves mostly to *spread the fire*. Because of this, you will probably *not* find a PW extinguisher in chemistry labs, and furthermore, you should avoid the temptation to throw a beaker of water on a Class B fire.

Class B fires are best extinguished by either CO_2 (BC) or dry chemical (BC or ABC) extinguishers and these should be the type of extinguishers that are located in, or near, the lab. Always determine where the fire extinguishers are, and what type they are, before working in any lab. (See *Chemical Connection 2.2.1.2* Applications of General Chemistry Principles in a CO_2 Extinguisher.)

CHEMICAL CONNECTION *2.2.1.2*

APPLICATIONS OF GENERAL CHEMISTRY PRINCIPLES IN A CO_2 EXTINGUISHER

Many features of the CO_2 fire extinguisher illustrate principles of molecular behavior.

In what state is the carbon dioxide inside the extinguisher? Most humans only experience CO_2 as either a gas (such as the bubbles in a carbonated beverage) or a solid, in the form of "dry ice." In fact, one of the few common experiences of sublimation is watching a chunk of dry ice "disappear" as it sublimes. Some folks might wonder why it's not melting, as most solids do at "high temperature." However, inside a CO_2 fire extinguisher the CO_2 is a liquid. The phase diagram in Figure 2.2.1.1 shows that at room temperature (about 25 °C) CO_2 can be condensed to its liquid form under high pressure. In fact, the vapor pressure of liquid CO_2 at 25°C is about 63 atm, which means that the pressure of the CO_2 gas over the surface of the liquid is 63 atm. Little wonder that the gas is very forcefully ejected when the extinguisher is discharged! CO_2 extinguishers do not have, or need, pressure gauges since the pressure inside, as long as some liquid is present, will be 63 atm whether there are 5 pounds of liquid CO_2 or 1 gram of liquid CO_2 present. The extinguisher is checked monthly by its weight.

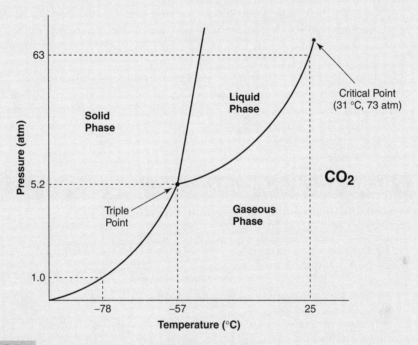

FIGURE 2.2.1.1   Phase Diagram for CO_2. The pressure inside a charged CO_2 fire extinguisher is about 63 atm at room temperature. Most of the CO_2 is in the liquid phase.

In what form is the CO_2 after the extinguisher is discharged? Only moments after discharge, much of the CO_2 appears as the solid, much like a "snow" that blankets the fire. The phase diagram indicates that at room pressure (1 atm) CO_2 can only exist as a solid if the temperature is less than −78 °C. (This is the normal sublimation temperature of CO_2.)

How did the CO_2 get so cold? The vapor inside the fire extinguisher is a room temperature. What caused the gas to cool to a temperature less than −78 °C? What caused a change in temperature of over 100 °C in less than a second? As the gas is ejected it expands considerably due the pressure change from 63 atm to 1 atm. The intermolecular forces between the gas molecules must be overcome and the potential energy of the system must increase. The only source of this energy is from the kinetic energy of the system. (This is an adiabatic process, with no heat being gained from or lost to the environment during the gas expansion.) The loss of kinetic energy causes a considerable decrease in temperature. This process is called Joule–Thomson expansion. Gas compression increases temperature and gas decompression decreases temperature. This same principle is used in the process of refrigeration. This is a good reason to hold the handle above the nozzle of the extinguisher and not the large wide-mouth nozzle—it will get surprising cold.

Why does the fire extinguisher get (a little) cooler? As the gas is discharged, the pressure drops to less than 63 atm inside the extinguisher. This causes the liquid to start to evaporate as it attempts to restore the equilibrium vapor pressure. As the liquid vaporizes the endothermic heat of vaporization self-cools the liquids. This, in turn, will cool the extinguisher a bit.

Class C Fires: Electrical Fires

Fires that involve energized electricity (Class C) will inevitably also be Class A fires. It is usually some electrical "short" that starts the fire but the materials that are burning will probably be some plastic or other combustible materials. If a fire appears to be Class A but you suspect that "live electricity" is present, you must not use water on such a fire. Water conducts the electricity and will pose the possibility of electrocution of a stream or pool of water. Firefighters are at risk of electrocution when using hose streams near high-voltage wires.

The best extinguisher for an electrical fire in a laboratory is the dry chemical ABC extinguisher since this will also handle the "Class A" part of the fire, as well as the "Class B" flammable liquids in the vicinity of the fire. A CO_2 extinguisher (BC) might work on some Class A/C fires but as the coding suggests, this is not a guarantee. We also note here that although uncommon in academic circumstances, rooms filled with computers or high-priced analytical instrumentation sometimes have automated fire suppression systems that use either CO_2 or, more commonly, another inert gas that displaces enough oxygen from the room to suppress the fire. These agents do not cause further harm to computers, unlike a dry chemical extinguisher. (See *Chemical Connection 2.2.1.3* What Chemicals Are Used in Fire Extinguishers?)

CHEMICAL CONNECTION *2.2.1.3*

WHAT CHEMICALS ARE USED IN FIRE EXTINGUISHERS?

The most common fire extinguishers[7] found in laboratories are ABC and BC. Some laboratories may use Type D extinguishers if they are using reactive metals or pyrophoric organometallic compounds, such as organolithium or organomagnesium compounds, such as Grignard reagents. A pressurized water portable fire extinguisher (Class A) is unlikely to be found in a lab since many lab fires will be Class B and a high-pressure stream of water will only make a Class B fire worse.

There are two main chemical components in a fire extinguisher: the propellant and the extinguishing agent. The extinguishing agents are described below. Nitrogen is often used as a propellant; that is, it pushes the extinguishing agent such as a dry powder out of the nozzle. Carbon dioxide is another common propellant. Extinguishers using solid chemicals and powders also contain additives to keep them free-flowing and resistant to "caking."

Dry chemical extinguishers. Dry chemical extinguishing agents are usually expelled by gas pressure from within the extinguisher. *ABC* extinguishers usually are filled with ammonium phosphate, $(NH_4)H_2PO_4$. Ammonium phosphate is mildly corrosive if it comes in contact with water or moisture, so areas must be cleaned carefully after their use. When heated, phosphate decomposes to metaphosphoric acid, which is a sticky polymer, $(HPO_3)_n$, that effectively seals the burning material from atmospheric oxygen and suffocates the fire. *BC* extinguishers usually are filled with a bicarbonate salt, where the cation is sodium, potassium, or urea potassium. Cleanup is much easier than the ABC type. BC extinguishers are ineffective on Class A fires.

Dry powder extinguishers. There are several types of chemicals used in these extinguishers. Sand is the most inexpensive version of a Class D extinguishing agent. Commercial Class D extinguishers may contain a variety of agents such as sodium chloride and thermoplastic additives, copper powder, graphite, or sodium carbonate.

Halon extinguishers. These are the so-called clean agents that leave no residue but are effective as ABC extinguishing agents. Halon® extinguishers, containing bromochlorodifluoromethane or bromotrifluoromethane, were used for many years; however, Halon use was phased out in 2000 as a result of the Montreal Protocol on Substances that Deplete the Ozone Layer. Halotron® extinguishers, containing dichlorotrifluoroethane or other similar agents, have replaced Halon, since these agents are deemed more environmentally friendly but are also very effective at extinguishing fires and leaving no residue. These extinguishers may use inert propellants, such as argon. These extinguishers are less common in laboratories because they are likely to be more expensive.

Carbon dioxide extinguishers. Carbon dioxide is a common extinguishing agent for Class BC fires. They are relatively inexpensive. In this instance, the propellant and the extinguishing agent are the same substance. It is not effective for Class A fires since these fires usually contain enough heat to reignite after the carbon dioxide has dissipated.

Class D Fires: Active Metals

"Active metals" are those metals, or metal-containing compounds, that are good reducing agents, or easily oxidized. Some examples are elemental sodium, potassium, and lithium and compounds such as sodium hydride, lithium aluminum hydride, or the Grignard reagent (Mg-R-X). These reagents are rarely used *except* in chemistry labs. Some industrial processes involving magnesium or titanium at high temperature, particularly under the condition of high surface area, also present a fire hazard.

For all active metal fires, water is the worst possible extinguishing agent. Water will accelerate the fire since the active metals react with water. *Never use water on an active metal fire.* The fire in an "active metal fire" actually results from the combustion of the hydrogen gas that is produced when the metal reduces the hydrogen in the water creating hydrogen gas and heat. As you may already know, the use of elemental sodium or any hydride compound in an experiment is always under the condition of the strict exclusion of water. In fact, these reagents are often used to remove trace amounts of water from organic solvents.

Active metal fires are very dangerous and frequently involve flammable organic solvents. Thus, they are often both Class B and Class D fires. Unfortunately, there is not an ideal, universal Class D extinguisher. Various commercial extinguishers are specified for specific metals, but these may or may not be available in the lab. The main action of the extinguisher, often powdered graphite or sodium chloride, is to exclude oxygen and water from the metal or compound and to absorb heat. (Although not perfect, a bucket of ordinary sand is usually a reasonable extinguishing agent.) While these agents will mitigate the Class D component of the fire it will likely be necessary to also extinguish any burning solvent using a BC or ABC extinguisher.

Laboratory Fires and Laboratory Hoods

So what should you do in the event of a fire in your laboratory? If the fire is small and you believe that you can safely extinguish the fire, then it may be appropriate to do so. But you should know exactly what you are doing; that is, you must be prepared to handle emergencies. You should know how to operate the extinguisher and you should know if the extinguisher that you have is appropriate for the class of fire. Whatever your decision, remember that fires can get out of control rapidly. If your safety might be compromised, then do not try to extinguish the fire. It is better to *evacuate immediately!*

If you are working at a hood and a fire occurs in that hood, if you can safely close the hood and exit, then it is appropriate to do this—remember your safety is most important. Do not put yourself at increased risk. *Alert* others of the emergency. *Exit* the laboratory and building. *Pull* the alarm as you leave the building.

Your laboratory hood may turn off if a fire alarm sounds. Laboratory hoods are ventilated cabinets designed to contain and exhaust away gases, vapors, or aerosols that might be released during laboratory operations. They operate by drawing air from the laboratory room, which in turn draws air from the adjacent corridors. This in essence means that hoods draw air into a building. Air or oxygen can feed a fire so it is not unusual for hood systems to be designed to automatically shut down in the event of a fire.

However, some buildings do not automatically shut off hoods when a fire alarm sounds. Nevertheless, you must be aware that if the alarm does go off and you are working in a hood that the hood may shut down so that it is no longer providing protection. If you can safely close your hood and immediately leave the laboratory and building, then do so.

RAMP

- *Recognize* there are different kinds of fire extinguishers for use in fighting the four classes of fire.
- *Assess* the class of fire so that the right extinguisher can be selected.
- *Minimize* your risks of injury by selecting the right fire extinguisher for a fire–if the fire is small and you can safely extinguish the fire without risk to yourself.
- *Prepare* for fire emergencies by knowing where extinguishers are located, knowing how to operate extinguishers, and knowing which extinguisher is needed to fight a specific fire.

References

1. RED ADAIR. BrainyQuote; available at http://www.brainyquote.com/quotes/quotes/r/redadair195669.html (accessed March 15, 2009).
2. American Industrial Hygiene Association Laboratory Health and Safety Committee. Laboratory Safety Incidents: Fires. Lithium Aluminum Hydride Fire; available at http://www2.umdnj.edu/eohssweb/aiha/accidents/fire.htm#Aluminium (accessed September 14, 2009).
3. American Industrial Hygiene Association Laboratory Health and Safety Committee. Laboratory Safety Incidents: Fires. Shelf-Collapse Causes Spills and Fire; available at http://www2.umdnj.edu/eohssweb/aiha/accidents/fire.htm#Shelf (accessed September 14, 2009).
4. National Ocean and Atmospheric Administrations. Cameo Chemicals; available at http://cameochemicals.noaa.gov/ (accessed September 14, 2009).
5. National Institute for Occupational Safety and Health. Pocket Guide to Chemical Hazards; available at http://www.cdc.gov/niosh/npg/ (accessed September 14, 2009).
6. National Fire Protection Association. *Fire Protection Handbook*, 18th ed., NFPA, Quincy, MA, 1997, Section 6, Chapter 18, pp. 281–296.
7. National Fire Protection Association. *Fire Protection Handbook*, 18th ed., NFPA, Quincy, MA, 1997, Section 6, Chapter 23, pp. 368–385.

QUESTIONS

1. Most of the fires that one would expect to encounter in an organic synthesis lab would be

    (a) Class A
    (b) Class B
    (c) Class C
    (d) Class D

2. The portable fire extinguisher of most use in an organic lab will be

    (a) A
    (b) BC
    (c) ABC
    (d) BC or ABC

3. Although rarely encountered in academic environments, some industrial labs may have whole-room, automated suppression systems. These systems use CO_2 or some inert gas and are designed mostly for

    (a) Class A fires
    (b) Class B fires
    (c) Class C fires
    (d) Class D fires

4. Active metal fires

    (a) Should not be extinguished using water
    (b) Are usually associated with strong reducing agents
    (c) Should be extinguished only with a Class D extinguishing agent
    (d) All of the above

2.2.2

CHEMICAL SPILLS: CONTAINMENT AND CLEANUP

Preview This section describes how you should respond to a chemical spill and how common chemical spills can be cleaned up.

> *One sometimes finds what one is not looking for.*
>
> Alexander Fleming, English scientist[1]

INCIDENT 2.2.2.1 MIXED SOLUTIONS SPILL[2]

In a qualitative analysis experiment, a student accidentally knocked over a rack of small test tubes containing a variety of solutions on an open bench. He notified the instructor, who was aware that each of the solutions by itself was not highly hazardous as a spill, but also knew that some combinations of solutions might be hazardous. The immediate area was vacated and two of the solutions were identified as containing cyanide and acid. All of the hoods near the spill were opened and other hoods in the lab were closed to produce the maximum air exhaust in the area of the spill. The total volume of liquid spilled was less than 15 mL. Paper towels were used to quickly soak up the spill (while wearing gloves) and they were transferred to a chemical hood. These were later placed in a very large beaker of water to effect dilution. Eventually, the highly diluted solutions were flushed down the sink and the paper towels discarded.

What lessons can be learned from this incident.

Prelude

Section 2.1.3 presented an introduction to how to handle spills of solids and liquids in most academic labs associated with courses. This section discusses the kinds of spills and chemical releases that are more likely to occur in advanced and research labs.

We again begin with the premise that, for the most part, students and researchers should *not* be the principal persons involved in cleaning up spills, except in the most trivial circumstances. Most organizations, academic and otherwise, have personnel with more training in these matters. For some hazmat incidents, OSHA requires special training known as HAZWOPER training (Hazardous Waste Operations and Emergency Response) although this is not universally required at colleges and universities.

However, in some circumstances it may be necessary or practical that you, as a student or employee, may need to respond to a chemical spill rapidly in a circumstance where waiting for someone else is unwise. With this in mind, we present more information about responding to chemical spills and other emergency situations.

General Procedures

Section 2.1.3 concluded with the general procedure for responding to chemical spills and releases:

- *Evacuate*. Depending on the size of the spill and the hazards posed by the chemical, it may be necessary to evacuate the immediate area or the entire building. If necessary, a fire alarm can be activated to effect evacuation of a building.

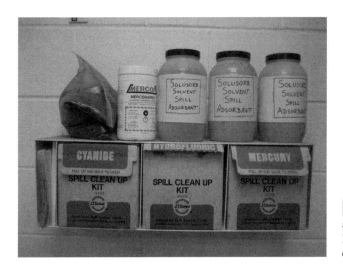

FIGURE 2.2.2.1   Chemical Spill Kit. There are many kinds of spill kits. It is best to know their location in laboratories since some spills should be contained as quickly as possible.

- *Communicate*. You should alert someone, preferably your laboratory instructor or assistant, that there has been a spill. If you know the nature of the spill, you should relay this information so that appropriate action can be taken. You should alert others of the possible hazard and advise them to avoid this area.

- *Isolate*. If possible, try to limit the extent of the spill (for liquids) by containing the spill with sand or other appropriate material.

- *Mitigate*. Under some circumstances you can participate in cleaning up the spill.

Also, as noted in Section 2.1.3, the treatment of any injuries associated with a spill and possibly calling 911 take top priority.

More About Spill Containment and Cleanup

If you have a spill, you will need to decide if this is a spill that you can take care of or if you need to get some help. For small spills, maybe up to about 100 mL, you may be able to do this yourself. But spills much larger than this probably require some additional experience—and even a small spill of a very hazardous material requires assistance. Never hesitate to ask for help!

Many laboratories have buckets or cabinets with supplies for cleaning up chemical spills. These are often called chemical spill kits (see Figures 2.2.2.1 Figure 2.2.2.2). While these kits are likely to contain written instructions and needed supplies and protective equipment, you should not attempt to clean up a spill unless you have received some instruction and demonstration about carrying out a cleanup operation. Depending on the nature of the chemical, it may be possible to clean up a spill with only a few paper towels that are then disposed of in an appropriate manner (ask what is appropriate). For larger spills you should seek assistance, which often comes from the institution's health and safety department.

For spills of liquids and solutions it is desirable to contain the spill—you don't want it to spread so, if you can, encircle it with appropriate materials. There are commercially available spill control "pillows," "socks," and "pads" that absorb many times their own weight in acids, bases, and organic solvents. These are an effective and safe way to minimize the spread of a spill and largely accomplish a cleanup as well. These containment devices often contain material such as shredded polypropylene filling, which can absorb acids, bases, solvents, and oils. They do not neutralize acids or bases, but only absorb them. They should not be used for highly reactive chemicals such as hydrofluoric acid, fuming nitric acid, fuming sulfuring acid, or strong oxidizing chemicals. Alternately, if available, a "dike" of sand around the spill can temporarily contain the spread of the spill.

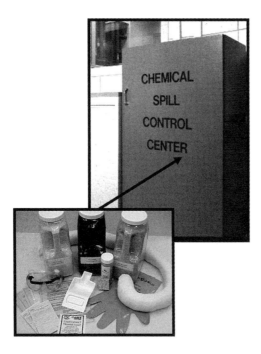

FIGURE 2.2.2.2   Chemical Spill Cabinet. Spill cabinets will contain useful equipment to handle many kinds of spills. The picture shows a "sock" that can be used to surround, contain, and absorb a spill.

Acid and Base Spills

Because acids and bases are used with some frequency in the laboratory, it is not uncommon to have to clean up an acid or base spill. Spills can be adsorbed using "universal chemical adsorbents"; however, these do not neutralize acids or bases. Vermiculite (hydrated silicates) or kitty litter (clay) are often used as absorbents. A "universal" adsorbent consisting of a 1:1:1 mixture of kitty litter, sodium bicarbonate, and sand has been used by some institutions.[3]

Neutralizing agents can be used for spills of acids or bases. There are also commercially available packets of materials that provide sorbent materials and scoops and plastic bags for containment and disposal.

- Acid spills are neutralized with sodium carbonate (also known as soda ash), sodium bicarbonate, or calcium carbonate. Commercially available neutralizing agents, such as *Neutrasorb®*, *Spill-X-A®, or Hazorb®acid neutralizer*, are also available and some have an indicator so that the color will be changed when the neutralization is complete. However, there are precautions for some of these cleanup materials indicating that they should not be used to clean up highly reactive acids, such as hydrofluoric acid, fuming sulfuric acid, or peroxy acids. You will need to read the precautions about using these.

- Base or caustic spills are neutralized with sodium bisulfate, citric acid, or dilute acetic acid (vinegar). There are also commercially available neutralizing agents for bases, such as *Hazorb®* caustic neutralizer, *Spill-X-C®*, or *Neutracrit ®*. Some of these commercial reagents contain indicators to help guide you in neutralizing the spill.

Solvent and Flammable Liquid Spills

You may encounter the spill of a solvent or flammable liquid in your work in the laboratory. Once again, if these spills are small you can probably take care of these with a universal absorbent. However, if you encounter a large spill you should seek assistance immediately and notify others to leave the area. This is especially true if this is a flammable liquid, since the chemical could encounter a source of ignition that could result in a fire.

- Solvents and flammable liquids can be cleaned up using vermiculite, kitty litter, or activated charcoal. Commercial sorbents are also available, such as Spill-X-S®, Solusorb®, or Chemsorb®. These sorbents can be used for spills of other organic chemicals but should not be used for oxidizing agents.
- It is important to act quickly and get a sorbent onto the solvent spill so that the vapor concentration will be reduced and the risk of fire can be reduced.
- If the air concentration becomes high, it may be necessary for the cleanup crew to wear respirators. Anyone wearing a respirator must be specifically trained and fit-tested for the respirator being used. If you have not been trained and fitted for a respirator, DO NOT decide that you can just put on a respirator and that you will be protected—it is likely to offer only a false sense of security. You will be putting yourself at increased risk.

Other Spills: HF, Hg, Reactive Chemicals

While there are spill kits for mercury, you should seek professional assistance in cleaning up mercury spills. Unless you have been specifically trained in cleaning up hydrofluoric acid, mercury, or other reactive chemicals, you should not attempt to deal with these spills yourself. Seek professional assistance, usually the environmental, health, and safety department of your institution. These spills are especially hazardous and require more experience.

Cleaning Up a Spill

If you decide that you can clean up a spill:

- You should put on appropriate protective equipment including chemical goggles, protective gloves, and a lab coat or protective suit. The cleanup of some spills may require respirators, and you must be medically cleared, fit-tested, and trained in the use of respirators before using these.
- You should first try to contain the spill by building a dike around it with spill tubes or sand.
- After this, you can add the appropriate spill cleanup reagent carefully and slowly—don't be in a hurry to do this.
- After the spilled material is absorbed, transfer the absorbent to a waste container—there is usually a bucket, scoops, and bags in the spill kit for this. Put the other contaminated materials in the bucket also.
- You may need to wipe up the area with some paper towels or other absorbent material to ensure that all the material is cleaned up. Put all of that material in the bucket too.
- Close up the bucket, put a label that describes the spilled materials, and call your EHS department to take care of this and to replenish the spill kit.

Leaking Gas Cylinders

When you use compressed gases, it is possible that a leak could develop. The first thing to determine is what gas is leaking. The most common laboratory gases are N_2, O_2, He, and Ar and small leaks of these gases pose no safety hazard. However, a leak from a tank of CO, HCN, NH_3, HCl, or any other toxic gas may require immediate evacuation rather than any remedial action. The size of the leak and the local ventilation may affect this decision.

Leaks are often located at the cylinder valve or at coupling or connectors. Leaks can be detected with soapy water and can often be remedied by tightening the connection with the proper wrench. You should turn off the pressure while you are making these adjustments. If you cannot stop the leak, then seek some assistance from a more experienced colleague or instructor or from your EHS department.

If gas is leaking around the neck of the cylinder where the valve screws into the cylinder, you will not be able to stop this. You should take action to remove the cylinder to the outside if it is a small leak and you can do this safely. You may want to call your EHS department to assist in taking care of this problem. The leak can only be repaired by the supplier and they will need to be called to retrieve the leaking cylinder. Similarly, if there is a leak around the pressure relief device that is on the main cylinder valve, you will not be able to fix this. Nevertheless, it will need to be moved outside. In some cases, there may be leaks in small cylinders that can be placed in a laboratory hood—a lecture bottle, for example.

If you do encounter a large leak that cannot be easily stopped, immediately alert everyone to leave the area and call emergency personnel. If the gas is flammable, it is possible that an explosion and fire can occur if a source of ignition is encountered; even turning off the lights can create a spark. Don't turn off anything, just evacuate. If this is a corrosive or toxic gas, anyone trying to deal with this leak will need to be protected with appropriate protective equipment and probably a self-contained breathing apparatus. You must be trained in doing these operations and it is best to allow emergency personnel to do this work. If you know what the situation is, you should be there when the emergency personnel arrive so that you can inform them about the nature of the emergency and what other hazards might be present.

Summary

You are likely to encounter chemical spills during your work in the laboratory. You should know what you will do before this happens—remember the "prepare" part of RAMP. You should know where the nearest spill kit is located and inspect it to understand what it contains and how it can be used to clean up spills. If possible, you should ask for a training session to practice cleaning up a spill.

When you do clean up a spill, follow the instructions provided in the spill kit. You should ask for assistance or for others to observe you doing this cleanup. Dispose of cleanup materials used for the spill as hazardous waste with the appropriate label.

RAMP

- *Recognize* that chemical spills can be hazardous and appropriate action is necessary.
- *Assess* the relative size of the spill and the nature of the materials to determine if you can clean up the spill safely or if professional assistance is needed.
- *Minimize* the effect of the spill by using available resources such as a spill kit and call for assistance.
- *Prepare* for spills that you might experience by learning how to clean up a spill, finding the nearest spill kit, and understanding how to use this effectively.

References

1. ALEXANDER FLEMING, BrainyQuote; available at http://www.brainyquote.com/quotes/quotes/a/alexanderf264431.html (accessed March 27, 2009).
2. DAVID C. FINSTER. Personal account of incident.

3. Indiana University, Environmental, Health, and Safety Department. Chemical Spill Response Guide, 2000; available at http://www.ehs.indiana.edu/em/spilweb.pdf (accessed April 1, 2009).

QUESTIONS

1. In general, in what order should the following steps be taken when responding to a chemical spill?

   (a) Communicate, isolate, mitigate, evacuate.
   (b) Isolate, evacuate, mitigate, communicate.
   (c) Evacuate, communicate, isolate, mitigate.
   (d) Evacuate, isolate, mitigate, communicate.

2. Commercially available "pillows" or "socks" are designed to contain the spread of liquid spills and

   (a) Generally "neutralize" the hazardous nature of most spills
   (b) Can be used with all chemicals
   (c) Absorb many times their own weight in liquid
   (d) All of the above

3. Which chemical is *not* a good neutralizing agent for an acid spill?

   (a) Na_2CO_3
   (b) $NaHCO_3$

(c) NaOH

(d) CaCO$_3$

4. Which substance is *not* a good agent for absorbing an organic solvent spill?

    (a) Kitty litter

    (b) Vermiculite

    (c) Activated charcoal

    (d) Sand

5. A spill kit may include personal protective equipment, buckets, scoops, absorbing agents, plastic bags, and paper towels. Which of these materials should be included as part of the material to be disposed of by trained personnel?

    (a) Only the absorbing agent and contaminated paper towels

    (b) The absorbing agent, contaminated paper towels, and plastic bags

    (c) The absorbing agent, contaminated paper towels, plastic bags, and the buckets and scoops

    (d) Everything that was potentially contaminated in the cleanup process

6. Leaks from high-pressure gas cylinders involving toxic gases

    (a) Require immediate evacuation of the lab

    (b) Require immediate evacuation of the lab unless they can be quickly detected and corrected by tightening a coupling or connection

    (c) Should only be fixed if the leak is large and the ventilation is poor

    (d) Needn't be fixed if the leak is very small and the lab ventilation is good

7. The best response to a chemical spill

    (a) Involves a quick recognition of the size and nature of the spill

    (b) Involves people who know how to properly clean up a spill

    (c) Usually requires a hazmat team from the local fire department

    (d) Both (a) and (b)

CHAPTER 3

UNDERSTANDING AND COMMUNICATING ABOUT LABORATORY HAZARDS

HOW CAN chemicals harm you? How can you know which chemicals harm in particular ways? This chapter presents the means by which the chemical community and others have developed systems to communicate the various dangers of chemicals. The good news is that there is plenty of information about the hazards of many of the chemicals that you are likely to encounter in chemistry laboratories, but sometimes it takes a little more than "common sense" to understand the hazards. And, sometimes it takes a little effort to find the information. This chapter helps students begin to understand how to find and interpret safety information. The World Wide Web has made this "finding" fairly easy but it takes some knowledge to do the "interpreting" correctly! This chapter is one of three chapters covering the first principle of safety: *recognizing hazards*.

INTRODUCTORY

3.1.1 Routes of Exposure to Hazards   Describes the potential pathways of exposure and explains how blocking these pathways can prevent exposure.

3.1.2 Learning the Language of Safety: Signs, Symbols, and Labels   Discusses hazard recognition, an essential skill needed by every laboratory worker, through the basics of understanding labels, signs, symbols, terms, and other sources of information.

3.1.3 Finding Hazard Information: Material Safety Data Sheets (MSDSs)   Describes some good resources with information about hazards of chemicals and how they might be used.

INTERMEDIATE

3.2.1 The Globally Harmonized System of Classification and Labelling of Chemicals (GHS)   Presents a brief overview of the United Nations' Globally Harmonized System of Classification and Labelling of Chemicals (GHS) for defining and communicating hazard information and protective measures related to the physical, health, and environmental risks of chemicals through product labels and safety data sheets (SDSs).

3.2.2 Information Resources About Laboratory Hazards and Safety   Describes information resources that are easily available about the hazards of chemicals and how to use this information to recognize the hazards of chemicals.

3.2.3   Interpreting MSDS Information   Provides an overview of the various parts of a typical MSDS and provides suggestions to help you find important information about the hazards of a chemical.

ADVANCED

3.3.1   Chemical Hygiene Plans   Describes OSHA's Laboratory Standard that requires laboratory management to develop a Chemical Hygiene Plan (CHP), to appoint a Chemical Hygiene Officer (CHO), and how a CHP can help in devising protection from exposure.

3.1.1

ROUTES OF EXPOSURES TO HAZARDS

Preview   This section describes the potential pathways of exposure and explains how blocking these pathways can prevent exposure.

Concern for man himself and his fate must always form the chief interest of all technical endeavors.... Never forget this in the midst of your diagrams and equations.

Albert Einstein[1]

INCIDENT 3.1.1.1   REUSING GLOVES[2]

The young chemist was working in his research laboratory, and he wanted to protect himself from exposure to the chemicals he was using. He used a chemical hood for his experiments and personal protective equipment—chemical goggles and gloves. He had selected rubber gloves and used them in the morning for handling his chemicals. When he returned in the afternoon, he put on the gloves again and went to work. Sometime later, he became aware that his skin was burning and had turned red. Then he realized that he had put on the gloves outside in, and now the contaminated outside with chemicals he was using was inside in direct contact with his skin. The exposure was so severe that he required medical treatment, he could not work in the laboratory for two months, and he finally had to stop working on this research project because he became allergic to one of the chemicals.

What lessons can be learned from this incident?

How Can I Be Exposed to Hazards?

A *hazard* is a potential source of danger or harm. The word *potential* means something that is capable of being dangerous or harmful. *Exposure* means coming in direct contact with a hazard or chemical in a fashion that can injure or harm. You could receive a dose of a chemical by ingesting it, breathing it, or allowing contact with your skin—that is, by providing the chemical access to your body. You might also receive a dose of another type, such as trauma from being struck directly by an object (shrapnel) from an explosion or being struck indirectly by a secondary object that was struck by something else moving. The dose, duration of exposure, and path of exposure play significant roles in the extent of injury or harm. The important thing to remember is that we want to minimize or eliminate exposure to hazards.

Routes of exposure are the pathways by which chemicals may reach or enter the body. To cause toxic effects in a living organism or body, a chemical must find a way to come in contact with and enter the body. To prevent exposure to harmful chemicals, it is only necessary to prevent entry or exposure to the body. If you know the potential routes of exposure and you seek to prevent chemicals from entering by those pathways, you will effectively prevent exposure to those chemicals and will be protecting yourself from harm. There are four routes of entry or exposure: ingestion, inhalation, skin and eye exposure, and injection.

Don't Eat That Here!

Ingestion, also known as the *oral* route, takes place through swallowing or allowing the toxic material to enter through the mouth and proceed through the alimentary canal (mouth, pharynx, esophagus,

Laboratory Safety for Chemistry Students, by Robert H. Hill, Jr. and David C. Finster
Copyright © 2010 John Wiley & Sons, Inc.

stomach, small intestine, large intestine). To be toxic by this route, a chemical must be absorbed in its passage through the alimentary canal, either in the stomach or intestines. Fortunately, the pathway of ingestion can be effectively eliminated or minimized with the strict application of two basic safety rules—prohibiting eating or drinking in the laboratory and never tasting any chemical in the laboratory (see Section 1.1.1, Table 1.1.1.2).

Sometimes ingestion occurs in an indirect fashion. If you lay your pen or pencil on the lab bench and it gets contaminated by a chemical and then you (perhaps by habit) hold that pen or pencil between your teeth, the chemical may transfer into your mouth. Similarly, placing a contaminated finger in your mouth would have the same effect.

A famous incident of unintentional ingestion was that of Constantine Falhberg who, in 1879, synthesized the saccharin molecule. Since lab safety procedures were less cautious at the time, he contaminated his hands with the powder, didn't wash his hands after leaving the lab, and later detected an unusually sweet and "off" taste when the saccharin was inadvertently transferred to his food. Since saccharin is about 300 times sweeter (per gram) than common table sugar, even a small amount was noticeable. In fact, the "off" taste is characteristic of saccharin in doses higher than the amounts now used, for example, in the sweetener Sweet and Low®.

The Eyes Have It!

As you read this, you should recognize that your eyes are a special set of sensitive organs exposed to the open environment. Eyes are very sensitive to chemical exposures and injuries from flying projectiles. The laboratory affords many opportunities for handling a variety of hazardous materials, including corrosives, irritants, toxicants, and perhaps even compounds that could under certain conditions explode. Additionally, there will likely be more than one person carrying out operations in the laboratory, and you will not know from moment to moment what these other laboratory workers might be doing. It is not uncommon for an experiment to go awry and for chemicals or projectiles from a splatter, spill, or perhaps even an explosion to come in contact with your eyes. People working in laboratories have suffered severe eye injuries and even blindness from these kinds of exposures.

It is not enough to wear eye protection only when you *think* there might be a chance of exposure. You cannot predict when an incident may occur that results in something being projected into your face and eyes, either from your own experiments or another person's activity in the same laboratory. To prevent exposure of your eyes, it is essential that you *always* wear eye protection in the laboratory. Chemical safety goggles are an example of personal protective equipment that you should wear in the laboratory. You will find more information on eye protection later in this book in Section 7.1.2. Your teacher will likely provide specific instructions on the required eye protection for the laboratory where you are working. However, it is ultimately up to you to protect your eyes by following the simple procedure of always wearing appropriate eye protection in the laboratory.

Did You Smell That?

Inhalation is a very important route of entry. By inhaling or breathing a toxic chemical, it can enter into the lungs, where it can be absorbed directly into the bloodstream with the exchange of air. We can prevent inhalation exposure in the laboratory by using appropriate containment measures. Liquid chemicals that can readily move into the gaseous phase at normal temperature and pressure from an open container are called *volatile* liquids. We can keep containers such as bottles, jars, or flasks closed when not in use and minimize the concentration of the vapor in the air, thus minimizing our exposure. We can also work with such liquids in fume hoods to virtually eliminate our exposure to the vapors.

Ordinary air consists of about 78% N_2, 21% O_2, 1% Ar, and traces of other gases such as CO_2, Ne, CH_4, and He. Since the process of respiration uses O_2 and generates CO_2, the air we exhale is about 78% N_2, 17% O_2, 4% CO_2, 1% Ar, and traces of other gases. (Thus, doing CPR, *cardiopulmonary resuscitation*, on someone who is not breathing is very helpful since even your exhaled, "used" air still contains plenty of oxygen.) A variety of contaminants can occur in air and we can conveniently divide the possibilities into two groups: homogeneous contaminants and heterogeneous contaminants.

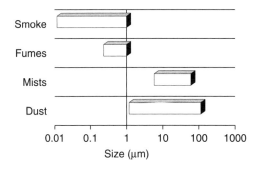

FIGURE 3.1.1.1   Ranges of Particles Sizes for Aerosols. Air can contain many sizes of particles that can be inhaled. There should be none of these in a laboratory. Any procedures that generate these aerosols should be conducted in chemical hoods. Industrial manufacturing facilities are a more likely site to encounter these particles.

A homogeneous mixture is one where both components are in the same phase. Thus, we can identify nonair gases as the contaminants in such a mixture. In some instances the gases will be elements or compounds that are ordinarily gases at room temperature and pressure such as ozone (O_3), chlorine (Cl_2), or methane (CH_4). More often, though, the gaseous contaminant will be the vapor form of a compound that is primarily a liquid at room temperature and pressure. (See *Chemical Connection 3.1.1.1*. Predicting the Vapor Pressure of a Liquid.) Some examples are the common organic solvents toluene and dichloromethane

A heterogeneous mixture in air is one where the contaminant is either a liquid or a solid. When the particle size ranges from about 0.01 to 100 μm, the mixture is called a colloid or an aerosol. (A colloid that is a heterogeneous mixture of one liquid in another liquid is called an emulsion: for example, milk is an emulsion.) Some common aerosols are mists, fumes, and dusts. *Mists* are tiny droplets of liquid suspended in the air. Clouds and fog are mists of water droplets in air. Figure 3.1.1.1 shows the ranges of particle sizes.

CHEMICAL CONNECTION *3.1.1.1*

PREDICTING THE VAPOR PRESSURE OF A LIQUID

Using a liquid may or may not involve the inhalation of the vapors. How can we predict how much of the liquid will vaporize? (In addition to this first question, we should also consider what concentration of vapor is safe to breath. This is discussed in Section 6.2.2.)

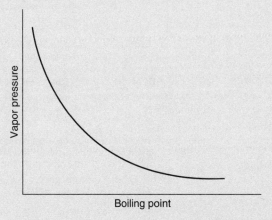

FIGURE 3.1.1.2   Relationship Between Vapor Pressure and Boiling Point. This sketch shows that liquids with low boiling points will have high vapor pressures. What are the safety concerns when using liquids with high vapor pressures?

Volatile liquids are those liquids that present a reasonable chance of vaporization under ordinary lab conditions of about 25 °C and 1 atm of pressure. The volatility of a liquid is indicated by its boiling point and its vapor pressure (see Figure 3.1.1.2). Liquids with low boiling points have higher vapor pressures and liquids with high boiling points have lower vapor pressures.   (continued on next page)

TABLE 3.1.1.1   Normal Boiling Points and Vapor Pressures of Selected Chemicals at 25°C

Chemical	Normal boiling point (°C)	Vapor pressure (mm Hg) at 25°C
Dichloromethane	39	423
Acetone	56	200
Isopropanol	82	41
Water	100	24
Toluene	111	32
Dimethylsulfoxide	189	0.6

We also know that vapor pressure increases as temperature increases, but this relationship is not linear (see Figure 3.1.1.3). This is predicted by the Clausius–Clapeyron equation:

$$P = B \exp(\frac{-\Delta H_{vap}}{RT}) \quad \text{or} \quad \ln P = \ln B - (\Delta H_{vap}/RT)$$

where P is the vapor pressure, ΔH_{vap} is the enthalpy of vaporization, R is the gas constant, T is temperature, and B is a constant that depends on the liquid. (This equation is typically introduced in most general chemistry textbooks in the chapter on "Liquids.") Thinking about safety, it may be that a liquid that has a relatively low vapor pressure at 25 °C and presents a negligible breathing hazard will have a much higher vapor pressure at elevated temperature. For example, dimethylsulfoxide (DMSO) has a tiny vapor pressure at 25 °C (0.6 torr) but at 150 °C (39 degrees below its boiling point) the vapor pressure is 238 torr. Thus, even a "nonvolatile" solvent becomes more volatile as temperature increases.

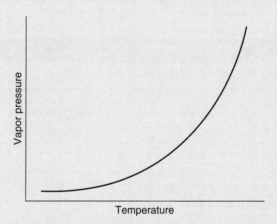

FIGURE 3.1.1.3   Relationship Between Vapor Pressure and Temperature. The Clausius-Clapeyron equation predicts that vapor pressure increases non-linearly as temperature increases. What are the safety concerns when using liquids at high temperatures?

It is possible to find vapor pressure curves for liquids on the Internet, or by using the enthalpy of vaporization it is easy to calculate a "second vapor pressure" if you already know one value using the "two-point" format of the Clausius–Clapeyron equation:

$$\ln(P_2/P_1) = -(\Delta H_{vap}/R)(1/T_2 - 1/T_1)$$

In the laboratory you may form mists or aerosols when a bottle of chemical is opened, when the contents of an open bottle are shaken, when chemicals are poured into other vessels, or when a chemical is spilled and hits a solid surface. *Fumes* are a colloidal suspension of solid or liquid particles. Grinding a metal can generate fumes when the metal gets hot enough to vaporize but then quickly cools and recondenses to a solid in the form of tiny small particles. "Fuming sulfuric acid" and "fuming nitric acid" produce clouds of these acids over the surfaces of the liquids. *Dusts* are solid particles suspended in air, usually created by some operation, such as handling a fine powder (perhaps during weighing), or crushing, grinding, or blending a solid material.

Very small aerosol particles, ≤ 5 μm, can remain suspended in air for extended time periods (minutes to hours) and provide additional opportunities for exposure. These particles are a health concern since they may be deposited in the smaller airways and alveoli of the lungs. Depending on their chemical nature and composition, some of these particles may be dissolved and absorbed into the body, or if they are insoluble may remain in the lungs permanently. When you are exposed to larger sized aerosol particles (>10 μm), they may be inhaled and are deposited in the secretions of the nose, the nasopharynx, or in the upper regions of the lungs, where they can be swallowed and ingested or coughed up from the upper parts of the lungs, swallowed, and ingested.

Another air contaminant with which you are already familiar is smoke. *Smoke* is a mixture of aerosol particles (dry particles and droplets) that result from incomplete combustion (burning) of carbon-based materials—from a fire or from a cigarette or other burning materials. Smoke will also contain many molecular-size contaminants such as CO_2, CO, NO, and NO_2 and, depending on what is burning, it may also contain HCN and halogen acids (HF, HBr, and HCl). Thus, smoke is a deadly mixture with many homogeneous and heterogeneous components.

When you use volatile chemicals or powdery solids on the open bench or in open containers, you will be exposing yourself to the possibility of inhalation of the chemicals. If you smell chemicals, it is an indication that you are being exposed. Some chemicals can be detected by odor at very low airborne concentrations; that is, they have very low odor thresholds and their odor gives you warning before concentrations reach levels that cause harm. However, there are other compounds that have relatively high odor thresholds, meaning that by the time you smell them, you may have already inhaled a significant dose. It is always important to minimize exposure to chemicals, so you should avoid breathing or inhaling chemicals in the laboratory as much as possible.

Such Lovely Skin—Our Personal Barrier to Exposure!

Skin exposure, also known as *dermal exposure*, is the most common route of exposure to chemicals. Most adverse occupational health exposures to chemicals have occurred through this route—and this is also true of laboratory exposures. The skin normally provides us with good protection from many, but not all, outside agents that might injure us if they were to gain entry into our bodies. When some chemicals come in contact with your skin, there is a possibility that they may stay on or near the surface of the skin, or they can react with skin at the site of exposure or they can be absorbed through the skin into the bloodstream and then proceed to your body's organs. When our skin is compromised with breaks or cuts, these injured sites can provide additional routes of entry into the body, and it is important to protect these vulnerable areas from exposure when using toxic chemicals.

One other area of concern for skin exposure is surface contamination produced by handling a fine powder or a liquid that forms fine particles when shaken. These particles—solid or liquid—can cover or contaminate a surface and may not be readily visible. This surface contamination does not move or leave the surface unless it is removed by cleaning. So, for example, if you are weighing out a solid material and you spill some material or the dust settles on the benchtop from your handling operations, it needs to be removed or you and the next person using this area could be exposed to these materials. Your hand or arm could come in contact with the chemical(s) and it can be absorbed through the skin or accidentally ingested when your contaminated hand touches your mouth. This kind of contamination could also be carried on your clothes. All of this can be prevented through careful attention to cleaning surfaces, wearing gloves, laboratory coats, or aprons, and perhaps covering surfaces with plastic-backed absorbent paper.

The best practice is to always minimize or avoid dermal exposure to chemicals. This can be accomplished in a number of ways. Protective or outer clothing, such as a lab apron or a laboratory coat, can be worn to protect your body. It's best to wear long sleeved shirts, pants or long skirts, and closed-toe shoes so that if you spill the chemical it will not come in direct contact with your skin. Your college probably has rules about what clothes you can and cannot wear in the lab.

If a significant exposure to a hazardous chemical is possible, then your hands should be protected by wearing protective gloves. This does not mean that you need to wear gloves at all times in the laboratory. However, it is critical if you do need to wear gloves that you select gloves that are specific for the chemical(s) that you will be using. There is no universal glove that protects in all situations. One kind of glove may protect against a chemical or a class of chemicals but may be ineffective against another chemical or class of chemicals. Some chemicals can permeate (pass through) the glove material and still reach the skin. Other chemicals may break down or dissolve the glove material to reach the skin. It is important to choose your gloves with some knowledge about the protective properties of the gloves toward those classes of chemicals you will be handling. You will find more on this topic in Section 7.1.3 on protecting your skin.

Exposure by Injection Is Possible, But Not Likely

Injection is not usually an important route of unintentional toxic exposure because chemists don't routinely use needles and syringes. However, this route is possible if a glass container breaks and creates a shard or sliver of glass contaminated with this toxic chemical. If this sticks into your body, some of the chemical could be injected. However, only the most toxic chemicals could be received in a dose sufficient to cause a toxic effect.

Some advanced techniques in chemical synthesis and analysis actually do involve the use of needles and syringes. Obviously, it is important to use this equipment with care to avoid an accidental "stick." And, chemists conducting analyses in clinical, hospital, or toxicology laboratories working with infectious substances such as blood or serum must take extra care to avoid accidental injection. This topic will be discussed more in Section 4.3.3 regarding bloodborne pathogens and infectious materials.

Working in the Lab: RAMP

- *Recognize* what routes of exposure are possible for a particular chemical based on its physical properties.
- *Assess* the level of risk for exposure to skin or eyes and/or the possibility for inhalation or ingestion.
- *Minimize* the risk by preventing exposure using good lab practices and personal protective equipment.
- *Prepare* for emergencies by knowing what to do if there is potentially damaging exposure to a chemical.

References

1. GORTON CARRUTH and EUGENE EHRLICH. *The Giant Book of American Quotations*, Gramercy Books, New York, 1988, Topic 207. Science, #14.

2. R. H. HILL. Personal account of incident.

QUESTIONS

1. What factor does *not* play a significant role in the extent of harm upon exposure to a chemical?

   (a) Dose
   (b) Length of exposure
   (c) State (gas, liquid, solid)
   (d) Path of exposure

2. What is the most effective method for avoiding exposure by ingestion?

   (a) Taste only chemicals that your instructor gives you permission to taste.
   (b) Taste only chemicals that you know are nontoxic.
   (c) Never eat or drink anything while in a chemistry lab.
   (d) Only eat food in a lab when you know that it cannot be contaminated with a toxic chemical.

3. In order to avoid exposure to your eyes in a chemistry lab you should wear eye protection

   (a) Only when working with corrosive chemicals
   (b) Only when your instructor requires it
   (c) Only when there are other students working nearby who might do something stupid
   (d) All of the time

4. Volatile chemicals are substances that

   (a) Are very reactive
   (b) Have a vapor pressure of at least 100 mm Hg

(c) Are liquids that vaporize readily at normal pressure and temperature

(d) Are gases at normal pressure and temperature

5. The best way to avoid inhalation of volatile chemicals while working with them is to

(a) Work in a chemical hood when using them

(b) Avoid breathing the vapors by holding beakers and flasks at arm's length

(c) Wear a common dust mask

(d) Always keep beakers, flasks, and bottles of the chemical capped or covered

6. Which definition is *not* correct?

(a) Mists are tiny droplets of liquid suspended in the air.

(b) Fumes are small solid particles that are formed when a solid is vaporized.

(c) Dusts are solid particles suspended in air.

(d) A heterogeneous mixture in air is one where the contaminant is a colloid.

7. Smoke

(a) Is a mixture of aerosol particles resulting from the incomplete combustion of carbon-based materials

(b) Contains many toxic, small-molecule compounds

(c) Is a colloid

(d) All of the above

8. It is generally true that

(a) Almost all chemicals have an odor threshold below their toxic threshold

(b) Almost all chemicals have an odor threshold above their toxic threshold

(c) Whenever you can smell a chemical you are being exposed to it

(d) None of the above

9. Most adverse occupational exposures to chemical are through

(a) Inhalation

(b) Dermal exposure

(c) Ingestion

(d) The eyes

10. The best way(s) to avoid chemical contact with the skin is to wear

(a) Clothes that protect most of your body

(b) A lab coat

(c) Shoes that cover your feet

(d) All of the above

11. Gloves worn in labs

(a) Are generally resistant to almost all chemicals

(b) Usually protect against only a limited range of chemicals

(c) Generally protect against most chemicals, but only for a short period of time

(d) Need to be selected carefully to protect against the particular chemicals you are using

12. Having a chemical "injected" through your skin

(a) Is fairly unlikely in labs because chemists use needles and syringes only rarely

(b) Is possible if there is an explosion that produces sharp glass that is contaminated

(c) Is not likely to produce a toxic dose

(d) All of the above

3.1.2

LEARNING THE LANGUAGE OF SAFETY: SIGNS, SYMBOLS, AND LABELS

Preview   Hazard recognition, an essential skill needed by every laboratory worker, is presented through the basics of understanding labels, signs, symbols, terms, and other sources of information.

The new circumstances under which we are placed call for new words, new phrases, and for the transfer of old words to new objects.

Thomas Jefferson[1]

INCIDENT 3.1.2.1   ACETIC ACID EXPLOSION[2]

A technician was carrying out a procedure with boiling acetic acid as the solvent. She needed to add some additional solvent, so she obtained what she thought was a bottle of acetic acid and added some of its contents to the flask. Immediately there was a loud noise and the boiling solution blew out hitting the ceiling. A cloud of brownish fumes was noticed after the eruption, and upon checking the bottle, the technician realized that she had picked up a bottle of nitric acid instead of acetic acid. Because the technician was wearing the correct personal protective equipment, there was no injury.

What lessons can be learned from this incident?

Why Do You Need to Know About Hazards?

"They're everywhere, they're everywhere!" This quote is familiar from an old movie but when it comes to hazards, it is close to the truth. A *hazard* is a possible source of danger, and our world holds many hazards.

- Swimming is great fun, but the water for nonswimmers or even good swimmers can be a great hazard under certain circumstances. Indeed, the chemical that kills more than 3500 Americans each year (by drowning) is water![3] Yet we cannot live without water—it is essential to our very being.

- Fire can be a great hazard if uncontrolled, but under our control, fire is a major tool that we use to supply us with heat or to cook our food.

- Flying to far destinations is very important for many of us, but aircraft can be involved in crashes, although rarely; air travel is actually safer than auto travel.

- Automobiles are the major mode of transportation for most of us here in the United States, but these cars can also be very dangerous if used improperly—there are 40,000 deaths from automobile incidents every year.[4]

- We use many medicines that provide comfort, cure our ills, or prevent disease, but if we don't use them properly they can pose serious risks to our health.

The point is that many things, including chemicals, that we use everyday have not only very substantial benefits but very real hazards. The positive side of these hazardous materials is that we can learn how to maximize the benefits while at the same time minimizing the hazards of these things by taking certain steps.

When working in the laboratory, as in any other workplace, you will encounter chemicals or other materials or processes that just like the rest of the world have both benefits and hazards. Your safety and the safety of others depend on your developing the skill to recognize hazards and to manage, control, or minimize the risks of those hazards. This section introduces some basics of recognizing hazards in the laboratory—the language of hazards. A laboratory is usually a very safe place to work. However, if you work in a laboratory, you are likely to encounter chemical, biologic, radiation, or physical hazards. This section focuses on recognizing chemical hazards, while other sections deal with other kinds of hazards.

Name that Chemical—Watching for Mistaken Identity

There are millions of chemicals in our world, and likely millions more to be identified. Long ago chemists recognized this and they developed systematic ways to name chemicals—a process known as systematic nomenclature. About 100 years ago the American Chemical Society's Chemical Abstracts Service (CAS) developed a system of identifying chemicals with unique numbers—these are known as CAS numbers. You will find that it is not unusual for chemicals to have multiple names, but there will always be only one CAS number. (See *Special Topic 3.1.2.1* Finding the CAS Number for a Chemical.) Let's look at an example.

SPECIAL TOPIC *3.1.2.1*

FINDING THE CAS NUMBER FOR A CHEMICAL

The CAS number for a chemical is probably listed on the label if you are retrieving a chemical from the bottle or container in which it was received from a chemical supply company. But, to more generally find a CAS number, it is easy to access reliable web sites. Examples of two such web sites are ChemIndustry.com (http://www.chemindustry.com/apps/chemicals) and the National Institutes of Standards and Technology (http://webbook.nist.gov/chemistry/). At these sites you can search by name or formula to find a CAS number.

The chemical diethyl ether is also known by the common name "ether" and is also sometimes called "ethyl ether." The CAS number for diethyl ether, ether, or ethyl ether is 60-29-7 and its chemical formula is $(C_2H_5)_2O$, also written as $C_2H_5OC_2H_5$. If you look at the name you can understand why it has been called by all three names. All ethers have two substituents with one on each side of the oxygen. This chemical has two ethyl groups, hence the *di*ethyl part of the name. Some chemists call it ethyl ether, knowing that there is an ethyl on both sides of the oxygen. In fact, IUPAC (International Union of Pure and Applied Chemistry), which determines the official names of chemicals, calls this ethyl ether since when both groups on either side of the oxygen atom are the same, the *di-* prefix is omitted. Lastly, the name used by many chemists is simply "ether" because when we speak of ether, we all understand that it is diethyl ether (see Figure 3.1.2.1).

This brief introduction illustrates the confusion that can arise in the naming of chemicals or using nomenclature. It is meant to give you a brief understanding about the importance of knowing which

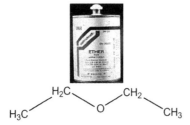

FIGURE 3.1.2.1   Ether Structure and Laboratory Chemical Container. Ethyl ether is a common laboratory chemical, also called "diethyl ether" and "ether." It is very flammable and tends to form dangerous peroxides if stored over extended periods of time.

specific chemical you are using so that you do not make a mistake in recognizing its hazards. As you can see, knowing the CAS number can help you unambiguously identify the correct chemical.

Look at the Label!

Chemicals come in bottles or containers that are labeled. OSHA requires that the label must contain information about the principal hazards of the chemical—that's the law! You should always examine the label to look for hazard information, and this information often comes in specific terms or symbols denoting the hazard or hazards of a chemical. Did you catch that we used "hazard or hazards" in the previous sentence? It is not unusual for a chemical to have *more* than one kind of hazard, but a label may only have information about the principal hazard of a chemical—it may not necessarily include *all* of its potential hazards. For example, you may find that the label on a bottle of *n*-hexane says "Flammable," "Highly Flammable," or "Extremely Flammable," but as you learn more about *n*-hexane you will find that it also has toxic properties.

Table 3.1.2.1 lists terms describing hazards with which you should be familiar. Hazards are sometimes described in more general terms by symbols or pictograms. Figure 3.1.2.2 shows some common pictograms used to describe hazards. The U.S. Department of Transportation uses some of these pictograms on its labeling and you can see these placards on containers or on trucks carrying hazardous materials.

You may be able to use some of this hazard information to judge the relative hazards of chemicals. For example, if a bottle of diethyl ether is labeled as "Extremely Flammable" and a bottle of ethyl benzoate is labeled as "Combustible," this will tell you that diethyl ether is a much greater fire hazard than ethyl benzoate. This means too that diethyl ether requires much more attention in handling, but there are very good procedures for using diethyl ether safely! The labels help us know what major hazards are posed by chemicals, but, as mentioned above, you need to keep in mind that some chemicals have more than one hazardous property. While diethyl ether's principal hazard is its extreme flammability, it is also slightly toxic and inhalation of ether can cause sedation and unconsciousness—it once was used to induce anesthesia for surgical operations. In addition to flammability and toxicity, ethers can form explosive peroxides as they age in storage.

The point here is to not let the principal hazard mask other important hazardous properties of a chemical—it is important to review the Material Safety Data Sheet to be sure that you understand all of the potential hazards of the chemicals that you are using, by examining available information about those chemicals. You should also know that it is possible that some hazardous properties of chemicals have not yet been recognized, studied, or documented, so we must treat all chemicals with care, minimizing your exposure to the greatest extent possible. This is easy to do in a well-equipped laboratory so you will be able to use most chemicals very safely. This will be discussed more in Chapter 7.

Hazard Rating Systems

Sometimes labels contain a hazard rating symbol, box, or diamond. These hazard rating systems are often very helpful because they help you recognize the significance of the hazard of that chemical. Some chemical companies have developed their own rating and you may find these on bottles of chemicals. Unfortunately, there is currently no universally agreed-upon system that is in common use. Labels need to be read carefully and any rating system needs to be understood before interpreting some number that is representing some hazard level.

The most well-known hazard rating system is by the National Fire Protection Association (NFPA) and is called the NFPA diamond. It is a multicolored diamond that covers three hazard classes as well as specific other hazards, and is shown is Figure 3.1.2.3. This diamond is actually designed for firefighters and it represents the hazard under conditions of a fire—it is not specifically designed for laboratory safety. (See *Special Topic 3.1.2.2* How Is a Chemical Different in a Fire?) The diamond is subdivided into four smaller diamonds—each with its characteristic color. If each point of the large diamond is treated like a clock, you will find at 9 o'clock a blue diamond for health hazards, at 12 o'clock a red diamond for flammability, at 3 o'clock a yellow diamond for instability, and at 6 o'clock a

TABLE 3.1.2.1   Terms Describing Hazards of Chemicals

Allergen	A chemical that causes an allergic reaction—that is, can evoke an adverse immune response in a person. See also sensitizer below. Example: toluene diisocyanate, CAS 584–84–9
Carcinogen	A chemical that causes cancer in humans or animals—indicating that it has potential to cause cancer in humans. Example: benzidine, CAS 92–87-5
Combustible	A chemical that burns under most conditions once ignited, but it does not ignite and burn as easily as a flammable (see below). Example: aniline, CAS 62–53-3
Compressed gas	A gas stored under pressure—under this pressure it might be in a gaseous or liquid state but under normal pressure it is a gas. Example: propane, CAS 74–98-6
Corrosive	A chemical that causes destruction of living tissue at the site of contact. Example: sodium hydroxide, CAS 1310–73-2
Cryogen	A chemical that is stored at extremely low temperatures. Example: carbon dioxide (normally as gas, but it is used in solid form, known as dry ice, formed when the gas is compressed and kept cold), CAS 124-38-9
Embryotoxin	A chemical that is harmful to a developing embryo—a developmental toxin. A teratogen (see below) is an embryotoxin that causes physical defects in a developing fetus. Example: carbon disulfide, CAS 75-15-0
Explosive	A chemical that can produce a sudden release of energy or gas when subjected to ignition, sudden shock, or high temperature. Example: 1,3,5-trinitrotoluene, aka TNT, CAS 118-96-7
Flammable	A chemical that is easily ignited and burns very rapidly. Sometimes this term is preceded by descriptors, such as Highly or Extremely (Flammable) to denote particularly flammable chemicals. A synonym is inflammable—means the same thing. Example: diethyl ether, CAS 60-29-7
Inflammable	A chemical that is easily ignited and burns very rapidly. A synonym is flammable—means the same thing. Example: acetone, CAS 67-64-1
Irritant	A chemical that causes irritation, either reversible or irreversible, to skin or living tissue at the site of contact—resulting in redness, swelling, itching, or a rash at the site of contact. Example: aluminum powder, CAS 7429-90-5
Lacrimator or lachrymator	A chemical that causes tears upon exposure. Example: α-chloroacetophenone, CAS 532-27-4
Mutagen	A chemical that causes mutagenic or genetic changes to DNA in persons or animals, with the potential to cause adverse effects in future generations of the exposed person or animal. Example: ethidium bromide, CAS 1239-45-8
Nonflammable	A chemical that does not ignite, does not burn, or is extremely difficult to burn. Example: carbon tetrachloride, CAS 56-23-5
Oxidizer or oxidizing agent	A chemical that can rapidly bring about an oxidation reaction by supplying oxygen or receiving electrons during oxidation. Example: nitric acid, CAS 7697-37-2
Peroxide former	A chemical that can form peroxides upon storage conditions—peroxides are highly reactive and unstable, and can explode upon concentration or sudden shock. Example: diisopropyl ether, CAS 108-20-3
Poison	A chemical that is known to be extremely toxic to humans or animals. Example: hydrogen cyanide, CAS 74-90-8
Pyrophoric	A chemical that readily ignites and burns in air spontaneously—without a source of ignition. Example: phosphorus (white or yellow), CAS 7723-14-0
Reactive	A chemical that readily reacts or decomposes rapidly (violently) due to shock, pressure, temperature, or contact with air, water, or an incompatible chemical. Example: acetyl chloride, CAS 75-36-5
Sensitizer	A chemical that upon repeated exposure causes severe adverse immune reaction—see also allergen. Example: formaldehyde, CAS 50-00-0
Stench	A chemical that has an extremely offensive odor. Example: ethane dithiol, CAS 540-63-6
Teratogen	A chemical that causes physical defects in a developing fetus or embryo. Example: thalidomide, CAS 50-35-1
Toxic	A chemical that causes adverse health effects in humans or animals upon exposure. Example: lead, CAS 7439-92-1
Water reactive	A chemical that upon contact with water or moisture reacts violently to catch fire. Example: sodium, CAS 7440-23-5

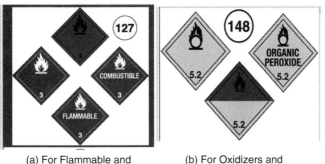

(a) For Flammable and
Combustible Materials

(b) For Oxidizers and
Organic Peroxides

FIGURE 3.1.2.2 DOT Pictograms and Placards for Flammable/Combustible Chemicals and Oxidizers.[5] These pictograms are most often seen on trucks on highways. (From *2008 Emergency Response Guide*, Department of Transportation; available at http://hazmat.dot.gov/pubs/erg/gydebook.htm.)

white diamond for special hazardous properties. Three of these hazard diamonds—health, flammability, instability—are rated with numbers ranging from 0 to 4, with 0 being no significant hazard and 4 being extremely dangerous. The fourth diamond is for special hazards, such as OX (means this is an oxidizer), W̶ (means this reacts violently with water—do not use water), or SA (for simple asphyxiant). Since each chemical has its own set of hazards, you will find that the NFPA diamond is distinct for each chemical. Figure 3.1.2.4 shows a laboratory chemical bottle with an NFPA diamond on the label.

FLAMMABLE (Red diamond)

4 Extremely flammable

3 Ignites at normal temperatures

2 Ignites when moderately heated

1 Must be preheated to burn

0 Will not burn

INSTABILITY (Yellow diamond)

4 May detonate – Vacate area if materials are exposed to fire

3 Strong shock or heat may detonate – Use monitors from behind explosive resistant barriers

2 Violent chemical change possible – Use hose streams from distance

1 Unstable if heated – Use normal precautions

0 Normally stable

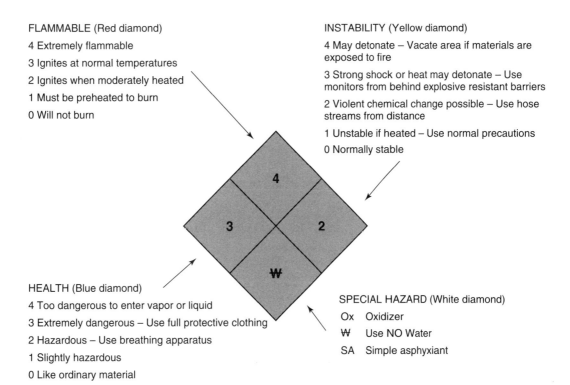

HEALTH (Blue diamond)

4 Too dangerous to enter vapor or liquid

3 Extremely dangerous – Use full protective clothing

2 Hazardous – Use breathing apparatus

1 Slightly hazardous

0 Like ordinary material

SPECIAL HAZARD (White diamond)

Ox   Oxidizer

W̶   Use NO Water

SA   Simple asphyxiant

FIGURE 3.1.2.3 NFPA Diamond. The fire diamond is frequently used in chemical laboratories. The ratings indicate the hazard level under fire conditions, not necessarily ambient laboratory conditions. (The NFPA diamond is reprinted with permission from NFPA 704–2007, *System for the Identification of the Hazards of Materials for Emergency Response*. Copyright © 2007 National Fire Protection Association, Quincy, MA. This reprinted material is not the complete and official position of the NFPA on the referenced subject, which is represented only by the standard in its entirety. The NFPA classifies a limited number of chemicals and cannot be responsible for the classification of any chemical whether the hazard of classifications are included in NFPA or developed by other individuals.)

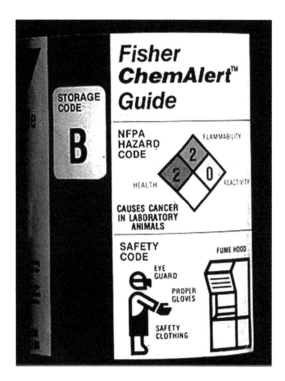

FIGURE 3.1.2.4   Laboratory Chemical Bottle with NFPA Diamond. It is important to know the meaning of the ratings (0–4) in the fire diamond to properly assess the various hazard levels of health, flammability, and reactivity.

SPECIAL TOPIC 3.1.2.2

HOW IS A CHEMICAL DIFFERENT IN A FIRE?

The NFPA diamond was designed for firefighters to quickly assess the hazards of a chemical under the conditions of a fire since that is how they are most likely to encounter chemicals. This is different from the conditions in laboratories. The NFPA diamond considers issues such as:

Feeding the fire.   Strong oxidizers may create bigger hazards under fire conditions since they may contribute to the combustion of flammable materials more than at ambient temperatures. Good examples of such chemicals are all nitrate salts.

Enhanced reactivity.   While magnesium is unreactive with water at room temperature, when it is hot it will behave like an "active metal" (similar to sodium) and reduce the hydrogen in water to H_2, which is extremely flammable.

Decomposition products.   Ammonium chloride is relatively safe under ambient lab conditions but at high temperatures it may decompose to HCl and NH_3, both of which are toxic gases.

Combustion products.   Sulfur is relatively safe under lab conditions but in a fire will produce SO_2, a highly toxic gas.

Because of the differences between "lab" and "fire" conditions, the NFPA hazards for various chemicals must be interpreted carefully. Some chemicals will be more dangerous in a fire situation than during laboratory use.

As noted above, there are many other systems for labeling chemicals in current use. For this reason, the United States is participating in a global effort to develop a uniform system to identify and label hazardous materials. This new international chemical labeling system is called *The Globally Harmonized System of Classification and Labelling of Chemicals* (GHS). This is a system that is being proposed under the auspices of the United Nations and will be designed, as the titled indicates, to harmonize the classification of chemicals of various countries with a standardized labeling system. The GHS is discussed more in Sections 3.2.1 and 6.2.1.

Working in the Lab: RAMP

- *Recognize* lab hazards by reading chemical labels carefully, reviewing and understanding hazard rating systems, and knowing the language of hazards from chemicals.

- *Assess* the level of risk based on likelihood of exposure during the course of your work.

- *Minimize* the risk by preventing exposure using good lab practices and personal protective equipment.

- *Prepare* for emergencies by understanding how chemicals can harm you and having a plan of action in the event of an exposure that causes harm.

References

1. Gorton Carruth and Eugene Ehrlich. *The Giant Book of American Quotations*, Gramercy Books, New York, 1988, Topic 132. Language, #31.
2. Fariba Mojtabai and James A. Kaufman. *Learning by Accident*, Vol. 1, Laboratory Safety Institute, Natick, MA, 1997; modeled after incident #177, p. 31.
3. Centers for Disease Control and Prevention; available at http://www.cdc.gov/HomeandRecreationalSafety/Water-Safety/waterinjuries-factsheet.htm (accessed July 12, 2009).
4. Car Accident Statistics; available at http://www.car-accidents.com/pages/stats.html (accessed July 12, 2009).
5. Department of Transportation. *2008 Emergency Response Guide*; available at http://hazmat.dot.gov/pubs/erg/gydebook.htm (accessed October 14, 2009).

QUESTIONS

1. For every chemical there is a unique
   - (a) Chemical formula
   - (b) Molar mass
   - (c) Safety warning
   - (d) CAS number

2. Most chemicals
   - (a) Are not very hazardous
   - (b) Have only one hazardous property
   - (c) Have multiple hazards, but not all of them are always listed on the label
   - (d) Have multiple hazards, with all of them listed on the label

3. Table 3.1.2.1 lists how many terms that describe hazards of chemicals?
   - (a) 5–9
   - (b) 10–14
   - (c) 15–19
   - (d) 20–24

4. What term describes a chemical that causes cancer in humans or animals?
   - (a) Carcinogen
   - (b) Embryotoxin
   - (c) Lacrimator
   - (d) Teratogen

5. What term describes a chemical that causes physical defects in a developing fetus or embryo?
   - (a) Carcinogen
   - (b) Embryotoxin
   - (c) Lacrimator
   - (d) Teratogen

6. What term describes a chemical that causes tears upon exposure?
   - (a) Pyrophoric
   - (b) Embryotoxin
   - (c) Lacrimator
   - (d) Mutagen

7. What term describes a chemical that is stored at extremely low temperatures?
   - (a) Pyrophoric
   - (b) Cryogen
   - (c) Lacrimator
   - (d) Compressed gas

8. Which statement is true with regard to the NFPA diamond?
   - (a) The blue diamond at the 9 o'clock position refers to special hazards.
   - (b) The white diamond at the 6 o'clock position refers to instability hazard.
   - (c) The red diamond at the 12 o'clock position refers to flammability.
   - (d) The yellow diamond at the 3 o'clock position refers to health hazard.

9. The hazards ratings in the NFPA diamond
   - (a) Describe how a chemical behaves in a fire, which is almost identical to how it behaves under "lab" conditions
   - (b) Describe how a chemical behaves in a fire, which is not always the same as how it behaves under "lab" conditions
   - (c) Was designed for firefighters and is of almost no practical use for chemists
   - (d) Refer to categories of chemicals, not individual chemicals

3.1.3

FINDING HAZARD INFORMATION: MATERIAL SAFETY DATA SHEETS (MSDSS)

Preview This section identifies some good resources with information about hazards of chemicals. It includes a description of these resources and how they might be used.

We don't know a millionth of one percent about anything.

Thomas Edison[1]

INCIDENT 3.1.3.1   CHEMICAL SENSITIVITY[2]

Three female students were using 1-fluoro-2,4-dinitrobenzene (FDNB). While working with this reagent over a 2-month to 2-year period, the tips of their fingers and nails exhibited a yellow staining characteristic of FDNB, and blisters developed on their fingers. Testing revealed that the students and their 39-year instructor had developed sensitivity to FDNB—investigation showed that it had been previously reported to be a potent allergen and a strong irritant. The students suggested that their protective gloves were permeable to FDNB, which led to the sensitization. Testing of their gloves revealed that FDNB passed through their protective gloves.

What lessons can be learned from this incident?

50,438,639 Chemicals in 2009—Now that's a Lot of Chemicals!

In September 2009 the American Chemical Society's Chemical Abstracts Service (CAS) reported that it had CAS numbers for 50 million chemical substances, and the list was continuing to grow, causing them to move to the use of 10-digit CAS numbers in the future.[3] Figure 3.1.3.1 shows how the number of chemicals has grown in recent years. Not all chemicals are tested for toxic properties, and we have relatively little to no information about the hazards of the vast majority of these compounds. In some instances, the information that exists must be sought after by looking for specific information for a specific compound, and even then, it may be hard to find. (See *Special Topic 3.1.3.1* Why Not Test the Health Effects of All Chemicals?)

SPECIAL TOPIC *3.1.3.1*

WHY NOT TEST THE HEALTH EFFECTS OF ALL CHEMICALS?

Money, money, money! It requires lots of money to pay for proper testing! To determine something about the health effects of a chemical requires that it be tested, usually in animals, and rats are the most popular model used for this. A typical toxicity study might cost $50,000 to $100,000 or even more. Unless there is real commercial value in a chemical, it is not likely to be tested for toxicity. Furthermore, this kind of testing is only for immediate effects. Longer term health effects from repeated smaller doses of a chemical are often only recognized by careful observation of people exposed to a chemical for a long period of time. While there are animal toxicity tests for long-term health effects, they are not always predictive of health effects in humans—and they are enormously expensive. Also keep in mind that while the CAS has identified 50 million chemicals (in 2009), many of those chemicals may have been reported and used in only one or a few research laboratories—they are only laboratory chemicals and not used by the public or even any significant number of other laboratories.

Laboratory Safety for Chemistry Students, by Robert H. Hill, Jr. and David C. Finster
Copyright © 2010 John Wiley & Sons, Inc.

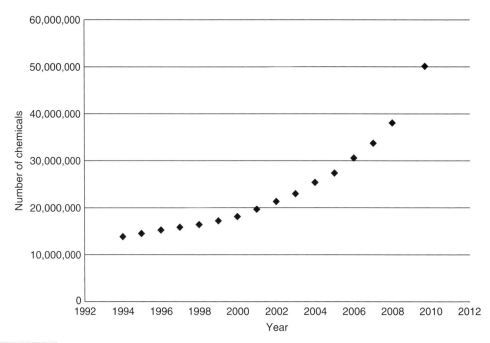

FIGURE 3.1.3.1   Number of CAS Registered Chemicals. For the current number of CAS chemicals access http://www.cas.org/cgi-bin/cas/regreport.pl. How many chemicals do you think will be cataloged by the time you graduate from college?

For the much smaller number of chemicals that are made and sold here in the United States, perhaps less than 0.1% of those reported by CAS, we rely on the chemical suppliers to provide us with known and reported hazard information. However, the chemical suppliers can only report the information that is available, and oftentimes only limited hazard information is available—this is particularly true of toxicological properties. Nevertheless, you should always strive to obtain all of the information you can find about the hazards of a chemical and never make the assumption that exposure to any chemical is safe.

So the bad news is that we don't know about the toxicological properties of >99% of the known compounds. The good news is that for most of the chemicals you will encounter in an undergraduate chemistry curriculum, we have at least some toxicological data. Thus, for most of the chemicals that you are likely to use as a student, you can learn about their health effects.

Material Safety Data Sheets—From Hieroglyphs to the Internet

If you want to know about the hazards of a chemical, one source is its Material Safety Data Sheet (MSDS). MSDSs, prepared by chemical suppliers, describe properties and information pertinent to the safety of specific chemicals. In 1980 the Occupational Safety and Health Administration (OSHA) wrote a regulation called the Hazard Communication Standard that required all chemical suppliers provide MSDSs to chemical users.[4] Some of the information found in MSDSs was recorded more than 4000 years ago in hieroglyphs by the Egyptians in their tombs and papyrus records.[5] In today's world you can find MSDSs at many sites on the Internet.[6] When any company or individual purchases a chemical for the first time, many suppliers provide a paper copy of the MSDS with the shipment of that chemical. Alternatively, some may only provide you with access to their electronic database of MSDSs.

MSDSs are documents that are intended to help workers by providing readily available information about the hazards of a chemical and they must include some basic information about chemicals. OSHA requires specific information be provided in MSDSs and recommends (but does not require) that all MSDSs follow the American National Standards Institute (ANSI) Z400.1 format.[7,8] This format is shown in Table 3.1.3.1. Furthermore, the newly proposed Globally Harmonized System of Classification and

TABLE 3.1.3.1 American National Standards Institute (ANSI) Material Safety Data Sheet Format[7–9]

1. Identification
2. Hazard(s) Identification
3. Composition/Information on Ingredients
4. First Aid Measures
5. Firefighting Measures
6. Accidental Release Measures
7. Handling and Storage
8. Exposure Controls/Personal Protection
9. Physical and Chemical Properties
10. Stability and Reactivity
11. Toxicological Information
12. Ecological Information
13. Disposal Considerations
14. Transport Information
15. Regulatory Information
16. Other Information

Labelling of Chemicals (GHS, see Section 3.2.1) proposes the same format for its worldwide Safety Data Sheets (SDSs).[9]

While MSDSs are very common and accessible, it is not legally required that they be accurate. Thus, while it is reasonable to assume that the data presented is intended to be accurate, the documents usually contain a disclaimer (in small print) that absolves the company from being held responsible for the accuracy of the information. Because of this, and other reasons discussed later, many safety professionals do not rely on MSDSs for the best safety information.

Why Do I Need to Know About MSDSs?

Section 3.2.3 will discuss each section of an MSDS in detail. We begin here with an overview of these documents.

Your primary use for an MSDS is to find information about the hazard of the chemical you are about to use, and to find any safety procedures or special handling instructions for this chemical. As you might guess from an earlier comment, not all MSDSs are created equal in terms of quality—some are good, some are not, some may have erroneous information, and a few may not provide enough information for you to judge the hazards associated with the chemical. You may also find inconsistencies in MSDSs, where, for example, one section might indicate that ventilation is needed to keep air concentrations of a chemical below its permissible exposure limit, but later the MSDS indicates that there is no established exposure limit.[10] Nevertheless, an MSDS is a good place to start looking for information about the hazards of a chemical. You may want to consider examining MSDSs for the same chemical using other manufacturers to find out if hazards are more clearly delineated.

MSDSs can also be a primary source of information for emergency responders—especially fire departments and medical emergency personnel, such as doctors. Many facilities try to have hard copies of these documents on hand to give to emergency responders in the event of a spill that requires outside assistance or an exposure injury requiring trained medical care.

Finding Hazard Information in MSDSs

Step #1:. When using an MSDS, first make sure that you have the right MSDS by comparing the label information with the product identification information at the beginning. (Reference 6 provides a link to many sources for MSDSs. You might wish to review an MSDS to see how they are structured. Section 3.2.3 discusses more about how to read an MSDS.) Ensure that the name is the same as the label and read the name carefully since many chemicals have similar names. If you can find a CAS number, make sure it is the same. (Special Topic 3.1.2.1 describes how CAS numbers are unique identification numbers assigned by the Chemical Abstracts Service.)

Step #2:. Find the hazard information about the chemical you will be using. Depending on the format used by the supplier, you may have to look in different sections, but it is usually found early in one or more sections. Look for sections that may be called "Hazard Identification," "Physical and Chemical Characteristics," "Fire and Explosion Hazard Data," "Reactivity and Stability Data," or "Health Hazard Data." You may find additional information in later sections of the MSDS that can also help you evaluate the hazard of this chemical.

You should look for terms similar to those discussed in Section 3.1.2, such as corrosive, flammable, toxic, or irritant. Some MSDSs will make the hazards very clear; for example, for *acetone* the MSDS might indicate "Danger! Extremely Flammable." Other MSDSs might describe acetone's hazards somewhat differently, such as "Flammable liquid and vapor. Vapor may cause flash fire." It may have the ever-popular phrase "Harmful if swallowed or inhaled."

In other cases, the hazard may not be as clearly explained. For instance, you may find under potential health effects, a section entitled "Inhalation," which explains that the inhalation of vapors irritates the respiratory tract or that exposure can cause respiratory tract and throat irritation. It is easy to deduce that this chemical is an irritant. But it may go further and say that overexposure or high concentrations may cause central nervous system depression, unconsciousness, respiratory failure, and death. These are toxic effects. It may also use medical terms that you may not understand or be familiar with, such as narcosis, dyspnea, asphyxia, or emphysema. Sometimes the use of these terms can be confusing and you may want to seek help to understand what these mean. A very nice online glossary of safety-related terms can be found at http://www.ilpi.com/msds/ref/index.html. Nevertheless, what you should notice is that the harmful effects progress from relatively minor to sometimes more serious as exposure increases.

Learning About Safe Handling Precautions

After learning about the hazards of the chemical you will want to learn about specific procedures for safe handling. For example, it may state: "Keep away from heat, sparks, flames, or other sources of ignition." MSDSs often contain special information for firefighters. For example, it may indicate that water is not effective in putting out fires involving this chemical. You will find a section that describes personal protective equipment (PPE) and exposure controls—eye protection, skin protection, and respiratory protection. A significant problem with many MSDSs is that the PPE is not specific enough to be useful. For example, to protect your skin from exposure, the MSDS may indicate that you should use appropriate protective gloves—but of course you want to know which gloves are appropriate. Each specific chemical requires a specific kind of glove—there is no universal glove! You will also find procedures that describe how to handle emergency situations, such as spills, but these directions are often not specific enough to be useful. For example, the recommendation to "follow state and local protocols" is accurate, but not helpful.

Reading MSDSs with a Critical Eye

Since MSDSs are very easy to access, they are used a great deal as a "first line of defense" in your investigation of the properties and hazards of a particular chemical. As noted above, they are not always as reliable or accurate as they could be, but they nevertheless are a good starting point for getting a general sense of the hazards posed by a particular chemical. Depending on your particular use of a chemical, you may well wish to further investigate its properties.

Many chemists also believe that MSDSs sometimes overstate the hazards of a chemical, potential adverse health effects, and the requirements for personal protective equipment. It is easy to guess that this practice is a result about concerns over potential liability from the use of this chemical and the corresponding advice from legal counsel. While this is understandable, it can make it much more difficult to discern the most important hazards associated with a chemical, the most important adverse health effects, and the appropriate PPE that would be required in a laboratory setting. Remember that MSDSs were not written for laboratory use, but rather focus on the use of that chemical in an industrial setting where large quantities of chemicals may be used.

You will learn more about other sources of information (besides MSDSs) in Section 3.2.2.

Using MSDSs in the Lab: RAMP

The RAMP acronym is easy to use with MSDSs. The MSDS can be a useful source of information in order to:

- *Recognize* the hazards of a particular chemical.
- *Assess* the level of risk posed by the chemical based on the physical properties of the chemical.
- *Minimize* the risk by preventing exposure using personal protective equipment and other safety equipment recommended in the MSDS.
- *Prepare* for emergencies based on the assessed hazard level and the possible effects of exposure.

References

1. GORTON CARRUTH and EUGENE EHRLICH. *The Giant Book of American Quotations*, Gramercy Books, New York, 1988, Topic 129. Knowledge, #3.
2. J. S. THOMPSON and O. P. EDMONDS. Safety aspects of handling the potent allergen FDNB, *Annals of Occupational Hygiene* **23**: 27–33 (1980).
3. Chemical Abstracts Service Registry. Registry Number and Substances Count; available at http://www.cas.org/expertise/cascontent/registry/regsys.html (accessed September 18, 2009).
4. U.S. Department of Labor, Occupational Safety and Health Administration. Part 29 Code of Federal Regulations, 1910.1200, Hazard Communication; available at http://www.osha.gov/pls/oshaweb/owadisp.show_document?p_table=standards&p_id=10099 (accessed January 19, 2008).
5. SAMUEL AARON KAPLAN. Development of Material Safety Data Sheets, presented at the 191st American Chemical Society Meeting, New York, NY, April 13–18, 1986; available at http://jrm.phys.

ksu.edu//Safety/kaplan.html (accessed November 9, 2007).
6. MSDSonline®, Where to find MSDS on the Internet; available at http://www.ilpi.com/msds/index.html (accessed November 9, 2007).
7. U.S. Department of Labor, Occupational Safety and Health Administration. Recommended Format for Material Safety Data Sheets (MSDSs), issued 07/22/2004; available at http://www.osha.gov/dsg/hazcom/msdsformat.html (accessed January 19, 2008).
8. American National Standards Institute (ANSI). Z400.1–2004, Hazardous Industrial Chemicals—Material Safety Data Sheets—Preparation, ANSI, New York, 2004.
9. United Nations. Globally Harmonized System of Classification and Labelling of Chemicals (GHS), 2nd revision, United Nations, New York, 2007.
10. ROBERT J. ALAIMO and LYNNE A. WALTON. Material Safety Data Sheets. In: Handbook of Chemical Health and Safety (ROBERT J. ALAIMO, ed.), American Chemical Society Oxford University Press, Washington, DC, 2001, pp. 317–321.

QUESTIONS

1. What fraction of the known chemicals have been tested for toxicity?

   (a) Almost none
   (b) Less than 0.1%
   (c) About 25%
   (d) About 65%

2. What fraction of the chemicals that you will use in undergraduate chemistry labs will have some safety information associated with them?

   (a) Almost none
   (b) A tiny fraction
   (c) About half
   (d) Most of them

3. An MSDS is a document that

   (a) Is required to be provided with a chemical when purchased from a chemical supply company
   (b) Has a wide range of safety information in a mostly standardized format
   (c) Is not legally required to be accurate
   (d) All of the above

4. MSDSs are commonly used by chemists because

   (a) They are easily accessible
   (b) They represent the most reliable information about chemicals
   (c) They contain safety information in "common language" that chemists with no safety training can understand
   (d) All chemists have had training on how to read MSDSs

5. When using an MSDS, the most important first step is to make sure that the MSDS

   (a) Is in the "standard format"
   (b) Refers to the chemical you are using
   (c) Has been approved by OSHA
   (d) All of the above

6. Hazard information about a particular chemical that is listed on MSDSs from various sources

   (a) Will always be identical, using standardized language
   (b) May vary

   (c) May be listed in different orders, but will always be identical
   (d) Is rarely consistent

7. Safety information on MSDSs is usually

   (a) Specific enough to be of use to chemists
   (b) Written mostly for "industrial" or manufacturing situations more than for chemists in research labs
   (c) Written mostly for chemists in research labs more than for "industrial" or manufacturing situations
   (d) Written to be useful for chemists, emergency responders, and physicians

3.2.1

THE GLOBALLY HARMONIZED SYSTEM OF CLASSIFICATION AND LABELLING OF CHEMICALS (GHS)

Preview This section presents a brief overview of the United Nations' Globally Harmonized System of Classification and Labelling of Chemicals (GHS) for defining and communicating hazard information and protective measures related to the physical, health, and environmental risks of chemicals through product labels and safety data sheets (SDSs).

Ignorance is preferable to error; he is less remote from the truth who believes nothing, than he who believes what is wrong.

Thomas Jefferson[1]

INCIDENT 3.2.1.1 ISOPROPANOL EXPLOSION[2]

A laboratory worker was preparing some samples using a procedure that involved isopropanol. When he finished he poured the excess isopropanol into a plastic bottle labeled "2-Propanol." There was an immediate reaction and the container burst and sprayed its contents on the laboratory worker. He was startled, but managed to call for help. Nearby colleagues came to his aid and when he reported that his skin was burning they used an emergency eyewash and emergency shower to wash away the splash. He was taken to the university medical center for an evaluation and later released after being treated for acid burns. Investigators learned that the bottle had been mislabeled and actually contained concentrated nitric acid and some copper waste.

What lessons can be learned from this incident?

What Is the Globally Harmonized System for Classification and Labelling of Chemicals (GHS)?

Today's marketplace is global and we receive and use foods, goods, and materials from countries all over the world. To improve the interactions of countries in the chemical industry, a consensus of nations under the purview of the United Nations has come together to develop a system for classifying and labeling hazardous chemicals so that all countries classify and label chemicals in the same way. This project is named the Globally Harmonized System for Classification and Labelling of Chemicals, or GHS.

In 1992 the *United Nations Conference on Environment and Development* in Rio de Janeiro adopted a mandate to develop GHS and by 2000 the initial system was formed. Designing an international system to classify and label chemicals has been a slow and complicated process and implementation of this system is still in progress. The first edition of GHS appeared in 2003. The document describing GHS is known as "The Purple Book"[3] and it underwent revisions in 2005, 2007, and 2009.

GHS is an international system that incorporates MSDS-like documents, a standardized labeling system for chemicals, and the protocols for the determination of the hazard ratings. Many parts of this classification system will seem familiar to those in the United States who use MSDSs, but some differences will exist since it has been fine-tuned and agreed upon by this large group of nations as an international system for improving cooperation for global trade. Compliance with GHS is voluntary.

For those countries that adopt GHS, its use allows those countries to sell and buy chemicals across their borders with confidence that the hazardous chemicals are identified and labeled in the same way as in other GHS countries, allowing transport and shipping companies to handle these chemicals safely. The labels inform the user of any special hazards associated with the chemical using common chemical pictograms, signal words, and hazard statements on labels to alert the handler, emergency personnel, and ultimately the user of the relative risk of a chemical being shipped. The task to establish GHS was an arduous one that involved the input of many countries and compromise to reach these consensus classifications.

Organization of the GHS Hazard Ratings

Table 3.2.1.1 shows the 28 hazard classes by which chemicals are assessed in the GHS.[3] The main classes are physical hazards, health hazards (for humans), and environmental hazards (that focus on aquatic systems). Each class has a unique relative rating system (categories) using letters, numbers, or other descriptions. In all cases the most hazardous classes have the lowest numerical rating or "A" and the least hazardous have higher numbers and subsequent letters. You should *not* be concerned with learning everything in Table 3.2.1.1 but rather it is presented to show you the breadth of hazards covered by GHS. GHS labeling can be used for recognition of hazards, so that you know that a low number or an early alphabetical character is a warning of a potentially very hazardous material. You will learn later that GHS can also be used a tool for risk assessment.

It is worth noting, in particular, that the flammability rating system[1-4] is deceptively similar to the NFPA rating system for flammable liquids except that *the rating systems are numerically reversed!* The NFPA system ranks the least flammable as "0" and most flammable as "4." This unfortunate lack of consistency will plague chemists and firefighters since the NFPA system is in widespread use in the United States.

The GHS ratings for flammability are presented in Section 5.1.2. The GHS ratings for health hazards are presented in Section 6.2.1. Hazards designated in GHS are described throughout this book (see Table 3.2.1.1).

Implementation

The original goal was to begin worldwide implementation of GHS in 2008. The United Nations maintains a web site that identifies the implementation progress for 67 countries.[4] GHS implementation may not be complete until 2015.

In the United States, the implementation involves four federal agencies: the Occupational and Safety Health Administration (OSHA), the Environmental Protection Agency, the Department of Transportation, and the Consumer Product Safety Commission, but the key agency affecting laboratories is OSHA. OSHA issued a notice of proposed rule-making in October 2009.

Using GHS

When you work in a laboratory, you can use GHS to help you assess the relative risk of handling hazardous chemicals. GHS will specify the format for MSDS-like documents that will contain the information shown below. These will be called Safety Data Sheets (SDSs) and will have the following information sections[5]:

1. Identification
2. Hazard(s) identification
3. Composition/information on ingredients
4. First aid measures

3.2.1 THE GLOBALLY HARMONIZED SYSTEM OF CLASSIFICATION AND LABELLING OF CHEMICALS (GHS)

TABLE 3.2.1.1   GHS Hazard Classes [3]

	Rating system categories
Physical hazards (sections with information about this hazard)	
Explosives (5.2.2)	Unstable, Divisions 1.1, 1.2, 1.3, 1.4, 1.5, 1.6
Flammable Gases (5.1.2)	1, 2
Flammable Aerosols (5.1.2)	1, 2
Oxidizing Gases (5.2.3, 5.3.2, 5.3.3)	1
Gases Under Pressure (5.3.1)	Four categories
Flammable Liquids (5.1.2, 5.2.2)	1, 2, 3, 4
Flammable Solids (5.2.2)	1, 2
Self-reactive Substances (5.2.3)	A, B, C, D, E, F, G
Pyrophoric Liquids (5.2.3)	1
Pyrophoric Solids (5.2.3)	1
Self-heating Substances and Mixtures (5.3.3)	1, 2
Substances Which on Contact with Water Emit Flammable Gases (5.2.3)	1, 2, 3, not classified
Oxidizing Liquids (5.2.3, 5.3.2, 5.3.3)	1, 2, 3
Oxidizing Solids (5.2.3, 5.3.2, 5.3.3)	1, 2, 3
Organic Peroxides (5.3.2)	A, B, C, D, E, F, G
Corrosive to Metals (5.1.1, 5.2.1)	1
Health hazards (sections with information about this hazard)	
Acute Toxicity (4.1.2)	1, 2, 3, 4, 5
Skin Corrosion/Irritation (5.1.1, 5.2.1)	1A, 1B, 1C, 2, 3
Serious Eye Damage and Eye Irritation (5.1.1, 5.2.1)	1, 2A, 2B
Respiratory Sensitization (4.1.2)	1
Skin Sensitization (4.1.2)	1
Germ Cell Mutagenicity (4.2.1)	1A, 1B, 2
Carcinogenicity (4.3.1)	1A, 1B, 2
Reproductive Toxicity (4.2.1)	1A, 1B, 2
Target Organ Systemic Toxicity (TOST): Single Exposure (4.1.2)	1, 2, 3
Target Organ Systemic Toxicity (TOST): Repeated Exposure (4.2.1)	1, 2
Aspiration Hazard (4.1.2)	1, 2
Environmental Hazards (sections with information about this hazard)	
a. Acute Aquatic Toxicity (4.1.2)	1, 2, 3
b. Chronic Aquatic Toxicity (4.2.1)	1, 2, 3, 4

5. Fire fighting measures

6. Accident release measures

7. Handling and storage

8. Exposure control/personal protection

9. Physical and chemical properties

10. Stability and reactivity

11. Toxicological information

12. Ecological information

13. Disposal considerations

14. Transport information

15. Regulatory information

16. Other information

These can be read like MSDSs (see Sections 3.1.3 and 3.2.3).

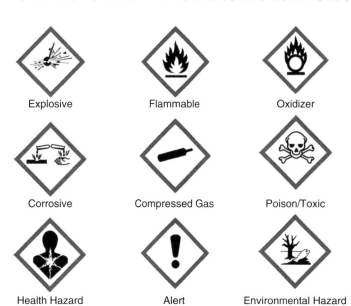

FIGURE 3.2.1.1   GHS Pictograms.[5] These pictograms will probably become commonplace in the next decade as the GHS is adopted and implemented in the United States. (Courtesy of the United Nations Economic Commission for Europe. Copyright © 2007 United Nations, New York and Geneva.)

Additionally, the format of labeling of chemicals will be standardized. Each label will contain a product identifier, a pictogram (symbol), a signal word, a hazard statement, and supplier information. Manufacturers or suppliers of chemicals can provide additional information on the label. The category designation will be the letter or number, shown in Table 3.2.1.1, and will be included in the SDS. However, if a transport pictogram appears on the package, the GHS pictogram for some hazards may not appear. There are nine pictograms[5,6] and two signal words—"danger" or "warning." Some examples of hazard statements are "May cause respiratory irritation," "Toxic in contact with skin," and "Heating may cause a fire or explosion." All of the pictograms and the hazards they represent are shown in Figure 3.2.1.1.[5]

RAMP

The GHS Safety Data Sheets and GHS labeling will be useful in RAMP.

- *Recognize* the hazards of a particular chemical.
- *Assess* the level of risk posed by the chemical based on the physical properties of the chemical.
- *Minimize* the risk by preventing exposure using personal protective equipment and other safety equipment recommended in the SDS.
- *Prepare* for emergencies based on the assessed hazard level and the possible effects of exposure.

References

1. THOMAS JEFFERSON. Notes on the State of Virginia. In: *Bartlett's Familiar Quotations*, 17th edition (J. KAPLAN, ed.), Little, Brown, and Co., Boston, 2002, p. 357, #5.
2. Office of Environment, Health, and Safety, University of California, Berkeley. Lessons Learned. Nitric Acid Spray—September 2004; available at http://www.ehs.berkeley.edu/lessonslearned.html (accessed September 14, 2009).
3. United Nations Economic Commission for Europe. Division of Transportation, Dangerous Goods; available at http://www.unece.org/trans/danger/publi/ghs/ghs_rev02/02files_e.html (accessed September 14, 2009).
4. United Nations Economic Commission for Europe. Division of Transportation, Dangerous Goods; available at http://www.unece.org/trans/danger/publi/ghs/implementation_e.html (accessed September 14, 2009).
5. United Nations Economic Commission for Europe. Division of Transportation, Dangerous Goods; available at http://www.unece.org/trans/danger/publi/ghs/pictograms.html (accessed September 14, 2009).
6. Occupational Safety and Health Administration. Guide to the Globally Harmonized System of Classification and Labelling of Chemicals (GHS); available at http://www.osha.gov/dsg/hazcom/ghs.html#3 (accessed September 14, 2009).

QUESTIONS

1. The GHS

   (a) Is a system for classifying and labeling hazardous chemicals so that all countries classify and label chemicals in the same way

   (b) Was adopted by the United Nations in 2003 after initially being designed in 1992

   (c) Allows participating countries to sell and buy chemicals across their borders with confidence that the hazardous chemicals are identified and labeled in the same way

   (d) All of the above

2. GHS labels will use

   (a) Chemical pictograms only since language barriers prevent the use of words

   (b) "Signal words" that will be understood regardless of what language is used

   (c) Chemical pictograms and signal words

   (d) Chemical pictograms, signal words, and hazard statements

3. The GHS is

   (a) Simple and uses consistent rating systems for various hazards

   (b) Simple but uses inconsistent rating systems for various hazards

   (c) Complicated but uses consistent rating systems for various hazards

   (d) Complicated and uses inconsistent rating systems for various hazards

4. What are the main categories of hazards used in the GHS?

   (a) Physical and health

   (b) Human and environmental

   (c) Physical, human, and environmental

   (d) Physical, health, and environmental

5. The rating systems used by the NFPA and the GHS for flammable liquids and solids are

   (a) Identical

   (b) Very similar

   (c) The "reverse" of each other

   (d) Similar, except the NFPA uses Arabic numbers and the GHS uses Roman numerals.

6. The original and the now-expected adoption dates for the GHS are

   (a) 2003 and 2008

   (b) 2003 and 2015

   (c) 2008 and 2015

   (d) 2008 and uncertain

7. What United States agencies are involved in the implementation of the GHS?

   (a) EPA, DOT, OSHA

   (b) EPA, DOT, CPSC

   (c) EPA, OSHA, CPSC

   (d) EPA, DOT, OSHA CPSC

8. The GHS will use

   (a) MSDS-like documents only

   (b) Labels

   (c) 29 pictograms that can be understood internationally

   (d) All of the above

3.2.2

INFORMATION RESOURCES ABOUT LABORATORY HAZARDS AND SAFETY

Preview This section will outline information resources that are easily available about the hazards of chemicals, enabling you to use this information to recognize the hazards of chemicals and protect yourself from those hazards.

Three weeks in the lab can save you 20 minutes in the library![1]

INCIDENT 3.2.2.1   DIMETHYL SULFATE AND SODIUM AZIDE EXPLOSION[2]

During the preparation of methyl azide from dimethyl sulfate and sodium azide, an explosion occurred. The same reaction had previously been carried out without incident. Upon investigation, the chemists found that *Bretherick's Handbook of Reactive Chemical Hazards* reported that this reaction can result in an explosion if the acidity of the solution drops to pH 5. It was surmised that sodium hydroxide, which was added to maintain the pH above 7, was not added at a sufficient rate to keep the solution at the required pH. Further steps (visible color indicator) were taken in subsequent preparations to ensure that the pH was maintained in the appropriate range.

What lessons can be learned from this incident?

Where Do I Begin?

In an ideal world, it would be nice to be able to find accurate information about chemicals very easily. And, happily, it is possible to find some accurate information about many chemicals with relative ease using a combination of printed materials and Internet resources. The danger is that some printed and Internet resources may be outdated, confusing, or wrong and there is often no easy way to identify these problems. Clearly, using a search engine like Google will likely lead to a mixture of reliable and unreliable information. While using Wikipedia is convenient, it is not reviewed on a systematic basis and the chance for even clerical errors is higher than in reviewed resources. So, the question facing someone looking for safety information is where to start, and whom to believe.

In the 21st century, it is likely that most laboratorians will be sitting at their computer as they are thinking about lab experiments and the facile use of the Internet makes this a tempting resource for information. However, most laboratories should also have a good collection of printed materials readily available and perhaps a short walk to a bookshelf or library can provide a reliable source of information. We will present this discussion with these two general categories in mind: Internet resources and printed materials.

Internet Resources

As you know, the Internet is a vast resource of information. But, the quality of that information can vary tremendously so Internet resources must be reviewed with a critical eye to be used wisely and judiciously.

The first inclination, because it is so easy, is to use a search engine like Google or to use Wikipedia to find information about a chemical. In many cases this may be sufficient, particularly if one is looking for simple information about a chemical such as boiling point or aqueous solubility. In general, though,

---

we discourage the use of unreviewed online resources to collect safety information since the possibility for error is higher than in reviewed materials and the consequences of inaccurate information are so high.

A Google search can be used to identify many MSDSs for a chemical, too. While these first seem like reliable sources, you should know that the federal law that mandates their existence does not require that they be accurate! Furthermore, since MSDSs are both technical and legal documents (because they carry liability for their authors) some comments about the hazards of chemicals are overstated or too vague to be useful. As with Wikipedia, "simple information" about a chemical is likely to be correct but some safety information will not be as useful as you might wish it would be. Reading MSDSs will be discussed more in Section 3.2.3. Table 3.2.2.1 identifies a handful of web sites that lead to good MSDS collections.

TABLE 3.2.2.1   Internet Reference Resources

Best[a1]	URL and description	Notes
	Books and Pamphlets	
1*	http://www.nap.edu/catalog.php?record_id=4911; 1995 *Prudent Practices in the Laboratory: Handling and Disposal of Chemicals*	Internet version of this popular and reliable safety reference book, geared to the research laboratory. A new edition of this book is expected in 2010 but it is not likely to be available on the Internet.
2*	http://membership.acs.org/c/ccs/pubs/SACL_Students.pdf; 2003 *Safety in Academic Chemistry Laboratories: Volume 1—Accident Prevention for College and University Student*, 7th edition	This volume is designed for use by students in higher education class laboratories.
3*	http://membership.acs.org/c/ccs/pubs/SACL_faculty.pdf; 2003 *Safety in Academic Chemistry Laboratories: Volume 2—Accident Prevention for Faculty and Administrators*, 7th edition	This volume is designed for use by teaching staff in higher education.
4	http://membership.acs.org/c/ccs/pub_1.htm; 2001 *Chemical Safety for Teachers and Their Supervisors*, by J Young	This book is designed for secondary school laboratories.
5	http://www.cdc.gov/niosh/docs/2007-107; 2006 *School Chemistry Laboratory Safety Guide*, NIOSH Publication 2007-107	This book is designed for secondary school laboratories.
	Chemical Profiles	
6*	http://www.cdc.gov/niosh/npg/ *NIOSH Pocket Guide to Chemical Hazards*	Written for industrial and emergency response exposures, but provides key safety information for several hundred chemicals and is reliable and easily accessible.
7*	http://books.nap.edu/openbook.php?record_id= 4911&page=237; 1995 Appendix B of *Prudent Practices in the Laboratory*: Laboratory Chemical Safety Summaries (LCSSs)	LCSSs are reliable and useful descriptions of 88 common chemicals. The general format provides the basis for a risk assessment of chemicals to be used in a laboratory process.
8*	http://jchemed.chem.wisc.edu/	Chemical Laboratory Information Profiles (CLIPs) are very useful synopses of the hazards of laboratory chemicals.
	Journals and Magazines	
9*	http://www.sciencedirect.com/science/journal/18715532 *Journal of Chemical Health and Safety*	Search archives for safety articles about particular chemicals and specific processes; subscription (individual or institutional) is required for full access.

TABLE 3.2.2.1 (Continued)

Best[a2]	URL and description	Notes
10	http://jchemed.chem.wisc.edu/ *Journal of Chemical Education*	Search archives for articles about particular chemicals; subscription required for full access.
11	http://pubs.acs.org/cen/ *Chemical & Engineering News*	Weekly, official magazine of the American Chemical Society. Often has letters and articles related to chemical health and safety.
Databases and MSDSs		
12*	http://toxnet.nlm.nih.gov/ National Library of Medicine, National Institutes of Health. *Toxnet*	Contains links for biomedical references for specific chemicals.
13*	http://www.cdc.gov/niosh/ipcs/icstart.html, several languages http://www.cdc.gov/niosh/ipcs/nicstart.html, in English NIOSH, *International Chemical Safety Cards*	Collection of International Chemical Safety Cards, which are MSDS-like safety database
14*	http://www.osha.gov/web/dep/chemicaldata/default.asp#target OSHA Chemical Database	Links to *NIOSH Pocket Guide to Chemical Hazards*, DOT Emergency Response Guidebook, NOAA CAMEO database based on chemical names.
15	http://www.ilpi.com/msds/index.html	Links to many MSDS web sites.
16	http://www.cdc.gov/niosh/topics/chemical-safety/	A homepage with many links.
17	http://www.sigmaaldrich.com/Area_of_Interest/The_Americas/United_States.html *Aldrich Handbook of Fine Chemicals*	MSDSs can be found by searching on chemical name.
Standards		
18*	www.osha.gov (main web site) http://www.osha.gov/html/a-z-index.html&S (site index, use "Standards" link) OSHA, 29 Code of Federal Regulations, Part 1910 Standards	Standards for chemicals found at 1910.1000 and higher.
19	http://www.nfpa.org/aboutthecodes/list_of_codes_and_standards.asp NFPA Standards online. The three most relevant NFPA standards related to laboratory safety are (1) Standard System for Identification of Hazards of Materials for Emergency Response, NFPA 704, NFPA, Quincy, MA, 2007; (2) Standard on Fire Protection for Laboratories Using Chemicals, NFPA 45, NFPA, Quincy, MA, 2004; and (3) Standard for Portable Fire Extinguishers, 2007 Edition, NFPA 10 , NFPA Quincy, MA, 2007	Need to follow a few links and register as a guest.
General Chemical Safety information		
20*	http://www.ilpi.com/msds/osha/index.html; OSHA links http://www.ilpi.com/msds/ref/demystify.html; the "MSDS-Demystifier" http://www.ilpi.com/msds/ref/index.html; glossary	Useful explanations of basic chemical safety information.
21	http://www.unece.org/trans/danger/publi/ghs/ghs_rev02/02files_e.html United Nations, *Globally Harmonized System of Classification and Labelling of Chemicals (GHS)*, 2nd revision	A description of the Globally Harmonized System for chemical classification and labels.

[a] The best ones are marked with an asterisk.

Beyond these simple searching techniques, there exist a number of excellent web sites for reliable safety information. In Table 3.2.2.1 we have sorted these into the following:

- **Books and pamphlets online.** Some publishers have graciously allowed their print versions of books and pamphlets to be accessible online as a public service.

- **Chemical profiles.** These are typically 1–2 page synopses about particular chemicals that have been written by safety professionals using language that appropriately trained chemists can understand in order to work safely with the chemicals. There are two online sources that contain very useful "profiles" of chemicals that include safety information. Jay Young has written a lengthy series of *Chemical Laboratory Information Profiles* (CLIPs) in the *Journal of Chemical Education*. The online version is viewable by subscription only, but the collection of CLIPs is searchable and hard copies can be located in most educational libraries. In Appendix B of the online *Prudent Practices*, all 88 of the *Laboratory Chemical Safety Summaries* (LCSSs) are viewable. The *NIOSH Pocket Guide to Chemical Hazards*, available from the Internet as well as in printed form, also offers detailed information about many chemicals.

- **Scientific journals.** Two journals are listed that can provide useful safety information to practicing chemists and chemical educators. (We have not listed other primary research journals, of which there are many, where the audience is highly specialized toward medicine, chemical hygiene, or epidemiology.)

- **Chemical databases and MSDSs.** There are many Internet sources of databases of chemicals and lists of MSDSs. We have listed resources that can easily be accessed and understood by chemists and chemistry students.

- **Standards.** Standards are rules or guidelines that establish performance levels. OSHA standards are federal law. NFPA standards only have the legal status of "recommendations" but are considered to be the guidelines against which behavior is measured.

- **Safety information.** This is a collection of web sites with generic and useful information about laboratory safety.

In Table 3.2.2.1 we have placed an asterisk next to the best resources and these are the web sites that we recommend for starting to learn about the toxicological information for a chemical.

Printed Materials

Printed materials include reference books and other books that are not generally available electronically. Since journals are now more commonly used online, they are listed in Table 3.2.2.1. We have listed some "traditionally printed materials" in Table 3.2.2.2 since that is how they are generally accessed. Our recommendations for the "best first places to start" are indicated with an asterisk.

Planning to Work Safely in the Lab

It is always a good idea to thoroughly research a topic using reliable reference materials before beginning lab experiments. This not only can prevent you from performing an unnecessary experiment or conducting an experiment that is destined to "fail," but more importantly it can also help you to conduct your experiments safely. Most nonresearch experiments in academic labs have already been reviewed for "success" and safety by professors. Any new experiment that is part of a research project needs to be approached with greater caution.

It is wise to collect information from several, hopefully reliable, sources and compare information. When sources disagree, it is important to try to resolve or understand the discrepancy. One way to do this might be through an examination of the date of the material since sometimes information gets updated. In situations where differences cannot be resolved, the prudent approach is to assume the most alarming information is the correct information. Reviewing what you have learned with a mentor or colleague is smart, too, since that conversation may lead to a better understanding of the information.

TABLE 3.2.2.2   Printed Safety Reference Materials

Best[a3]	Title	Author	Publisher	Year
	Books of General Interest			
1*	*Prudent Practices in the Laboratory: Handling and Disposal of Chemicals*; New edition in June 2010	National Research Council	National Academy Press, Washington, DC	2010
	New edition in early 2010			
2*	*Bretherick's Handbook of Reactive Chemical Hazards*, 7th edition, Vols. 1 and 2	P. Urben (editor)	Elsevier, New York	2006
3*	*The Merck Index*, 14th edition	M.J. O'Neil (editor),	Merck & Co., Inc., Whitehouse Station, NJ	2006
4*	*NIOSH Pocket Guide to Chemical Hazards*, available online at http://www.cdc.gov/niosh/npg/	National Institute for Occupational Safety and Health	U.S. Government Printing Office, Washington, DC	2005
5	*Handbook of Chemical Health and Safety*	R. J. Alaimo (editor)	Oxford University Press, New York; ACS. Washington, DC	2001
6	*Sittig's Handbook of Toxic and Hazardous Chemicals and Carcinogens*, 5th edition, Vols. 1 and 2	R. P. Pohanish	William Andrew, Norwich, NY	2008
7	*Patty's Toxicology*, 5th edition, Vols. 1–9	E. Bingham, B. Cohrssen, and C. H. Powell (editors)	John Wiley & Sons, Hoboken, NJ	2001
8	*Proctor and Hughes' Chemical Hazards of the Workplace*, 5th edition	Gloria J. Hathaway and Nick H. Proctor	John Wiley & Sons, Hoboken, NJ	2004

(continued)

TABLE 3.2.2.2 (Continued)

Best[a]	Title	Author	Publisher	Year
9	*A Comprehensive Guide to the Hazardous Properties of Chemical Substances*, 3rd edition	Pradyot Patnaik	John Wiley & Sons, Hoboken, NJ	2007
10	*Hazardous Chemicals Desk Reference*, 6th edition	Richard J. Lewis, Sr.	Wiley - Interscience, Hoboken, NJ	2008
11	*Sax's Dangerous Properties of Industrial Materials*, 11th edition; available online for one-time fee ($1200) or annually at cost (2008)	Richard J. Lewis, Sr.	Wiley - Interscience, Hoboken, NJ	2004
	Specialized Books and References			
12	*CRC Handbook of Chemistry and Physics: A Ready-Reference Book of Chemical and Physical Data*, 88th edition	David R. Lide (editor)	CRC Press, Boca Raton, Fl	2008
13	*TLVs® BEIs® Threshold Limit Values for Chemical Substances and Physical Agents and Biological Exposure Indices*; available to members and some libraries	American Conference of Governmental Industrial Hygienists (ACGIH)	ACGIH, Cincinnati, OH	yearly
14	*Catalog of Teratogenic Agents*, 12th edition	Thomas H. Shepard and Ronald J. Lemire	Johns Hopkins University Press, Baltimore, MD	2007
15	*Reproductive Hazards of the Workplace*	Linda M. Frazier and Marvin L. Hage	John Wiley & Sons, Hoboken, NJ	1998

[a]The best first places to start are indicated with an asterisk.

RAMP

- *Recognize* hazards by conducting appropriate searches of information resources.
- *Assess* the level of risk for exposure to chemicals and the use of equipment by using these resources.
- *Minimize* the risk by using appropriate personal protective equipment and engineering controls as presented in the resources.
- *Prepare* for emergencies by knowing what equipment to use and how to use it.

References

1. This is an amusing reversal of a quote frequently attributed to Frank Westheimer, a chemist, who said "A couple of hours in the library can save you weeks in the laboratory."

2. M. E. BURNS and R. H. SMITH. The instability of azides. *Chemical & Engineering News* **62** (2): 2 (1984).

QUESTIONS

1. The most reliable information about the hazards of a particular chemical can be found

   (a) In Wikipedia
   (b) Using Google
   (c) In the MSDS
   (d) In other printed or online materials that have been reviewed

2. Which printed resources are also available online (for free)?

   I. *Prudent Practices in the Laboratory (2010)*
   II. *Bretherick's Handbook of Reactive Chemical Hazards*
   III. *NIOSH Pocket Guide to Chemical Hazards*
   IV. NFPA Standards

   (a) I, III, and IV
   (b) II, III, and IV
   (c) II and III
   (d) III and IV

3. In which of the following resources can "chemical profiles" be found?

   I. *Prudent Practices in the Laboratory*
   II. *Safety in Academic Chemistry Laboratories*
   III. *NIOSH Pocket Guide to Chemical Hazards*
   IV. *Journal of Chemical Education*
   V. NFPA Standards

   (a) I, II, III, and V
   (b) II, III, IV, and V
   (c) I, III, and IV
   (d) All of the above

3.2.3

INTERPRETING MSDS INFORMATION

Preview This section provides an overview of the various parts of a typical MSDS and provides suggestions to help you find important information about the hazards of a chemical.

In the fields of observation chance favors only the prepared mind.

Louis Pasteur[1]

INCIDENT 3.2.3.1 ISOPROPYL ETHER DETONATION[2]

A student working in a laboratory needed a chemical for an experiment. While looking for this chemical he came across a bottle of isopropyl ether. He remembered that isopropyl ether was well known for explosions from peroxide formations. As he looked at it, it appeared to be filled with large crystals and he knew that it should be a liquid, not a solid. He reported this to his professor who came to look at the bottle too. The professor ordered everyone out of the stockroom and called local emergency response officials who sent out the bomb squad. The bomb squad removed the bottle for controlled detonation in an open field.

What lessons can be learned from this incident?

Gleaning Useful Information for Hazard Assessment from MSDS

Material Safety Data Sheets (MSDSs) are required by OSHA's Hazard Communication Standard[3] and chemical suppliers are required to provide copies of these for each chemical that it sells to its customers. MSDSs are written to inform workers who use chemicals in workplaces about the hazards associated with the chemicals. Their general structure and contents were described in Section 3.1.3. MSDSs have proved useful for many people, including those working in industrial workplaces who handle large quantities of chemicals, and also to emergency response personnel. Most MSDSs, however, do not focus on the use of chemicals in laboratories, where small quantities of many different chemicals are used in normal operations. Nevertheless, there can be much useful information found in these documents—particularly when making safety plans to work with a particular chemical.

Although there is a legal requirement that chemical manufacturers produce MSDSs for chemicals that they sell, there is no specific required format. However, the American National Standards Institute designed a 16-part format[3] that was recommended by OSHA[4] and a similar format has been adopted by the United Nations for its Safety Data Sheets in its Globally Harmonized System of Classification and Labelling of Chemicals (GHS).[5]

The various sections of a typical MSDS written in the ANSI/GHS formats are discussed below with an emphasis on how this can be useful to you and what the limitations of the information might be. Keep in mind that the format is suggested and not required, so it may vary with manufacturer. Even if the ANSI format is used, there may be variations with the types of information reported in a given section. There is an online OSHA document (http://www.osha.gov/dsg/hazcom/ghs.html#7.0) that compares the MSDS formats for ANSI, SIO, OSHA, and GHS formats.

We will describe each of the 16 ANSI sections below. However, we first want to caution you about some MSDSs that may be inaccurate, misleading, or incomplete.

Laboratory Safety for Chemistry Students, by Robert H. Hill, Jr. and David C. Finster
Copyright © 2010 John Wiley & Sons, Inc.

Skepticism About MSDSs

In an ideal world, the information in an MSDS would be accurate, reliable, and useful. Unfortunately, there are a handful of reasons why MSDS information may be incorrect or misleading. Let's explore these situations so that you can use these documents intelligently and critically.

MSDSs have two principal types of information in them: data and recommendations. The data are various characteristics of the chemical (or mixture of chemicals). This will include "simple" features such as boiling point and aqueous solubility that have likely been known accurately for decades. There should be little reason to be suspect about this kind of data, although any document is susceptible to clerical error and evaluations of MSDSs have shown significant errors. One study of 150 MSDSs showed that 11% of dangerous materials had been misidentified, and more than half of the MSDSs had incorrect health information and exposure level information.[6] In a 1997 report commissioned by OSHA, an expert panel found that only 11% of MSDSs were accurate in four critical areas: health effects, first aid, personal protective equipment, and exposure limits.[7] The panel found that health effects data were frequently incomplete. Even more alarming was a study of MSDSs for 11 degreasing agents from Korea in which the authors report that most of the information on composition, safety classification, LD_{50} values, and exposure levels was wrong.[8] There have been other reports of criticism and skepticism of MSDSs. The chairman of the U.S. Chemical Safety and Hazard Investigation Board reported to a congressional panel that defective MSDSs were contributing factors in 10 of 19 major chemical accidents.

The point here is not that all or even the majority of MSDSs are faulty, but rather you should always view information from MSDSs with some skepticism and seek to further investigate or corroborate any information that is critical for your safety. It is prudent to look for other sources to verify information. Section 3.2.2 provides additional, reliable sources of information about the hazards of chemicals.

MSDSs also include recommendations about safety procedures, medical treatment, and chemical disposal. It is not unusual to find that these recommendations are so broad as to be of little or no use. Since these recommendations were made with the industrial workplace in mind, rather than the laboratory, it would not be unusual to encounter recommendations that are not useful for laboratory applications. This is particularly true of safety equipment recommendations. For example, you probably don't need to wear rubber boots or steel toed shoes in a laboratory, although this might be wise in a factory. In the same vein, however, recommendations such as wearing protective gloves are not very useful if they don't tell you what kind of gloves to use. Lastly, recommendations could easily change as more information comes to light and if MSDSs are not updated, they will contain incorrect information. New laws and new techniques don't fundamentally change the "data" about a chemical but can change how we need to manage risk and exposure with new updated recommendations.

It is also important to understand the process by which MSDSs are written. Ideally, we might hope that a trained safety professional has studied the scientific literature about a chemical and thoughtfully produced a scientifically accurate and "lab-useful" document. While scientists and physicians might sometimes author some parts of the MSDS, it is also true that lawyers representing the company that sells the chemical (for which the MSDS is written) will advise authors to write in a fashion that limits the liability of the manufacturer. MSDSs typically have legal disclaimers included to protect the manufacturer.

Also, and perhaps surprisingly, MSDSs are not always written by humans! When the Hazard Communication Standard was enacted in 1970 and implemented in 1986, many chemical manufacturers already were selling thousands of chemicals for which MSDSs were now "suddenly" required. It is estimated that about 650,000 chemicals are now used in the United States for which MSDSs are required. Having humans write these MSDSs was financially unviable. Computer programs were constructed that allowed information databases to be searched to craft MSDSs automatically. This understandably generated a particular "style" of MSDS, including many generic phrases and warnings that are "correct" but not always useful.

The result of this process of authorship of an MSDS sometimes leads to less-than-useful information. For example, although is it correct to be advised to "dispose in compliance with local, state, and federal regulations," this hardly has the specificity that helps a chemist actually dispose of a chemical responsibly. Other examples of perhaps computer-generated recommendations are listed in *Special Topic 3.2.3.1* Silly and Not-So-Silly MSDSs.

SPECIAL TOPIC 3.2.3.1

SILLY AND NOT-SO-SILLY MSDSs

An incident that circulated on the Internet some years ago involved the city council of Aliso Viejo, California. The council was considering banning the use of Styrofoam cups since a local paralegal had done some research on the manufacturing process and came upon the MSDS for dihydrogen oxide at http://www.dhmo.org/. This web site has links to various MSDSs for dihydrogen oxide (H_2O) that list various hazards associated with its use. Sometime before the vote it was determined that dihydrogen oxide was water and the council avoided further embarrassment.

This incident highlights what, in fact, are some silly interpretations and entries on some MSDSs. We cite three more examples.

An MSDS for a buffered saline solution indicates that upon exposure to the eye one should "immediately flush with potable water for a minimum of 15 minutes and seek assistance from MD" and for skin exposure, "flush with copious amount of water and call MD." Since this product is likely designed as an irrigating agent for eyes, the cautions seem silly. This is probably a good example of a computer-generated MSDS where the "universal" recommendation for eye and skin exposure is to flush with water. (Actually, this is generally good advice for most chemical exposures to the eyes and skin.)

Another classic example of somewhat dubious laboratory advice comes from an MSDS for deionized water, where the recommended laboratory protective equipment consists of "safety glasses and lab coat" and it is further recommended that the product be stored in a "tightly closed container." Again, these are generically good recommendations for most chemicals but tend to make deionized water sound somewhat more hazardous than we know it to be.

Similarly, an MSDS for sucrose (table sugar) recommends that you "practice good chemical hygiene after using this material, especially before eating, drinking, smoking, using the toilet, or applying cosmetics."

The danger of these kinds of entries arises when what seems to be ridiculous advice gets ignored or causes someone to too-quickly dismiss sound cautions. The MSDS for sucrose also has a long list of cautions about using various engineering controls, avoiding breathing dust, and wearing a respirator, chemically protective gloves, boots, aprons, and gauntlets. However, some of this is actually good advice. When you think of sucrose, you might think of a teaspoon of a sweet granular substance to dissolve in iced tea. However, a large sucrose-processing facility will likely generate considerable sucrose dust that may in fact be an eye or skin irritant upon extended exposure. And such a cloud of sucrose can also explode (as many combustible dusts and powders do) if an ignition source is supplied. MSDSs for laundry detergents are similarly scary and their entries would make one cautious about touching the substance.

The key point to remember is that "ordinary" chemicals and substances with which we are familiar may sound overly dangerous in the MSDS when, in fact, they do present hazards under some circumstances. Unfortunately, some MSDSs likely overstate some hazards, particularly when they are not written by a knowledgeable safety professional or when lawyers convince authors to "err on the side of excess warning" for legal reasons.

Format of a Material Safety Data Sheet or Safety Data Sheet[3–5]

Chemical and Hazard Identification—MSDS Sections 1, 2, and 3

Section 1: Chemical Product and Company Identification

Information. Identifies the specific chemical product by principal name and other names, Chemical Abstracts Service number (CAS No.), product code (catalog number), and company name with emergency contact information.

Use. Allows comparison with label to ensure that this is the correct MSDS for the product; contact information may be needed to get additional information for an emergency, accident, or for recommendations for specific personal protective equipment (PPE).

Section 2: Composition/Information on Ingredients (GHS lists Hazards Identification before Composition/Information on Ingredients.)

Information. This reports the chemical's purity and CAS number. If this is a mixture of chemicals this will be noted and the components and their approximate percentages will be identified.

Use. This information may be needed in the event of an emergency, for example, to clean up a spill or for medical treatment.

Section 3: Hazards Identification (GHS lists Hazards Identification before Composition/Information on Ingredients.)

Information. Emergency overview—warnings that appear on the label; potential health effects usually given by route of exposure; known chronic effects. The new GHS lists hazard classifications and pictograms in this section.

Use. Physical hazards are usually clearly identified. The user is alerted to potential health effects with a range of symptoms and signs resulting from exposure. However, it is more difficult to determine the relative importance of these health effects unless some relative ranking or classification is given here. In some cases a broad range of health effects are given, such as mild irritation, coughing, headache, central nervous system depression, narcosis, unconsciousness, and death. While this covers the whole gamut of possible effects that could result from exposure, this type of presentation leaves readers to determine how important exposure might be in regard to their health. The new GHS hazard classifications clarify this quandary by assigning appropriate relative hazard ratings.

Emergency Measures—Sections 4, 5, and 6
Section 4: First Aid Measures

Information. Provides general procedures for first aid treatment in the event of exposure, usually by route of exposure. Some MSDSs may give more information for medical treatment by physicians.

Use. This information can be very useful in developing a safety plan to deal with emergencies in the event of an exposure. First aid measures need to be taken within minutes of exposure to ensure that further damage is prevented. For example, a splash in the eyes may call for immediate washing for several minutes under an emergency eyewash or a sink. This immediate action could mean the difference between minor eye irritation and severe eye damage from delayed treatment.

Section 5: Firefighting Measures

Information. Provides flammable properties; fire extinguishing media; unusual fire or explosion hazards; and special firefighting instructions.

Use. You can use the fire extinguisher information in developing a safety plan. Firefighters can use this information if they are called to handle a fire with this chemical. Firefighters may ask for MSDSs for all chemicals if they are called to your facility in an emergency. They may have computers on their trucks that have access to this kind of information.

Section 6: Accidental Release Measures

Information. Describes what should be done in the event of a spill.

Use. This information can be used in developing a safety plan. Laboratories are sometimes equipped with spill kits and you may be able to take care of small spills. As part of your plan you may need to identify special materials for the cleanup, and prepare for that potential event. Larger spills probably need to be managed by safety professionals or emergency response personnel.

Managing This Hazard—Sections 7 and 8
Section 7: Handling and Storage

Information. This is where you will notice that the focus is not on the laboratory use of the chemical. The information usually focuses on handling and storage of the container, such as grounding the container during transfer of the contents. Here you will find general warnings such as keep away from flames and ignition sources or use in ventilated areas. It also describes the conditions for storage of the chemical and perhaps may provide a list of incompatible chemicals. There may be general personal hygiene recommendations here.

Use. This section may only be useful from a very general view, since there are likely to be no specific recommendations for handling this chemical in the laboratory. Additionally, the recommendations may focus on handling large quantities in metal drums—hence the need for grounding.

Section 8: Exposure Controls/Personal Protection

Information. This describes measures to prevent exposure such as ventilation, engineering controls, or administrative controls. Recommendations for personal protective equipment (PPE) will be found in this section. If there are occupational exposure limits or guidelines, they will be described here.

Use. The information about exposure controls is usually very general and is often directed to the employer or the engineering or safety professional staff. The recommendations for PPE may be only marginally useful since it is described in generic terms and specific information about PPE is needed to be of use. For example, this section might explain that you need to "use impervious gloves, boots, overalls" but it does not say what kind of gloves are impervious, and clearly the recommendation for boots and overalls is not for laboratory personnel. The occupational exposure limits and guidelines can be useful in your assessment of the relative risk of handling this compound (see Section 6.2.2).

Data Used for Hazard Recognition/Assessment Information—Sections 9, 10, 11, and 12

Section 9: Physical and Chemical Properties

Information. This lists physical and chemical properties of the chemical.

Use. This information can be useful in helping you assess the relative risk of handling the chemical. It may also be useful in your experiments—solubility, boiling points, or melting points are listed.

Section 10: Stability and Reactivity

Information. The stability of the chemical is described as well as conditions to avoid when working with this chemical. Other information relative to reactivity is described including incompatibility with other chemicals; hazardous products that can result from decomposition or burning; and if hazardous polymerization is likely to occur.

Use. This information can be very useful to the laboratory worker—knowing about a chemical's reactivity is essential. Knowing about incompatibles to avoid during use and storage is also essential. The formation of hazardous products is particularly useful to emergency personnel, especially firefighters. The possibility of hazardous polymerization is also important to know for the laboratory worker and emergency personnel.

Section 11: Toxicological Information

Information. This section reports known toxicity testing and may also have some information about human exposure. It should list both acute and chronic toxicities if these are known.

Use. This information is very useful in determining the relative risk of health effects from exposure to this chemical. It can be used to help you classify a chemical's relative risk using GHS or some other risk rating system. However, you must keep in mind that there is often only very limited information available on toxicity testing, and chronic toxicities, in particular, may not have been evaluated or detected by testing methods. This means that a chronic health effect such as cancer or sensitization may not have been identified, but could still be a possibility—hence, always avoid chemical exposures. One must also examine the date when a particular MSDS was written. (See Section 16 below.) While information like a boiling point is not likely to change (!), toxicity data, in particular, is continually open to reexamination and any MSDS more than a few years old might not have the best current toxicological data.

Section 12: Ecological Information

Information. Data regarding the behavior and effects of this chemical on the environment are described here.

Use. This information will be especially useful to environmental protection specialists in case of a spill that gets into our waterways or landfills. Information about a chemical's stability in the environment can be very useful in deciding what to do if the chemical is released accidentally.

Regulated Activities—Sections 13, 14, and 15

Section 13: Disposal Considerations

Information. All hazardous waste is regulated in the United States under the Resource Conservation and Recovery Act (RCRA) administered by the U.S. Environmental Protection Agency or state environmental authorities. This section describes in a general way considerations concerning disposal.

Use. This section is useful to environmental specialists within your organization, but you must understand that you cannot just pour hazardous chemicals down the drain—there are legally acceptable ways to dispose of hazardous waste and these must be followed. You will likely receive very specific instructions from your environmental specialists.

Section 14: Transport Information

Information. Transport or shipping of hazardous materials is regulated by the U.S. Department of Transportation. This section provides general information about shipping and labels.

Use. This section is principally useful to personnel within your organization who ship or transport hazardous materials to other places. You should know that you just cannot send a hazardous material to someone unless you use authorized shipping procedures—special containers, packaging, and labeling requirements must be met.

Section 15: Regulatory Information

Information. This section describes various U.S. regulations that are applicable to this chemical. It includes requirements for reporting if spills exceed a certain quantity. This section also may list specific right-to-know laws in U.S. states that are applicable to this chemical.

Use. This information is useful to the employer and the environmental and safety staff in helping them comply with legal requirements. While you may not focus on this at all as a student, someday there is a good chance that you will be involved in making decisions related to the use of chemicals.

Other Information—Section 16

Section 16: Other Information

Information. This section may list the date of MSDS preparation or last revision; references may be cited; some MSDSs list hazard ratings (such as NFPA ratings) here as well as label and hazard warnings. A disclaimer may be given here.

Use. This information can be helpful in finding the latest information about the hazards of a chemical or in ensuring that you have the latest MSDS available. The hazard ratings will be useful in assessing the risk of handling this chemical.

MSDSs can provide some very useful information to anyone working in a laboratory, particularly in developing safety plans or in assessing hazards. Students in laboratories should understand that most MSDSs are written for general industrial use, not for laboratory use, and some information is written so generally as to be of little practical value for the laboratorian.

Using MSDSs in the Lab: RAMP

We repeat here the advice listed at the end of Section 3.1.3. The RAMP acronym is easy to use with MSDSs. The MSDS can be a useful source of information in order to:

- *Recognize* the hazards of a particular chemical.

- *Assess* the level of risk posed by the chemical based on the physical properties of the chemical.

- *Minimize* the risk by preventing exposure using personal protective equipment and other safety equipment recommended in the MSDS.

- *Prepare* for emergencies based on the assessed hazard level and the possible effects of exposure.

References

1. J. BARTLETT. *Bartlett's Familiar Quotations*, 17th edition (J. KAPLAN, ed.), Little, Brown, and Co., Boston, 2002, p. 533, line 8.
2. N. V. STEERE. *Chemical Safety in the Laboratory*, Vol. 1, Division of Chemical Safety, American Chemical Society, Washington, DC, 1974, p. 69.
3. American National Standards Institute. American National Standard for Hazardous Industrial Chemicals—MSDS Preparation (ANSI Z400.1–2004).
4. U.S. Department of Labor, Occupational Safety and Health Administration. Recommended Format for Material Safety Data Sheets (MSDSs), issued July 22, 2004; available at http://www.osha.gov/dsg/hazcom/msdsformat.html (accessed January 19, 2008).
5. United Nations. Globally Harmonized System of Classification and Labelling of Chemicals (GHS), 2nd revision, Annex 4, United Nations, New York, 2007.
6. P. W. KOLP, P. L. WILLIAMS, and R. C. BURTAN. *American Industrial Hygiene Association Journal* **178A**: 183 (1995).
7. U.S. Department of Labor, Occupational Safety and Health Administration. Hazard Communication: A Review of the Science Underpinning the Art of Communication for Health and Safety; available at http://www.osha.gov/SLTC/hazardcommunications/hc2inf2.html (accessed August 27, 2008).
8. C. G. YOON, T.-W. JEON, C.-K. CHUNG, M.-H. LEE, S.-I. LEE, S.-E. CHA, and I.-J. YU. *Korean Industrial Hygiene Association Journal* **10**: 18 (2000).

QUESTIONS

1. MSDSs are required by

   (a) The OSHA Lab Standard
   (b) The OSHA Hazard Communication Standard
   (c) The OSHA General Duty Clause
   (d) The EPA

2. It is estimated that about what percent of MSDSs have incorrect information in them?

   (a) Less than 2%
   (b) 11%
   (c) 37%
   (d) At least 61%

3. Which of the following descriptions apply to MSDS recommendations about safety procedures, medical treatment, and chemical disposal?

   I. They are usually specific and accurate.
   II. They are designed more for industrial workplace circumstances than research labs.
   III. They are sometimes so broad in nature as to be of little use.
   IV. They are regularly updated.

   (a) I and IV
   (b) I, II, and IV
   (c) II, III, and IV
   (d) II and III

4. Which of the following are the contributing sources for MSDSs?

   I. Scientists
   II. Physicians
   III. Lawyers
   IV. Computer programs

   (a) I and II
   (b) I and III
   (c) I, II, and III
   (d) I, II, III, and IV

5. What information is usually in Section 1: Chemical Product and Company Identification?

   I. Principal name of the chemical
   II. CAS number
   III. Synonyms and other names for the chemical
   IV. Company name and emergency contact number

   (a) I and II
   (h) I and III
   (c) I, II, and III
   (d) I, II, III, and IV

6. What information is *not* in Section 3: Hazards Identification?

   (a) Environmental hazards
   (b) Potential health effects on humans
   (c) Physical hazards such as flammability and reactivity
   (d) Route of exposure information

7. The information in Section 4: First Aid Measures is mostly

   (a) Useful only to paramedics
   (b) Useful only to physicians
   (c) Useful only to paramedics and physicians
   (d) Useful for both medically trained and nontrained persons

8. Section 7: Handling and Storage is mostly written for

   (a) Chemistry students in academic labs
   (b) Laboratory instructors
   (c) Lab managers and stockroom personnel
   (d) Truck drivers who deliver chemicals

9. Which of the following describe Section 8: Exposure Controls and Personal Protective Equipment?

   I. It is written more for industrial situations than for research labs.
   II. It often contains very general information about PPE that is of limited use.
   III. It contains exposure limits information.

      (a) I and II
      (b) I and III
      (c) II and III
      (d) I, II, and III

10. Section 10: Stability and Reactivity contains information useful to

    (a) Chemists in research labs
    (b) Emergency responders
    (c) Chemistry students
    (d) All of the above

11. Section 11: Toxicological Information may have inaccurate information because

    (a) We simply don't know much about the toxicity of most chemicals
    (b) It is usually generated from computer searches that are notoriously unreliable
    (c) Many errors are made in the animal testing that leads to these data
    (d) This information is more subject to being updated compared to most MSDS information

12. Section 13: Disposal Considerations is usually

    (a) Accurate and specific with regard to federal, but not state and local, laws and regulations
    (b) Accurate and specific with regard to federal, state, and local laws and regulations
    (c) Accurate and specific with regard to federal and state, but not local, laws and regulations
    (d) Not very specific

3.3.1

CHEMICAL HYGIENE PLANS

Preview   This section describes OSHA's Laboratory Standard, which requires that laboratory management develop a Chemical Hygiene Plan (CHP) and appoint a Chemical Hygiene Officer (CHO), and also describes how a CHP can help protect you from exposure.

> *Out of this nettle, danger, we pluck this flower, safety.*
>
> *King Henry IV*, Shakespeare[1]

INCIDENT 3.3.1.1   HYDROFLUORIC ACID EXPOSURE[2]

A scientist working in a laboratory was using hydrofluoric acid, when he inadvertently put his finger in the solution. It did not hurt immediately but after a short time he began to experience severe pain. He washed off his finger and went to the clinic. The clinic physician learned that the scientist was using hydrofluoric acid, but since the doctor did not know that anyone was using this chemical on the campus, he did not have the appropriate treatment for hydrofluoric acid burns—a specialized first aid supply, calcium gluconate gel. The scientist was sent to a city hospital, where he finally received appropriate treatment. While his finger was severely burned, he did not lose it.

What lessons can be learned from this incident?

Planning to Succeed

Laboratories are different from many other workplaces since in many cases a variety of chemicals are used in relatively small quantities for short periods of time. In contrast, industrial chemical workplaces handle large quantities of a few chemicals consistently over many years. Thus, a chemist or a laboratory scientist may be exposed to many different chemicals, but usually for shorter periods of time and in quantities that are below those exposures found in industrial situations. However, this does not necessarily mean that there is less risk in the laboratory or that there are no opportunities for serious exposures, but rather that a different approach to limiting exposures from industrial practices is necessary.

Under what is called the "general duty clause" of the Occupational Safety and Health Act of 1970,[3] OSHA requires that each employer "furnish to each of his employees ... a place of employment ... free from recognized hazards that are likely to cause death or serious physical harm." Employees are required to "comply with occupational safety and health standards and all rules." Although this OSHA regulation does not cover students while they are in class, it does apply to all employees of colleges and universities, including faculty, instructors, maintenance, custodial staff, and students who are employed by the university as instructors or laboratory assistants. Thus, it is for moral reasons that college and universities work to ensure that their students are protected from hazardous chemicals in general.

In 1990 the Occupational Safety and Health Administration (OSHA), recognizing that laboratories require a different approach from industrial settings in managing exposures, issued the regulation *Occupational Exposures to Hazardous Chemicals in Laboratories*, known today as the "Lab Standard."[4] This Lab Standard is a performance standard, meaning that OSHA does not tell an employer how to prevent exposure to its employees, but rather each employer can use his/her own methods to control exposures

Laboratory Safety for Chemistry Students, by Robert H. Hill, Jr. and David C. Finster
Copyright © 2010 John Wiley & Sons, Inc.

TABLE 3.3.1.1   Components of a Chemical Hygiene Plan (29 CFR 1910.1450 [4])

Regulation section (3)	Topic
(e)(3)(i)	Standard Operating Procedures Involving Use of Hazardous Chemicals
(e)(3)(ii)	Criteria to Determine and Implement Control Measures to Reduce Employee Exposure to Hazardous Chemicals
(e)(3)(iii)	Requirements to Ensure Control Measures Perform Properly
(e)(3)(iv)	Employee Information and Training
(e)(3)(v)	Operations Requiring Prior Employer Approval
(e)(3)(vi)	Medical Consultation and Examinations
(e)(3)(vii)	Designation of Chemical Hygiene Officers
(e)(3)(viii)	Requirements for Handling Particularly Hazardous Chemicals
(e)(3)(viii)(A)	Designated Areas
(e)(3)(viii)(B)	Containment Equipment
(e)(3)(viii)(C)	Procedures for Safe Removal of Contaminated Waste
(e)(3)(viii)(D)	Decontamination Procedures

as long as the methods meet certain requirements to limit exposure. Foremost is the requirement for a Chemical Hygiene Plan (CHP). This is a general safety plan written by management for employees working in their laboratories.

The Chemical Hygiene Plan or CHP

The CHP is the principal tool used by your university or college in its efforts to protect its laboratory workers. The CHP must be easily available to all employees and it is usually used for laboratory students as well. You should review the CHP of your own college/university and become familiar with the general contents. It should be pretty easy to find the CHP online at your institution. A CHP is required for any institution that has a laboratory. OSHA's general requirements for CHPs are shown in Table 3.3.1.1. However, because of the latitude in the Lab Standard, CHPs take a variety of approaches to compliance with OSHA requirements by using a varied CHP format that reflects the specific hazards and resources available at the institution. All CHPs will have the same basic components, but the components may vary in detail, content, and presentation. Good CHPs provide general guidance in critical areas of handling chemicals. However, for individual experiments, particularly in research, it is important that you have a more specific safety plan (see Sections 6.3.3 and 7.3.6). As a student, any experiment should be planned in cooperation with your faculty mentor, working under the overall guidelines of your local CHP. Furthermore, there are other personnel available in your institution who can also help develop and recommend safe work practices with chemicals, particularly the Chemical Hygiene Officer identified in the CHP.

Chemical Hygiene Officer (CHO)

The Lab Standard that requires the CHP also requires that each institution designate a Chemical Hygiene Officer, commonly called the CHO. The Lab Standard requires that the CHO have the training and experience to provide guidance in developing and implementing the CHP. The management of each institution is responsible for identifying the specific duties of its CHO. Although the Lab Standard does not require specific duties for the CHO, Appendix A of the Lab Standard contains guidance for CHO duties:

- Developing the CHP and appropriate measures for handling chemicals
- Overseeing the purchase, use, and disposal of chemicals in the lab

- Ensuring that there are adequate safety surveys, inspections, and audits of chemical and laboratory operations
- Assisting and working with investigators to develop safety measures and appropriate safe facilities for handling chemicals
- Regular review and improvement of the institution's chemical safety program
- Understanding the current legal requirements for laboratories, including laboratory chemicals and materials that are regulated

While most students in the early years of undergraduate study may not know of the CHO for their institution, as your work proceeds and you, as an advanced student, become more independent and conduct chemical research, you will likely want to seek out the CHO. The CHO can be a good resource for information about the specific safety procedures that should be used in handling chemicals. In larger universities, there is likely an Environmental Health and Safety (EHS) office on campus and the CHO likely is part of that institutional structure and has special training. In smaller colleges the CHO might be an EHS person or perhaps a member of the chemistry faculty (who has far less direct EHS training). If your institution has a safety committee (as many do), the CHO is likely to be a member of this committee. Several papers go into greater depth about CHOs and their recommended responsibilities.[5-8] You may find these references valuable if you are taking on this role as the CHO in a future job.

RAMP

- The CHP and CHO may be a resource for *recognizing* hazards in your laboratory.
- The CHO may be a resource to assist you in *assessing* the risk of the hazards in your laboratory.
- The CHO should be able to assist you in *minimizing* exposures by recommending equipment and safe practices.
- The CHP should have components that will assist you in *preparing* for emergencies and the CHO can help in the process.

References

1. *Bartlett's Familiar Quotations*, 17th edition, (J. KAPLAN, ed., Little, Brown, and Co., Boston, 2002, p. 185, line 16.
2. R. H. HILL. Personal account of an incident.
3. Occupational Safety and Health Act of 1970. Section 5. (a) (1). 29 U.S. Code 654; available at http://www.osha.gov/pls/oshaweb/owadisp.show_document?p_table=OSHACT&p_id=3359 (accessed June 18, 2009).
4. Occupational Safety and Health Administration. *Occupational Exposures to Hazardous Chemicals in Laboratories*, 29 Code of Federal Regulations Part 1910.1450; available at http://www.osha.gov/pls/oshaweb/owadisp.show_document?p_table=STANDARDS&p_id=10106 accessed March 25, 2008).
5. R. J. ALAIMO and K. P. FIVIZZANI. Qualifications and training of chemical hygiene officers. *Journal of Chemical Health and Safety* **3**(6): 11–13 (1996).
6. J. A. KAUFMAN, R. W. PHIFER, and G. H. WAHL. What every CHO must know. *Journal of Chemical Health and Safety* **4**(3): 10–13, 44 (1997).
7. H. J. ELSTON and W. E. LUTRELL. The CHO's spectrum of responsibilities. *Journal of Chemical Health and Safety* **5**(3): 15–18 (1998).
8. H. J. ELSTON. A whistle-stop tour of OSHA for laboratories: regulations and standards for the Chemical Hygiene Officer. *Journal of Chemical Health and Safety* **5**(6): 9–11 (1998).

QUESTIONS

1. Chemical Hygiene Plans are required by

   (a) The OSHA Lab Standard
   (b) The OSHA Hazard Communication Standard
   (c) The OSHA General Duty Clause
   (d) The EPA

2. Legally, what group of people is *not* bound by the Chemical Hygiene Plan?

   (a) Students who are not employees
   (b) Researchers in chemistry labs
   (c) Researchers who have a B.A./B.S., M.S., or Ph.D. in chemistry
   (d) Faculty who have federal grant money

3. The CHP is a "performance standard," which means that

   (a) OSHA specifies how to prevent exposure to chemicals in labs
   (b) OSHA does not specify how to prevent exposure to chemicals in labs

(c) OSHA requires that employers establish the methods by which employees meet certain requirements to limit exposures

(d) Both (b) and (c)

4. What is *not* necessarily part of a CHP?

   (a) Standard Operating Procedures Involving Use of Hazardous Chemicals

   (b) The CHO duties

   (c) Medical Consultation and Examinations

(d) Designation of Chemical Hygiene Officers

5. Which is generally *not* a duty of a CHO?

   (a) Developing the CHP and appropriate measures for handling chemicals

   (b) Lab inspections

   (c) Understanding legal requirements for labs

   (d) Assisting lab personnel in conducting particularly dangerous experiments

CHAPTER 4

RECOGNIZING LABORATORY HAZARDS: TOXIC SUBSTANCES AND BIOLOGICAL AGENTS

WHAT MAKES something toxic or poisonous? Most chemistry students are not likely to consider ingesting laboratory chemicals, but understanding the toxicology of chemicals can heighten such awareness of avoiding this and also help you understand other means of exposure besides ingestion. Although most chemical exposures, if they occur, are likely to be brief in an academic lab, other industrial settings and some career paths can provide opportunities for the risk of exposure to small amounts of chemicals over longer periods of time. So, it's good to know the full range of the ways chemical might harm your health, including carcinogens and biological agents. Recognizing these toxic hazards is an important skill in laboratory safety.

INTRODUCTORY

4.1.1 Introduction to Toxicology   Presents the basic principles of toxicology, acute and chronic toxicities, metabolism, the language of toxicology, and factors that influence toxicity.

4.1.2 Acute Toxicity   Discusses the types of acutely toxic chemicals that present immediate hazards in chemistry courses and provides examples of each type.

INTERMEDIATE

4.2.1 Chronic Toxicity   Provides a brief introduction to chronic toxicity and why it is important to understand this type of toxicity when working with chemicals over longer periods of time.

ADVANCED

4.3.1 Carcinogens   Presents a brief overview of the causes of cancer and an overview of carcinogens. This section is a companion to the discussion about chronic toxicants found in Section 4.2.1.

4.3.2 Biotransformation, Bioaccumulation, and Elimination of Toxicants   Discusses the fate of toxicants in the body, including their biotransformation, bioaccumulation, and elimination.

4.3.3 Biological Hazards and Biosafety   Explains the hazards of biological agents, where and how they might be encountered, and some general approaches to prevent exposures.

4.1.1

INTRODUCTION TO TOXICOLOGY

Preview   You will learn about the basic principles of toxicology, acute and chronic toxicities, metabolism, the language of toxicology, and factors that influence toxicity.

> *What is it that is not poison? All things are poison and nothing is without poison. It is the dose only that makes a thing not a poison.*
>
> Paracelsus[1]

INCIDENT 4.1.1.1   CHEMICAL DERMATITIS[2]

A student carried out a "new" laboratory preparation called the "Dipoles Experiment." The student synthesized in three separate phases over a period of one week a chemical similar to a commercially used optical whitener. The student developed an acute dermatitis on her face, chest, arms, and thighs. Six other students, who had also performed the same experiment, developed varying degrees of dermatitis. Skin testing showed that the five of these seven students had positive response to CBPH (chlorobenzaldehyde phenylhydrazone), a chemical used in the last step of the experiment.

What lessons can be learned from this incident?

Prelude

Section 4.1.1 presents an introduction to the basic science and terminology of toxicology. Section 4.1.2 discusses toxic hazards (acute toxicants) about which you should learn fairly early in your academic career. Section 4.2.1 is mostly about chronic toxicants, which pose less of an immediate hazard in academic laboratories, and Section 4.3.1 discusses carcinogens. This thread of toxicology will finish in Chapter 4 with Sections 4.3.2 about the biotransformation of toxicants and Section 4.3.3 about biological hazards in chemistry labs. Responding to emergencies involving toxic substances is presented in Chapter 2.

What Is Toxicology and Why Do I Need to Know About It?

Toxicology is the study of the adverse systemic effects of chemicals. Some dictionaries define toxicology as the study of poisoning. While poisoning is studied in toxicology, toxicology encompasses more than poisons and most chemicals are not classified as poisons (see more below). *Toxicity* is the ability of a chemical to cause systemic changes or harm to a living organism by damaging an organ or adversely modifying a biochemical system or process. "Systemic" means that the effect is produced remotely from the site of exposure because the toxic material moves from the site of attack by absorption into the bloodstream. All chemicals have potential toxic properties depending on the dose rate, and it is important you have a basic understanding of toxicology and its terms so you will be able to understand the hazardous properties of chemicals.

A chemical producing toxic effects is termed a *toxicant*, or in the case of a naturally occurring toxic compound, it may be called a *toxin*. It is common to use the word toxin mistakenly to describe

---

Laboratory Safety for Chemistry Students, by Robert H. Hill, Jr. and David C. Finster

all toxic materials. Toxicant is an all-encompassing term that includes all toxic chemicals, while toxin refers only to a subset of toxicants produced as natural products—such as *botulinum toxin*, one of the most toxic chemicals known to humans.

It's the Dose that Matters—What Was Your Level of Exposure?

Once a toxic chemical is in contact with or enters the body, it can exert its effects. Whether a chemical produces any toxic effects is determined by the dose received over a period of time. This is called the dose rate. The quote by 16th century physician Paracelsus at the beginning of this section embraces the basic concept of toxicology that is commonly phrased as "the dose makes the poison." Whether a chemical exerts its toxic effects depends on the amount or dose of the chemical that is received by the organism. Another critical factor that determines if a chemical is toxic is the time period over which the dose was received.

The dose–response relationship is illustrated by the graph in Figure 4.1.1.1. In this figure it can be seen that up to a certain dose there is no toxic response observed. At a certain point, called a *threshold*, toxic effects begin to be observed and then as the dose increases the toxic effects become more pronounced up to a maximum.

Acute toxicity is the ability of a chemical to do systemic damage to an organ(s) or a biological system or process with a single dose. The result is often, but not always, very rapid and produces immediate effects. The time frame for acute toxicity is usually considered to be seconds, minutes, hours, or days. Examples of acutely toxic chemicals are shown in Table 4.1.1.1.

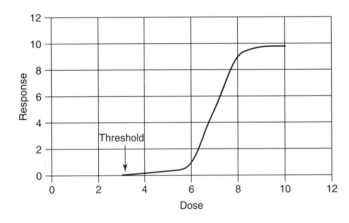

FIGURE 4.1.1.1   Dose–Response Curve. This is a generic dose–response curve for a toxic substance. The threshold is the dose below which no effect is detected. The x-axis can range from micrograms (for very toxic substances) to grams (for less toxic substances). The y-axis is the percentage of a population that exhibits some illness or death.

TABLE 4.1.1.1   Acute and Chronic Toxicities for Selected Chemical Toxicants Encountered in Chemistry Laboratories[a]

Toxic class	Acutely toxic chemicals (see Section 4.1.2)	Chronically toxic chemicals (see Section 4.2.1)
Irritants	Al_2O_3, CS_2, EDTA	
Sensitizers/allergens	Formaldehyde, diazomethane, aluminum trichloride	
Asphyxiants	CO, N_2, He, CO_2, CH_4, H_2	
Teratogens	Ethanol, acrylonitrile, nitrogen mustard	
Carcinogens		Ethylene oxide, Cr(IV) compounds, benzene, chloroform, vinyl chloride
Organ toxicants	Ethanol, aniline, toluene, DMF, CCl_4	Hg, hexane, phenol
Neurotoxins	Dimethylmercury, acetone, CS_2, ethanol	Acrylamide, lead, methyl iodide
Poisons	Cyanide, osmium tetroxide, phosphorus, sodium azide	Arsenic compounds, phosphorus, osmium tetroxide

[a]Cells that are blank indicate the absence of known chemicals in those classes.

Chronic toxicity is the ability of a chemical to do systemic damage to an organ(s) or a biological system or process with multiple smaller doses over a prolonged period of time, usually months or years. In relative terms the dose for acute toxicity is much higher than the dose for chronic toxicity in which small continuous or repeated doses are received. A chemical with chronic toxicity may be an accumulative poison in situations where the body retains the small doses until it accumulates enough of the chemical to exhibit its toxic effects. Examples of chemicals that exhibit chronic toxicity are shown in Table 4.1.1.1.

It is important to distinguish these because acute toxic effects are not predictors of chronic toxic effects—that is, the effects of acute toxicity or chronic toxicity may be different because they may affect different organ systems or different biological systems or processes.

Metabolism—The Body's Defense Against Toxic Chemicals

Fortunately, your body has considerable ability to deal with toxic chemicals through a process in which chemicals are *metabolized*. Through metabolism, toxicants are changed to nontoxic, less toxic, or perhaps even more toxic products that are excreted from the body. Metabolism is a *dynamic* process that can be accomplished in many ways. The word dynamic means that it is always ongoing, so if you get small doses of many chemicals, they may be metabolized and either used in the body for some purpose or excreted from the body before they can exert any significant toxic effect. This is why you might be able to consume some toxic chemicals such as ethyl alcohol in beverages or drugs like acetaminophen but only experience some short-term immediate effects. In that same way, we may be exposed to very small amounts of toxic chemicals in our food or water, but because the amounts are very small the body's metabolic system can eliminate these chemicals without causing any adverse effect. If it were not so, we would likely experience problems from the many chemicals that enter our bodies every day. Metabolism takes place primarily in the liver, so if your liver is damaged, it may affect your ability to metabolize chemicals, including drugs. The process of toxicant metabolism is discussed further in Section 4.3.2.

Some chemicals are resistant to metabolism, and instead of being excreted, they may be stored or accumulated in the body. Many of us have trace amounts of certain chlorinated chemicals that were used frequently many years ago. (DDT, a long-lasting pesticide, is an example.) These substances remain in our environment even years later, and when ingested they are stored in our fat tissue. While these chemicals may exert some toxic effects at higher doses, they are believed to be relatively nontoxic in the amounts stored in our fat. This process is known as *bioaccumulation*. (Not all scientists agree about this, however; see *Special Topic 4.3.2.1* Controversy in Science—Epidemiological Challenges and Degrees of Certainty.)

It is also possible for chemicals to become more toxic when metabolized. A good example of this is ethylene glycol, the major component of antifreeze. When ethylene glycol is ingested it is metabolized to products that accumulate and cause toxic effects, including metabolic acidosis (a condition of excess acid in the body) and severe damage to the kidneys and brain from the formation of crystals of calcium oxalate. Death can occur from ethylene glycol poisoning.

The Language of Toxicologists—Lethal Doses (LD$_{50}$ and LC$_{50}$ Values)

Over many years scientists and physicians have gathered information about the toxic effects of chemicals from reports of accidental poisonings, homicides, suicides, and occupational exposures to humans. However, this experience and information is very limited and subject to limitations of estimations of dosages received. A larger amount of information about toxicity comes from experimental animal toxicological studies. Toxicologists often evaluate the toxicity of chemicals through testing with laboratory animals, such as rats or mice. In their studies they have developed some important terms used to evaluate the relative risk of a chemical.

LD$_{50}$ or "lethal dose fifty" is a term used to describe the acute toxicity of a chemical and is usually expressed in milligrams (mg) of chemical given per kilogram (kg) of the animal. By expressing doses in units like mg/kg it makes it easier to compare variations in dosages between species. LD$_{50}$ is the single dose that is lethal to 50% of the experimental animals dosed. In practice, this is an extrapolated

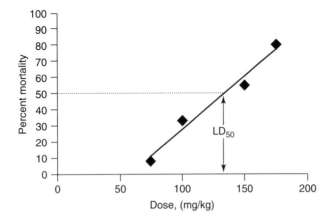

FIGURE 4.1.1.2 Dose–Response Curve to Estimate an LD_{50}. This graph indicates the methodology for the determination of an LD_{50}. Several data points are measured above and below the LD_{50} value, which is then determined graphically. The data are not always linear, and the slopes of the plots can be steep or shallow. What are the safety implications of a steep, or shallow, curve?

or estimated dose, rather than a measured one. Tests are conducted with a range of doses so that at each dose some animals survive but others do not. For example, at one dose 90% of the animals dosed survive, while at another dose 70% survive, at another dose 25% survive, and at another dose 10% survive. The LD_{50} is then estimated from a graph of the data points for these experiments (see Figure 4.1.1.2). The LD_{50} is used primarily in oral or dermal dosing experiments. Figure 4.1.1.2 is a generic plot; some chemicals have steep slopes and some chemicals have shallow slopes.

If a chemical is tested through inhalation, the term for toxicity is LC_{50} or "lethal concentration fifty." The LC_{50} is the lethal dose that causes death in 50% of the tested animal population. As with an LD_{50}, this is an extrapolated dose and is reported in terms of the air concentration of the toxic chemical, and is often expressed as parts per million (ppm) or mg/m^3 or $\mu g/L$. (See Section 6.2.2 to learn about ppm.) Determining LC_{50} values is expensive since it requires extra equipment (inhalation chambers and careful measurements of air concentrations) not required in measuring lethal doses by oral, dermal, or other routes. Since LD_{50} values are much easier and less expensive to determine than LC_{50} values, most of the information we have in experimental toxicology is reported in terms of LD_{50} values. Since keeping smaller animals is less expensive, rats or mice are often used in toxicity studies.

Extremely Toxic Chemicals — The Poisons

What is a poison? Are all chemicals poisons? There is a wide variation in the understanding of the term poison, especially in the media and literature, and even among scientists. This variation has resulted in a misunderstanding of the term poison. Perhaps this is because we commonly use the term "poisoned" or "poisoning" to describe incidents where people have taken a dose of a chemical into their body (intentionally or unintentionally) that resulted in serious illness or even death. Nevertheless, even though someone is "poisoned" by a chemical does not mean that the chemical was a poison (see *Special Topic 4.1.1* Our Understanding of a Poison—A Little on the Cloudy Side).

SPECIAL TOPIC *4.1.1.1*

OUR UNDERSTANDING OF A POISON—A LITTLE ON THE CLOUDY SIDE

Let's examine some definitions of "poison."

- *The American Heritage College Dictionary* defines poison as "a substance that causes injury, illness, or death, esp. by chemical means."[3]
- *Taber's Cyclopedic Medical Dictionary* describes a poison as "any substance which, taken into the body whether by ingestion, inhalation, injection, absorption, etc., produces a deleterious, injurious, or lethal effect. Virtually any substance can be poisonous if consumed in sufficient quantity so that the term poison more often implies an excessive degree of dosage rather than any list of substances."[4]

- The United States Court of Appeals' decision in a criminal poisoning defines a poison as "any substance introduced into the body by any means which by its chemical action is capable of causing death."[5]

- *Casarett & Doull's Toxicology* comments that "one could define a poison as any agent capable of producing a deleterious response in a biological system, seriously injuring function or producing death. This is not however a useful working definition for the very simple reason that virtually every known chemical has the potential to produce injury or death if it is present in a sufficient amount."[6] However, the authors do not go on to provide a better working definition of a poison.

- M. Alice Ottoboni in *The Dose Makes the Poison* explains that "a chemical that causes illness or death is often referred to as a poison. This concept is erroneous and has resulted in a great deal of confusion in the public mind about the nature of the toxic action of chemicals. Poisons are chemicals that produce illness or death when taken in very small quantities. Legally, a poison is defined as a chemical that has an LD_{50} of 50 milligrams (mg), or less, of chemical per kilogram (kg) of body weight."[7] The legal definition comes from a child protection act.[8]

In our judgment, it is not useful to consider everything that can cause adverse effects as a poison. Imagine if we labeled salt or milk as poisons! This would not only be inappropriate but dangerous. How would we know what is "really poisonous"? Defining a poison is best described by Ottoboni and Poison Control Centers using LD_{50} and LC_{50} values. You will also learn later that a new system of labeling hazardous chemicals, known as GHS (see Sections 3.2.1 and 6.2.1) identifies the most toxic chemicals using the terms "Danger" and "Fatal if ingested" along with the "Skull and Crossbones" symbol—they do not use the word poison.

A poison is a chemical that, when taken or being exposed to it in very small quantities, can cause severe illness or death. In this book we define a poison using the language of toxicologists using LD_{50} and LC_{50} values. A poison ("highly toxic" chemical) is defined as a chemical having[7,8]

- Oral $LD_{50} \leq 50$ mg/kg
- Dermal $LD_{50} \leq 200$ mg/kg
- Inhalation $LC_{50} \leq 200$ ppm or 2 mg/L (dust)

All poisons are toxic chemicals, but all toxic chemicals are not poisons. This does not mean that you can't be poisoned by other chemicals, because if the dose is great enough over a period of time, you can be poisoned, have serious effects, or even die from chemicals that are toxic, but not poisons. However, poisons are extremely toxic, so much so that they merit special attention, usually with the label containing "Poison," "Danger," and the old "Skull and Crossbones" symbol. Relatively few chemicals are poisons.

While there is often focus on the poisonous effects of synthetic chemicals, the most poisonous chemicals are bacterial toxins, marine toxins, fungal toxins, mycotoxins, venoms (from snakes, insects, arachnids, and other animals), and some plant toxins. Table 4.1.1.2 shows the LD_{50} values for toxicants including poisons and nonpoisons. This can help you judge the relative acute toxicity of a chemical, but it does *not* indicate chronic or long-term toxicity. Whether a substance is fatal, toxic, or harmful also depends on the dose, of course. There is also no relationship between toxicity and whether a compound is "simple" or "complex." The mechanism of toxicity varies greatly, and structure–activity relationships are not simple (see *Chemical Connection 4.1.1.1* Structure–Activity Relationships).

CHEMICAL CONNECTION *4.1.1.1*

STRUCTURE–ACTIVITY RELATIONSHIPS

You may have already learned that there are many clear relationships between physical properties and chemical structures. For example, there is a relationship of boiling point with molar mass in a series of structurally similar organic compounds such as simple aromatic hydrocarbons: benzene (b.p. 80 °C), toluene (b.p. 111 °C), and *p*-xylene (b.p. 138 °C).

TABLE 4.1.1.2   Poisons and Nonpoisons

Poison	LD_{50} (mg/kg, oral), rat	Natural	Synthetic
Botulinum toxin	0.000001	X	
Aflatoxin B1	0.048	X	
Tetrodotoxin	0.3 (mice)	X	
Arsenic trioxide	1.5		X
Phosphorus	3		X
Sodium cyanide	6.5		X
Hydrogen cyanide	10		X
Trimethyltin chloride	13		X
Osmium tetroxide	14		X
Acrolein	42		X
Sodium azide	27		X
Nicotine	50	X	
Toxic, but not poison			
Ammonium dichromate	67.5		X
Barium chloride	118		X
Caffeine	192	X	
Copper nitrate	940		X
Hydrogen peroxide (8–20%)	1518		X
Sodium chloride	3,000	X	
Calcium hydroxide	7,340		X
Ethanol	10,600	X	
Vitamin C	11,900	X	

When assessing the risk of chemicals, it would be helpful if we could identify chemicals that have similar structure and biological activity (toxic) relationships, often called *structure–activity relationships* or *SARs*. For example, if we knew that benzene had a particular toxicity, we might predict that toluene or xylene had a similar effect. Unfortunately, this is a much more complicated matter than just looking at boiling points. Benzene causes aplastic anemia and leukemia, while toluene and xylene do not exhibit these adverse properties but do have other toxic properties of their own. Why?

Interactions between chemicals and biological systems within humans or animals are almost always complex. Chemicals' specific structures, even when very similar, often result in their being metabolized completely differently in the body. Furthermore, metabolism is often accomplished by enzymes, and enzymes often require very specific chemical structures (the "key and lock" concept). This means that very small, subtle changes in structure can have dramatic effects on biological activity. As a result, biological activity is often exceedingly difficult to predict. In fact, the SAR paradox is that not all similar compounds have similar biological activities. Nevertheless, drug companies continue to find ways to explore improvements in therapeutic properties while seeking to avoid toxic properties.

Pharmaceutical companies and other researchers trying to identify new drugs for the marketplace often use computational chemistry *quantitative structure–activity relationships* (QSARs) studies to examine the relationship of activity and small structural changes. This is a complicated, complex, and expensive method but this is almost always cheaper than using test compounds in whole animals and has been successful in limited cases. It continues to remain a highly active area of research.

Contrary to popular understanding in our culture, no generalization whatsoever can be drawn about the toxicity of a substance based on whether it is naturally occurring or synthetic (man-made). The phrase "naturally occurring" should not imply any particular level of relative safety nor should "synthetic" imply a relative hazard. In fact, the most toxic substances known are naturally occurring substances. A good example of this is botulinum toxin, which has an estimated lethal dose for humans of about 0.000001 mg/kg. Botulinum toxin is produced by *Clostridium botulinum* bacteria and is sometimes produced when some alkaline or low-acidity foods are improperly prepared and canned. The result can be poisoning and the fatality rate in known cases is 5–10%. Paradoxically, it is also the active ingredient in the cosmetic treatment Botox®.

Multiple Factors Influence Toxicity

Besides the dose of the substance, there are many other factors that can affect the degree of toxic effect a substance may exhibit. These are summarized in Figure 4.1.1.3.

The *route of exposure* can have a profound effect on a toxic chemical's effect on the animal or person (see Section 3.1.1). In experimental toxicology, usually the oral route is used for comparisons. Doses by the oral route may be given by feeding the toxic chemical in food or water or it may be given by gavage to animals in which the toxic chemical or its dilute form is inserted directly via a tube into the animal's stomach. Examples of the variation in toxicity with route are given in Table 4.1.1.3. A toxic chemical may be given by the dermal route on the skin by painting or applying a patch with the chemical. Another route is via inhalation in which an animal is exposed to a chemical at a given concentration in the air over a specified period of time.

Table 4.1.1.3 shows that generally the oral route is more effective in delivering a toxic dose than the dermal route. It is more difficult to compare the oral or skin routes with the inhalation route since the units are so different. To find out how much an animal received by the inhalation route requires more information, including how much of the airborne concentration was inhaled by the animal over a period of time. This varies by animal size and physiology of the animal as well as the activity of the animal. This is complicated and it is not easy to compare inhalation dose with other dose routes.

The *species* affected by the toxicant is an important factor in toxicity. While humans are usually of primary concern to us with regard to toxicity, as noted above data about human toxicity are relatively

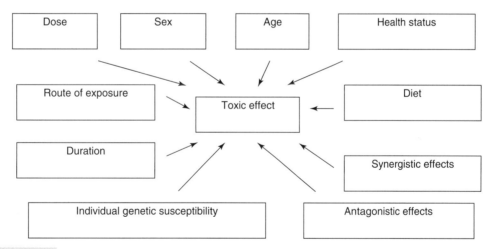

FIGURE 4.1.1.3   Factors Affecting Toxicity. Many factors determine the toxic effect of a substance for a particular individual. It is not surprising that not all people exhibit the same effect from a particular exposure.

TABLE 4.1.1.3   Variability of Toxicity by Route

Substance	Oral LD$_{50}$	Skin LD$_{50}$	Inhalation LC$_{50}$
Acrylonitrile	78 mg/kg (rat)	148 mg/kg (rat)	333 ppm/4 h (rat)
Methanol	14,200 mg/kg (rabbit)	15,800 mg/kg (rabbit)	81,000 ppm/14 h (rabbit)
Aniline	250 mg/kg (rat)	1,400 mg/kg (rat)	250 ppm/1 h (rat)
Triiodomethane	355 mg/kg (rat)	1184 mg/kg (rat)	183 ppm/7 h (rat)

TABLE 4.1.1.4 Variability of Toxicity by Species

Substance	Species	Oral LD_{50}
TCDD	Hamster	1157 μg/kg
TCDD	Mouse	114 μg/kg
TCDD	Monkey	2 μg/kg
TCDD	Guinea pig	0.5 μg/kg

scarce and we rely on animal toxicity as a crude guide. However, experimental animals are only approximations (perhaps even poor ones in some cases) for potential toxic effects in humans. Among animal species, for example, mice, rats, guinea pigs, rabbits, and monkeys, there can be considerable variability in LD_{50} doses and toxic effects (see Table 4.1.1.4). Chemicals may cause an effect in one or two species but not in other species. From Table 4.1.1.4 you will see that the guinea pig is more than 2000 times more sensitive to TCDD than is the hamster. TCDD (2,3,7,8-tetrachlorodibenzo-*p*-dioxin) is infamous as the dioxin found at various contamination sites at Seveso, Italy (1976), Love Canal, New York (1979), and Times Beach, Missouri (1985). It was also the toxicant used in the nonfatal poisoning of President Viktor Yushencko of Ukraine (2004).

There is no one animal model that best approximates potential hazards to humans. Toxicologists consider information from animal experiments using multiple species when possible. Even then, it is possible that humans can experience different toxic responses to chemicals than observed in animals and this is particularly true of long-term effects like cancer. Nevertheless, experience has shown that frequently animal toxicity testing can predict effects upon humans. If this were not so, we would have not been able to develop today's drugs. It is upon good scientific judgment among toxicologists and pharmacologists that we can make reasonable approximations of the effects of chemicals on humans.

Age can be a factor in toxicity. Generally, the very young and very old are more susceptible to the toxic effects of chemicals. The biological systems of children are not fully developed and these immature systems can leave them more vulnerable to toxicity. Old age brings about breakdowns or weaknesses in mature protective systems. This also relates to the *health status* of the individual. As we age we develop more chronic health conditions and diseases, such as diabetes, heart conditions, cancer, or arthritis. These conditions weaken and compromise our overall health status, making us more susceptible to exposures to chemical or biological agents in our environment. *Diet* has also been shown to have an effect on the toxicity of substances in animal studies.

Sex can result in clearly recognized differences in toxicity. The most important area of concern related to sex is that of toxic effects upon the reproductive process. While some facts are known about this area, there has been relatively little research in this area, except with respect to chemicals that interrupt the conception process (i.e., birth control). However, it has been recognized that some chemicals cause effects in the reproductive process. For example, dibromochloropropane causes sterility in men exposed in occupational settings.

Genetics plays a role in individual susceptibility to chemicals, but its role in causing adverse effects in humans is often poorly understood. Over the years there has been debate about whether exposure to chemicals in our environment caused certain adverse events or whether the events were actually related to our genetic makeup. And it turns out that these two views are very difficult to distinguish since both are plausible explanations of adverse events. It has long been proposed that chemicals can act as triggers to set off adverse effects in "genetically susceptible" people. This is particularly true in regard to cancer. Over the years we have come to recognize that genetics plays a significant role in our own personal health status and also plays a significant role in each person's susceptibility to toxic chemicals.

Almost all information about toxicants also assumes that a given substance is acting alone. This is sometimes not the case. When two substances have a greater toxic effect than the simple "addition" of the effects of each alone, we say that there is a *synergistic effect*. In essence, the whole is greater than the sum of the parts. It is also possible that there can be an *antagonistic effect*, where the combined toxic effect is less than expected from the components of a mixture of toxic substances. Physicians and pharmacologists try to understand these two effects as best as possible when prescribing or administering medications.

Working with Toxicants

Given the wide variety of toxicants and routes of exposure, and considering both immediate and delayed effects, it is extremely important to be aware of the toxicity of the chemicals used in labs. In the event of a known or suspected toxic exposure, it is best to contact the lab instructor or supervisor immediately to determine the best course of action.

RAMP

- *Recognize* what chemicals are toxic.
- *Assess* the level of risk based on likelihood of exposure by various routes.
- *Minimize* the risk by preventing exposure using good lab practices and personal protective equipment such as gloves and using chemical hoods for inhalation hazards.
- *Prepare* for emergencies by knowing the toxic hazards of the chemicals that you are using.

References

1. M. A. OTTOBONI. *The Dose Makes the Poison: A Plain-Language Guide to Toxicology*, 2nd edition, Van Nostrand Reinhold, New York, 1997, p. 31.
2. D. F. BARRETT. Outbreak of dermatitis among chemistry students—due to alpha-chlorobenzaldehyde phenylhydrazone. *Transactions St. John's Hospital Dermatological Society* 59:52–55 (1973).
3. *The American Heritage Dictionary*, 2nd college edition, Houghton Mifflin Company, Boston, 1985.
4. Clarence Wilbur Taber. *Taber's Cyclopedic Medical Dictionary*, 10th edition, F. A. Davis Company, Philadelphia, 1968.
5. United States Court of Appeals. Ninth Circuit, No. 97–16830, *Stanton v. Benzler*, decided June 17, 1998; available at http://bulk.resource.org/courts.gov/c/F3/146/146.F3d.726.97–16830.html (accessed July 28, 2009).
6. CURTIS D. KLAASSEN (editor). *Casarett & Doull's Toxicology: The Basic Science of Poisons*, 6th edition, McGraw-Hill, Medical Publishing Division, New York, 2001, p. 13.
7. M. A. OTTOBONI. *The Dose Makes the Poison: A Plain-Language Guide to Toxicology*, 2nd edition, Van Nostrand Reinhold, New York, 1997, pp. 26–27. From Federal Hazardous Substances Act (Public Law 86–613; July 12, 1960). See Consumer Product Safety Commission at http://www.cpsc.gov/BUSINFO/fhsa.pdf (accessed September 8, 2009).
8. Poison Control Centers. Poison Control Centers and Emergency Facilities, Pesticide Safety, Pesticide Poisoning; available at http://ipm.ncsu.edu/peach/poisoncontrol.pdf (accessed July 31, 2009).

QUESTIONS

1. *Toxicity* refers to the capability of a substance to
   - (a) Cause systemic harm to a living organism
   - (b) Damage an organ in a living organism
   - (c) Adversely modify a biochemical system
   - (d) All of the above

2. A *toxin* is a toxicant that
   - (a) Is a single, pure chemical instead of a mixture
   - (b) Is a naturally occurring toxic compound
   - (c) Is a part of a chemical warfare agent
   - (d) Is a low molecular weight molecule

3. With regard to toxic dose, the *threshold* is the level at which
   - (a) Permanent damage is done to an organism
   - (b) A person starts to feel pain or discomfort
   - (c) Toxic effects begin to occur
   - (d) Some antidote can have an effect

4. Acute toxicants are characterized by what features?
   - (a) Single, relatively high dose
   - (b) Single, relatively low dose
   - (c) Long-term exposures of relatively high doses
   - (d) Long-term exposures of relatively low doses

5. Chronic toxicants are characterized by what features?
   - (a) Single, relatively high dose
   - (b) Single, relatively low dose
   - (c) Long-term exposures of relatively high doses
   - (d) Long-term exposures of relative low doses

6. The process of metabolism explains why
   - (a) Almost no chemicals have a toxic effect on humans
   - (b) Many toxicants can accumulate in the body
   - (c) Humans can ingest small amounts of toxic substances with little or no effect
   - (d) Acute exposures to toxicants are usually less harmful than chronic exposures

7. When molecules are metabolized they
   - (a) Usually become less toxic molecules
   - (b) Usually become more toxic molecules
   - (c) Can become either more toxic or less toxic
   - (d) Always become less toxic

8. An LD_{50} represents the
   - (a) Amount of a substance that kills 50% of a test population
   - (b) Concentration of a substance that kills 50% of a test animal population
   - (c) Amount of a substance that is known to have killed at least 50 people due to accidental exposure

(d) Concentration of a substance that is known to have killed at least 50 people due to accidental exposure

9. The methodology used to determine an LD_{50} involves

(a) Slowly increasing the dose until 50% of a test population dies

(b) Slowly increasing the dose until 5% of a test population dies and then multiplying the dose by 10

(c) Dosing animals to get a wide range of death rates and interpolating the data to 50%

(d) Finding the minimum dose that kills 100% of a population and then dividing by 2

10. What dimension is *not* used to represent an LC_{50}?

(a) ppm

(b) mg/m^3

(c) mg/mL

(d) µg/L

11. Poisons are compounds with an LD_{50}

(a) ≤ 50 mg

(b) ≤ 50 mg/kg

(c) ≥ 50 mg

(d) ≥ 50 mg/kg

12. What percentage of compounds can be classified as poisons?

(a) A small percentage

(b) About 25%

(c) About 50%

(d) Almost 80%

13. The toxicity values shown in Table 4.1.1.2 cover a range of how many orders of magnitude?

(a) 3

(b) 5

(c) 8

(d) 11

14. Which statement is *true*?

(a) In general, synthetic compounds are more toxic than naturally occurring compounds.

(b) In general, naturally occurring compounds are more toxic than synthetic compounds.

(c) There is no relationship between the source of a compound and its toxicity.

(d) More complex molecules tend to be more toxic.

15. Table 4.1.1.3 shows data that allows one to conclude which of the following?

I. The toxic dose is largely independent of route of exposure.

II. Most compounds have remarkably similar toxic doses.

III. LC_{50} data is often reported with different time exposures.

IV. Oral LD_{50} doses are generally lower than dermal (skin) LD_{50} doses.

(a) I and III

(b) II and III

(c) I and II

(d) III and IV

16. Table 4.1.1.4 shows which of the following?

I. Most animal species react about the same to a particular dose (per kg) of a toxicant.

II. LD_{50} values vary widely according to species.

III. Generally, the larger the animal, the lower the LD_{50}.

IV. Generally, the larger the animal, the higher the LD_{50}.

(a) I and III

(b) II and III

(c) II and IV

(d) Only II

17. To extrapolate animal testing to humans, it has been found that

(a) Using monkeys provides very similar toxicity results as those shown for humans.

(b) There is no single animal species that simulates human responses well, but testing several species generally leads to reasonably good estimates of human toxicity.

(c) In general, other animals are mostly more sensitive to toxicants than are humans.

(d) In general, other animals are mostly less sensitive to toxicants than are humans.

18. Figure 4.1.1.3 shows that the toxic effect of a substance on an individual depends

(a) On about 10 factors

(b) Almost entirely on factors that are beyond the control of the individual

(c) On factors that are remarkably consistent in a population of individuals

(d) Mostly on two main factors

19. Figure 4.1.1.1 shows that

(a) the dose–response curve is linear enough to generally assume that an LD_{50} is 10 times smaller than an LD_5

(b) Most compounds have a threshold nearly equally to zero

(c) The dose–response curve is not linear over a wide range of doses

(d) Generally, higher doses ironically lead to smaller responses

20. Synergistic effects explain which of the following?

I. The effect of being exposed to two toxicants is sometimes less than the predicted "sum" of the exposures.

II. The effect of being exposed to two toxicants is sometimes more than the predicted "sum" of the exposures.

III. There can be variation in the response of a group of people to the same dose of a toxicant.

IV. Many people have a delayed reaction to a toxicant.

(a) I and III

(b) II and III

(c) II and IV

(d) IV

21. Antagonistic effects explain which of the following?

I. The effect of being exposed to two toxicants is sometimes less than the predicted "sum" of the exposures.

II. The effect of being exposed to two toxicants is sometimes more than the predicted "sum" of the exposures.

III. There can be variation in the response of a group of people to the same dose of a toxicant.

IV. Many people have a delayed reaction to a toxicant.

   (a) I and III
   (b) II and III
   (c) II and IV
   (d) IV

22. Which is *true*?

   (a) Synthetic compounds are more toxic than naturally occurring compounds.

   (b) Synthetic compounds are less toxic than naturally occurring compounds.

   (c) We can draw no general conclusion about the toxicity of a compound based on whether is it synthetic or naturally occurring.

   (d) If a naturally occurring compound is also prepared synthetically, it will *not* have the same toxicity behavior.

4.1.2

ACUTE TOXICITY

Preview Types of acutely toxic chemicals that present immediate hazards in chemistry courses are explained and examples of each type are given.

Salt is white and pure—there is something holy in salt.

Nathaniel Hawthorne[1]

INCIDENT 4.1.2.1 AZIDE IN THE TEA[2]

Five technicians working in a laboratory ate lunch in a small nearby storeroom. After drinking cups of tea, three of them felt ill with a "fuzzy" head and pounding heart. Another technician drank two cups and developed violent cramp-like pains in his chest, and he was taken to the hospital believing he was suffering from a heart condition. A fifth technician drank only half of a cup of tea, but also felt ill. Finally, one of the technicians remembered that she had filled the tea kettle with laboratory distilled water that had been treated with sodium azide to prevent growth of bacteria, a common practice in some laboratories. All technicians recovered.

What lessons can be learned from this incident?

Toxic Chemicals—A Plethora of Effects

This section describes acutely toxic chemicals that, upon significant exposure, can produce immediate toxic effects. Because these immediate effects are easier to recognize and study, we know much more about the effects of acute toxicants. Nevertheless, we may not know *all* of the toxic effects of chemicals, particularly for long-term or chronic effects. In Section 4.1.1 you learned that the effects that chemicals exhibit depend on many factors, including dose and time. Salt, that "holy" substance praised by Hawthorne, is toxic with a LD_{50} of 3000 mg/kg, and children have died of salt poisoning. But salt is required in small doses for our well-being and we often use it liberally in our foods. Caffeine is toxic and has an oral LD_{50} in rats of 192 mg/kg., but you would have to drink about 100 cups of coffee to receive a lethal dose in a few hours. As discussed in Section 4.1.1, "the dose makes the poison." Some chemicals may have more than one toxic effect, and the mechanisms of these effects vary markedly.

Our knowledge of acute toxicity is based on what we have learned over the last few centuries, and the information about toxic effects is relatively scant in light of the millions of compounds known to exist. Not more than a few thousand chemicals have been investigated to any significant degree. It is not unusual to suddenly discover toxic properties of some chemical that were not previously recognized simply because it had not been tested. For example, around the mid-1970s vinyl chloride (Figure 4.1.2.1), a chemical used for many years to make polyvinyl chloride (PVC) polymer, was recognized as the cause of a rare cancer, known as angiosarcoma. Sometimes toxic compounds are discovered by accident and sometimes through systematic testing. It is not surprising to learn that a compound once believed to be "safe" in the absence of any evidence to the contrary is discovered upon investigation to be toxic in a previously unrecognized way. For example, diacetyl, a component of butter flavoring in microwave popcorn, was recently found to cause a rare severe lung disease called bronchiolitis obliterans that can be fatal.

Laboratory Safety for Chemistry Students, by Robert H. Hill, Jr. and David C. Finster
Copyright © 2010 John Wiley & Sons, Inc.

4-15

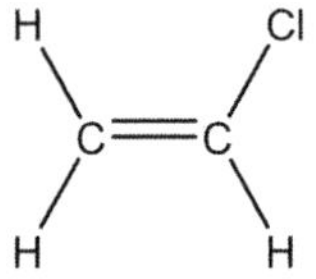

FIGURE 4.1.2.1 Structure of Vinyl Chloride. The EPA has classified vinyl chloride as a Group A, human carcinogen.

While some chemicals can have adverse effects, we need to avoid becoming *chemophobic* (having an abnormal fear of chemicals) since our world is made up of chemicals, and even you are made of chemicals. When handling chemicals, it is prudent, and always possible, to prevent and minimize chemical exposures. In today's laboratories we can and do handle many toxic chemicals safely and the methods and procedures to do so have evolved over many years of experience. If you follow these prudent procedures, you will greatly reduce risks of having toxic effects from chemical exposures, even when you don't know all of a chemical's possible toxic effects.

Acutely Toxic Chemicals Found in Laboratories

Table 4.1.2.1 lists several types of acute toxicants and examples of each. You will find that some acute toxicants fall neatly into one type, but you will find many that exhibit more than one type of toxicity and have multiple hazardous properties. For example, phenol is moderately toxic (with an LD_{50}, mouse, oral, of 270 mg/kg), is an eye and skin irritant, exhibits mutagenic and teratogenic effects, affects the central nervous system, is corrosive (even as a solid), and is flammable!

It's Really Irritating How Sensitive You Can Be!

Irritants are noncorrosive chemicals that cause reversible inflammation (redness, thickening, rash, scaling, and blistering) at the point of contact with skin. They can cause itching and the effects are displeasing, particularly from a cosmetic perspective. *Contact dermatitis* is the single most prevalent occupational skin disease in the United States[3] since irritants are very common. In tests, about one-third of 1200 chemicals were found to be irritants, by far the leading adverse effect for toxicants. That means your chances of encountering an irritant in the laboratory are pretty good.

The two types of contact dermatitis are irritant and allergic contact dermatitis, and they are clinically indistinguishable. Irritants involve a nonimmune response caused by the direct action of the irritant on the skin. Related, but different, is allergic contact dermatitis that is caused by a delayed hypersensitivity reaction of the immune system (see allergens below). Many chemicals have irritant properties and some examples are shown in Table 4.1.2.1. Dilute concentrations of corrosives can have

TABLE 4.1.2.1 Examples of Acute Toxicants Found in Chemistry Laboratories

Acute Toxicants	Examples
Irritants	Al_2O_3, CS_2, EDTA, acetone, heptane, sodium carbonate, tetrahydrofuran, ethyl acetate, isopropanol
Sensitizers/allergens (classification of sensitizers as acute toxicants based on reactions after sensitization has occurred)	Formaldehyde, toluene diisocyanate, 1-fluoro-2,4-dinitrobenzene, diazomethane, bichromates, dicyclohexyldicarbodiimide, latex rubber
Asphyxiants	CO, N_2, He, CO_2, methane, H_2
Teratogens, fetotoxicants, developmental toxicants	Ethanol, ethylene oxide, methylmercury, lead compounds
Organ toxicants	Carbon tetrachloride, toluene, vinyl chloride, ethanol, arsenic, caffeine, aflatoxin, chlorine, arsenic compounds
Neurotoxicants	Methylmercury, mercury, lead, nicotine, acetone, carbon disulfide
Poisons	Cyanides, phosgene, Cl_2, osmium tetroxide, arsenic compounds

irritant effects, but strongly corrosive chemicals or reactive chemicals can produce chemical burns and tissue destruction at the point of contact (see Sections 5.1.1 and 5.2.1).

Sensitizers or *allergens* are chemicals that produce their effects by evoking an adverse response in the body's immune system called *hypersensitivity*. In grouping sensitizers/allergens among acute toxicants, it is understood that there is a delay of a few weeks between initial exposure and a demonstrated hypersensitive reaction upon reexposure. Nevertheless, we list these as acute toxicants because of their typical severe acute *response* to the toxicant after sensitization. The hypersensitive reaction can occur only after a prior exposure has resulted in sensitization, and a subsequent reexposure occurs.

In developing sensitivity, small molecules such as some laboratory chemicals may act as *haptens*—compounds that react with larger proteins in your body to form *antigens*. These antigens may be recognized by the body as "foreign," and antibodies to these antigens are formed. Upon reexposure to this antigen, these antibodies form rapidly, evoking their natural reaction to protect the body. The antibodies react with the antigen, causing adverse reactions such as swelling, redness, and itching. Once a person is sensitized to a chemical, further exposure, even a very small amount, can result in an allergic reaction, such as allergic contact dermatitis, hives, runny nose, sneezing, swollen red eyes, and headache. A respiratory sensitizer can evoke asthma, difficulties in breathing, and rarely anaphylactic shock (a rapidly occurring condition in which a person's breathing system becomes severely restricted, resulting in death if not treated immediately). There is a genetic component to sensitivity. Not all persons exposed to a particular allergen develop sensitivity to that compound. We do not clearly understand what makes some people sensitive to a chemical and others not.

Perhaps a sensitizer that you know best either by reputation or by personal experience is poison ivy (oak or sumac). The chemical that produces the sensitizing reaction is known as urushiol. However, allergic reactions to laboratory chemicals are not uncommon, for example, see Incident 4.1.1.1. Sensitization to a chemical can be very debilitating and can result in requirements for dramatic changes in lifestyle to avoid further exposure to the sensitizer. Prevention of exposure is essential when dealing with sensitizers. Examples of known sensitizers are shown in Table 4.1.2.1.

You Don't Do Well Without Oxygen

Asphyxiants interfere with the body's uptake and/or transport of oxygen to vital organs, the result being oxygen deprivation leading to cell death, organ death, and finally death of the individual. Asphyxiants that function by simply displacing oxygen from the air are called *simple asphyxiants*. Inert gases such as helium, carbon dioxide, nitrogen, and methane are simple asphyxiants. The degree of hazard is related to the density of the gas (see *Chemical Connection 4.1.2.1* How Do We Know the Density of a Gas?). Lighter-than-air common asphyxiants (such as He) will "pool" near the ceiling in a room. Heavier-than-air molecules (such as CO_2) can "pool" near the floor or in low areas. Persons working in confined spaces need to be particularly mindful of these hazards.

CHEMICAL CONNECTION *4.1.2.1*

HOW DO WE KNOW THE DENSITY OF A GAS?

The ideal gas law, $PV = nRT$, is very useful to scientists because it can readily be applied to all gases (under most conditions) without regard to the identity of the gas. (P = pressure, V = volume, n = number of moles, R = the gas law constant, and T = temperature.) This equation can be rearranged to $n/V = P/RT$. Then, using $\mathcal{M}$ = molar mass and m = mass, we can write

$$n/V = P/RT$$

Multiplying both sides of the equation by $\mathcal{M}$ = molar mass, we have

$$n\mathcal{M}/V = P\mathcal{M}/RT$$

Dimensionally, $n\mathcal{M}/V$ is density (in g/L):

$$n \cdot \mathcal{M}/V = d$$

$$\text{mol} \cdot (\text{g/mol})/\text{L} = \text{g/L}$$

So

$$d = P\mathcal{M}/RT$$

Thus, the density of a gas depends directly on the molar mass. The density of air is approximately $(0.790)(28.0$ g/mol$) + (0.21)(32.0$ g. mol$) = 28.8$ g/mol if we take air to be about 79% N_2 and 21% O_2. If we choose the density of air to be 1.0, as a relative value, we can easily calculate the relative density of other common gases, as shown in Table 4.1.2.2.

TABLE 4.1.2.2   Molar Masses and Relative Densities of Common Gases

Gas	Molar mass (g/mol)	Relative density
He	4.00	0.139
CH_4	16.0	0.556
N_2	28.0	0.972
CO	28.0	0.972
Air	28.8	1.00
O_2	32.0	1.11
Ar	40.0	1.39
C_3H_8	44.1	1.53
CO_2	44.0	1.53
Cl_2	70.9	2.46
THF, C_4H_8O	72.0	2.50
Ethyl ether, $(C_2H_5)_2O$	74.0	2.57
Hexane, C_6H_{14}	86.0	2.99

This table can also serve as a crude guide to help us think about the placement of smoke detectors and carbon monoxide detectors in the home. Smoke detectors should be placed on ceilings since, although smoke is relatively "heavy" in term of its constituents, it will always be very hot, which will always produce a relatively low density. "Hot air rises" is the rule of thumb in this matter. Carbon monoxide detectors, on the other hand, can be placed at any location since the CO has the same density as air. "Leaking CO" will likely be slightly warmer than ambient air since it is the product of some incomplete combustion but mixing and convection effects will not allow it to "pool" at the ceiling. CO detector manufacturers recommend "eye level" placement of CO detectors more to keep them away from pets and children than for any reasons related to density.

Convection and mixing of gases must always be considered. Otherwise, in a "stagnant" room we would expect the nitrogen to collect in the upper 80% of the room and the oxygen to collect in the lower 20%. The *kinetic molecular theory* explains easily why this does not occur: gases are in constant, random motion.

The last three entries in the table are common organic solvents. Their relative densities illustrate that they can "creep" along the surface of a benchtop, perhaps eventually finding a source of ignition as warned about in Section 5.1.2 on *flammables*.

Nitrogen gas, which is about 79% of air, is a simple asphyxiant if the concentration in air becomes large enough to "dilute" the level of O_2 to a dangerous level. Tanks of nitrogen gas are common in laboratories and a leaking tank of this otherwise harmless gas can pose a hazard in a confined space. Table 4.1.2.3 lists the effects of decreased O_2 concentrations on humans. Liquid oxygen and liquid nitrogen tanks are strikingly similar in design and deaths at nursing homes have been recorded when someone inadvertently and tragically connected a liquid nitrogen tank to the oxygen system that supplies supplemental O_2 to rooms in nursing homes. Due to this hazard, nitrogen and oxygen regulators have

TABLE 4.1.2.3   Effects of Oxygen Deficiency on the Human Body[4,5]

Percent O_2	Possible results
20.9	Normal
19.0	Some unnoticeable adverse physiological effects
16.0	Increased pulse and breathing rate, impaired thinking and attention, reduced coordination
14.0	Abnormal fatigue upon exertion, emotional upset, faulty coordination, poor judgment
12.5	Very poor judgment and coordination, impaired respiration that may cause permanent heart damage, nausea, and vomiting
10	Inability to move, loss of consciousness, convulsions
< 6	Death

"opposite threads" exactly for the purpose of avoiding this kind of mistake. Events like this are caused by *two* mistakes: not reading the label on the tank *and* using the wrong regulator.

Asphyxiants that function by chemically interfering with oxygen uptake are called *chemical asphyxiants*. Carbon monoxide, the most common chemical asphyxiant, is the leading agent of poisoning in the United States, causing more than 500 deaths and more than 15,000 hospital visits each year. Exposure to carbon monoxide in homes and businesses is usually the result of products from some source of combustion such as gas furnaces, water heaters, or open burning sources such as propane grills. Carbon monoxide is a colorless, odorless, tasteless gas and exhibits a moderate toxicity with a reported human LD_{LO} of 4000 ppm or 4570 mg/m^3 (30 min). Symptoms begin to appear at lower doses of 500 to 1000 ppm with headache, dizziness, nausea, and fatigue, which may be confused with flu. (The LD_{LO} is the lowest observable lethal dose for a given compound and most LD_{LO} values are derived from reported accidental or intentional poisonings.)

The mechanism of its toxicity is easily understood at the molecular level and involves interference with a vital function of the blood. In the blood a large (molar mass = 64,000 g/mole!) molecule, hemoglobin, takes up oxygen in the lungs. Oxygen attaches itself reversibly to the iron atoms in hemoglobin and it is transported throughout the body to cells where respiration occurs. Oxygen and carbon monoxide are of similar molecular shape and size. Due to its polarity, carbon monoxide binds about 240 times more effectively than oxygen does with hemoglobin. (See *Chemical Connection 4.1.2.2 Using Lewis Acid–Base Theory to Predict Chemical Interactions*.) This produces *carboxyhemoglobin*, which decreases the blood's capacity to carry oxygen. With significant concentrations of carboxyhemoglobin formed, the body's cells cannot receive enough oxygen, and they eventually begin to die of asphyxiation.

CHEMICAL CONNECTION *4.1.2.2*

USING LEWIS ACID–BASE THEORY TO PREDICT CHEMICAL INTERACTIONS

There are many important enzymes and other large biological molecules in the body that have transition metal atoms at their active sites. Co, Ni, Fe, Cu, and Zn are very common in this regard and for the sake of convenience, we usually regard these atoms as being in some positive oxidation state. Since Lewis acids are electron pair acceptors, the presence of empty d, s, and p orbitals make these ions good Lewis acids. Thus, molecules with lone pair electrons such as O_2, CO, halide ions, or O-, N-, and S-containing compounds will be good Lewis bases and we can predict that they might attach to the metal atoms in biological molecules. This simple model is a good starting point for analysis but not sufficient to explain all interactions, or lack of them. Molecular polarity will also play an important role in Lewis acid–base interactions. Even with this consideration, though, explaining why O_2 and CN^- attach to the iron atom in hemoglobin but N_2, CO_2, and Cl^- do not requires more sophisticated bonding models.

Carbon monoxide is rarely used in the laboratory, and you are more likely to encounter risk from carbon monoxide poisoning at home. Carefully examine all potential sources of carbon monoxide to ensure that you and others are not exposed. Carbon monoxide detectors are relatively inexpensive, and, along with smoke detectors, every home that uses any gas or fuel-fired appliance should have one.

Other chemicals can ultimately interfere with breathing, but they are not asphyxiants due to their mode of action. Breathing reactive toxicants, such as hydrogen cyanide, phosgene, chlorine, ammonia, nitrogen dioxide, or methylfluorosulfate, can cause severe pulmonary edema (among other toxic effects), a condition that can be life-threatening if not treated quickly. As these gases are inhaled, they come in contact with the air sacs. While the exact mechanism is unknown, the effects of these gases are to damage the air sac–capillary interface by making it more permeable so that after some period, from a few minutes to several hours depending on dose and toxicant, serum begins to leak from the capillaries into the air sac. The fluid in the air sac, a condition called edema, disrupts the normal oxygen–carbon dioxide exchange and the exposed person begins having difficulty breathing. If this condition is not treated rapidly, the lungs "fill up" with fluid and, without adequate oxygen, the person eventually lapses into unconsciousness that may become life-threatening and fatal. The lesson here is to avoid exposures to reactive chemical toxicants. Always use chemical hoods when handling reactive gases.

Developmental Toxicants—Teratogens and Fetotoxicants

There are a variety of compounds that interfere with the genetic process. Most of these compounds present long-term, but not short-term, hazards in the lab. Section 4.3.1 will discuss chemicals that cause cancer, called carcinogens. Here, we briefly present a group of compounds that is of immediate concern: you should know about them even though they are not commonly encountered in academic labs.

Chemicals that can interfere with the normal development of the fetus are called *developmental toxicants*. *Teratogens* can cause damage to the developing fetus, resulting in birth defects and abnormalities in the baby. Teratogens can cause birth defects with exposure in the first trimester of pregnancy. Fortunately, there are relatively few known teratogens. The most well-known teratogen is thalidomide, which caused limb malformations in the children of women who took the prescribed drug as a sleep aid and for nausea and vomiting during their pregnancy.[6] This is not a chemical that you will likely encounter in the laboratory.

Some chemicals can cause effects on the developing fetus in other ways and are termed *fetotoxicants*. The most well-known fetotoxicant, ethyl alcohol, causes fetal alcohol syndrome (FAS), a condition in which the child is born potentially with a range of possible abnormalities. The Centers for Disease Control and Prevention issued the following statement about FAS:

> One of the most severe effects of drinking during pregnancy is fetal alcohol syndrome (FAS). FAS is one of the leading known preventable causes of mental retardation and birth defects. If a woman drinks alcohol during her pregnancy, her baby can be born with FAS, a lifelong condition that causes physical and mental disabilities. FAS is characterized by abnormal facial features, growth deficiencies, and central nervous system (CNS) problems. People with FAS might have problems with learning, memory, attention span, communication, vision, hearing, or a combination of these. These problems often lead to difficulties in school and problems getting along with others. FAS is a permanent condition. It affects every aspect of an individual's life and the lives of his or her family.[7]

While some very limited exposure to alcohol in the lab is possible, significant exposure is highly unlikely in a laboratory setting. However, the frequent and sometimes excessive use of alcohol on college campuses is an important public health issue, and we include the comments above since they provide a striking example of the results of a common fetotoxicant. The rate of FAS in the United States is estimated to be between 0.2 and 1.5 per 1000 live births. Simply put, women who drink alcohol while pregnant put the fetus at considerable risk for birth defects.

Organ Toxicants—Toxic Chemicals That Affect Specific Organs or Body Systems

Organ toxicants primarily affect one or more organs or body systems. For example, many chlorinated hydrocarbon compounds are hepatotoxic (i.e., liver toxicants). Carbon tetrachloride and vinyl chloride, sometimes used in the laboratory, target the liver. Other hepatotoxic laboratory chemicals include ethanol, allyl alcohol, dimethylformamide, dichloroethylene, and methylene dianiline.

Ethylene glycol is a well-known kidney toxicant (nephrotoxicant) that causes severe damage upon ingestion by formation of calcium oxalate crystals. This compound has been implicated in several high visibility intentional poisonings and murders. Elemental mercury and mercury salts can be encountered in the laboratory, are highly toxic, and accumulate as Hg^{2+} bound to sulfhydryl groups within the kidney. Cadmium, another heavy metal, is also nephrotoxic. Another laboratory chemical, chloroform, causes nephrotoxicity by targeting the proximal tubule within the kidney.

Another group of acute toxicants are neurotoxicants, sometimes called neurotoxins. Neurotoxicants interfere with the peripheral and central nervous systems. We know more about neurotoxicants than most other classes of toxic compounds because studies of the mechanisms of these toxicants have been largely successful and have produced a variety of drugs and other useful chemicals. The most hazardous neurotoxicants include pesticides, natural venoms, and nerve gases (chemical warfare agents). Chemical warfare agents are restricted by laws and when permission is granted for use of these agents, they require special handling, facilities, and security measures. These are substances that you are not likely to encounter in early undergraduate laboratories but could encounter in undergraduate or graduate research laboratories.

Some organic solvents, such as *n*-hexane and carbon disulfide, are known neurotoxins and others are suspected neurotoxins. The route of exposure can be ingestion, inhalation, or absorption through the skin. The effects on the central nervous system (CNS) can range from apparent intoxication, psychomotor impairment, dizziness, headache, and nausea. Fortunately, reports of laboratory intoxication incidents from neurotoxicants are rare.

Other categories of organ toxicants are pulmonary (lungs), dermal (skin), cardiovascular, immune system, and endocrine system.

RAMP

- *Recognize* what chemicals are toxic, using the categories in this section as a guide for analysis.
- *Assess* the level of risk based on likelihood of exposure by various routes. Be mindful of "confined spaces" (small labs) with regard to asphyxiants.
- *Minimize* the risk by preventing exposure using good lab practices and personal protective equipment such as gloves and using ventilation hoods for inhalation hazards.
- *Prepare* for emergencies knowing the toxic hazards of the chemicals that you are using.

References

1. Nathaniel Hawthorne. From G. CARRUTH and E. EHRLICH, *The Giant Book of American Quotations*, Gramercy Books, New York, 1988, Section 92. Food, #9.
2. O. P. EDMONDS, and M. S. BOURNE. Sodium azide poisoning in five laboratory technicians. *British Journal of Industrial Medicine* **39**:308–309 (1982).
3. DAVID E. COHEN, and ROBERT H. RICE. Toxic Responses of the Skin. In: *Casarett & Doull's Toxicology: The Basic Science of Poisons*. (CURTIS D. KLAASSEN, ed.), McGraw-Hill, New York, 2001, Chapter 19, p. 653.
4. Compressed Gas Association, 2001. Found in U.S. Chemical Safety Hazard Investigation Board. Safety Bulletin. Hazards of Nitrogen Asphyxiation; available at http://www.csb.gov/assets/document/SB-Nitrogen-6-11-03.pdf (accessed September 14, 2009).
5. Centers for Disease Control and Prevention. National Institute for Occupational Safety and Health. NIOSH Respirator Selection Logic 2004; available at http://www.cdc.gov/niosh/docs/2005-100/chapter5.html (accessed September 15, 2009).
6. JOHN M. ROGERS, and ROBERT J. KAVLOCK. Developmental Toxicology. In: *Casarett & Doull's Toxicology: The Basic Science of Poisons*. (CURTIS D. KLAASSEN, ed.), McGraw-Hill, New York, 2001, Chapter 10, p. 353.
7. Centers for Disease Control and Prevention. Fetal Alcohol Spectrum Disorders (FASDs); available at http://www.cdc.gov/ncbddd/fas/fasask.htm (accessed September 15, 2009).

QUESTIONS

1. Acute toxicants are characterized by what features?

   (a) Single, relatively high dose
   (b) Single, relatively low dose
   (c) Long-term exposures of relatively high doses
   (d) Long-term exposures of relatively low doses

2. The most common toxic effect of chemicals is the category of

   (a) Allergens

(b) Poisons

(c) Asphyxiants

(d) Irritants

3. There is often a delay between initial exposure and the response for allergens, but they are considered to be acute toxicants because

(a) A high dose is required

(b) After sensitization has occurred, the reaction is immediate

(c) They involve the immune system

(d) All of the above

4. "Simple asphyxiants"

(a) Are chemically unreactive

(b) Displace oxygen from the air

(c) Can be lighter or heavier than the average molar mass of air

(d) All of the above

5. "Chemical asphyxiants"

(a) Interfere with oxygen uptake through a chemical reaction in the body

(b) Are actually the same as "simple asphyxiants"

(c) Are usually not gases

(d) Cause the blood to coagulate

6. Pulmonary edema

(a) Is the condition of excess fluid in the lungs

(b) Can be caused by nonasphyxiant chemicals that react with lung air sacs

(c) Can occur rapidly or slowly

(d) All of the above

7. Teratogens and fetotoxicants

(a) Have immediate and obvious effects when ingested

(b) Interfere with the genetic process

(c) Can have an effect on a fetus, even though the mother may not know this

(d) Are frequently encountered in introductory chemistry labs

8. Substances that affect the central nervous system are called

(a) Hepatotoxicants

(b) Neurotoxins

(c) Nephrotoxicants

(d) Teratogens

9. What compounds often used as pesticides are also neurotoxins?

(a) Cholinesterases

(b) Nephrotoxicants

(c) Organophosphates

(d) Tetratogens

4.2.1

CHRONIC TOXICITY

Preview   This section provides a brief introduction to chronic toxicity and why it is important to understand this type of toxicity when working with chemicals over longer periods of time.

> *There can be no intelligent control of the lead danger in industry unless it is based on the principle of keeping the air clear from dust and fumes.*
>
> Alice Hamilton, occupational health scientist[1]

INCIDENT 4.2.1.1   XYLENE POISONING[2]

Five cytotechnologists, each working in separate laboratories, were preparing slides for microscopic examination. Their work involved using xylene as a solvent in laboratories without ventilation in the last step of a process for cleaning and staining cytological/histological slides. Each woman became ill, with the severity of symptoms varying from very severe to moderate, but generally they experienced chronic headache, fatigue, chest pain, and other symptoms. One woman, after being hospitalized several times, was finally diagnosed with xylene poisoning but was permanently disabled. Three others changed to jobs that did not use xylene and their conditions improved. After investigation of the laboratory in the fifth case, a recommendation was followed and implemented installing a laboratory hood. The technologist resumed work using the hood and her health improved.

What lessons can be learned from this incident?

Chronic Toxicants and the Laboratory

In your work in the laboratory, you will encounter chemicals that have documented or suspected chronic effects—chronic toxicants. While the incident described above illustrates an adverse outcome from exposure to a chronic toxicant, these cases are rare in laboratories. The nature of effects from exposures to chronic toxicants usually requires receiving doses over an extended period, usually many months or years, and this is an unlikely scenario in a laboratory situation. Most laboratories use very small amounts of a wide variety of chemicals over many years of time, and reports of chronic toxic effects from laboratory exposures have been sparse and rare. Nevertheless, a few epidemiologic studies have found that chemists, as a group, were at excess mortality risks for suicide and cancer but the causes of these observations are not clear.[3-9]

As in Section 4.1.2 about acute toxicants, we will organize the discussion here using various categories of chronic toxicants. However, the experience from these chronic toxicants has been derived from "industrial exposure," where typical exposures are much greater than typical laboratory exposures. Nevertheless, this does not mean that there is no risk from handling chronic toxicants in the laboratory—as in all work where chemicals are handled you should seek to minimize exposures. Table 4.2.1.1 lists several types of chronic toxicants and examples of chronic toxicants that might be encountered in a laboratory. Carcinogens, toxicants that cause cancer, will be discussed in Section 4.3.1.

Laboratory Safety for Chemistry Students, by Robert H. Hill, Jr. and David C. Finster
Copyright © 2010 John Wiley & Sons, Inc.

TABLE 4.2.1.1   Chronic Toxicants that May Be Encountered in Laboratories

Chronic toxicant types	Examples of chronic toxicants
Carcinogens	Ethylene oxide, Cr(IV), vinyl chloride, formaldehyde, arsenic, benzidine
Organ toxicants	Hg, ethanol, beryllium, benzene, carbon tetrachloride, chloroform, arsenic
Neurotoxicants	Methylmercury, carbon disulfide, hexane, acrylamide, lead, nicotine, arsenic

Understanding Chronic Toxicity — The Basics

Chronic toxicity is the production of systemic damaging effects from exposure to repeated low-level doses of a chemical over a long period of time resulting in chronic poisoning, cancer, or other effects. Acute toxicity does not predict chronic toxicity. We know much less about chronic toxicity than acute toxicity because it is much more difficult to recognize and study. Studies evaluating chronic toxicity are also very expensive. Symptoms of chronic toxicity often appear slowly and subtly upon its subject, and the relationship between exposure and illness may not be easily recognized over a long period of time. Connections between symptoms and exposure may be obscured by other chronic diseases and the breakdown of the body with time.

Although chronic toxicity may be suspected from an exposure based on serendipitous observations, it is quite another thing to establish a definitive link between exposure and effect. Sometimes chronic toxicities from exposure to chemicals are recognized through large, expensive epidemiological studies of large numbers of people so that relationships between exposures and outcomes can be correlated by statistical analysis and inferred by probability. Usually most chronic toxicants are identified through chronic animal studies and through follow-up studies may be correlated with human exposures and outcomes (see *Special Topic 4.2.1.1* Epidemiology and Its Role in Discovering Chemically Induced Disease).

SPECIAL TOPIC *4.2.1.1*

EPIDEMIOLOGY AND ITS ROLE IN DISCOVERING CHEMICALLY INDUCED DISEASE

Epidemiology is the study of disease frequency among populations and the disease's relationship to factors that might be possible causes of the disease. Epidemiologists use statistics to identify associations between exposure and disease. Statistical associations can be powerful tools in helping find the probable cause of a disease, but statistical associations do not by themselves identify causes with certainty. To find the cause, epidemiologists must clearly define the disease (case definition), demonstrate statistically significant differences between exposed and unexposed populations, identify a potential causative agent, and link exposure of the causative agent to the disease. These studies are difficult because critical data may be missing, particularly adequate exposure information.

Only strong associations with a causative agent, such as a chemical, can lead to further studies to establish definitive cause. The news media often cite studies that show or imply an association as the cause of disease, but these reports are often misleading and incorrect because the hard work to identify the cause has not been done. For example, there is probably a strong association between wearing skirts and breast cancer but skirts don't cause breast cancer. Rather, most breast cancer is among women and skirts are more commonly worn by women.

Epidemiologic studies helped establish links between certain chemicals and cancers shown in Table 4.2.1.2. Some of these associations were discovered through epidemiologic investigations of "cancer clusters." Clusters are groups of health-related events, such as cancer, occurring in a greater than expected number of cases within a group of people in a geographic area over a specified time period. These clusters are often reported or detected by the public or an observant clinician. While most clusters do not result in significant associations, a few cluster cases such as those associated with rare diseases (e.g., angiosarcoma) or diseases not usually associated with a certain age group are more likely to be verified as "real."

There are also many famous case studies where epidemiology has either reasonably linked a causative agent to a disease or reasonably proved that a suspected causative agent was not linked to a disease. These are called "false positives" since at first an association is detected using statistics but later some methodological flaw reveals the absence of the statistical connection. Some examples that you can further read about using an

Internet search are Legionnaires' disease, Agent Orange, breast cancer and silicone implants, thalidomide, and Bendectin. The National Cancer Institute has many "fact sheets" on a wide range of topics related to cancer at http://www.cancer.gov/cancertopics/factsheet/Risk. Two other examples[14] are a false positive in early studies of breast cancer associated with serum level of DDE [1,1-dichloro-2,2-bis(p-chlorophenyl)ethylene] and a false positive between acrylonitrile exposure and lung cancer in occupational settings.

TABLE 4.2.1.2   Chemicals Epidemiologically Linked to Cancer

Chemical	Type of cancer
Arsenic	Skin and lung
Asbestos	Mesothelioma
Benzene	Leukemia
Benzidine	Bladder
2-Naphthylamine	Bladder
Chromates	Respiratory tract
Nickel	Respiratory tract
Vinyl chloride	Angiosarcoma

The human body has developed over the thousands of years of its evolution a myriad of defense systems that can, in most people, metabolize the many chemicals that enter our bodies through our food and drink, our skin, and our lungs in small, inconsequential amounts and concentrations. As a result of this, it is possible for us to be exposed to low levels, or even occasional moderate levels, of toxic chemicals without exhibiting any adverse effects. The body's defenses take care of these through removal by elimination from the body or storage in body tissues. These metabolic processes will be described more in Section 4.3.2.

Let's consider what happens if you are exposed to a chemical that exhibits chronic toxicity. Keep in mind that there are millions of chemicals, and only a fraction of these have been tested and are known to exhibit significant chronic toxicity. Some chemicals can accumulate with storage in the body over time—meaning they are stored in the body and as exposure continues the concentration of the chemical increases with time. Exposures to low levels of toxicants over a long period of time are usually taken care of by your body's defenses, but at some point in time your body's defensive systems may break down or become overwhelmed as the toxicant dose increases. Some of these chemicals are toxic enough so that as the concentration reaches a certain level adverse effects from chronic exposure begin to appear. This follows the same sort of dose–response curve as acute toxicity, except with low dose levels it takes a much longer time to reach this point because the toxicant must reach a certain toxic level in the body or cause accumulated damage to cells, tissues, or organs. See *Chemical Connection 4.2.1.1* Solubility, Storage, and Elimination.

CHEMICAL CONNECTION 4.2.1.1

SOLUBILITY, STORAGE, AND ELIMINATION

Xenobiotic substances are those that do not naturally occur in an organism. They may be other naturally occurring substances or they may be synthetic molecules. When a person is exposed to some xenobiotic substance, what happens? This question will be answered at greater length in Section 4.3.2 (Biotransformation, Bioaccumulation, and Elimination of Toxicants), but let's take a quick look at the application of some general chemistry principles to this topic first.

There are three key ideas presented in general chemistry that play an important role in this discussion: solubility, equilibrium, and kinetics. Let's examine an oversimplified, but useful, scheme to think about the fate of a xenobiotic in your body.

First, we need to think about solubility. The two main categories are lipophilic and hydrophilic. Lipophilic molecules are those that prefer to be dissolved in fatty tissue. These will be molecules that are mostly nonpolar just like fatty tissue. "Lipo" refers to lipids, which is another name for fats. "Philic" refers to "-loving" so lipophilic means "fat-loving." We could also call these hydrophobic molecules as they are nonpolar and they are "water-hating." Hydrophilic means "water-loving" and these will be polar molecules or ionic substances.

The bloodstream is an aqueous environment so when a lipophilic molecule enters the bloodstream it will prefer to reside in fatty tissue. We can look at Figure 4.2.1.1 to think about this process. A lipophilic molecule will set up a dynamic equilibrium between the blood and fatty tissue (B-F), probably with the equilibrium heavily favoring the fatty tissue. This is equilibrium "B-F" in the figure. If more exposure occurs, then the concentration of the molecule in the blood increases and applying the *LeChatelier effect* we know that this will set up a new equilibrium with higher concentrations in both the blood and fatty tissue.

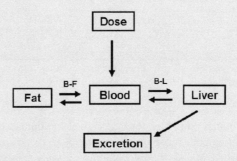

FIGURE 4.2.1.1   Pathways for Xenobiotic Substances. The arrows represent pathways for the movement of a substance in an organism. A complete understanding of this involves both a detailed analysis of multiple, connected equilibria and various rate constants.

A polar xenobiotic molecule will not enter the fatty tissue but will be absorbed by the liver via the blood–liver (B-L) equilibrium. The function of the liver is to metabolize the xenobiotic molecule, probably to a more polar molecule, for eventual excretion in the urine and feces.

The rates of all the reactions in the figure are important, too. If the rate of the toxic mechanism in the blood and the rates at which the toxicant affects other organ systems are faster than the rate at which the toxicant moves from the blood to the liver, then the toxic effect may be exerted before the liver can perform its heroic function. The same could be said of the B-F equilibrium. If the B-F equilibrium is fast enough and the toxicant can be stored in the fatty tissue (where even at high levels it cannot exert its toxic effect while "safely" stored in the fatty tissue), then even though there may be an "acutely toxic" level in the body overall, since most of the toxicant is stored in fat, the level in the blood would be at a subacute level.

Some lipophilic molecules will also set up equilibrium with the liver, where they can get metabolized to less lipophilic forms for eventual excretion. In this fashion, depending on the various combinations of equilibrium and rate constants, a lipophilic molecule in the fatty tissue could slowly be excreted by being in equilibrium with the blood, which is also in equilibrium with the liver.

At a point called the threshold, when the chronic toxicant reaches dose levels that can cause adverse effects, the effects of the chronic exposure begin to appear (Figure 4.2.1.2). As exposure continues, the effects begin to build and increase until subtle but overt symptoms begin to appear—keep in mind that this chronic exposure can take years or even decades to manifest itself. Because we often experience various aches, pains, and discomforts over our lifetimes, we may not associate an overt symptom, at least not at first, with a significant problem. At some point, however, the symptoms become unmistakable, perhaps painful or clearly uncomfortable, and we usually seek medical care at this point although a physician may have a difficult time establishing causality and must treat the illness based largely on symptoms.

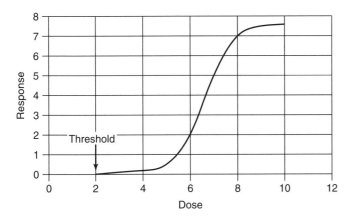

FIGURE 4.2.1.2 Dose–Response Curve for Chronic Toxicity. This is a generic dose–response curve for chronic toxicity. The *x*-axis can refer to amounts of a substance inhaled or ingested. In this example, the threshold is about 3, below which no effects are detected. The *y*-axis is the incidence of some disease or death. The smooth line may imply many data points, but sometimes only a relatively small data set is available to estimate the curve.

Risks of Chronic Toxicity

As indicated above, laboratory work is a relatively low-risk operation if you know and follow safety principles. However, this does not mean that you are not at risk of chronic chemical exposures because they come from places other than the laboratory.

Many experts believe that the major causes of chronic diseases come from genetics, lifestyles, and age. Each of us has a different and unique genetic makeup—a broad variance has evolved in biological defense systems with our individual bodies exhibiting strengths in some areas and weaknesses in other areas. This means that not everyone reacts to a chronic toxic exposure in the same way. Some people may fend off the toxic effects, while others may exhibit symptoms early. Thus, there may be a spectrum of responses to an exposure.

Your lifestyle plays a significant role in your risk of chronic toxicity or chronic disease—depending on whether you smoke, drink or eat to excess, skip exercise, abuse drugs, or engage in other risky behaviors. Smoking is the leading preventable cause of disease in the United States. The Centers for Disease Control and Prevention have reported the following:

> Smoking harms nearly every organ of the body; causing many diseases and reducing the health of smokers in general. The adverse health effects from cigarette smoking account for an estimated 438,000 deaths, or nearly 1 of every 5 deaths, each year in the United States. More deaths are caused each year by tobacco use than by all deaths from human immunodeficiency virus (HIV), illegal drug use, alcohol use, motor vehicle injuries, suicides, and murders combined.[10]

These genetic and lifestyle factors and the normal aging process further complicate a complex problem when it comes to recognition and study of chronic toxicity. It is possible to recognize and understand chronic toxicities, but it can be very difficult.

Mercury, a Potential Chronic Poison with Laboratory Exposures

There have been relatively few reports of chronic toxicity from laboratory work and most of these are now historical case studies from times when chemists had little understanding of risks from chemicals. Perhaps one of the most significant concerns, at least from past experience, has been for mercury, because it is very toxic by inhalation and readily vaporizes.

The German chemist Alfred Stock did pioneering work in the chemistry of boron hydrides in the early 20th century. Stock invented the glassware and equipment that allowed him to work with these air-sensitive compounds and elemental mercury was used as an essential component of these systems, most often in pressure measuring devices called manometers. After working in unventilated labs for decades with mercury pools and mercury spills, Stock developed severe mercurialism. In fact, as he realized that the source of his illness was exposure to mercury, he began a lifelong campaign to document the effects of this disease and warn other scientists of the dangers of working with mercury. Symptoms included intellectual deterioration, loss of memory, slurred speech, and various nerve disorders. Despite these challenges, Stock made significant discoveries not only in boron hydride chemistry, but also the chemistry of silicon hydrides and arsenic and phosphorus chemistry. W. N. Lipscomb, Nobel Laureate in 1976 for his work in boron hydrides, quoted Sidgwick in his Nobel Laureate address: "All statements about the hydrides of boron earlier than 1912, when Stock began to work on them, are untrue."[11]

While most modern laboratories have done much to minimize the use of mercury in the laboratory, it would not be unusual for you to encounter mercury in your laboratory work. Mercury metal is a heavy liquid that is used in laboratory equipment, such as thermometers and manometers, and is also used in electrical switches, thermostats, and fluorescent lights. Breakage of thermometers or other mercury-containing equipment or spills of mercury create opportunities for long-term, chronic exposure to mercury if these spills are not cleaned up immediately. Because liquid mercury is dense, it gets into cracks and crevices and is often difficult to remove. Some don't even consider it important enough to clean up if they experience or see mercury spill. The mercury seemingly "disappears" as it moves into these low-lying places to become diffused throughout the floor boards and tiles—but it still is likely to be present. Mercury, being toxic by inhalation and having a small, but significant, vapor pressure, can evaporate into the laboratory air so that chemists can be exposed through inhalation over long periods of time. This is the perfect recipe for chronic poisoning.

A 1972 report on contamination of chemical laboratories by mercury revealed that several students, professors, technicians, and employees of a university experienced mercury chronic toxicity or had elevated urinary mercury levels.[12] This was recognized when a student reported to a clinic experiencing chronic fatigue, headaches, and mental lethargy and it was learned that he had been potentially exposed to mercury in his laboratory. Further investigation revealed elevated mercury concentrations in his urine and in the air of these laboratories. Only three of 28 people working in these laboratories exhibited symptoms, but 26 of 28 had elevated mercury levels in their urine. The widespread contamination of many laboratory rooms resulted from spilled mercury from broken thermometers and manometers over the many years of use. The laboratories were decontaminated, dropping mercury air concentrations below acceptable levels. Urinary concentrations of the exposed individuals eventually dropped back within normal range. This report highlights the potential hazards of mercury if spills or broken thermometers are not promptly taken care of. Furthermore, the reduction of mercury-containing equipment, when possible, will reduce your chances of mercury intoxication. Whenever possible, replace mercury-containing devices with nonmercury devices, such as thermometers.[13]

Working with Chronic Toxicants

As noted above, it is not likely that you will be working with many chronic toxicants in the lab setting over long periods of time, and even if you are, modern safety protocols reduce this hazard dramatically. However, since you may someday be supervising labs and/or industrial manufacturing facilities, it is very important to be aware of this category of toxicants and the OSHA rules that govern their exposure. OSHA requires under the Laboratory Standard (29 CFR 1910.1450) that laboratorians and their supervisors be aware of the risks and exposure hazards associated with chronic toxicants. Several governmental and private organizations are involved in setting exposure limits. For example, OSHA mandates limits as *permissible exposure limits* (PELs). The American Conference of Governmental Industrial Hygienists (ACGIH) recommends *threshold limit values* (TLVs). The National Institute of Occupational Safety and Health (NIOSH) lists levels that are *immediately dangerous to life and health* (IDLH). And there are more. This myriad of regulations, recommendations, and acronyms (related to both acute and chronic toxicants) will be explained in greater detail in Section 6.2.2.

RAMP

- *Recognize* what substances are chronic toxicants.
- *Assess* the level of risk based on likelihood of exposure by various routes.
- *Minimize* the risk by preventing exposure using good lab practices and personal protective equipment such as gloves and using ventilation hoods for inhalation hazards.

- Since chronic toxicity represents a long-term episode it is less critical to *prepare* for emergencies as one should for episodes of acute toxicity. Being knowledgeable about the signs and symptoms of chronic exposures is the best "preparation" and this involves the steps above: recognize, assess, and minimize.

References

1. Alice Hamilton, BrainyQuote; available at http://www.brainyquote.com/quotes/authors/a/alicenh_amilton.html (accessed June 12, 2009).
2. R. N. Hipolito. Xylene poisoning in laboratory workers: case reports and discussion. *Laboratory Medicine* **11**(*9*):593–595 (1980).
3. F. P. Li. Suicide among chemists. *Archives of Environmental Health* **19**:518–520 (1969).
4. F. P. Li, J. F. Fraumeni, N. Mantel, and R. W. Miller. Cancer mortality among chemists. *Journal of the National Cancer Institute* **43**:1159–1164 (1969).
5. G. R. Olin. The hazards of a chemical laboratory environment—a study of the mortality in two cohorts of Swedish chemists. *American Industrial Hygiene Association Journal* **39**:557–562 (1978).
6. C. E. Searle, J. A. H. Waterhouse, B. A. Henman, D. Barlett, and S. McCombie. Epidemiological study of the mortality of British chemists. *British Journal of Cancer* **38**:192–193 (1978).
7. G. R. Olin and A. Ahlbom. The cancer mortality among Swedish chemists graduated during three decades. *Environmental Research*, **22**:154–161 (1980).
8. S. K. Hoar and S. Pell. A retrospective cohort study of mortality and cancer incidence among chemists. *Journal of Occupational Medicine* **23**:485–494 (1981).
9. J. Walrath, F. P. Li, S. K. Hoar, M. W. Mead, and J. F. Fraumeni. Causes of death among female chemists. *American Journal of Public Health* **75**:883–885 (1985).
10. Centers for Disease Control and Prevention. Smoking and Tobacco Use: Health Effects of Cigarette Smoking; available at http://www.cdc.gov/tobacco/data_statistics/fact_sheets/health_effects/effects_cig_smoking/ (accessed August 3, 2009).
11. N. V. Sidgwick. *The Chemical Elements and their Compounds*, Clarendon Press, Oxford, UK, 1950, p. 338.
12. K. D. Rose, E. W. Simpson, and D. Weed. Contamination by mercury in chemical laboratories. *Journal of the American College Health Association* **20**:197–199 (1972).
13. E. H. Rau. NIH's laboratory mercury campaign and its potential applications in health initiatives to reduce public exposure. *Journal of Chemical Health and Safety* **13**(*3*):5–15 (2006).
14. P. Boffetta, J. K. McLaughlin, C. La Vecchia, R. E. Tarone, L. Lipworth, and W. J. Blot. False-positive results in cancer epidemiology: a plea for epistemiological modesty. *Journal of the National Cancer Institute* **100**:988–995 (2008); available at http://jnci.oxfordjournals.org/cgi/content/full/djn191 (accessed July 30, 2009).

QUESTIONS

1. The likelihood of suffering an adverse, chronic effect while working with chemicals in undergraduate laboratories is

   (a) Small, because almost no chronic substances are allowed in labs
   (b) Small, because chronic exposure in academic labs is unlikely
   (c) Moderate, because academic labs take few precautions against exposure to chronic toxins
   (d) Large, but we don't know how large since all of the effects are delayed

2. Which statement is true?

   (a) Acute toxicity does not predict chronic toxicity.
   (b) We know much more about acute toxicity than we do about chronic toxicity.
   (c) Studies that evaluate chronic toxicity are very expensive.
   (d) All of the above are true.

3. Chronic toxicity studies

   (a) Are sometimes not conducted because we don't know what symptom or effect to measure
   (b) Sometimes occur based on the accidental exposure of a group of people to a chemical
   (c) Are usually more definitive in establishing a cause–effect relationship than acute toxicity studies
   (d) Both (a) and (b) are true.

4. It is known that we ingest small amounts of large numbers of synthetic and naturally occurring chronically toxic compounds regularly

   (a) And this accounts for the relatively high cancer rate in the United States
   (b) But since the toxic effects of these substances are not detected it is clear that the body's defense system can usually handle these chemical assaults
   (c) But the delay time between ingestion and illness is predicted to be over 80 years so the effects are rarely seen
   (d) But the effect of these compounds is masked by other illnesses

5. Which of the following factors are the primary causes of chronic disease?
   I. Genetics
   II. Lifestyle choices
   III. Age
   IV. Chemicals

   (a) I, II, and IV
   (b) II, III, and IV
   (c) I, II, and III
   (d) I, and IV

6. The element mercury

   (a) Is much more toxic by ingestion than inhalation
   (b) Produces symptoms as soon as blood level concentrations rise to detectable levels

(c) Was widely used in laboratories in the 20th century where spills of the liquid Hg occurred and exposed chemists to Hg vapors

(d) Is the preferred liquid in thermometers because it is so chemically inert

7. The Laboratory Standard (OSHA 29 CFR 1910.1450)

(a) Ignores the issue of chronic toxicity because it is so rare in chemistry labs

(b) Regulates the specific levels of known chronic toxicants

(c) Requires that scientists be aware of risks and hazards associated with chronic toxicants

(d) Requires that the ACGIH determines the TLVs for all chronic toxicants.

4.3.1

CARCINOGENS

Preview This section presents a brief overview of the causes of cancer and an overview of carcinogens. This section is a companion to Section 4.2.1 where we discussed the chemistry of chronic toxicants.

The public appears to have little concern for chemicals that destroy lung tissue, or nerve tissue, or kidneys, but chemicals suspected of causing cancer are viewed with great apprehension or alarm.

M. Alice Ottoboni, *The Dose Makes the Poison*[1]

INCIDENT 4.3.1.1   LABORATORY EXPOSURE TO CARCINOGENS[2]

A chemist worked in a laboratory where he inhaled bis(chloromethyl)ether, chloromethyl methyl ether, and a small amount of dimethyl sulfate. He developed extensive pulmonary carcinoma (lung cancer) from this 7-year exposure and in early 1975, at 42 years of age, died. Around the same time period as his laboratory work, experiments by other investigators revealed that bis(chloromethyl)ether was a powerful carcinogen in animals. Dimethyl sulfate was also shown to be an animal carcinogen. In 1974 bis(chloromethyl)ether, chloromethyl methyl ether, and dimethyl sulfate were included in the list of 13 carcinogens closely regulated in the United States by OSHA.[3] Today, bis(chloromethyl)ether is recognized as perhaps the most potent human carcinogen known (see Figure 4.3.1.1).

What lessons can be learned from this incident?

Cancer and Its Causes

We categorize cancer-causing agents as chronic toxicants, as indicated in Table 4.2.1.1. Many of the features of these chemicals have similar characteristics to chronic toxicants, including nonlinear dose–response curves with thresholds, as generically shown in Figure 4.1.1.1.

Cancer is one of the most feared words in our vocabulary, but often the real risks of cancer are overshadowed by sensational stories and headlines about the tragic occurrence of cancer in an individual, family, or group of people and that suggests that there are purported links to chemical exposure. Here are some facts about cancer that will help place the discussion of cancer risk from chemicals in a helpful context:

- One in four people will get some sort of cancer in their lifetime.

- Cancer is not a single disease but actually a large group of diseases—each having a different cause (etiology) and a different prognosis (predicted outcome).

- Cancers are malignant neoplasms—an abnormal growth or tumor. Some tumors may be classified as benign—these are not cancer. Malignant tumors have cells that grow more rapidly than normal cells, cells that lose the typical features of normal cells, and cells that invade adjacent tissues or migrate to distant locations from the site of the cancer.

- There are four large groups of cancers—leukemias, lymphomas, sarcomas, and carcinomas.

- The majority of cancers are caused by lifestyle and genetic background, a lesser percentage is attributable to sun and radiation exposure, and cancer from occupational (workplace) exposures to chemicals accounts for about 5% of all cancers.

Laboratory Safety for Chemistry Students, by Robert H. Hill, Jr. and David C. Finster
Copyright © 2010 John Wiley & Sons, Inc.

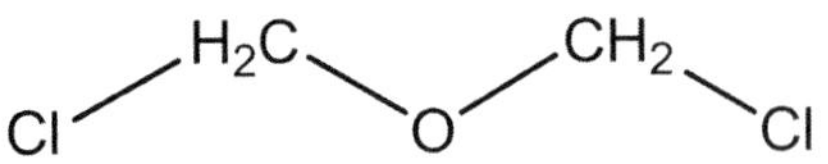

FIGURE 4.3.1.1 Structure of Bis(chloromethyl)ether. This compound is a regulated carcinogen. The EPA recommends <0.0000038 parts per billion (!) in lakes and streams. OSHA set a limit of 1 ppb as the highest acceptable level in workplace air.

Perhaps the most important observation for our discussion is the last entry in the list above: the majority of cancers are caused by lifestyle and genetic background. Lifestyle causes include smoking, lack of exercise, being overweight, and abusing drugs. Smoking, the leading cause of preventable disease in the United States, involves inhalation of chemicals produced when tobacco is burned; thus, cancer from smoking is chemically induced—but it is a lifestyle choice. If you live or work closely with someone who smokes in your presence, you are also being exposed to these chemicals, although to a lesser extent than the smoker. Exposure to excessive sunlight, radiation, and alcohol are also associated with cancer.

It is estimated that 5% or perhaps less of cancers are attributable to chemical exposure in the workplace—most of this being in industrial workplaces where there has been continuous and long-term exposure to specific chemicals. This means that your chances of contracting cancer from working in a laboratory are probably very, very low—particularly if your experience in the laboratories involves a large variety of chemicals rather than one or two chemicals over a long period of time. Nevertheless, being prudent in preventing exposure to chemicals will protect you from this risk.

What chemicals merit some concern when working with them over long periods of time, usually at relatively low levels of exposure? A definitive answer to this is not readily at hand but given how little we know about, and how difficult it is to determine, the long-term effects of exposure to low levels of toxicants, the prudent posture to take is to assume that any chemical might exhibit chronic toxicity. This realization needn't induce a lifelong panic about working with chemicals but instead should urge us to work with them in a safe fashion by minimizing exposures to chemicals.

In fact, OSHA mandates the limitation of occupational exposure to chemicals both through a series of specific regulations for some chemicals and more generally through CFR 1910.1200 ("Right to Know") and CFR 1910.1450 (the "Lab Standard"). Generally, OSHA sets exposure limits through *permissible exposure limits* (PELs) and these data are easily found on MSDSs. The PELs are usually *threshold limit values* (TLVs) that have been measured for many hundreds of substances by the American Conference of Governmental Industrial Hygienists (ACGIH).

Chemical Carcinogens

Chemicals that are known to cause cancer in humans or animals are called *carcinogens*. Those that are known to cause cancer in humans are regulated in the United States by OSHA—meaning there are special requirements when using these chemicals. All of the chemicals that are known to cause cancer have been identified through epidemiological studies of industrial workplaces and have been the result of repeated, significant exposure to these chemicals in these industrial workplaces over time. Table 4.3.1.1 lists chemicals regulated by OSHA as carcinogens.

There are other recognized organizations that have identified chemicals as known carcinogens or probably carcinogenic to humans, including the International Agency for Research on Cancer (IARC)[4] and the National Toxicology Program (NTP).[5] IARC critically reviews and publishes information about carcinogenic properties of chemicals and processes. IARC monographs are available in printed form and also in electronic form from the IARC web site.[4] NTP publishes its *Report on Carcinogens* that lists known and suspected carcinogens and provides details about each of these.[5] There are several good discussions and listings of carcinogens found elsewhere.[1,6,7]

While it is possible that you could encounter some of these carcinogens in the laboratory, most laboratory workers use small amounts of a large variety of chemicals intermittently over their careers—thus reducing their chances of long-term exposure to carcinogens. Nevertheless, even if a chemical has not been reported to have toxic effects, including chronic ones, you should always seek to prevent exposures in your laboratory work. There are markedly few reports of laboratorians contracting cancer from chemical exposures, except where there have been significant repeated exposures over time. Incident

4.3.1 CARCINOGENS

TABLE 4.3.1.1 Chemical Carcinogens Regulated by the Occupational Safety and Health Administration[3]

Carcinogen name	CAS number	OSHA standard citation
Asbestos		29 CFR 1910.1001
13 Carcinogens		29 CFR 1910.1003
4-Nitrobiphenyl	92-93-3	29 CFR 1910.1003
1-Naphthylamine	134-32-7	29 CFR 1910.1004
Methyl chloromethyl ether	107-30-2	29 CFR 1910.1006
3,3'-Dichlorobenzidine	91-94-1	29 CFR 1910.1007
Bis(chloromethyl)ether	542-88-1	29 CFR 1910.1008
2-Naphthylamine	91-59-8	29 CFR 1910.1009
Benzidine	92-87-5	29 CFR 1910.1010
4-Aminobiphenyl	92-67-1	29 CFR 1910.1011
Ethyleneimine	151-56-4	29 CFR 1910.1012
β-Propiolactone	57-57-8	29 CFR 1910.1013
2-Acetylaminofluorene	53-96-3	29 CFR 1910.1014
4-Dimethylaminoazobenzene	60-11-7	29 CFR 1910.1015
N-nitrosodimethylamine	62-75-9	29 CFR 1910.1016
Vinyl chloride	75-01-4	29 CFR 1910.1017
Inorganic arsenic		29 CFR 1910.1018
Chromium(VI) compounds		29 CFR 1910.1026
Cadmium and its compounds		29 CFR 1910.1027
Benzene	71-43-2	29 CFR 1910.1028
Coke oven emissions		29 CFR 1910.1029
1,2-Dibromo-3-chloropropane	96-12-8	29 CFR 1910.1044
Acrylonitrile	107-13-1	29 CFR 1910.1045
Ethylene oxide	75-21-8	29 CFR 1910.1047
Formaldehyde	50-00-0	29 CFR 1910.1048
Methylenedianiline	101-77-9	29 CFR 1910.1050
1,3-Butadiene	106-99-0	29 CFR 1910.1051
Methylene chloride	75-09-2	29 CFR 1910.1052

4.3.1.1 lists one such report. This individual likely was exposed over several years to some powerful, now regulated carcinogens, and contracted cancer. In today's modern laboratories we strive to minimize all chemical exposures and while you will never be able to eliminate the risk of cancer if you handle a carcinogen, you will surely be able to greatly reduce your personal risk by taking prudent precautions with all chemicals.

What fraction of chemicals are carcinogens? This question has been researched a great deal in recent decades and several hundred chemicals have been tested for carcinogenicity in mice and rats. Overall, 57% of the naturally occurring chemicals tested were carcinogens and 59% of the synthetic chemicals tested were carcinogens.[8] These tests are run using relatively high doses of chemicals, so for this reason and also including the animal-to-human extrapolation uncertainty we must be cautious about overinterpreting these data. Humans are more efficient at repairing damage to DNA so these percentages might well be lower for human exposure. In general, too, we are exposed to many more naturally occurring chemicals in foods that we eat as compared to synthetic chemicals. These exposures are generally far below levels at which they are likely to actually induce cancer.

Mutagens

Mutagens are chemicals that produce genetic changes or mutations to genes that are composed of molecules of deoxyribonucleic acid, known as DNA. Some of these changes are regarded as the events that initiate adverse effects such as cancer or birth defects. Not all mutations, however, produce adverse reactions, such as cancer. In fact, while all carcinogens are mutagens, only about 50–60% of known mutagens are carcinogens. Many mutagens cause no effect and some are beneficial. A potent mutagenic chemical commonly used in many biochemical laboratories is ethidium bromide.[9,10] Ethidium bromide has not been shown to cause mutagenic effects in humans; nevertheless, minimizing exposure in the laboratory to this compound would be prudent.

Screening compounds to determine if they are carcinogens usually begins with tests for mutagenicity using bacteria or fruit flies. Compounds that are mutagenic are then tested for carcinogenicity using rats, mice, or hamsters. There are two types of cells—somatic cells (body cells) and germ cells (reproductive cells). Most tests for mutagenic effects utilize somatic cells. While there are known somatic cell mutagens, there are no known human germ cell mutagens. For mutagenic compounds to elicit mutagenic effects in humans, they must be germ cell mutagens. For a mutation to be inherited it must occur in the germ cell and that cell must survive to propagate its effects. Detecting these germ cell mutagens requires that a parent or parents have their reproductive cells affected (mutated), that these cells survive, and that offspring of these parents develop mutagenic changes as a result of these mutated germ cells. Correlating mutagenic changes and chemical exposures is an extremely difficult process to prove or establish any such connection. It seems unlikely that any large number of compounds will be identified as human mutagens. Nevertheless, that does not mean that a mutagenic compound could not elicit a response upon exposure. All of this suggests that it is best to be prudent when working with mutagenic substances.

Working with Carcinogens

As noted in Section 4.2.1 in reference to chronic toxicants, it is not likely that you will be working with many chronic toxicants or carcinogens in the lab setting over long periods of time, and even if you are, modern safety protocols reduce this hazard dramatically. However, since you may someday be supervising labs and/or industrial manufacturing facilities, it is very important to be aware of carcinogens and the OSHA rules that govern their exposure. OSHA closely regulates chemicals that are recognized carcinogens in the workplace. These are listed in Table 4.3.1.1. The practices listed in these regulations are designed to protect those people who handle these compounds in industrial settings; nevertheless, it is always prudent to handle these compounds with care in the laboratory. In fact, OSHA requires under the Laboratory Standard (29 CFR 1910.1450)[11] that special practices be established and followed when handling any select carcinogens (see *Special Topic 4.3.1.1* Select Carcinogens). These special practices include the following:

- Establishing a designated area where these select carcinogens can be handled
- Using containment devices such as chemical hoods or glove boxes
- Establishing procedures for safe removal of contaminated waste
- Establishing procedures for decontamination

SPECIAL TOPIC *4.3.1.1*

SELECT CARCINOGENS

OSHA defines the term "select carcinogen" in its Laboratory Standard found in 29 CFR 1910.1450(b)[11] Definitions in the following manner:

Select carcinogen means any substance that meets one of the following criteria:

    (i) It is regulated by OSHA as a carcinogen; or

    (ii) It is listed under the category "known to be carcinogens" in the Annual Report on Carcinogens published by the National Toxicology Program (NTP)(latest edition); or

    (iii) It is listed under Group 1 ("carcinogenic to humans") by the International Agency for Research on Cancer Monographs (IARC)(latest editions); or

(iv) It is listed in either Group 2A or 2B by IARC or under the category "reasonably anticipated to be carcinogens" by NTP, and causes statistically significant tumor incidence in experimental animals in accordance with any of the following criteria:

    a. After inhalation exposure of 6–7 hours per day, 5 days per week, for a significant portion of a lifetime to dosages of less than 10 mg/m^3;

    b. After repeated skin application of less than 300 mg/kg of body weight per week; or

    c. After oral dosages of less than 50 mg/kg of body weight per day.

OSHA intends to incorporate the Globally Harmonized System of Classification and Labelling of Hazardous Chemicals (GHS). GHS has established hazard classes and hazard categories for all chemical hazards, including carcinogens. It seems likely that OSHA will also include GHS definitions of carcinogens in its listing of select carcinogens (see Sections 3.2.1 and 6.2.1 about GHS).

This standard is a performance standard, meaning that OSHA does not tell you how to accomplish these requirements: you have the latitude to devise and follow your own guidelines—but you must follow the guidelines that you identify. In the Laboratory Standard OSHA provides a nonbinding Appendix A—National Research Council Recommendations Concerning Chemical Hygiene in Laboratories [Nonmandatory]—as an example of a Chemical Hygiene Plan. This CHP provides some good suggestions as to how these special practices for handling carcinogens might be developed. Thus, if you look at that appendix you will find some of these suggested practices for chemicals of moderate chronic toxicity or high acute toxicity or high chronic toxicity:

- Minimize exposure by any route through reasonable precautions.
- Prepare a plan for use and disposal of these materials and obtain approval of this plan.
- Use and store these materials only in areas of restricted access with special warning signs.
- Use a chemical hood when handling these materials; glove boxes may also be used for these materials.
- Use protective gloves and long sleeves to prevent skin exposure.
- Conduct all transfers in a "controlled area."
- Always wash hands and arms immediately after working with these materials.
- Be prepared for spills and accidents.
- Protect surfaces from contamination with trays or plastic-backed absorbent paper.
- Thoroughly decontaminate or incinerate contaminated clothing.
- Protect vacuum pumps from contamination.
- Decontaminate "controlled areas" before resuming normal work in that area.
- Remove all protective apparel before exiting this restricted area, and thoroughly wash your hands, arms, face, and neck.
- Use a wet mop and HEPA-filtered vacuum cleaner for removing any dry powders.
- Consult a physician for advice about medical surveillance if you regularly work with significant amounts of these materials.
- Maintain records of amounts used and stored, including the names of persons using these materials.

RAMP

- *Recognize* what substances are carcinogens and suspected carcinogens.
- *Assess* the level of risk based on likelihood of exposure by various routes.
- *Minimize* the risk by preventing exposure using good lab practices and personal protective equipment such as gloves and using ventilation hoods for inhalation hazards. Follow specific requirements stated in compound-specific OSHA standards.

- Since carcinogens represent a long-term episode it is less critical to *prepare* for emergencies as one should for episodes of acute toxicity.

References

1. M. A. OTTOBONI. *The Dose Makes the Poison*, 2nd edition, Van Nostrand Reinhold, New York, 1997, p. 101.

2. U. BETTENDORF. Berufbedingte Lungenkarzinome nach Inhalation alkylierender Verbindungen: Dichlorodimethyläther, monochlorodimethyläther, and dimethylsulfat [Occupational lung cancer after inhalation of alkylating compounds: dichlorodimethyl ether, monochlorodimethyl ether, and dimethyl sulfate]. *Deutsche Medizinische Wochenschrift* **102**:396–398 (1977).

3. Occupational Safety and Health Administration. Carcinogens, *Federal Register* **39**:3756 (January 29, 1974) (see 29 CFR 1910.1003). See other individual standards 29 CFR 1910.1001–1052; available at http://www.osha.gov/SLTC/carcinogens/standards.html (accessed August 28, 2009).

4. International Agency for Research in Cancer. *Monographs on the Evaluation of Carcinogenic Risk to Humans*, Vols. 1–92 (as of May 2008); available at http://monographs.IARC.fr/index.php (accessed May 13, 2008).

5. National Toxicology Program. *Report on Carcinogens*, 11th edition; available at http://ntp.niehs.nih.gov (accessed May 13, 2008).

6. E. K. WEISBURGER. Carcinogenesis. In: *American Chemical Society's Handbook of Chemical Health and Safety*, (R. J. ALAIMO, ed.), Oxford University Press, New York, 2001, pp. 141–147.

7. H. C. PITOT and Y. P. DRAGAN. Chemical Carcinogenesis. In: *Casarett & Doull's Toxicology: The Basic Science of Poisons*, 6th edition (C. D. KLAASSEN, ed.), McGraw-Hill, New York, 2001, Chapter 8, pp. 241–319.

8. L. S. GOLD, T. H. SLONE, and B. N. AMES. What do animal cancer tests tell us about human cancer risk? Overview of analyses of the carcinogenic potency database. *Drug Metabolism Reviews* **30**(2):359–404 (1998).

9. National Toxicology Program. Ethidium Bromide, CAS Registry Number: 1239-45-8 Toxicity Effects; available at http://ntp.niehs.nih.gov/index.cfm?objectid=E87CFE2D-BDB5-82F8-F93649545FEBF468 (accessed August 28, 2009).

10. National Research Council. *Prudent Practices in the Laboratory: Handling and Disposal of Chemicals*. Appendix B: Laboratory Chemical Safety Summaries—Ethidium Bromide, National Academy of Sciences, Washington, DC, 1995, pp. 310–311.

11. Occupational Safety and Health Administration. Occupational Exposure to Hazardous Chemicals in the Laboratory—29 CFR 1910.1450; available at http://www.osha.gov/pls/oshaweb/owadisp.show_document?p_table=STANDARDS&p_id=10106 (accessed August 28, 2009).

QUESTIONS

1. What estimated percentage of cancers in the United States are attributable to chemical exposure in the workplace?

   (a) <5%
   (b) About 10%
   (c) About 18%
   (d) About 26%

2. What percentage of chemicals known to cause cancer in humans is regulated by OSHA?

   (a) Less than 10%
   (b) About 25%
   (c) About 60%
   (d) 100%

3. What element, and its compounds, are regulated by OSHA?

   (a) Asbsestos
   (b) Arsenic
   (c) Cadmium
   (d) Mercury

4. For most chemists who work in labs,

   (a) It is probable that they will use and handle some carcinogens, but the likelihood of contracting cancer based on lab activity is very small if appropriate steps are taken to limit exposure
   (b) It is very unlikely that they will use and handle any carcinogens since carcinogens are so highly regulated
   (c) It is very likely that they will use and handle carcinogens and some may develop cancer but establishing the causal link to chemical exposure is almost impossible

   (d) They will avoid using anything carcinogenic since OSHA regulations make it illegal to require a chemist to use carcinogens without signing a release statement

5. What percentage of chemicals tested for carcinogenicity actually turn out to be carcinogens?

   (a) Less than 5%
   (b) About 10%
   (c) About 45%
   (d) About 60%

6. The percentage of chemicals found to be carcinogens in lab tests is probably higher than the actual percentage of human carcinogens in all known chemicals because

   (a) These expensive tests are more likely to be run on compounds that are already suspected to be carcinogens
   (b) Very high doses are used and the high-dose-to-low-dose extrapolation probably overestimates the actual chance of the chemical being a carcinogen in humans
   (c) Humans have a better DNA repair mechanism than some animals
   (d) All of the above

7. Which statement is *true*?

   (a) Most carcinogens are mutagens.
   (b) All mutagens are harmful.
   (c) About 50–60% of known mutagens are carcinogens.
   (d) Mutagenicity tests are only performed on known carcinogens.

8. Which statement is *true*?

(a) OSHA regulates about 30 substances as known carcinogens.

(b) OSHA regulates almost 300 substances as known carcinogens.

(c) OSHA recommends, but does not require, that special practices be established when handling select carcinogens.

(d) OSHA regulates mutagens more strictly than carcinogens.

9. A substance is a "select carcinogen" if it

   (a) Is regulated by OSHA as a carcinogen

   (b) Is "known to be a carcinogen" in the Annual Report on Carcinogens published by the NTP

   (c) Is listed as Group 1, Group 2A, or Group 2B by the IARC and causes statistically significant tumor incidence under specified conditions

   (d) All of the above

4.3.2

BIOTRANSFORMATION, BIOACCUMULATION, AND ELIMINATION OF TOXICANTS

Preview   This section discusses the fate of toxicants in the body, including their biotransformation, bioaccumulation, and elimination.

I know nothing except the fact of my ignorance.

Socrates[1]

INCIDENT 4.3.2.1   DIOXIN EXPOSURE IN THE LAB[2]

In 1956 a chemist synthesized a powerful chloracnegic agent, 2,3,7,8-tetrabromodibenzo-*p*-dioxin (TBDD). He initially contracted chloracne, a chemically induced acne, after an initial synthesis. The toxicity of this compound was unknown but the chemist did not use a chemical hood or personal protective equipment during this work. Later that same year he synthesized 2,3,7,8-tetrachlorodibenzo-*p*-dioxin (TCDD) and this work resulted in a very severe case of chloracne—he was hospitalized for evaluation and later released. He recovered from the two incidents and was in good health in 1991. At that time the chemist volunteered to have his blood sampled for the measurement of TBDD in the blood sample itself and also in the blood adjusted for its lipid content—an indication of the compound in his fat tissues. The analysis revealed that the concentration of TBDD in blood lipids in 1991 was 625 parts per trillion. The concentration of TCDD in blood lipids was 16 parts per trillion. Based on this observation as well as other measurements, the initial body burden of halogenated dioxins in 1956 was estimated to have been between 13,000 and 150,000 parts per trillion.

What lessons can be learned from this incident?

What Happens When Toxicants Enter Your Body

Chemical toxicants, sometimes called xenobiotics, enter the body through various routes usually in very small doses—ingestion frequently from our foods, inhalation from our air, and through the skin (dermal absorption). Sometimes we intentionally take xenobiotics in the form of drugs, medicines, or therapy for illness, other medical conditions, or prevention of adverse medical conditions. Even in small amounts the body has evolved a process that seeks to detoxify and eliminate these toxicants. At one time this process was called detoxification (meaning to make less toxic); however, because this process sometimes results in the production of more toxic products it is now more commonly referred to as biotransformation or metabolism. This is a dynamic and continuous process—meaning that it is always going on in your body. The purpose of this process is to transform lipid-like (hydrophobic) toxicants into more polar (hydrophilic) forms so that they can be eliminated from the body. One of the body's defense mechanisms to protect cells and their internal workings involves making these toxicants more polar and more water soluble to prevent these compounds from entering cells through their lipid membranes.

Once toxicants enter the body, most pass through the liver where they undergo biotransformation with the purpose being to enable their elimination. The principal route of elimination is through the

Laboratory Safety for Chemistry Students, by Robert H. Hill, Jr. and David C. Finster
Copyright © 2010 John Wiley & Sons, Inc.

kidney and out in the urine. However, most nonpolar toxicants cannot be eliminated in their native form. They must be transformed into more polar, less lipid-like forms that can more easily be eliminated. The body has evolved a three-step system to accomplish this—Phase I Biotransformation, Phase II Biotransformation, and Elimination.

Phase I Biotransformation of Chemical Toxicants

Phase I Biotransformation is the first step in this process to break down chemical toxicants into products (metabolites) that are not harmful to the body and that can easily be excreted. In Phase I Biotransformation, your body, using selected enzymes such as the P450 series, seeks to convert the molecule into a more polar form by (1) removing some part of the toxicant, (2) modifying some part of the molecule, or (3) adding a polar functional group. These transformations are accomplished by enzymes that *hydrolyze, reduce, or oxidize* the toxicant, thereby producing a polar, reactive functional group that makes it more likely to be further transformed to a water-soluble molecule. See *Chemical Connection 4.3.2.1* Enzymes—The Cells' Tools for Biotransformations.

CHEMICAL CONNECTION *4.3.2.1*

ENZYMES—THE CELLS' TOOLS FOR BIOTRANSFORMATIONS

Enzymes are made of proteins and are the organic catalysts produced by living cells. Each cell contains a multienzyme system, and this enzyme system is complex and interrelated in substrates and products—that is, the product of one enzymatic reaction may become the substrate (reactant) for another enzymatic reaction. The protein structure of each enzyme is stereochemically specific and forms complexes with substrates to carry out transformations—an enzyme–substrate complex. This complex is then transformed with a coenzyme or cofactor to the biotransformed product.

Most xenobiotics are biotransformed by enzymes in the liver, although there is also significant xenobiotic enzyme activity elsewhere in the body. These hepatic enzymes have broad specificity and can metabolize a wide variety of xenobiotics. Within the liver the biotransforming enzymes are located in the microsomes (smooth endoplasmic reticulum), due to the solubility of lipophilic xenobiotics in lipid membranes. There is also additional enzymatic activity in the cytosol and to a smaller extent in other areas of the cell.

There are variations in the structures of these enzymes among individuals which can produce different rates of biotransformation of xenobiotics—hence, some people are more susceptible to some toxicants while other people are more resistant to those toxicants. This may be because some people lack key enzymes, or, if present, these enzymes do not work well. This is why, for example, medications work well for some people but not others.

Cytochrome P450 enzymes are the most widespread, active, and versatile in their xenobiotic Phase I transformation activity. These enzymes are composed of heme-containing enzymes in the ferric ion state. In transformations the ferric ion is reduced to the ferrous ion that can bind O_2 and CO. These enzymes basically add oxygen or remove hydrogen in a stepwise process to generate Phase I Biotransformation products. Most cytochrome P450 transformations require an additional enzyme (coenzyme) to assist in the transfer of electrons. Cytochrome P450 enzymes carry out many kinds of oxidations—hydroxylations, epoxidations, heteroatom oxidations, N-hydroxylations, dealkylations, ester hydrolysis, and dehydrogenation.

Those seeking further information about this topic should consult Andrew Parkinson, Chapter 6, Biotransformation of Xenobiotics, in *Casarett & Doull's Toxicology: The Basic Science of Poisons*.[3]

Hydrolysis involves breaking down compounds, such as carboxyl esters, amides, epoxides, or ethers, and this process usually produces free carboxylic acids or hydroxyl groups. *Oxidation* is accomplished in a number of ways—adding a hydroxyl group or an epoxide group, oxidizing an alcohol to an aldehyde, or oxidizing an aldehyde to an acid. For example, toluene is oxidized in stages through benzyl alcohol, then benzaldehyde to benzoic acid—a less toxic, more polar product. *Reduction* is the common process for converting nitro compounds to water-soluble compounds that can more easily be eliminated. Thus, the nitro group in nitrobenzene is reduced to a hydroxylamine and finally to an amine to produce aniline—both products are more polar than the parent molecule. The purpose of all of these

TABLE 4.3.2.1 Potential Phase I Biotransformation Products of Various Organic Functional Groups

Functional group[a]	Phase I process: intermediate products	Phase I Biotransformation products[a]
Acids: —CO_2H		Acids: —CO_2H
Alcohols: —OH	*Oxidation*: Aldehydes: —CHO	Alcohols: —OH; Acids: —CO_2H
Aldehydes: —CHO	*Oxidation*	Acids: —CO_2H
Ketones: —CO—	*Reduction*	Alcohols: —OH
Esters: —CO_2R	*Hydrolysis*	Alcohols: —OH; Acids: —CO_2H
Epoxides: —COC	*Hydrolysis*	Alcohols: —OH {Diols}
Ethers, —COC—	*Hydrolysis*	Alcohols: —OH
Amides: —$CONH_2$	*Hydrolysis*	Acids: —CO_2H
Amines, aliphatic: —CNH_2	*Oxidation:* Ketones: —CO—	Alcohols: —OH
Amines, alkylaliphatic: —CNR_2	*Oxidation:* Hydroxyalkyl: —$CNRCH_2OH$; Dealkylated: —CNHR; CNH_2; Ketones: —CO—	Alcohols: —OH
Amines, aromatic: R—NH_2		Amines, aromatic: R—NH_2
Nitroaliphatics: —CNO_2	*Reduction:* Amines, aliphatic: —CNH_2; Ketones: —CO—	Alcohols: —OH
Nitroaromatics: R—NO_2	*Reduction:* Aromatic hydroxylamines: R—NHOH	Amines, aromatic: R—NH_2
Aromatics with aliphatic groups: R—CHR′R″	*Oxidation:* Aromatic with hydroxy side chains: R—CHR′—OH	Aromatic acids: R—CO_2H
Aromatics without aliphatic groups	*Oxidation*	Phenols: R—OH
Phenols: R—OH		Phenols: R—OH
Thiols: —SH		Thiols: —SH
Thiol ethers: —S—R	*Hydrolysis*	Thiols: —SH

[a] Same entries in both columns 1 and 3 indicate that these compounds ordinarily do not change in Phase I of this multistep process.

transformations is to provide a more polar functional group that can be successfully transformed in another step to a compound that can be eliminated in the urine.

Table 4.3.2.1 shows Phase I Biotransformations for various organic functional groups. Thus, if you identify the functional groups within a molecule, you will probably be able to generally predict the Phase I Biotransformation product. While this table can predict in a general way what kind of Phase I Biotransformations may take place, it may not predict all products in some cases. You should look for information about the biotransformation or metabolism of specific compounds if you need to know more specifics. The point of this table is to illustrate how all of these varied functional groups are transformed to more polar groups, such as —CO_2H, —OH, —NH_2, and —SH, so that the products can be directly eliminated or further transformed in Phase II Biotransformations to water-soluble products.

While the goal of Phase I Biotransformation is to convert chemical toxicants into less toxic forms, sometimes the result is in fact a more toxic product. Ethylene glycol ($HOCH_2CH_2OH$) is a solvent commonly used in antifreeze products and deicers. Ethylene glycol is toxic by ingestion and about 100 people die each year from ethylene glycol poisoning. However, ethylene glycol is not the direct cause of the poisoning. When ethylene glycol is absorbed it is metabolized (biotransformed) to glyceraldehyde ($HOCH_2CHO$) and then to glycolic acid ($HOCH_2CO_2H$). Glycolic acid builds up to cause a condition called metabolic acidosis—a disturbance of the acid–base balance in the body with accumulation of excess acid that can cause hyperventilation. Glycolic acid is further converted to glyoxylic acid ($OHCCO_2H$) and this is finally converted to oxalic acid (HO_2CCO_2H)—a very toxic product. Oxalic acid reacts with calcium forming calcium oxalate, and this compound crystalizes in the kidney, causing serious damage to its vital function. These crystals also damage and clog small blood vessels in the brain, leading to neurological effects. Eventually, if untreated, acidosis and damage from calcium oxalate crystals cause sufficient damage to organ systems to be fatal. This example illustrates how the body's defense against chemical toxicants using Phase I Biotransformation is not always effective and may actually result in producing much more toxic products.

Phase II Biotransformation—Conjugation

Following Phase I Biotransformation, the next step in the process to eliminate chemical toxicants is called Phase II Biotransformation. In this process additional enzymes carry out reactions on the Phase I biotransformed products by adding functionalities that will make them very water soluble. This process is called conjugation because it links (or conjugates) the Phase I transformed product with another compound to produce a highly water-soluble product. The major kinds of conjugation in Phase II Biotransformation are:

- **Amino acid conjugates.** Carboxylic acids react with amino acids, such as glycine or glutamine, to form amides. For example, toluene is biotransformed to benzoic acid, and benzoic acid undergoes conjugation in Phase II with glycine to form hippuric acid ($C_6H_5CONHCO_2H$).

- **Glucuronic acid conjugates.** Glucuronic acid, $C_6H_{10}O_7$, is an oxidation product of glucose and it is commonly used in conjugations with phenols, alcohols, and carboxylic acids to form glucuronides. For example, 1,4-dichlorobenzene is transformed in Phase I Biotransformation to 2,5-dichlorophenol. This phenol in Phase II Biotransformation is conjugated with glucuronic acid to form 2,5-dichlorophenylglucuronide—an O-glucuronide (an ether).

- **Sulfate conjugates.** A series of enzymes called sulfotransferases add sulfate groups to the same set of compounds that undergo glucuronidation. Thus, phenols would be conjugated to produce sulfates—phenol is converted to phenylsulfate ($C_6H_5OSO_3^-$).

- **Glutathione conjugates.** Glutathione is a tripeptide (from glycine, cysteine, and glutamic acid). The thiol function of glutathione reacts with Phase I biotransformed products to produce Phase II conjugates that can be excreted directly in the bile or after further conversion to mercapturic acid derivatives can be excreted into the urine. In the formation of mercapturates, the glutathione derivative of the Phase I metabolite reacts with other enzymes to remove glutamic acid, then glucine, to produce a cysteine conjugate. The cysteine conjugate undergoes N-acetylation to produce the mercapturic acid product—this product is excreted into the urine. Bromobenzene is converted to a phenol in Phase I Biotransformation, followed by glutathione conjugation, and ending in formation of 4-bromophenylmercapturic acid [$BrC_6H_4SCH_2CH(NHCOCH_3)CO_2H$].

- **Acetylated conjugates.** Phase I Biotransformation products that contain amino, hydroxyl, or sulfur groups may be acetylated, producing conjugates that can be excreted in the urine. An example of a Phase I product that would be acetylated is p-aminobenzoic acid, producing p-(N-acetamido)benzoic acid.

Bioaccumulation of Chemical Toxicants

Most chemical toxicants are biotransformed and eliminated from the body in urine or bile; however, some xenobiotics are not amenable to these transformations—that is, the body's enzyme systems are ineffective in transforming these compounds to more polar forms. Since these compounds are often very lipid soluble and cannot be quickly eliminated, the body stores them in our fat tissues throughout the body—a process known as bioaccumulation. In order for bioaccumulation to occur, a chemical toxicant must be absorbed faster than it is eliminated. These chemical toxicants are stored in the body's fat or lipids until a dynamic state is reached where the rates of absorption and elimination become about the same. The concentrations of these xenobiotics in fat tissues can be estimated from concentrations in blood that normally carries a small amount of circulating lipid, as reported in Incident 4.3.2.1. (See also in Section 4.2.1, *Chemical Connection 4.2.1.1* Solubility, Storage, and Elimination.)

The fact that these chemical toxicants are being stored in our fat tissue should not be considered good or bad but rather a reflection of exposure—it is simply the way the body deals with these kinds of compounds because the body cannot eliminate or use them quickly. In fact, many of us have these kinds of xenobiotics stored in our fat tissues at very low concentrations. Chlorinated or brominated hydrocarbon pesticides or pollutants often fit into the group of compounds that bioaccumulate.

The amount of time that a chemical remains in the body is measured as the biological half-life—also known as the half-life of elimination, and designated as $T_{1/2}$. This is the amount of time it takes for half of the concentration to be eliminated from the body. For compounds that are quickly metabolized and eliminated, this biological half-life is short—a matter of hours or days. For example, proteins in your body's liver have a biological half-life of around 10 days. Many reactive xenobiotics have half-lives of only a few hours. Metallic mercury is a chronic poison in low, continuous doses and has a half-life of 20–90 days. Compounds such as the chlorinated hydrocarbons can be stored for long periods of time, often for years.

The compound that is best known and most well studied for its bioaccumulation is the pesticide known as DDT, or dichlorodiphenyltrichloroethane. It was a very popular and valued pesticide for many years in the first part of the 20th century and contributed significantly to the improvement of public health, especially for eradicating mosquitoes—carriers of parasites that cause diseases such as malaria. Rachel Carson, writing in *Silent Spring*,[4] caught the public's attention when she reported DDT and other pesticides' long-term effects on our environment and particularly our bird population. This book is often credited with starting the "environmental" movement in today's world. Particularly because of concerns for its effects on bird populations, DDT and other chlorinated hydrocarbon pesticides were banned in the 1970s. Nevertheless, DDT, particularly in the form of its environmental degradation product (and metabolite) DDE, still remains in our food supply and in our fat tissues even many years after its use was discontinued. The biological half-life for DDT (as metabolites: 70% DDE, 30% DDD) in humans is dependent on the concentration (see Figure 4.3.2.1). At higher levels, $T_{1/2}$ is about 2 years but as the concentration gets lower $T_{1/2}$ can be around 8 years. (*Note:* The nonconstant value of $T_{1/2}$ suggests that this decay is not a first-order reaction, unlike most situations where half-lives are quoted.)

Another group of compounds that are similar in their propensity to be stored in body fat are the polychlorinated biphenyls, aka PCB, which theoretically have 209 congeners. These compounds were

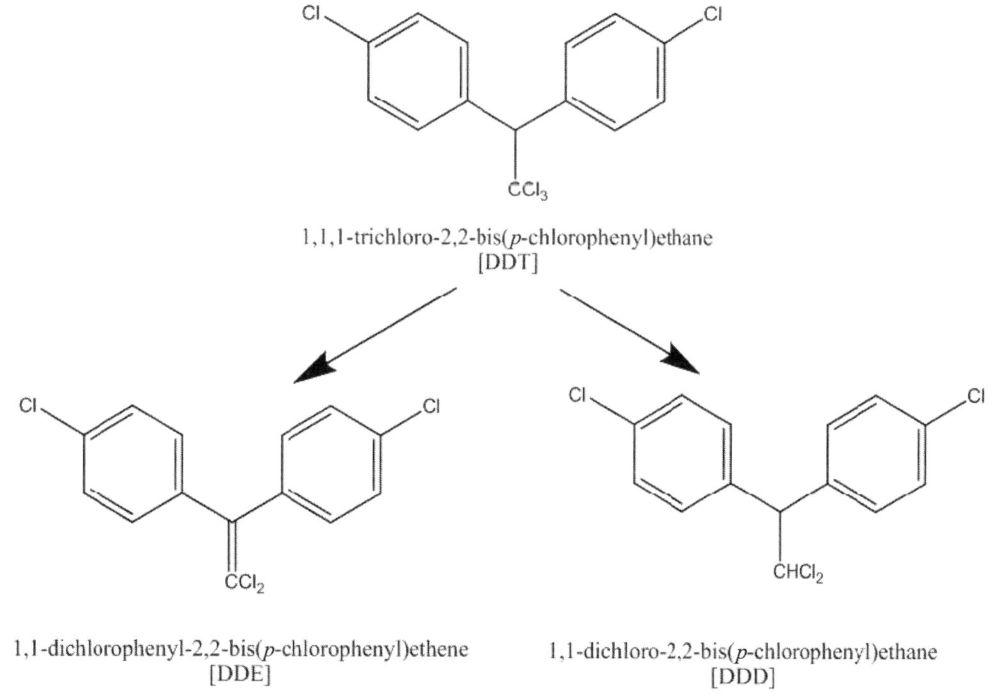

1,1,1-trichloro-2,2-bis(*p*-chlorophenyl)ethane
[DDT]

1,1-dichlorophenyl-2,2-bis(*p*-chlorophenyl)ethene
[DDE]

1,1-dichloro-2,2-bis(*p*-chlorophenyl)ethane
[DDD]

FIGURE 4.3.2.1 DDT and Its Metabolites. DDT was a widely used and effective pesticide employed to treat body lice infestations and to prevent the spread of malaria. However, its persistence in the environment and its damage to the ecosystem led to its ban in many places, including the United States.

widely used in electrical transformers for many years, until concerns about their long-term stability and exposure leading to potential health effects led to their discontinued use. Nevertheless, these compounds also remain in our environment and in our fat tissues. The biological half-life for PCBs in humans is dependent on the concentration and degree of chlorination—$T_{1/2}$ is probably around 2–4 years but as the concentration gets lower $T_{1/2}$ can become longer.

Although these compounds are stored in our body fat in low concentrations, many scientists believe that there is not strong evidence indicating that these compounds produce any significant health effects resulting from exposure to these chemicals in the low levels from our normal daily food supply; however, this is a subject of a great deal of controversy (see *Special Topic 4.3.2.1* Controversy in Science—Epidemiological Challenges and Degrees of Certainty).

SPECIAL TOPIC 4.3.2.1

CONTROVERSY IN SCIENCE—EPIDEMIOLOGICAL CHALLENGES AND DEGREES OF CERTAINTY

The presence in our fat tissues of xenobiotics with long half-lives, such as complex chlorinated hydrocarbons, is a well-established fact. In fact, it is unlikely to find any humans on Earth without some of these chemicals in their bodies. The concentration of these chemicals in the fat tissue is in continuous equilibrium with the fat (lipids) in the blood; thus, as the body slowly eliminates these chemicals from the blood, more of the chemical is "pulled" from the fat (i.e., to maintain the equilibrium). This raises the question of what effects these chemicals might have on individuals and populations, and how these effects might be tested.

Epidemiologically, the "gold standard" of testing is to identify two statistically large-enough and otherwise "matched" populations of humans, who either have or do not have at measurable concentrations the xenobiotic chemicals in their bodies and then determine if any biological or medical differences exist. Unfortunately, in this instance because we can no longer find the "control" group (without xenobiotics), this kind of epidemiological study cannot be done. Less definitively, but perhaps still useful, might be to identify populations with "high" versus "low" concentrations and look for response differences. However, this would require sampling an enormous number of individuals, carrying out an enormous number of complex analyses, and also then knowing what "response" to assess. Using extensive medical histories (which also requires much work), it might be possible to identify differences between these groups. And it would be good to involve a wide range of ages since many diseases might have long induction periods. As far as we know, this kind of study has not been done and, for practical reasons, might never be done due to its complexity and enormous expense.

Our best hope of understanding possible effects of these long-term low-level exposures is to examine studies of various groups that have been highly exposed through accidents, occupations, or gross environmental contamination. While these studies may have shown some effects, their applicability to populations with the much lower concentrations, as in most people, is unclear. And when the observed effects are considered in light of the normal aging and chronic disease processes, it makes for a very difficult and complex area to evaluate or interpret.

Some scientists also believe that it simply seems very unlikely that at the very low levels for some of these xenobiotics, there could even be an observable effect. In a practical sense, the huge amount of money that would be spent on these kinds of studies would be better spent on many other areas of public health, where we know the outcomes would be favorable.

Not all scientists agree with this argument, however. There is evidence that some xenobiotic substances have clearly been identified as interfering with animal endocrine (hormonal) systems. These molecules are called "endocrine disruptors." The effect in humans is more controversial and poorly understood. An Internet search of this topic will yield many web sites with opinions and "facts" in dispute.

As an historical note, the book *Our Stolen Future* was published in 1996[5] and launched much of the controversy surrounding the topic of endocrine disruptors. It has been compared to *Silent Spring* by Rachel Carson, published in 1962, a book about pesticide use that is widely credited for starting the environmental movement in the United States.[4]

Books like these often cause environmentalists and those in the chemical industry to adopt opposing positions because the proposition to ban the use of some chemicals has enormous economic impact. Studies that question the deleterious effects of chemicals are sometimes funded by industry groups whose objectivity and methodologies have come into question at times. Finding "the simple, scientific truth" is sometimes problematic. Methodological questions and interpretations of results often lead to uncertainties in conclusions.

Elimination of Chemical Toxicants

The pharmacokinetic process following exposure of chemicals is absorption into the blood, distribution of the chemical within the body, potential biotransformation (metabolism) of the chemical, and elimination. This process is remembered by the acronym ADME (absorption, distribution, metabolism, and elimination). The body is usually very effective in eliminating chemical toxicants, although some may take longer than others to make their exit. The kidneys are a major exit route for the elimination of the more polar chemical toxicants or metabolites. After the kidney has filtered the blood, needed nutrients are reabsorbed into the blood but hopefully unwanted products are eliminated in the urine. Exhaled breath from the lungs is another major route for elimination of volatile chemicals. Other chemical toxicants may be metabolized and eliminated into the bile that empties into the small intestine and later is eliminated (if not reabsorbed) in the feces—these are generally the more nonpolar, more lipophilic chemicals. It should be remembered that some chemicals, depending not only on the chemical but the matrix it is in, are not absorbed into the blood at all or are absorbed in small amounts and are eliminated directly into the feces.. Overall, the body does well in handling exposures to chemical toxicants, especially at low levels of exposure.

Working with Xenobiotic Substances

Chemistry students and chemists recognize the hazards of working with chemicals and take prudent steps to avoid unnecessary exposures. It is important to neither underestimate nor overestimate the risks. Underestimation may lead to risky exposures and overestimation may lead to taking unnecessary actions that cost time and money. In fact, it is human nature to become *less* fearful of events and situations that become "common" in our lives and this leads chemists to perhaps underestimate the risks from chemicals because we do not fear substances just because they have long, complicated names. So, we must constantly assess the actual risk level of any chemical or chemical reaction.

When working in laboratories, exposures to xenobiotic chemicals can occur in very limited ways—inhalation, skin, or eye exposures. (Ingestion is not a potential route because there is no eating or drinking in the lab.) We can minimize exposure through the use of containment (i.e., chemical hoods), through personal protection, and through safe practices. While these procedures minimize exposures, exposures probably cannot be eliminated completely.

It is not unusual for pregnant women to express concerns about working in the laboratory. For those students who might be expecting, you should know the following:

- There are very few known teratogens (chemicals that can cause damage to the fetus), and it is very unlikely that your institution would allow the use of a known teratogen in structured laboratory sessions.

- Many institutions can provide consultations with health and safety professionals to evaluate any potential exposures in laboratories. So if one has a concern, it might be possible to receive this kind of evaluation.

- Pregnant women concerned about working in the laboratory should always consult their personal physician to decide if working in the laboratory is safe for the baby.

- Laws prohibit employers from preventing a pregnant woman from working in any workplace. While this ruling does not apply directly to students, it is not likely that the institution that you are attending would take any steps to prevent you from working in the lab. Rather, the institution would probably take the view that the decision to work in the laboratory is up to the individual and her physician.

- Since most academic labs, except for research labs, are usually very structured and use experiments that have been tested and used by many students in the past, it is not likely that there would be experiments that have significant risks to anyone, including pregnant women.

RAMP

We apply the RAMP acronym to perform this assessment on a daily basis.

- *Recognize* that xenobiotics, including laboratory chemicals, are normally and continuously processed and eliminated from our bodies by an established mechanism. The level of risk posed by these hazards is related to the toxicity and dose of the substance.

- *Assess* the risk of the hazard of exposures to laboratory chemicals or xenobiotics by learning more about potential adverse effects and assessing the likelihood of exposure.

- *Minimize* exposures to laboratory chemicals or other xenobiotics, particularly those involving repeated or occupational exposures.

- *Prepare* for emergencies such as chemical spills so that you can avoid and minimize exposures to laboratory chemicals. "Preparation" involves having a plan to respond to these emergencies.

References

1. Socrates. BrainyQuote; available at http://www.brainyquote.com/quotes/authors/s/socrates.html (accessed September 14, 2009).
2. A. SCHECTER and J. J. RYAN. Persistent brominated and chlorinated dioxin blood levels in a chemist 35 years after dioxin exposure. *Journal of Occupational Medicine*, **34**:702–707 (1992).
3. Andrew Parkinson. Biotransformation of Xenobiotics. In: *Casarett & Doull's Toxicology: The Basic Science of Poisons* (CURTIS D. KLAASSEN, ed.), McGraw-Hill, New York, 2001, Chapter 6, pp. 133–224.
4. R. CARSON. *Silent Spring*, Fawcett Publications, 1962.
5. T. COLBORN, D. DUMANOSKI, and J. P. MYERS. *Our Stolen Future: How We Are Threatening Our Fertility, Intelligence, and Survival—A Scientific Detective Story*. Dutton, New York, 1996.

QUESTIONS

1. A xenobiotic substance is one that
   - (a) Has an unusually high toxicity
   - (b) Is usually eliminated before it can metabolize
   - (c) Does not normally occur in the body
   - (d) Was synthesized by Zena the Warrior Princess

2. The purpose of biotransformation of a xenobiotic substance in the body is to change it into a
   - (a) Naturally occurring, less toxic substance
   - (b) More hydrophilic form that can be eliminated more easily
   - (c) More hydrophobic form that can be eliminated more easily
   - (d) More hydrophobic form that is less toxic

3. In Phase I Biotransformation, enzymes are used to effect which chemical reaction?
   - (a) Hydrolysis
   - (b) Reduction
   - (c) Oxidation
   - (d) All of the above

4. In Phase I Biotransformation, epoxide function groups become
   - (a) Alcohols
   - (b) Carboxylic acids
   - (c) Phenols
   - (d) Ethers

5. In Phase I Biotransformation, chemical toxicants
   - (a) Are always converted into less toxic substances
   - (b) Are always converted into more toxic substances
   - (c) Can be converted into substances that can be either more or less toxic
   - (d) None of the above

6. In Phase II Biotransformation, what class of molecules is used to effect conjugation?
   - (a) Amino acids
   - (b) Carboxylic acids
   - (c) Alcohols
   - (d) Enzymes

7. In Phase II Biotransformation, xenobiotic substances are converted to
   - (a) More hydrophobic forms that more easily react with enzymes
   - (b) More hydrophilic forms
   - (c) Smaller molecules that are more hydrophilic
   - (d) Larger molecules that are more hydrophobic

8. In order for bioaccumulation to occur, a chemical toxicant must be
   - (a) Absorbed faster than it is eliminated
   - (b) Absorbed slower than it is eliminated
   - (c) Converted to a more hydrophilic form
   - (d) Converted to a more hydrophobic form

9. The half-life for the elimination of bioaccumulated substances is
   - (a) On the order of hours
   - (b) On the order of days
   - (c) On the order of years
   - (d) All of the above

4.3.3

BIOLOGICAL HAZARDS AND BIOSAFETY

Preview   This section discusses the hazards of biological agents, where and how they might be encountered, and some general approaches to prevent exposures.

A large number of laboratory workers handle such [hazardous biological] materials as part of their daily routine. The number has been estimated to be about 500,000 in the United States, but that number is probably a gross underestimate.

Committee on Hazardous Biological Substances in the Laboratory,[1] National Research Council

INCIDENT 4.3.3.1   EXPOSURE TO HUMAN SERUM[2]

A researcher was preparing to do some analytical measurements on human serum samples. She used long Pasteur pipettes to transfer the serum samples to other tubes. When she was done, the researcher put the used pipettes in a metal pan for decontamination via autoclaving. Another researcher used the metal pan to dispose of his pipettes but when he lifted the lid on the autoclave pan and reached in to put in his pipettes, one of the pipettes in the box stuck him in the lower part of his wrist. He learned that the pipette had been used in transferring human serum. He washed this immediately with soap and went to a clinic. The physician advised him that all he could do was to try and learn if the serum had any infectious materials—this could not easily be determined. Otherwise, he would have to be monitored to see if he would develop any symptoms of an infectious disease or markers for developed immunity to viruses.

What lessons can be learned from this incident?

Viruses and Bacteria—Agents of Infection!

Biological, or microbiological, agents are another class of hazards that you may encounter in the laboratory. As you can see from the quote above, there are opportunities for people working in clinical, hospital, or physician-office laboratories to be exposed to these microbiological infectious materials. The Occupational Health and Safety Administration (OSHA) estimated that about one-quarter of those opportunities occur in academic laboratories.[3] Biological agents may also exist in chemical or microbiological research and teaching laboratories. Additionally, microbiological agents are ubiquitous in our world—they can occur everywhere but fortunately only a very few are really dangerous to humans (pathogenic). This section provides an introduction to the potential hazards of microbiological agents in the laboratory, but it is not comprehensive. If you find that your laboratory research offers the potential for exposure to microbiological agents, it is important that you learn more about this area.[1,4,5] *Biosafety in Microbiological and Biomedical Laboratories*, 5th edition, is an outstanding guide to laboratory safety with infectious agents.[4]

While there are several classes of pathogenic microbiological agents, we will focus on the two major classes that are of greatest concern to laboratorians: viruses and bacteria.

- *Viruses* are subcellular and submicroscopic agents of varied shapes that have genes made of a nucleic acid (DNA or RNA) central core. This core is enclosed in a protective protein, and it can also be enveloped in a lipoprotein membrane. Viruses are not considered to be living organisms.

Laboratory Safety for Chemistry Students, by Robert H. Hill, Jr. and David C. Finster
Copyright © 2010 John Wiley & Sons, Inc.

They are very small in size, ranging between 0.02 and 0.3 μm. Viruses cannot sustain themselves; they rely on invading a host cell and using that cell's energy and biosynthesis mechanisms to replicate themselves under appropriate conditions. Antibiotics are not effective in treating viral infections. Immunizations are used to prevent illnesses from certain viruses.

- *Bacteria* are cellular organisms of varied shapes (spheres, cones, cylinders, rods, curved rods, filaments) and sizes—most being 0.5–1.0 μm wide by 2–5 μm long. They have cell walls that have some components that react with various dyes—enabling them to be classified as gram-positive (retaining the dye) or gram-negative (not retaining the dye). Bacteria have all the internal structures of normal cells and are able to reproduce under appropriate conditions within a host. Bacteria can be grown as "colonies" in appropriate solid culture media. Antibiotics are often used for bacterial infections. See Figure 3.1.1.1 to compare the size of biological organisms such as viruses or bacteria with aerosols.

Opportunities for Exposures to Biological Agents in Laboratories

In the early part of the 20th century, there were many laboratory-acquired bacterial infections that resulted from laboratorians being carelessly exposed to aerosols of infectious materials, especially human specimens. These aerosols were generated as a result of manipulating infectious materials, such as blood or serum, in open operations, such as open centrifuge tubes, mixing open vessels containing infectious materials, or spilling infectious materials. As this was recognized and more emphasis was placed on containment as a strategy to prevent the uncontrolled release of infectious aerosols, these laboratory-acquired infections (LAIs) began to drop. LAIs still happen today through accidental spills and dropped or broken tubes or flasks, but there is clearly a strong attempt to prevent exposures to aerosolized infectious materials.

In the latter part of the 20th century, laboratory-acquired viral infections became an important concern. Human immunodeficiency virus (HIV) was discovered and the concern for handling blood specimens from HIV-positive or suspected samples became a driving force for more protections for the laboratory worker. This was especially true since there was no vaccine to prevent HIV infections and no known cure for acquired immunodeficiency syndrome (AIDS). Laboratory-acquired HIV infections are extremely rare.

However, infections from other viruses present significant risks. The hepatitis viruses (A, B, C, and D) became recognized as very real hazards for laboratory workers who handled human specimens, particularly blood. The risk of hepatitis infections from exposure to materials containing these viruses is much greater than the risk from HIV. Remarkably, a recent report[6] indicated dramatic reductions in acute hepatitis infections, principally due to strong vaccination programs for children for hepatitis A and B—a clear illustration of the need for hepatitis vaccination for laboratory workers exposed to human specimens. See *Special Topic 4.3.3.1* Chemical Analysis of Human Specimens.

SPECIAL TOPIC *4.3.3.1*

CHEMICAL ANALYSIS OF HUMAN SPECIMENS

An important part of the chemical enterprise is the chemical analysis of human specimens, often called clinical chemistry. Our medical system depends tremendously on these clinical tests, which have been developed by chemists over many years. Clinical chemists routinely handle human specimens, such as blood, serum, plasma, urine, saliva, spinal fluid, or feces, to provide data that help physicians and other health care providers determine medical treatments. There are many hundreds of these types of analytical methods. Some are automated and some have to be performed individually by a chemist or other trained professional.

A well-known example of this is the analyses for cholesterol and lipids that are used to determine risks of heart disease. Phlebotomists collect a sample of blood, which is then sent to a laboratory for analysis. A standard sample volume called an *aliquot* is extracted with a nonaqueous solvent to isolate the fatty portion of the blood. Specific tests for cholesterol and other lipids are then performed. Everyone who handles the sample must follow proven procedures to prevent exposures to bloodborne pathogens that might coexist in the blood. Clinical chemists have refined these methods so that they are reproducible and accurate, which is exactly what is desired, of course, when making a decision about prescribing drugs that (hopefully) have positive effects but also have negative side effects.

The level of total cholesterol is an indication of the risk of heart disease. Levels over 200 mg/dL are generally taken to indicate that there may be increased risk of heart disease. Similarly, HDL (high density lipoprotein), also called "good" cholesterol, should have levels of 60 mg/dL or higher. LDL (low density lipoprotein) is called "bad" cholesterol and levels of less than 100 mg/dL are desired. It is interesting to note that some dimensions arise from an artifact of habit. You probably have not encountered the "deciliter" in any chemistry labs but this is the customary unit used to report cholesterol levels.

You probably know students who play college sports and have had their urine tested for illicit drugs. These are the same tests used in the Olympics and in various professional sports. Everyone who handles these urine samples must follow universal precautions with regard to BBP exposure. Samples are first tested with a "screening procedure" that is relatively quick and inexpensive but also has a known rate of false positives and false negatives. Samples that test positive are subjected to a second analysis that is more accurate, and more expensive, so that false positives from the first test do not incorrectly identify drugs in a sample from a drug-free individual.

Bloodborne Pathogens—Hazards from Handling Human Blood Specimens

The very nature of infectious agents makes these hazards different from chemical hazards. A seemingly small exposure can result in a significant infection since an infectious agent has the propensity to multiply once it has entered the host (you). It is therefore important that, as with hazardous chemicals, you use a strategy that will prevent exposure to these agents.

In 1991 OSHA published a regulation—"Occupational Exposure to Bloodborne Pathogens." OSHA has since revised this standard in 1999 and again in 2001. You will note from the title that while there are other infectious materials, the emphasis is on infectious agents in blood, aka bloodborne pathogens or BBPs.

As with chemical hazards there are several potential routes of exposure to infectious agents in the laboratory—ingestion, dermal exposure, eye exposures, inhalation, and injection.

- Ingestion should not be an important route of exposure in the laboratory since safe laboratory practices prohibit eating, drinking, smoking, application of cosmetics, taking medications, and so on in the laboratory. Mouth pipetting, used extensive in the early part of the 20th century, is now a prohibited practice in the laboratory.

- Dermal exposure can be especially important if you have any breaks or compromises in your skin—cuts, scrapes, abrasions, dermatitis, or other skin conditions—that could be exposed to splashes, aerosols, or other inadvertent exposures and provide entry points for microbiological agents.

- Eye exposures have resulted in dangerous infections from inadvertent splashes or spills of infectious materials that have entered the eyes.

- Inhalation is an important route of exposure to infectious agents that have become airborne through agitation from various laboratory operations and methods of containment are discussed below.

- Accidental injection is a real concern when handling potential infectious materials, such as blood samples, so methods to prevent these exposures from "sharps" are discussed below.

Preventing Exposures to Bloodborne Pathogens and Other Infectious Materials

Hepatitis B virus (HBV) infections can seriously damage the liver. The good news is that HBV infections can be prevented by immunization. It is recommended that all laboratory workers who routinely handle blood samples take the HBV immunizations to ensure that if they are inadvertently exposed they will be protected.

For BBPs, in general, however, special lab procedures need to be followed to avoid exposure. You may hear the term "universal precautions," which means that you should consider all human specimens (body fluids and tissues) as potentially infectious. Universal precautions involve several steps discussed below.

Eye Protection   The best eye protection is to wear splash goggles. Many folks in the medical field wear some form of glasses with some limited degree of side protection, but in the event of a splash these may not be completely effective.

Gloves   Wearing gloves is standard practice when working with BBPs. Latex gloves are the historical standard in the medical profession, but allergic reactions to latex are now common enough that nitrile gloves are now the most commonly used gloves. (In the medical settings, even if a doctor, nurse, or EMT is not sensitive to latex, the patient might be!)

When handling blood or other infectious materials, you must take extra precautions to prevent exposures. Cover all wounds, cuts, abrasions, and scrapes with bandages that have a nonabsorbent top cover, such as a Band-Aid® or tape. Find out if any nonaqueous chemicals are used in the process and ensure that these gloves are adequate for protection from these as well. (See Sections 7.1.3 and 7.2.2 for more about gloves.)

Since the hands are such a likely area of potential exposure, gloves are especially important. Any exposed skin presents a risk so many workers in this field commonly wear a lab coat as an extra layer of protection against splashes and other skin contact.

Washing Hands   In theory, wearing gloves should eliminate the need to wash your hands. However, let's think about how germs can "travel." Someone wearing gloves needs to constantly remember that anything they touch while wearing (potentially contaminated) gloves will also become contaminated. A researcher wearing gloves who then writes some notes in a notebook will potentially have contaminated the pen and the notebook. Even raising or lowering a hood sash contaminates the hood sash. After removing the gloves (which also needs to be done in a manner that avoids contact with the *outside* of the glove), it is easy for forget what has subsequently been contaminated. Touching these items with bare hands provides the opportunity for skin exposure. Needless to say, frequently washing ones hands is simply prudent behavior. You should wash your hands after handling any potentially infectious material and again wash your hands before eating, before going to the restroom, or before taking breaks.

Disposal of Sharps   Unintentional injection is a very real possibility when handling infectious materials if you are using needles or if you come in contact with other sharps from broken glassware that contained infectious materials. You should never try to recap, bend, or break needles; they should be discarded in a sharps container. Glass Pasteur pipettes, scalpels, box cutter blades, and any other items that have the potential to cause a stick or cutting injury should also be discarded in a sharps container. You should receive special training on work practices and disposal procedures for needles and other sharps with the goal of preventing injuries.

Disinfecting Work Areas   Since contamination can spread so easily and inadvertently, when handling potentially infectious agents, such as blood, it is important to frequently clean and disinfect the work surfaces where these materials might be used. A 1% dilution of common household bleach (5% sodium hypochlorite) is an effective disinfectant—these solutions must be made daily. Equipment that may be potentially contaminated with blood should also be wiped down to ensure disinfection. Not all bacteria or viruses can be inactivated by exposure to chlorine (bleach) so it is best to seek information about the correct disinfectant for the agents in use. Public Health Agency of Canada has developed MSDSs for infectious agents and these MSDSs contain information about disinfection and laboratory safety.[7]

Finally, containment is a critical part of the process to prevent infectious agent exposures. You should use chemical hoods or laminar flow biological safety cabinets (BSCs), depending on what you are doing. All pipetting should be carried out with a mechanical pipettor—*never* mouth pipette. You need to focus on prevention of exposures to aerosols of potentially infectious materials. You should not mix bloodborne pathogen solutions in open tubes—always ensure that they are capped and when uncapped they should be in a chemical hood or BSC. Centrifuge tubes should be capped and if a tube is suspected of breaking during centrifugation, you must take precautions to ensure that the aerosols are not points of exposure. Internal chambers of centrifuges should be periodically disinfected and should be disinfected if a tube has broken during processing.

Biosafety Levels — Protections When Using Microbiological Agents in the Laboratory

As a chemistry student, most of you will not encounter laboratories where microbiological agents in their purified states are commonly used. However, these agents are often used in microbiological laboratories in academic settings and with the large amount of crossover of scientific disciplines, you should have at least a very brief introduction to standard practices for microbiological and biomedical laboratories.

The Centers for Disease Control and Prevention (CDC) and the National Institutes of Health (NIH) have established guidelines for safely using microbiological agents in laboratories. Their publication known as *Biosafety in Microbiological and Biomedical Laboratories* (also known as the BMBL) describes various levels of practices for microbiological agents and these are called *biosafety levels* (BSLs).[4]

- BSL-1 practices are used for agents that are not known to cause disease in normal, healthy humans.
- BSL-2 practices are used for agents that, when exposed via ingestion, dermal exposure, or mucous membrane exposure, can produce human disease of varied severity. There are usually medical treatments or immunizations available for these agents.
- BSL-3 practices are used for agents that are infectious via aerosol exposure and inhalation. These agents can cause serious and even fatal diseases; however, there are treatments or immunizations available for these agents.
- BSL-4 practices are used for agents that are infectious via aerosol exposure and inhalation. Disease from these agents can be life-threatening and there are no known treatments or immunizations available against infections from these agents.

These practices emphasize containment of the biological agent. The practices and levels of containment increase as BSLs increase. So minimal precautions are used in BSL-1 designated laboratories, BSL-2 designated laboratories use increased precautions, BSL-3 laboratories require the use of BSCs, and BSL-4 laboratories require full containment—body suits, glove boxes, and so on. If you want to learn more about these practices go to http://www.cdc.gov/od/ohs/biosfty/bmbl5/bmbl5toc.htm.

RAMP

- *Recognize* that infectious agents can be very real hazards if you handle human specimens or other tissues or fluids from living organisms.
- *Assess* the level of risk for exposure based on lab procedures.
- *Minimize* the risk for exposure by using methods from the Bloodborne Pathogen Standard or the BMBL. You may want to consider immunizations to protect you against exposures to some viruses.
- *Prepare* for an unexpected exposure by knowing the correct response procedure.

References

1. National Research Council. *Biosafety in the Laboratory: Prudent Practices for the Handling and Disposal of Infectious Materials*, National Academy Press, Washington, DC, 1989.
2. R. HILL. Incident described by a colleague.
3. U.S. Department of Labor. *Profile of Laboratories with the Potential for Exposure to Toxic Substances*. Occupational Safety and Health Administration, Washington, DC, U.S. Government Printing Office [2, B, p. 8], 1983; cited in reference 1, p. 8: took 127,000/500,000.

4. Centers for Disease Control and Prevention (CDC)/National Institutes of Health (NIH). Biosafety in Microbiological and Biomedical Laboratories, 5th edition (or most current one), U.S. Government Printing Office, Washington, DC, 2007; available at http://www.cdc.gov/od/ohs/biosfty/bmbl5/bmbl5toc.htm (accessed October 30, 2008).
5. American Industrial Hygiene Association. *Biosafety: Reference Manual*, 2nd edition, AIHA, Fairfax, VA, 1995.
6. Centers for Disease Control and Prevention. Surveillance for Acute

Hepatitis—United States, 2007, *Morbidity and Mortality Weekly Report (MMWR)*, Vol. 58, Surveillance Summaries (SS-3), May 22, 2009; available at http://www.cdc.gov/mmwr/PDF/ss/ss5803.pdf (accessed September 3, 2009).

7. Public Health Agency of Canada. Material Safety Data Sheets (MSDS) for Infectious Agents; available at http://www.phac-aspc.gc.ca/msds-ftss/index-eng.php (accessed September 3, 2009).

QUESTIONS

1. Viruses

   (a) Have a central core of DNA or RNA
   (b) Have a protective protein coating
   (c) Are not alive
   (d) All of the above

2. Bacteria

   (a) Are alive
   (b) Are all easily dyed
   (c) Cannot be killed with antibiotics
   (d) Contain no DNA

3. Which poses the largest risk of infection upon exposure?

   (a) Hepatitis viruses
   (b) HIV
   (c) Hepatitis and HIV have the same risk
   (d) This depends on the individual

4. When comparing infectious agents with chemical toxicants,

   (a) For chemical toxicants, "the dose makes the poison," while for infection agents, the dose is·largely irrelevant since even a tiny exposure can multiply in the body
   (b) For both substances, "the dose makes the poison"
   (c) Using PPE for infectious agents is much more important since exposure usually leads to fatality

   (d) Different kinds of PPE are used for infectious agents as compared to chemical toxicants

5. Which practice is generally *not* needed when using infectious agents?

   (a) Using gloves
   (b) Avoiding mouth pipetting
   (c) Using chemical splash goggles
   (d) None of the above

6. Which two practices are *more* important to consider when handling infectious agents as compared to handling chemical toxicants?

   (a) Washing hands and disposing of sharps
   (b) Washing hands and disinfecting work areas
   (c) Disposing of sharps and disinfecting work areas
   (d) Disposing of sharps and using gloves

7. How many biosafety levels are established by the CDC and NIH?

   (a) Two
   (b) Three
   (c) Four
   (d) Five

CHAPTER 5

RECOGNIZING LABORATORY HAZARDS: PHYSICAL HAZARDS

WHAT MAKES chemicals burn or explode? What are the dangers associated with high-pressure gas cylinders or glass vacuum lines? How safe is it to use liquid nitrogen? Recognizing hazardous physical properties is an important skill needed to keep you safe. Chemists use a wide variety of chemicals with various physical properties and many types of uncommon equipment in laboratories. Many lab incidents occur because students, or even experienced chemists, don't learn enough about chemicals and equipment before using them. This chapter reviews the many ways we can get hurt in a laboratory, and ways to stay safe in the lab.

INTRODUCTORY

5.1.1 Corrosive Hazards in Introductory Chemistry Laboratories Describes the hazards of corrosive chemicals that most likely will be encountered in the first year of chemistry.

5.1.2 Flammables—Chemicals with Burning Passion Discusses flammable chemicals, why we use them, what starts a fire, what chemical properties tell us about flammability, what makes chemicals flammable, and how chemical structure is related to flammability.

INTERMEDIATE

5.2.1 Corrosives in Advanced Laboratories Describes the chemistry of corrosive action and corrosives found in advanced labs.

5.2.2 The Chemistry of Fire and Explosions Follows the topic of flammable chemicals in Section 5.1.2 with a discussion of the nature of fire and explosions as chemical reactions.

5.2.3 Incompatibles—A Clash of Violent Proportions Describes incompatibles, why they should not be mixed, and steps to prevent incompatibles from coming in contact with each other.

ADVANCED

5.3.1 Gas Cylinders and Cryogenic Liquid Tanks Discusses the hazards of gas cylinders and cryogenic liquid tanks, and safety measures to take when handling or working with these containers.

5.3.2 Peroxides—Potentially Explosive Hazards Discusses potentially explosive peroxides as hazardous contaminants in some chemicals, including their formation, detection, and care in handling.

5.3.3   Reactive and Unstable Laboratory Chemicals   Discusses the hazards of unstable chemicals that are encountered in research and other laboratories.

5.3.4   Hazards from Low- or High-Pressure Systems   Describes the hazards associated with laboratory equipment at either low or high pressures.

5.3.5   Electrical Hazards   Describes common electrical hazards encountered in the laboratory.

5.3.6   Housekeeping in the Research Laboratory—The Dangers of Messy Labs   Discusses hazards arising in the lab from poor housekeeping.

5.3.7   Nonionizing Radiation and Electric and Magnetic Fields   Presents an overview of regions of nonionizing electromagnetic radiation and electric and magnetic fields and their hazards in chemistry labs.

5.3.8   An Array of Rays—Ionizing Radiation Hazards in the Laboratory   Presents an overview of radiation properties and hazards in chemistry labs.

5.3.9   Cryogenic Hazards—A Chilling Experience   Describes the hazards of cryogens, and procedures to minimize the risk in handling these chemicals.

5.3.10   Runaway Reactions   Describes runaway reactions, their impact, their causes, and steps to prevent these incidents.

5.3.11   Hazards of Catalysts   Discusses the special hazards of catalysts, where you might encounter these, and how you can safely handle these materials in the laboratory.

5.1.1

CORROSIVE HAZARDS IN INTRODUCTORY CHEMISTRY LABORATORIES

Preview The hazards of corrosive chemicals most likely encountered in the first year of chemistry are discussed.

The art of life is the art of avoiding pain.

Thomas Jefferson [1]

INCIDENT 5.1.1.1 SULFURIC ACID SPILL[2]

Four 2.5-L bottles of sulfuric acid were being carried down the hall by students. As one student turned to the other, the bottles banged together and broke. Both students fell on the slippery acid, and another bottle broke. Another person came to help and also slipped and fell. All three suffered serious burns from the sulfuric acid and cuts from the broken glass.

What lessons can be learned from this incident?

Corrosives Destroy Things—Especially Your Skin!

Corrosive chemicals are common hazards in the laboratory and also in our homes in the form of cleaning agents. Corrosives are defined as chemicals that cause harm or injury by damaging and destroying tissue, such as eyes or skin, at the point of contact or the exposure site (see Figure 5.1.1.1). Table 5.1.1.1 lists some corrosives that may be encountered in the first year of chemistry laboratory experiments. Corrosives can be gases, liquids, solids, or solutions.

Most common corrosives, such as strong acids in solution and strong bases as solids or solutions, can cause severe burns and damage to skin or eyes upon contact. When using corrosives, minimizing exposure is obviously very important. By using proper precautions, however, you can safely use corrosives in your laboratory work.

As discussed in Section 3.1.2, you should always learn about the hazardous properties of the materials you will be using in the laboratory. Identify pathways for potential exposures, and determine in advance (requiring that you *plan*) how you are going to control or manage any potential exposures through protective measures. For example, wear eye protection, such as chemical goggles, hand protection, such as protective gloves, and body protection, such as an apron or laboratory coat, long-sleeved shirts, long pants or skirts, and closed-toe shoes to protect against direct contact with corrosives. If you need to carry bottles of concentrated acids or bases, use bottle carriers. These carriers are usually rubber or some sort of heavy plastic designed to contain the corrosives should the bottle break. Be sure that you know the location of nearby eyewash stations and safety showers, and how to use them. If you are using corrosives that might be inhaled in the form of a gas or a powder, or that might form a fume or mist, use these chemicals in a chemical laboratory hood. More thorough discussions of the use of this safety equipment can be found in Chapters 2 and 7.

The key to safety is to recognize the hazard (e.g., corrosive properties), assess the risk of exposure (how you might be exposed), minimize the hazard (use personal protective equipment and safety equipment), and prepare for emergencies (spills, splashes).

FIGURE 5.1.1.1   GHS Corrosive Pictogram. Pictograms effectively communicate a hazard without using words. This pictogram suggests that a particular liquid is corrosive to solids and to human skin. (Courtesy of the United Nations Economic Commission for Europe. Copyright © 2007 United Nations, New York and Geneva.)

TABLE 5.1.1.1   Corrosive Chemicals that May Be Encountered in First Year Chemistry Laboratories

Acids	Bases (alkalis)	Gases	Oxidizing agents
Hydrochloric acid (HCl)	Sodium hydroxide (NaOH)	Nitrogen dioxide (NO_2)	Hydrogen peroxide (H_2O_2)
Sulfuric acid (H_2SO_4)	Potassium hydroxide (KOH)	Ammonia (NH_3)	Potassium permanganate ($KMnO_4$)
Nitric acid (HNO_3)	Ammonium hydroxide (NH_4OH)		Nitric acid (HNO_3)

Acids—Safe Handling Prevents Burns

Acids can be highly corrosive and cause severe burns upon contact. The potential for injury by an acid depends on its chemical structure, the area of the exposure, the concentration of the solution, the duration of exposure, and the temperature of the solution. The greater the concentration, duration, and temperature encountered, the greater is the potential for harm. This means that you can reduce your risk when handling acids (and other corrosives) by using dilute solutions (if possible), taking action to rinse any corrosives away quickly if they come in contact with your skin, and limit using corrosives at elevated temperatures. Sometimes you cannot avoid the use of concentrated corrosives or elevated temperatures for reactions, but you should take greater precautions as the risks increase.

The three main strong acids in common use in general chemistry labs are hydrochloric acid, sulfuric acid, and nitric acid. When a "simple" strong acid is desired, HCl is usually the best choice since it is cheap. HCl is also called muriatic acid in some commercial applications. Sulfuric acid is used when a strong acid is needed but one wants to avoid the presence of chloride from HCl, which may precipitate some cations, or in the presence of strong oxidizers (such as permanganate), which can oxidize chloride to chlorine gas. At high concentrations sulfuric acid is a very strong dehydrating agent, which is also part of its corrosive action in contact with skin. Of these three strong acids, nitric acid is the most hazardous since it is also a powerful oxidizing agent (see below). Strong solutions of nitric acid contain and will generate toxic nitrogen oxide gases (NO_x).

Strong acids (and strong bases) in concentrations greater than 1 molar are usually corrosive. While hydrochloric acid causes severe damage at pH 1 (= 0.1 M), its effects are less severe above pH 3 (= 0.001 M). Acids generally cause damage to proteins, forming a substance called coagulum. As this substance accumulates it can block or prevent further damage to underlying tissues. This is why acids may be less damaging than bases (the action of bases is described below), but nevertheless all corrosives have the potential to be very damaging.

The acids listed in Table 5.1.1.1 are usually found in stockrooms in concentrated solutions (see Figure 5.1.1.2) that are diluted to the desired concentrations (see *Chemical Connection 5.1.1.1* What Do Concentrated and Dilute Mean?). These may be diluted by a laboratory technician prior to use in the laboratory, or as part of your learning experience as a student, you may prepare these dilutions yourself. When diluting an acid, you should always *pour* acid into water. While the reverse process, pouring water into acid, may sound like an equivalent procedure, it is not. (See *Chemical Connection 5.1.1.2* Why Does Adding a Concentrated Strong Acid to Water Cause a Violent Reaction?) Pouring water into acid can cause a violent thermal reaction that could result in spattering acid, greatly elevating your risk of exposure. So, *pour acid into water*.

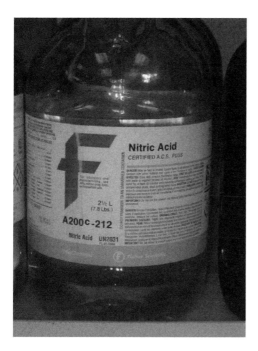

FIGURE 5.1.1.2 Concentrated Nitric Acid. Concentrated nitric acid is 15.8 M and 70% nitric acid by weight. It is very corrosive and produces toxic vapors. Exercise extreme caution if using this reagent.

CHEMICAL CONNECTION *5.1.1.1*

WHAT DO CONCENTRATED AND DILUTE MEAN

Chemists often use language and words in ways that are more specific than the same words used in a nonscientific conversation. While concentrated and dilute may just sound like relative terms to describe the molarity of solutions, in acid–base chemistry these have specific definitions. *Concentrated* means that the acid or base has the molarity found in Table 5.1.1.2. In fact, these are saturated solutions. *Dilute* is a less well-defined term and means that the concentration is lower than *concentrated*. This could be 6 M, 1 M, or less. So, while a "dilute solution" may sound like a "safe" one, 6 M is still pretty concentrated and probably quite hazardous.

TABLE 5.1.1.2 Molarities of Some Common Concentrated Acids and Bases at 25 °C

Acid	Formula	Molarity	Percentage, by weight	Base	Formula	Molarity	Percentage, by weight
Acetic (glacial)	$HC_2H_3O_2$	17.4	99%	Sodium hydroxide	NaOH	19.4	50%
Hydrochloric	HCl	11.6	36%	Potassium hydroxide	KOH	11.7	45%
Nitric	HNO_3	15.8	70%	Ammonium hydroxide	NH_4OH	14.2	56% as NH_4OH, 28% as NH_3
Phosphoric	H_3PO_4	14.7	85%				
Sulfuric	H_2SO_4	17.6	95%				

CHEMICAL CONNECTION *5.1.1.2*

WHY DOES ADDING A CONCENTRATED STRONG ACID TO WATER CAUSE A VIOLENT REACTION?

The process of mixing a strong acid with water will be very exothermic since the strong acid will be forming many strong hydrogen bonding interactions with water molecules and the formation of these strong interactions will release energy. Why does it matter if you add the acid to the water, or the water to the acid?

If we slowly add the acid to the water, we will limit the rate at which the heat is released since the acid will act as the limiting reagent. The water has a very high specific heat capacity (4.184 J/g-K) so it is very effective at absorbing the heat that is released. Finally, since the acid is almost certainly denser than the water, it will "sink" as it gets diluted and this will help to disperse the release of energy in the bulk of the solution.

If we add the water to the acid, the "excess" acid will react with all of the water quickly and the extensive hydrogen bonding will release a considerable amount of heat. Mostly, the solution is still the strong acid which will have a smaller specific heat capacity than water, and this will cause a larger increase in temperature. Since the water is less dense than the acid, the mixing is not very effective and the solution being heated is just the top layer of the solution. So, along with a smaller specific heat capacity, the volume of solution absorbing the heat is relatively small and this causes a much larger change in temperature. All of these factors can lead to sufficient increase in temperature that the solution boils at the surface and this takes the form of "spattering." What gets spattered is still quite a high concentration of hot, strong acid!

For similar reasons adding a strong base, such as NaOH pellets, to water must be done carefully. Add the pellets slowly to the water, with stirring. Sometimes this is best done in an ice bath to help absorb the heat that is released.

Common Corrosive Bases

The three most common bases that have corrosive properties are listed in Table 5.1.1.1: NaOH, KOH, and NH_4OH.

In times past sodium hydroxide was known as caustic soda, and potassium hydroxide was known as caustic potash, and both were called lye and both are highly corrosive. These bases are also known as alkalis, because they are products of the alkali metals Na and K.

Sodium hydroxide and potassium hydroxide are white solids that can be dissolved in water to make the solutions that you may encounter in the laboratory. These solutions are usually prepared by a laboratory assistant in order to save time in conducting experiments, but sometimes students may make their own NaOH or KOH solutions from the solid pellets. Don't be fooled by the pellets' appearance—they can be very harmful. Both liquid and solid forms are extremely corrosive and exposure can cause severe burns. Dissolving these hydroxide pellets in water generates a lot of heat, and a hot corrosive can be even more hazardous (see *Chemical Connection 5.1.1.3* Why Is the Heat of Solution So Exothermic for NaOH and KOH?). If you have some of these pellets in the palm of your hand, they will quickly absorb moisture from your hand and/or the air and can generate a very concentrated (saturated) solution. This will be very corrosive.

CHEMICAL CONNECTION *5.1.1.3*

WHY IS THE HEAT OF SOLUTION SO EXOTHERMIC FOR NAOH AND KOH?

When a salt dissolves in water, we can conceptually separate this into the processes of 1) separating the cations and anions from each other in the solid state into the gas phase, 2) and then letting these gas phase ions interact with the water as they go into solution:

(1) $\quad\quad\quad\quad\quad NaOH(s) \rightarrow Na^+(g) + OH^-(g) \quad \Delta H = +800\,\text{kJ/mol (estimated)}$
(2a) $\quad\quad\quad\quad\quad\quad\quad\quad Na^+(g) \rightarrow Na^+(aq) \quad \Delta H = -405\,\text{kJ/mol}$
(2b) $\quad\quad\quad\quad\quad\quad\quad\quad OH^-(g) \rightarrow OH^-(aq) \quad \Delta H = -453\,\text{kJ/mol}$

These three equations can be summed, using Hess' law:

(3) $\quad\quad\quad\quad\quad NaOH(s) \rightarrow Na^+(aq) + OH^-(aq) \quad \Delta H = -44.5\,\text{kJ/mol}$

Equation 1 shows a very endothermic process, which is what we would expect for the process of separating a +1 cation from a −1 anion. The Coulomb attraction between these ions will be very strong. The reverse of this equation shows the formation of the salt from the gaseous ions and the enthalpy term associated with this is called the lattice energy.

Equations 2a and 2b show the attraction of the +1 and −1 ions for the polar water molecule. In fact, it is clearly the exothermicity of these strong ion–dipole interactions that "outweigh" the endothermicity of the reverse of the lattice energy. So, when trying to account for why NaOH dissolves so endothermically, we can attribute this largely to the very strong ion–dipole interaction between the ions and water molecules.

This is understood a little more by appreciating that Equations 2a and 2b can be written more explicitly:

(2a′) $Na^+(g) + 6H_2O(l) \rightarrow [Na(H_2O)]^+(aq)$
(2b′) $OH^-(g) + 6H_2O(l) \rightarrow [OH(H_2O)]^-(aq)$

The equations show that about six water molecules surround the ion, so that there are actually six ion–dipole interactions per ion. This is called the *primary coordination sphere*. Beyond that there will still be more oriented water molecules around the primary coordination sphere in the *secondary coordination sphere*. All of these interactions contribute to the values of −405 and −453 kJ/mol.

Ammonium hydroxide, another base commonly used in the laboratory, is a weaker base than the alkali bases, but at high concentrations is also highly corrosive and irritating to the skin, eyes, and mucous membranes. Additionally, ammonia gas is released from ammonia hydroxide solutions and this ammonia can be very irritating to the eyes and mucous membranes.

In general, bases can cause severe damage because exposures may not be immediately painful, but harmful reactions with body tissues begin immediately if they are not removed immediately from the skin or point of contact. Bases act in a different way than do acids. Hydroxide ions saponify (break up) fats and proteins, and the bases continue to penetrate deeper into the skin or site of exposure, causing damage unless they are consumed or thoroughly washed away. We previously discussed how acids initially damage tissue, but the damaged tissue eventually forms a protective layer that prevents further damage. However, damage by bases does not form protective layers, so corrosive action continues until the base is washed away or used up. Furthermore, bases are not easily removed and it takes continuous washing for some time to remove them. The key here is to prevent exposure, but if you are exposed, you must take action quickly to avoid tissue damage.

Bases digest the fat molecules in our skin through a process called saponification. One of the products of this reaction will be the sodium (or potassium or ammonium) salt of a long-chain carboxylic acid, which is essentially a molecule of soap (see Figure 5.1.1.3).

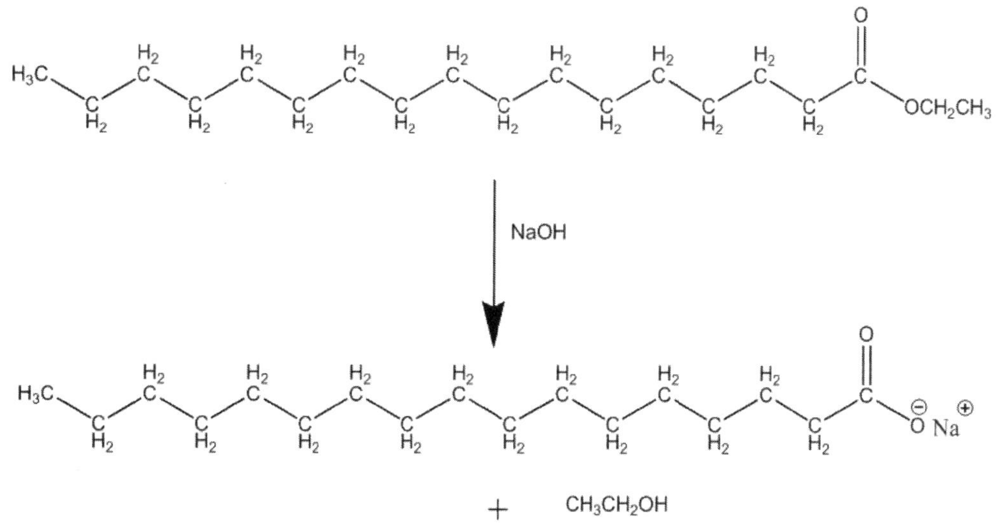

FIGURE 5.1.1.3   Saponification Reaction (also see Figure 5.2.1.1). This reaction will occur in your skin if you spill some NaOH solution on yourself.

The Eyes Are Particularly Sensitive to Corrosives, Especially Bases

The eye is especially susceptible to severe damage from exposure to corrosives. The concentration and length of exposure determine the amount of damage or injury. The eye is resistant to changes from solutions in the range of pH 3–10; however, outside this pH range (< pH 3 or > pH 10), the eye's epithelium is rapidly destroyed. [3] Eye exposure to concentrated sulfuric acid can be extremely serious due to the affinity of sulfate ions for corneal tissue, its dehydrating effect, and its high heat of hydration. Reducing exposure time is critical to prevent permanent damage or blindness—every second is important. Any exposure of base or acid should immediately be treated with thorough washing with water for at least 15 minutes, followed by an immediate examination by a physician in a nearby clinic or hospital emergency room. You can see why it's important to know the exact location of the eyewash station(s) in your lab and how to use these devices.

Again, the key here is to prevent exposure by wearing chemical splash goggles to protect those eyes. Remember RAMP?

Recognize:	Learn the dangers of corrosives.
Assess:	How skin and eyes could be exposed.
Minimize:	Wear splash goggles, protective gloves, and clothing.
Prepare:	Locate eyewash stations.

Corrosives Can Be Inhaled Too!

Some corrosive chemicals pose hazards via the inhalation route in gaseous, fume, mist, or powder forms that can cause damage to mucous membranes or lung tissue. Corrosive gases are more likely to be encountered in upper-level undergraduate laboratories and are classed as very hazardous. These gases can be inhaled and can cause severe damage to skin, eyes, nose, and the sensitive lining of the lungs that in some cases can lead to delayed fluid buildup in the lungs called pulmonary edema—a dangerous medical condition that can be fatal.

Let's take a look at two gases that you might encounter in first-year chemistry that illustrate the hazards of corrosive gases. You might encounter other gases, but the general principles of safety will be the same.

Some general chemistry programs include an experiment that digests the element copper with strong nitric acid:

$$Cu(s) + 4HNO_3(aq) \rightarrow Cu(NO_3)_2(aq) + 2NO_2(g) + 2H_2O(l)$$

There are two corrosive substances to be aware of here: a reactant, strong nitric acid, and a product, gaseous nitrogen dioxide. This red gas is a strong oxidizing agent and will react with water (moisture) in your lungs to produce nitric acid. Brief exposure to as little as 250 ppm will cause frothy sputum, difficulty breathing, and increased respiration and heart rates. And these symptoms may persist for 2–3 weeks. (In nonlab settings, NO_2 is a main component of smog in urban areas although concentrations are closer to 0.1 ppm. The main source of NO_2 is automobile engine exhaust. Catalytic converters use elemental rhodium to reduce the NO_2 to N_2.) *Any reaction producing NO_2 should be conducted in a chemical hood*.

While it is unlikely that you will use *gaseous* ammonia directly from a tank of NH_3, the common use of ammonium hydroxide solutions means that exposure to gaseous NH_3 is possible. In fact, "ammonium hydroxide" does not exist in solution; all of the ammonia exists as hydrated NH_3, with some of it dissociated as NH_4^+ and OH^-:

$$NH_3(aq) + H_2O(l) \rightleftharpoons NH_4^+(aq) + OH^-(aq)$$

And the dissolved ammonia is in equilibrium with the vapor over the surface of the liquid:

$$NH_3(aq) \rightleftharpoons NH_3(g)$$

Henry's law describes the solubility of a gas in water as a function of the partial pressure of the gas over the surface of the solution. It is described by the equation

$$S = k_H P$$

where S is the solubility of gas (as molarity), P is the partial pressure of the gas over the surface of the solution, and k_H is the Henry's law constant for a particular gas.

We can use Henry's law to calculate the vapor pressure of ammonia over the surface of solutions of ammonium hydroxide. The Henry's law constant for ammonia is 58 atm/M. Let's assume that we are using 6 M NH_4OH:

$$P = S/k_H$$
$$= (6\ \text{M})/(58\ \text{M/atm})$$
$$= 0.1\ \text{atm}$$

A concentration of 0.1 atm is equal to 100,000 ppm. The IDLH (*i*mmediately *d*angerous to *l*ife or *h*ealth) value for ammonia is 300 ppm (see Section 6.2.2 for more information about IDLH values). So if we take a good strong whiff of the vapor over the surface of 6 M ammonia, we are smelling ammonia at over 300 times the IDLH value! But one whiff is not 30 minutes. Alternatively, we can look at the LC_{lo} for mammals, which is 5000 ppm for 5 minutes. This is the lowest concentration known to cause death. So, if we are breathing 100,000 ppm we are breathing *20 times* the concentration that is lethal in 5 minutes for mammals.

All of this should convince you to *use 6 M NH_3* (and other high-concentration ammonia/ammonium hydroxide solutions) *in a chemical hood!* Even though it is a solution, the vapor over the surface of this solution is quite harmful.

Oxidizing Agents

Another category of corrosive chemicals that you might encounter in first-year chemistry is an oxidizing agent. These are less common than acids and bases, which is why it is important to take special note of them in advance. We'll discuss three likely suspects for exposure. The principles underlying these discussions would apply to all oxidizing agents, of course.

Nitric acid is probably the oxidizing agent that you are most likely to use—and we've already discussed this in the consideration of strong acids! So, it can be dangerous in two fashions. Its corrosive effect is due mainly to its oxidizing power, in fact, more so than its capability as a strong acid. In fact, nitrate salts such as KNO_3 and NH_4NO_3 are also good oxidizing agents. It's the nitrate that is the oxidizing species.

Hydrogen peroxide is another common oxidizing agent. It is also a fairly common household chemical that can be used for disinfecting wounds (3%) or decolorizing hair (15%). Laboratory solutions can be as high as 30%, which is extremely corrosive. The by-product of this oxidizing agent is water, which makes it convenient to use. Solutions with a concentration of > 8% are considered corrosive.

Some laboratory titrations are conducted using potassium permanganate ($KMnO_4$) since it is a strong oxidizing agent and the disappearance of the purple permanganate acts as an endpoint indicator. Dilute solutions that you are likely to use are only mildly irritating to the skin but high concentrations and the solid salts are very corrosive.

The bottom line here is that any good oxidizing agent will likely be able to oxidize you, too! Your skin and eyes will become the targets of these oxidizing agents unless appropriate precautions are taken.

Working with Corrosives

As described above, the main hazards are skin exposure and inhalation. Chemists should wear gloves when working with corrosive solutions; the type of glove depends on the specific corrosive and the concentration in solution. For acids and bases, nitrile gloves will be adequate for skin protection.

If there is a spill of a corrosive, it is important to isolate the area, keep other students away, report the spill to an instructor immediately, and use a spill kit if appropriate to do so. Responding to such a situation is discussed more in Sections 2.1.3 and 2.2.2.

More on Corrosives

This has been only an introduction to the common corrosives likely encountered in first-year chemistry labs. We shall return to this topic to discuss more about the chemistry of corrosives and other likely corrosive compounds in Section 5.2.1.

RAMP

- *Recognize* chemicals that are corrosive.
- *Assess* the level of risk based on likelihood of exposure and concentration of the corrosive.
- *Minimize* the risk by using a noncorrosive or less-corrosive chemical (if possible), eliminating exposure possibilities by using chemical goggles and gloves and/or working in a chemical hood. Add acid to water, when diluting concentrated solutions.
- *Prepare* for emergencies by learning locations of eyewashes, safety showers, sinks, and spill kits.

References

1. GORTON CARRUTH and EUGENE EHRLICH. *The Giant Book of American Quotations*, Gramercy Books, New York, 1988, Topic 4. Advice, #38.
2. N. V. STEERE. *Chemical Safety in the Laboratory*, Vol. 1, Division of Chemical Safety, American Chemical Society, Washington, DC, 1967, p. 116–.
3. DONALD E. CLARK. Chemical injury to the eye. *Journal of Chemical Health and Safety* 9 (2): 6–9 (2002).

QUESTIONS

1. Corrosives are chemicals that
   (a) Have a pH between 0 and 14
   (b) Are acids or bases
   (c) Are rarely encountered in introductory chemistry classes
   (d) Cause harm or injury by damaging tissue at the point of contact

2. Corrosives can be in which of the following forms?
   I. Liquids
   II. Solids
   III. Gases
   IV. Solutions
   (a) Only I and IV
   (b) Only I, II, and III
   (c) Only I and II
   (d) I, II, III, and IV

3. The potential for injury from an acid depends on which of the following?
   I. The chemical structure of the acid
   II. The concentration of the acid in solution
   III. The area of exposure (such as skin vs. eyes)
   IV. The duration of the exposure
   V. The temperature of the solution
   (a) I, II, and IV
   (b) II, III, and IV
   (c) I, II, III, and IV
   (d) I, II, III, IV and V

4. Which of the following are the three corrosive acids most likely to be encountered in introductory chemistry courses?

   I. Acetic acid
   II. Sulfuric acid
   III. Hydrochloric acid
   IV. Nitric acid
   (a) I, II, and III
   (b) I, III, and IV
   (c) II, III, and IV
   (d) I, II, and IV

5. Which acid is also a strong oxidizing agent?
   (a) Nitric acid
   (b) Sulfuric acid
   (c) Hydrochloric acid
   (d) All of the above

6. At what concentration are acids and bases generally considered to be corrosive?
   (a) Greater than 0.1 M
   (b) Greater than 1 M
   (c) Greater than 3 M
   (d) Greater than 6 M

7. What is the correct procedure for preparing the dilution of a solution of an acid?
   (a) Add the acid to the water as quickly as possible.
   (b) Add the acid to the water slowly with stirring.
   (c) Add the water to the acid as quickly as possible.
   (d) Add the water to the acid slowly with stirring.

8. Which base is *not* likely to be used in an introductory chemistry course?
   (a) NaOH
   (b) KOH

(c) NH$_4$OH

(d) LiOH

9. Which base, at high concentration, will also have a corrosive gas associated with it?

(a) NaOH

(b) KOH

(c) NH$_4$OH

(d) LiOH

10. Which statement is *true*?

(a) Corrosive bases are generally more dangerous than corrosive acids because corrosive acids damage tissue in a fashion whereby a somewhat protective layer is formed that prevents further damage.

(b) Corrosive acids are generally more dangerous than corrosive bases because corrosive bases damage tissue in a fashion whereby a somewhat protective layer is formed that prevents further damage.

(c) Corrosive bases are generally more dangerous than corrosive acids because the body is better able to neutralize corrosive acids.

(d) Corrosive acids are generally more dangerous than corrosive bases because the body is better able to neutralize corrosive bases.

11. Corrosive gases are not used directly in introductory chemistry experiments, but

(a) They are sometime encountered when using solutions of nitric acid or ammonia

(b) They are sometimes produced as a by-product of many common reactions

(c) They can be produced when chemicals are inadvertently mixed

(d) Even when used they present little hazard since it is common to wear respirators when using them

12. Of the following, which are the three most likely oxidizing agents to be used in an introductory chemistry experiment?

  I. Nitric acid

 II. Hydrogen peroxide

III. Potassium permanganate

IV. Chlorine

(a) I, II, and III

(b) I, II, and IV

(c) I, III, and IV

(d) II, III, and IV

13. What is the best strategy to avoid contact with corrosive chemicals?

(a) Wear chemical splash goggles.

(b) Wear appropriate gloves.

(c) Wear chemical splash goggles and appropriate gloves.

(d) Always work in a chemical hood.

5.1.2

FLAMMABLES—CHEMICALS WITH BURNING PASSION

Preview We discuss flammable chemicals, why we use them, what starts a fire, what chemical properties tell us about flammability, what makes chemicals flammable, and how chemical structure is related to flammability.

[The] fire resulted in a loss that cannot be assigned a monetary value.

Stephen Martin, commenting on a laboratory fire[1]

INCIDENT 5.1.2.1   SODIUM-SOLVENT FIRE[1]

On Saturday morning a postdoctoral fellow was working in his laboratory destroying excess sodium metal by treating it with alcohol. Believing that the sodium had been destroyed, he poured the residual material into the sink. To his surprise some sodium had not been destroyed and upon contact with water it caught fire. This fire then ignited some other solvent mixtures stored near the sink, and this larger fire destroyed three laboratories. Furthermore, the students working in these laboratories lost their notebooks and chemical products—a severe loss in research and in progress toward their degrees. Initially, costs of the fire were estimated around $400,000 to repair the laboratories. However, the local fire department reported that other incidents in the same location made them question the adequacy of the safety program. The fire department presented a list of recommendations that the university adopted and the modifications to make requested upgrades brought the cost to $30 million.

What lessons can be learned from this incident?

When Do We Use Flammables?

You frequently use flammable chemicals in your everyday activities, for example, gasoline for your car, natural gas for heating and cooking, liquefied petroleum gas (LPG; bottled gas) in grills, butane in lighters, acetone in nail polish, isopropyl alcohol in rubbing alcohol, and ethanol in alcoholic beverages. You could think of flammable chemicals as just another set of tools such as screwdrivers, cars, or computers. Chemists and other scientists also frequently use flammable chemicals in their laboratory operations, and it is likely that as a student you will also encounter these chemicals. While flammable chemicals can be very hazardous, they are very useful to us, and through past experiences we have learned how to use them safely so we can benefit from their useful chemical properties.

Flammable chemicals are chemicals that easily ignite and rapidly burn, releasing large amounts of energy, mostly in the form of heat. (Sometimes you see the term *inflammable*—it is a synonym of flammable.) Once ignited, they continue to burn as a self-sustaining reaction until the chemical is consumed or it is extinguished. Indeed, they have a passion for burning, and they come in all forms: gases, liquids, and even solids. In the laboratory we can use flammables in many ways, but most commonly we use flammable liquids as solvents for reactions and procedures in analytical and synthesis laboratories. A *solvent* is usually a liquid that is used to dissolve reactants (chemicals that react with each other) so a reaction can be carried out; it does not have to be flammable, but often is.

Flammable chemicals constitute an important hazard class found in laboratory operations or other situations using these materials. You are very likely to use flammables at some point in the laboratory or in your own daily activities, and it is prudent to understand what makes chemicals flammable and how you can manage, control, or minimize the risk when working with these chemicals. Common flammable laboratory solvents include acetone, ethanol, diethyl ether, ethyl acetate, hexane, and toluene.

While the focus of this section is on flammables, there is another category of burnable substances called combustible chemicals. These compounds don't catch fire as readily as flammable liquids. Combustible chemicals become flammable if they are heated so that they give off sufficient vapors to be easily ignited. While combustibles are more difficult to ignite, once ignited they also readily burn.

Characterizing Flammable and Combustible Chemicals

We can experimentally measure properties that help us determine the potential flammability of a chemical, and it is important to understand the terminology describing properties of flammable chemicals. The *boiling point* is the temperature at which a liquid's vapor pressure is at or just very slightly higher than atmospheric conditions so that bubbles readily appear in the liquid as it is vaporized rapidly. (The boiling point at $P = 1$ atm is the *normal* boiling point.) The boiling point of a flammable liquid is usually relatively low.

The *flash point* of a chemical is the lowest temperature at which its vapors near the liquid surface can be ignited under controlled conditions. For a liquid this is the lowest temperature at which vapors, above its surface mixing with air, can form an ignitable mixture. The *autoignition temperature* is the temperature at which a flammable chemical ignites in air spontaneously under controlled conditions. The lower the autoignition temperature, the greater the potential risk for a fire. Autoignition temperatures are generally quite high, as seen in Table 5.1.2.3. You needn't generally worry about vapors spontaneously burning.

Each chemical has a range of concentrations of its vapors called *flammability limits* within which a fire or explosion can occur, while fires will not occur below or above those limits. The lower explosive limit (LEL), sometimes called the lower flammability limit (LFL), is the lowest concentration of vapor in air, expressed in percent by volume, at which a fire can be started resulting in an explosion, when a source of ignition is present. Below the LEL, the concentration of the vapor is insufficient to support burning. The upper explosive limit (UEL), sometimes called the upper flammability limit (UFL), is the highest concentration of vapor in air expressed in percent by volume, at which a fire will be propagated. Above this limit the concentration of the vapor is too high to support burning. These limits become wider as temperatures are increased and as oxygen content is increased. The result of these variations in LEL and UEL is that flammability limits are not very useful. For example, when a flammable liquid is spilled its LEL is quickly achieved as the broad surface allows large amounts of chemical to vaporize, and if this occurs within the presence of an ignition source a fire or explosion can occur (see *Chemical Connection 5.1.2.1* Does Stoichiometry Matter?)

CHEMICAL CONNECTION 5.1.2.1

DOES STOICHIOMETRY MATTER?

Two important concepts in introductory chemistry are stoichiometry and limiting reactants. Let's look at the combustion of acetone as an example with the ideas in mind.

The balanced equation, assuming complete combustion to carbon dioxide and water, is

$$CH_3COCH_3 + 4O_2 \rightarrow 3CO_2 + 3H_2O$$

Since air is 20.8% oxygen (by volume) the stoichiometric (4:1) concentration of gaseous acetone will be 5.2%. However, the concept of "limiting reactant" suggests that *any* combination of acetone and oxygen would burn as long as both are present. So, the acetone should burn in air in some closed container, assuming excess O_2 in air, until *all* of the acetone is consumed. However, the LFL for acetone is 2.5% and the UFL is 12.8%. Thus, a match will ignite a mixture of air and 7% acetone, but not a mixture of air and 1.5% acetone, contrary to our notion of limiting reactant. Mixtures where the flammable concentration is less than the LFL are called too "lean" and mixtures where the flammable concentration is greater than the UFL are called too "rich."

The existence of LFLs and UFLs for all flammables suggests that the exact manner by which these reactions occur is more complicated than just the stoichiometry. When the fuel concentration is less than the LFL or greater than the UFL, the imbalance in the stoichiometry reduces the amount of heat energy (per mole of flammable or oxygen) released by the reaction. The gases present dissipate the heat due to their heat capacity. If this rate of heat dissipation is fast enough the temperature of the flame drops by several hundred degrees Celsius. Not enough of the released heat feeds back into getting the reactants over the activation energy barrier and the reaction stops.

LFLs are in the range of 1–4% and UFLs are usually 6–20%, with some exceptions in the 30–60% range. A most notable exception is acetylene with an LFL of 2.5% and a UFL of 81%. This extraordinary wide range makes this gas exceptionally dangerous since nearly any mixture in air will be explosive.

Stoichiometry is a very powerful tool in chemical calculations but doesn't always tell the whole story!

Fire Hazard Rating Systems

The United Nation's Globally Harmonized System (GHS) defines flammable and combustible liquids in terms of measurable chemical properties (Table 5.1.2.1). This hazard rating system uses a 1 for the highest level of flammability hazard and a 4 for the lowest level of flammability hazard. The other most commonly used system of categorizing fire hazard is the NFPA diamond. The NFPA flammability hazard criteria[3] are shown in Table 5.1.2.2. You will see that the NFPA and GHS ratings are "opposites" of each other. The NFPA rating increases in hazard from 0 to 4 (highest hazard by NFPA), while the GHS rating increases in hazard from 4 to 1 (highest hazard by GHS).

The NFPA and GHS define combustibles differently. The NFPA defines combustibles as compounds with flash points >38 °C (100 °F) but <93 °C (200 °F). The GHS defines combustibles as compounds with flash points >60 °C but ≤93 °C. These differences result in some chemicals being

TABLE 5.1.2.1 Properties of Flammable and Combustible Liquids as Defined by the Globally Harmonized System for Classification and Labelling of Chemicals[2]

Hazard category	Hazard description	Signal word	Flash point (°C)	Boiling point (°C)
HC 1	Extremely flammable	Danger	< 23	≤35
HC 2	Highly flammable	Danger	< 23	> 35
HC 3	Flammable	Warning	≥ 23 to ≤ 60	—
HC 4	Combustible	Warning	> 60 to ≤ 93	—

TABLE 5.1.2.2 The NFPA Categorization System for Flammable Substances[a]

Hazard rating	Description	Criteria
0	Will not burn	Will not burn at 816 °C for 5 min
1	Burns if preheated	Flash point (fp) >94 °C
2	Burns if moderately heated	fp between 38 °C and 94 °C
3	Burns under ambient conditions	fp <23 °C and boiling point (bp) >38 °C or fp between 23 °C and 38 °C
4	Burns rapidly under ambient conditions; vaporizes easily	fp <23 °C and bp <38 °C

[a] NFPA uses degrees Fahrenheit (°F) in its standards rather than degrees Celsius (°C) used in this table.

Source: Reprinted with permission from NFPA 704–2007 *System for the Identification of the Hazards of Materials for Emergency Response*. Copyright © 2007 National Fire Protection Association, Quincy, MA. This reprinted material is not the complete and official position of the NFPA on the referenced subject, which is represented only by the standard in its entirety. The NFPA classifies a limited number of chemicals and cannot be responsible for the classification of any chemical whether the hazard classifications are included in NFPA documents or developed by other individuals.

TABLE 5.1.2.3   Properties Used to Evaluate Flammability of Laboratory Solvents [2,4-6]

Chemical name—formula	Boiling point (°C)	Vapor pressure (mm Hg at 20 °C)	Lower and upper flammability	Autoignition (°C)	Flash point (°C)	GHS rating	NFPA rating
Acetic acid—CH_3CO_2H	118	11	4–16	463	39	3	2
Acetone—CH_3COCH_3	56	180	3–13	465	−18	2	3
Acetonitrile—CH_3CN	82	73	4–16	524	6	2	3
1-Butanol—C_4H_9OH	118	6	1.4–11	365	29	3	3
2-Butanone (methyl ethyl ketone)—$CH_3COC_2H_5$	80	71	2–10	515	−6.1	2	3
Carbon disulfide—CS_2	46.1	300	1–44	90	−30.0	2	4
Chloroform—$CHCl_3$	61	160	None	None	None	No rating	0
Dichloromethane—CH_2Cl_2	40	440		556	None	No rating	1
Diethyl ether—$(C_2H_5)_2O$	35	442	1.85–48	160	−45.0	1	4
Dimethylformamide—$(CH_3)_2NCHO$	153	2.6	2.2–15	445	58	3	2
Ethanol—C_2H_5OH	78.3	43	3.3–19	365	12.8	2	3
Ethyl acetate—$CH_3CO_2C_2H_5$	77	76	2.18–9	427	−4	2	3
Hexane—C_6H_{14}	68.9	124	1.1–7.5	225	−21.7	2	3
Methanol—CH_3OH	64.9	96	6.7–36	385	11.1	2	3
1-Propanol (n-propanol)—C_3H_7OH	97	15	2.1–13.5	433	25	3	3
2-Propanol (isopropanol)—C_3H_7OH	82.8	33	2.3–12.7	398	11.7	2	3
Tetrahydrofuran—C_4H_8O	66	132	2–11.8	321	−14	2	3
Toluene—$C_6H_5CH_3$	110.6	22	1.4–6.7	480	4.4	2	3

classified as "combustible" under the NFPA system but classified as "flammable" under the GHS, where those compounds having flash points >38 °C but <60 °C are flammable. So under the NFPA system, acetic acid and dimethylformamide are combustible and under the GHS these compounds are flammable. This should make little difference to your work with these compounds since you should already be taking precautions to avoid ignition sources. In the discussion below we will give you more information about flammable and combustible substances.

Table 5.1.2.3 shows the chemical and flammability properties of many solvents that are commonly used in laboratories. You'll note that most of the ratings of these chemicals put them in the "extremely flammable" or "highly flammable" categories.

What Starts a Fire?

Fires happen when three conditions are present in the same place at the same time: (1) a fuel, such as a flammable chemical, is present in gaseous form or is vaporized in a concentration that is within flammable limits (this will be discussed below); (2) a source of ignition is present such as a spark, flame, static electricity, or even a hot surface, depending on the properties of the chemical; and (3) oxygen (in air) or an oxidizing atmosphere is present. These three elements form what has been called the fire triangle in Figure 5.1.2.1. *What is important to know is that if you can remove one of the elements of this fire triangle you will prevent a fire*. STOP! Read that sentence again. Most laboratory work is done in the normal atmosphere of air, so you will need to concentrate on removing one of the other two sides of the fire triangle, the fuel or ignition source, to prevent fires. If your work requires using flammable chemicals and if you can't identify nonflammable substitutes, then you must focus on removing or preventing sources of ignition from coming in contact with vapors of the flammable chemicals you will

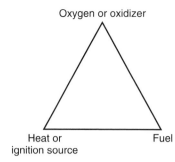

FIGURE 5.1.2.1 The Fire Triangle. The fire triangle helps explain how fires work and how to prevent fires.

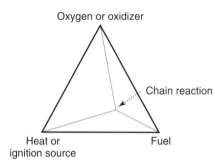

FIGURE 5.1.2.2 The Fire Tetrahedron. The fire tetrahedron helps explain how fires can be extinguished.

be using. The good news is that this is relatively easy to do, but the challenge is to keep up your guard and search for ignition sources.

We introduced the concepts of the fire triangle and fire tetrahedron in Section 2.1.2. The fire triangle gives us a simple model to help us think about the ways to prevent fires. There is a fourth element called the chain reaction added to the fire triangle to make the fire tetrahedron (see Figure 5.1.2.2) that explains how some fires are extinguished. Once started, a fire can be extinguished in any one of four ways: remove oxygen, remove the fuel, remove the heat, or break the chain reaction. The fire tetrahedron is a model used to think about ways in which to extinguish fires.

When using a flammable solvent, you must search to identify any potential sources of ignition prior to starting work with these materials. Sources of ignition can be flames, such as Bunsen burners, other burners, or torches, and sparks from electrical sources, such as electrical switches, any electrically powered equipment including hot plates, stirring motors, refrigerators, heat guns (hair dryers), and static electricity. Even ignition sources some distance from your operations should be considered, since vapors of laboratory solvents are generally heavier than air and can invisibly creep along tabletops to these remote ignition sources, where they can be ignited and flash back to the primary solvent source. Spills of flammable laboratory solvents are subject to this kind of remote ignition, where they might run along the floor to a hot motor or spark source. When pouring or transferring solvents from metal containers, it is good practice to ensure the container is grounded to prevent sparks from static electricity, especially in very dry environments (see Figure 5.1.2.3).

What Makes Chemicals Flammable?

Perhaps surprisingly, flammable *liquids* themselves do not burn but rather it is the *vapor* above the liquid that burns as the molecules move from the liquid to the gas phase. Flammable chemicals have relatively low molecular weights and high vapor pressures. If spilled on a benchtop or the floor, these flammable chemicals quickly vaporize and air currents can spread them into a wide area, where an ignition source could start a fire.

We can think of a chemical reaction profile to help us understand fire better. Figure 5.1.2.4 shows an exothermic reaction with a moderately high activation energy barrier. To start the fire, some ignition source is needed to overcome E_{act}, but once the fire is burning, the heat released can supply the

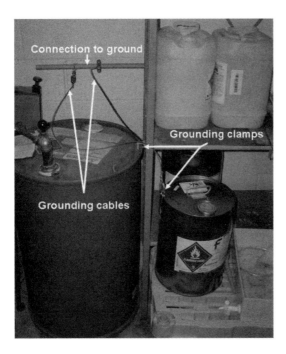

FIGURE 5.1.2.3   Proper Grounding for Metal Solvent Drum. Spring-loaded clips make it easy to connect a metal container to a grounded metal pipe.

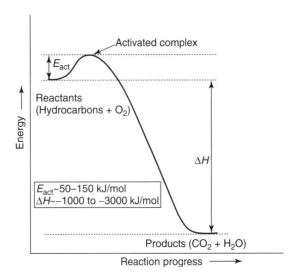

FIGURE 5.1.2.4   Exothermic Chemical Reaction Profile. For small hydrocarbons E_{act} will be 50–150 kJ/mol and ΔH_{rxn} will be -1000 to -3000 kJ/mol.

energy needed to overcome E_{act}. Most fires have this self-perpetuating characteristic. Thus, ΔH (heat, or enthalpy, of reaction) represents the exothermic heat released that perpetuates the reaction.

How Can I Recognize a Flammable Chemical?

Laws require that flammable chemicals be labeled as such. The GHS has devised a hazard class (HC) rating system that includes flammables and this system can help you recognize the relative flammable hazard of a chemical. These ratings range from HC 1 to HC 4 with HC 1 being extremely flammable and HC 4 being combustible liquids. The ratings are made on the basis of flash points and boiling points, but autoignition temperatures also play a role in flammability. Table 5.1.2.3 provides information about some of the most common solvents you may encounter when you are working in the laboratory.

Based on flash point and boiling point, the most flammable solvent in Table 5.1.2.3 is diethyl ether and it is also the one with the highest HC rating of 1. Ether, the common term for diethyl ether or

TABLE 5.1.2.4   Flammable and Combustible Fuels Used in Our Daily Activities [2,4,5]

Common name—formula	Vapor pressure (mm Hg at 20 °C)	Boiling point (°C)	Autoignition temperature (°C)	Flash point (°C)	GHS rating/ (NFPA rating)
Natural gas (methane)—CH_4	> 760	Gas	NA	NA	1/(4)
Liquefied petroleum gas (LPG)— $C_3H_6/C_3H_8/C_4H_8/C_4H_{10}$	>760	Gas	NA	NA	1/(4)
Gasoline—C_5–C_{12} hydrocarbons	38–300	38–204	280–456	−43	2/(3)
Kerosene—C_9–C_{16} hydrocarbons	5 (at 38 °C)	170–300	229	38–66	3/(2)
Fuel oil &4—C_{12}–C_{18} hydrocarbons		>275	263	64	4/(2)
Olive oil		300	343	225	No rating/(1)

ethyl ether, has been responsible for many laboratory fires. With such a low flash point, far below room temperature, even hot surfaces or static electrical sparks can set ether on fire. Based on the flash point, the next most flammable solvent in Table 5.1.2.3 is carbon disulfide, but its boiling point is higher than ether and that earns it an HC rating of 2. Take special note, however, that the autoignition temperature for this solvent, 90°C, is lower than any other chemical in Table 5.1.2.3, and that means it is more readily ignited than ether around any source of ignition such as a hot surface. In practice, there is little difference in the flammable hazard of these two compounds, and many flash fires have occurred with both solvents.

Only one solvent in Table 5.1.2.3, dichloromethane, is nonflammable. Acetic acid is a lower risk combustible as reflected by its HC rating of 3. All of the others chemicals in Table 5.1.2.3 have HC ratings of at least 2 and have very real potential to catch fire in the presence of a source of ignition. For comparison of chemical properties some chemicals that you are likely to encounter in your own daily activities are shown in Table 5.1.2.4. These hydrocarbon products come from oil or petroleum.

As noted above, you can now see that the GHS and NFPA ratings are the opposite of each other (see the last column in Table 5.1.2.4). (This is also true with regard to the NFPA ratings of health and reactivity hazards.) It seems likely that scientists will have to live with the NFPA ratings as the GHS also becomes standard.

Does Chemical Structure Influence Flammability?

Yes, it does! Flammable liquids are sometimes termed "volatile," which means they are easily vaporized at ambient temperatures. Many volatile chemicals are flammable, but there are also nonflammable volatile chemicals. The difference is in the structures of the molecules.

For example, in Table 5.1.2.3, methanol (CH_3OH) is a highly flammable liquid with a boiling point of 65 °C and a vapor pressure of 96 mm Hg at 20 °C, while dichloromethane, also known as methylene chloride (CH_2Cl_2), is a noncombustible liquid with a lower boiling point, 40 °C, and a higher vapor pressure, 440 mm Hg at 20 °C. The substitution of chlorine atoms (or other halogens) for hydrogen on a hydrocarbon molecule tends to make chemicals less flammable or nonflammable. For further discussion of this issue, see *Chemical Connection 2.2.1.1* Flammability of Halogenated Solvents and Halons.

The series of alcohols in Table 5.1.2.3—methanol, ethanol, 1-propanol, and 1-butanol—illustrates the relationship between molecular structure and chemical properties. Among these four alcohols, the boiling points, autoignition temperatures, and flash points increase while vapor pressures decrease, as the molecular weights of these compounds increase within this chemical class with the same functional groups (alcohol or —OH). In this series the vapor pressure decreases with increasing molecular weight.

These examples illustrate how chemical structure can influence flammability and other properties. Mostly, flammability is strongly related to volatility.

Working with Flammable Chemicals: RAMP

Given the obvious hazards of flammable chemicals, it makes sense to limit their use when possible. We can use RAMP to work safely with flammable chemicals.

- *Recognize* the appropriate flammability hazard associated with the solvent.
- *Assess* the level of risk under the circumstances by considering means by which the vapor might be generated and released and determine if any sources of ignition are available.
- *Minimize* the risk by substituting a less-flammable solvent (if possible), using less solvent (if possible), working in a ventilated area to limit vapor concentrations, and eliminating sources of ignition.
- *Prepare* for emergencies by locating and knowing how to use the appropriate fire extinguisher.

References

1. P. ZURER. Fire guts University of Texas chemistry lab. *Chemical & Engineering News*, pp. 10–11 (October 28, 1996); M. B. BRENNAN. Laboratory fire exacts costly toll. *Chemical & Engineering News*, pp. 29–34 (June 23, 1997).
2. *Globally Harmonized System for Classification and Labelling of Chemicals*, 2nd revision, United Nations, New York, 2007; available at http://www.unece.org/trans/danger/publi/ghs/ghs_rev02/02files_e.html (accessed August 28, 2009).
3. NFPA 704–2007. *System for the Identification of the Hazards of Materials for Emergency Response*, National Fire Protection Association, Quincy, MA, 2007.
4. National Institute for Occupational Safety and Health. *NIOSH Pocket Guide to Chemical Hazards*, NIOSH Publication 2005-149, U.S. Government Printing Office, Washington, DC, 2005; available at http://www.cdc.gov/niosh/npg (accessed August 28, 2009).
5. Manufacturing Chemists Association. *Guide for Safety in the Chemical Laboratory*, 2nd edition, Van Nostrand Reinhold Company, New York, 1972, Section V, Reference Charts, pp. 331–442.
6. National Research Council. *Prudent Practices in the Laboratory: Handling and Disposal of Chemicals*, Appendix B: Laboratory Chemical Safety Summaries, National Academy Press, Washington, DC, 1995, pp. 235–413.

QUESTIONS

1. Flammable chemicals are most commonly found in which category of chemicals?

   (a) Organic solutes
   (b) Organic solvents
   (c) Inorganic solutes
   (d) Inorganic solvents

2. Which statement is *true*?

   (a) Flammable chemicals and combustible chemicals have similar boiling points but different flash points.
   (b) Flammable chemicals and combustible chemicals have similar flash points but different boiling points.
   (c) Combustible chemicals burn more easily than flammable chemicals.
   (d) Flammable chemicals burn more easily than combustible chemicals.

3. The most flammable chemicals are identified by

   (a) "HC 1" in the GHS and "4" in the NFPA rating systems
   (b) "HC 1" in the GHS and "0" in the NFPA rating systems.
   (c) "HC 4" in the GHS and "4" in the NFPA rating systems.
   (d) "HC 4" in the GHS and "0" in the NFPA rating systems.

4. Which of the following are the three components of the fire triangle?

   I. A fuel
   II. A source of ignition
   III. An oxidizing agent
   IV. A chain reaction

   (a) I, II, and III
   (b) I, II, and IV
   (c) I, III, and IV
   (d) II, III, and IV

5. What is the most common oxidizing agent for fires?

   (a) Chlorine
   (b) Nitric acid
   (c) Elemental oxygen
   (d) Methane

6. As a general rule, organic solvents are

   (a) Lighter than air and not flammable
   (b) Heavier than air and not flammable
   (c) Lighter than air and flammable
   (d) Heavier than air and flammable

7. Which statement is *true* regarding the state of a chemical as it is burning?

   (a) Only liquids and gases can burn.
   (b) Only liquids can burn.
   (c) Only gases can burn.
   (d) Solids, liquids, and gases can burn.

8. Which statement is *true*?

   (a) Flammable substances have higher activation energies than combustible substances and both substances react exothermically.
   (b) Flammable substances have lower activation energies than combustible substances and both substances react exothermically.
   (c) Flammable substances have higher activation energies than combustible substances and both substances react endothermically.
   (d) Flammable substances have lower activation energies than combustible substances and both substances react endothermically.

9. The autoignition temperature for most organic solvents is

   (a) Fairly large
   (b) The temperature at which a flammable chemical ignites spontaneously in air
   (c) The temperature at which the solvent burns in an automobile
   (d) Both (a) and (b)

10. Which set of values best characterizes a flammable liquid?

    (a) Low BP, high vapor pressure, low flash point
    (b) Low BP, high vapor pressure, high flash point
    (c) Low BP, low vapor pressure, low flash point
    (d) High BP, low vapor pressure, low flash point

11. Which is *true*?

    (a) The LEL is lower than the UEL.
    (b) The UEL is lower than the LEL.
    (c) The UEL and LEL are often the same value.
    (d) The LEL and LFL are different numbers.

12. Which chemical is the most dangerous in the lab with regard to potential fire hazard?

    (a) Dichloromethane
    (b) Ethanol
    (c) Diethyl ether
    (d) Toluene

13. Which chemical is the least dangerous in the lab with regard to potential fire hazard?

    (a) Acetone
    (b) Ethanol
    (c) Dichloromethane
    (d) Butanol

14. Using the GHS criteria, gasoline is categorized as

    (a) Extremely flammable
    (b) Highly flammable
    (c) Flammable
    (d) Combustible

5.2.1

CORROSIVES IN ADVANCED LABORATORIES

Preview The chemistry of corrosive action is presented and several more corrosives found in advanced labs are discussed.

A very Faustian choice is upon us: whether to accept our corrosive and risky behavior as the unavoidable price of population and economic growth, or to take stock of ourselves and search for a new environmental ethic.

E. O. Wilson, American scientist[1]

INCIDENT 5.2.1.1 TRIFLUOROACETIC ACID/HYDROFLUORIC ACID BURN[2]

While working with trifluoroacetic acid, a chemist handled a container with an ungloved hand, but unnoticed was a small liquid residue on the outside of the glass. After a few hours the chemist began to experience severe pain on the palm and the inside of the thumb from a serious hydrofluoric acid burn. The burn required a skin graft. (Trifluoroacetic acid reacts with water to produce hydrofluoric acid.)

What lessons can be learned from this incident?

A Chemical Overview of Corrosive Compounds

The corrosive action at the molecular level may be an acid–base reaction, a redox reaction, or both. Let's take a look at each of these before a review of some common corrosives found in advanced laboratories.

Chemicals that are corrosive are often *strong oxidizing agents*. While nitric acid is obviously an acid, it is not the acidity *per se* that is the main part of the chemical action. Nitric acid will react with many elemental metals to dissolve the metal by increasing the oxidation state from zero (in elemental form) to +1, +2, or +3 (in most instances). For example, nitric acid reacting with elemental copper is shown by

$$Cu(s) + 4HNO_3(aq) \rightarrow Cu(NO_3)_2(aq) + 2NO_2(g) + 2H_2O(l)$$

Nitrate, with nitrogen in the +5 oxidation state, get reduced to nitrogen dioxide where nitrogen is in the +4 oxidation state. The solution becomes blue due to the Cu^{2+} ion and a toxic red gas, NO_2, is generated. Vapors from leaking bottles of oxidizing agents can similarly destroy metal shelves. For this reason many stockrooms use wood shelves to store corrosive chemicals.

The concentration of the acid is quite important here but not just because higher concentration provides more oxidizing compound. The oxidizing power of the acid is influenced by the concentration of the species according to the Nernst equation, shown below for the copper and nitric acid reaction:

$$E = E^o - (0.0591)\log([Cu^{2+}](P_{NO_2})^2[NO_3^-]^2[H^+]^4)$$

E is the oxidation potential of the solution, E^o is the standard state oxidation potential (or the negative of the reduction potential, as it is usually listed), and the other terms represent concentration and partial pressure. The value of E becomes *more positive* as the concentration of nitric acid is *increased*. Thus,

Laboratory Safety for Chemistry Students, by Robert H. Hill, Jr. and David C. Finster
Copyright © 2010 John Wiley & Sons, Inc.

although a higher concentration of acid *does* provide more oxidizing chemicals, the Nernst equation shows that the oxidizing power of the acid is increased. In some reactions the exponent in the $[H^+]$ term is even larger than 4, causing a dramatic increase in E as acid concentration increases. This also explains part of the extraordinary corrosive power of *aqua regia*, a 3:1 mixture of concentrated HCl and HNO_3. This acid mixture is able to dissolve metals such as gold and platinum due not only to both the extremely high concentrations of H^+ and NO_3^-, but also due to the complexing ability of the Cl^- ion to form $AuCl_3^-$ and $PtCl_6^{2-}$.

Acids such as HCl and H_3PO_4 owe their corrosive behavior primarily to the high concentration of acid, or the power of the H^+ cation (see *Chemical Connection 5.2.1.1* Why Are Some Acids *Strong* and Some *Weak* as Aqueous Solutions?). In human tissue, the combination of hydronium cation and low pH destroys protein molecules and catalytically causes the destruction of other organic materials. With regard to the destruction of metals, the hydronium ion in concert with O_2 can oxidize elemental metals to their cationic forms. Depending on the metal, H_2 gas might also get produced, which presents a flammability hazard.

CHEMICAL CONNECTION 5.2.1.1

WHY ARE SOME ACIDS *STRONG* AND SOME *WEAK* AS AQUEOUS SOLUTIONS?

For oxyacids such as HNO_3, H_2SO_4, H_3PO_4, and H_2CO_3, the simplest, but crude, rule of thumb is that if there are two or more oxo groups, the acid will be a strong acid (Figure 5.2.1.1). HNO_3 and H_2SO_4 illustrate this principle. This predicts that acids having only one oxo group will be weak acids, such as H_3PO_4 or H_2CO_3. Carboxylic acids have one oxo group, and almost all are weak acids unless there are overriding inductive effectives such as in trifluoroacetic acid ($K_a = 0.59$). Linus Pauling, and others, proposed simple equations such as $pK_a \approx 9 - 7n$ and $pK_a \approx 8 - 5n$, where n = the number of oxo groups. Remember that as pK_a is smaller, the acid is stronger. More sophisticated equations exist and can be found in most inorganic chemistry textbooks. These also involved a consideration of the electronegativity of the central atom in the molecule.

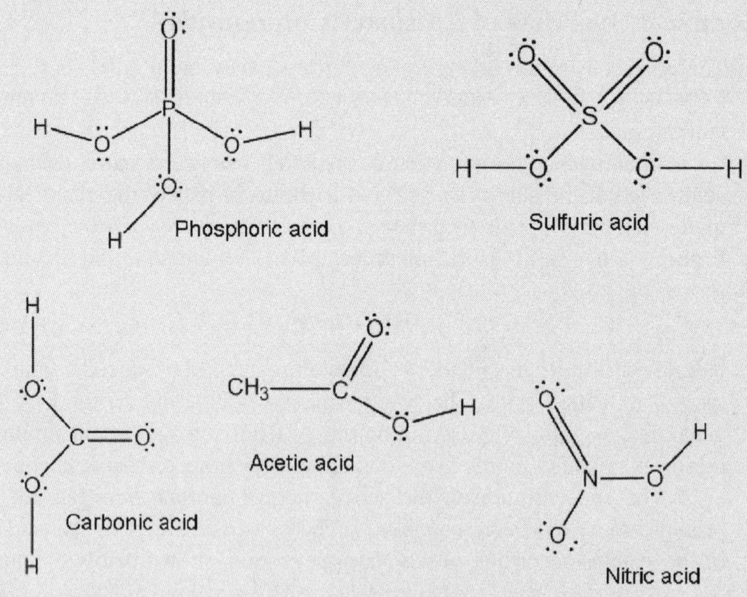

FIGURE 5.2.1.1   Lewis Diagrams of Oxyacids. The strength of the acid is related to the number of oxo groups.

For the hydrogen halides, HF (weak acid) seems curiously different from the strong acids, HCl, HBr, and HI. A simple explanation for this is that the HF bond is unusually strong (565 kJ/mol) and this inhibits dissociation compared to the other acids: HCl (431 kJ/mol), HBr (364 kJ/mol), and HI (298 kJ/mol).

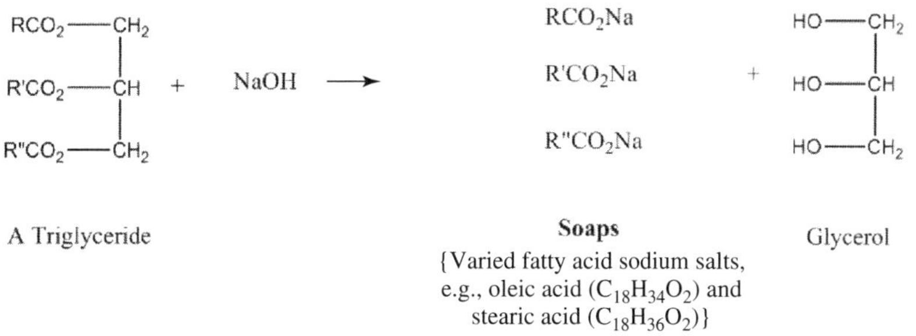

FIGURE 5.2.1.2 Saponification of a Triglyceride. Treating a fat with a strong base frees fatty acids that can be isolated as sodium salts that can be used as soaps.

A Triglyceride + NaOH → **Soaps** {Varied fatty acid sodium salts, e.g., oleic acid ($C_{18}H_{34}O_2$) and stearic acid ($C_{18}H_{36}O_2$)} + Glycerol

Bases such as NaOH are less destructive to metals but attack living tissue quite readily. In this instance, the reaction is a base-catalyzed hydrolysis reaction (*hydrolysis*—splitting apart a molecule using water). For example, a protein molecule with an amide bond can be split into a carboxylic acid and an amine by base-catalyzed hydrolysis:

$$\underset{\text{protein}}{RCONHR'} + H_2O \rightarrow \underset{\text{carboxylic acid}}{RCOOH} + \underset{\text{amine}}{R'NH_2}$$

Similarly, ester linkages found in molecules of fat can be split into a molecule of alcohol and one of carboxylic acid:

$$\underset{\text{ester}}{RCOOR'} + H_2O \rightarrow \underset{\text{carboxylic acid}}{RCOOH} + \underset{\text{alcohol}}{R'OH}$$

Lipids and fats contain three ester linkages as shown in Figure 5.2.1.2. When a strong base is used to destroy these ester linkages, the process is called saponification. One of the products of this saponification of fats is a salt of a fatty acid, such as sodium octadecenoate (oleic acid sodium salt)—this chemical can be used as a simple soap (see Figure 5.2.1.2).

Finally, some compounds react violently with water, which is the principal means by which they would react with skin and other tissue. Examples of these will be discussed below.

Remember that some chemicals have multiple modes of action. They may be oxidizing agents, acids, and/or dehydrating agents. The discussion below will list chemicals in particular categories but many chemicals will have multiple modes of action and it is important to know the overall reactivities of these chemicals. For example, the principal danger of glacial acetic acid is not so much as an acid but rather as a dehydrating agent. See Table 5.2.1.1 (the italicized entries were discussed in Section 5.1.1).

Acids

HCl, H_2SO_4, and HNO_3 are discussed in Section 5.1.1. Here, we consider three other acids commonly used in undergraduate chemistry labs.

Since acetic acid is a weak acid ($K_a = 1.7 \times 10^{-5}$) and you probably know it also as the principal component of vinegar, it may seem to be a reasonably safe compound. In low concentrations it is. However, acetic acid is sometimes used in its "glacial" form that is 99.8% acetic acid. Not only is acetic acid at concentrations >50% flammable (which is not a common danger associated with most acids), glacial acetic acid has a very high affinity for water and is a strong dehydrating agent. The exothermicity of this reaction is sufficient to cause burns and this strong acid is corrosive. Vapors of glacial acetic acid are understandably quite dangerous if inhaled.

Phosphoric acid is a weak acid ($K_{a1} = 7 \times 10^{-3}$) but can also be found in pure, 100% form. Phosphoric acid is also known as ortho-phosphoric acid and it solidifies below 21 °C. It is hygroscopic and a strong corrosive. In high concentrations it can cause severe burns to the skin. At lower concentrations it is an irritant. While phosphoric acid as a mist can be irritating to the eyes, nose, throat, and respiratory system, it is unlikely to cause pulmonary edema.

While all acids are very hazardous in concentrated form, hydrofluoric acid (HF) is extremely hazardous, and HF solutions ≥ 0.01 molar concentrations are very corrosive. While you are not likely to encounter HF in your early laboratory sessions, you should know something of its especially hazardous nature and that it requires special treatment upon exposure.

TABLE 5.2.1.1   Corrosive Substances Encountered in Undergraduate Chemistry Labs

Substance	Gas	Liquid	Solid	Solution	Acid	Base	Oxidant	Reductant	Water reactive
HCl	X			X	X				
H_2SO_4				X	X				X
HNO_3	X			X	X		X		
HF	X			X	X				
$HC_2H_3O_2$		X		X	X				X
H_3PO_4		X		X	X				
NaOH			X	X		X			
KOH			X	X		X			
NH_4OH				X		X			
F_2	X						X		
Cl_2	X			X			X		
Br_2		X					X		
I_2			X				X		
H_2O_2				X			X		
$KMnO_4$				X			X		
Na			X					X	X
NaH			X					X	X
$LiAlH_4$			X					X	X
Phenol			X		X				
P_2O_5			X						X
CaO			X						X

HF solutions must be handled with great care, and exposures to HF require special attention with flushing for only 5 minutes rather than the normal 15 minutes, then immediate, specific medical treatment, typically with benzalkonium chloride (Zephiran) or calcium gluconate.[3] No work should be done with HF unless these pharmaceuticals for treating HF burns are present, are readily available to the laboratory or to a nearby medical clinic or hospital, and a plan has been made with medical staff to treat these HF burns in the event of an exposure. See Incidents 5.2.1.1 and 3.3.1.1 for examples of HF burns. A publication titled *Recommended Medical Treatment for Hydrofluoric Acid Exposure* is available online from Honeywell.[4] As with any chemical, hydrofluoric acid can and is used in laboratories everyday without incident, but there have been reports of severe injuries from mishandling hydrofluoric acid. If you use this acid, be particularly diligent in learning about precautions in handling.

The Halogens—All Oxidizing Agents

The diatomic halogens (F_2, Cl_2, Br_2, and I_2) are all oxidizing agents. Fluorine is an exceptionally strong oxidizing agent, being able to oxidize almost anything that it contacts. The dangers to humans and physical structures are obvious. Cl_2 is also a very strong oxidizer. Its danger to humans is revealed by the fact that in WWI it was the first chemical warfare agent. Exposure to as little as 1000 ppm in airborne concentration (= 0.0010 atm = 0.76 mm Hg) for a few breaths can be fatal. This greenish colored gas should be used only in closed systems or chemical hoods.

Chlorine is used to prepare bleach—dilute solutions of sodium hypochlorite from the reaction with sodium hydroxide and chlorine. Chlorine forms explosive mixtures with ammonia due to the formation of nitrogen trichloride.[5]

Bromine is a dark red-brown liquid (BP = 59 °C) with a vapor pressure of 175 mm Hg at 20 °C. It is highly corrosive to the skin and either liquid or vapor contact with eyes is painful and destructive. Lachrymation, or tearing, begins at around 1 ppm, which functions as a good warning sign of exposure. Like chlorine, bromine can be fatal at 1000 ppm for short exposures.

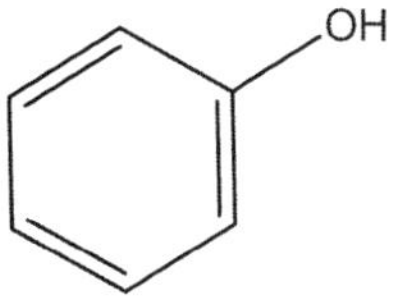

FIGURE 5.2.1.3 Structure of Phenol. The primary use of phenol is in the synthesis of bisphenol-A, which is used to make polycarbonate substances such as compact disks (CDs) and CD-ROMs.

You may know that the solution "tincture of iodine" is used as a disinfectant or to sanitize water. This solution is about 8% in ethanol and it is the oxidizing power of I_2 that makes it useful. At this concentration it is safe for exposure to skin and open wounds. Pure I_2, however, is quite hazardous. This dark purple solid has a melting point (MP) of 185 °C and a vapor pressure of 0.3 mm Hg (= 400 ppm) at 20 °C. Humans experience eye irritation at 1.6 ppm within 2 minutes. Iodine is more toxic than corrosive (2–3 grams is fatal) and it is a severe irritant to the respiratory tract.

All four of the elemental halogens represent very serious hazards in the lab. In the lungs they will all react to produce various oxyacids.

Other Dehydrating Agents and Water-Reactive Compounds

Advanced laboratory experiments sometimes use reducing agents. Elemental sodium reacts violently with water and produces hydroxide ions and highly flammable H_2 gas in the process. When sodium comes in contact with skin it reacts with moisture (e.g., sweat) to produce hydroxide, and the hydroxide will be at very high concentration and is very corrosive. Hydrides, such as sodium hydride and lithium aluminum hydride, are also common reducing agents. They react with water very readily in an exothermic fashion and produce H_2 gas. They are corrosive to the skin, eyes, and mucous membranes.

Sometimes "drying agents" are used to remove trace amounts of water from an enclosed atmosphere or from organic solvents. As you might guess, these substances are very reactive with water—which is exactly their mode of action as a drying agent. In addition to concentrated sulfuric acid, elemental sodium, hydride compounds, phosphorus pentoxide (P_2O_5, actually P_4O_{10}), and calcium oxide (CaO) are examples of other compounds that react readily with water. All of these will be corrosive substances in contact with the skin and eyes.

While accidental exposure to the skin for most chemicals is almost always remediated by flushing with water, since these chemicals react with water the best first remedy is to remove the solid, usually scraping it away with something (e.g., a credit card). Then, flushing with copious amounts of water is effective since only traces of the compound will remain and the exothermicity of the reaction will be tempered by the heat capacity of the water.

Phenol

A compound that defies easy categorization is phenol. Its structure (Figure 5.2.1.3) perhaps does not suggest its hazard: a simple benzene ring with a hydroxyl substituent. It is also less commonly known as carbolic acid.

However, this white, solid compound (MP = 41 °C) is rapidly absorbed through the skin and has both corrosive and toxic properties. It is perhaps more hazardous because it has a local anesthetic action at the site of contact, and exposure may not be immediately recognized. As with other corrosives, exposure to the skin, eyes, and mucous membranes should be avoided. An oral dose of as little as 1 gram can be fatal. However, it has good warning properties with an odor threshold of 0.06 ppm. (Odor thresholds are discussed in *Chemical Connection 7.1.4.1* "If I Can Smell It, Am I in Danger?")

Knowing of these hazards, you may be surprised to learn that phenol is the active compound in the commercial throat spray Chloraseptic®. This 1% solution has antiseptic properties.

Working with Corrosives

The main hazards of working with corrosives are skin exposure and inhalation. Chemists should wear gloves when working with corrosive solutions. The type of glove depends on the specific corrosive and the concentration in solution. For common acids and bases, nitrile gloves will be adequate for skin protection. A glove compatibility chart should be consulted to determine the appropriate gloves for other corrosives. Most corrosives should be handled in a chemical hood.

If there is a spill of a corrosive, it is important to isolate the area, keep other students away, report the spill to an instructor immediately, and use a spill kit if appropriate to do so. Responding to such a situation is discussed more in Sections 2.1.3 and 2.2.2.

RAMP

- *Recognize* chemicals with corrosive properties.
- *Assess* the level of risk based on likelihood of exposure and concentration of the corrosive.
- *Minimize* the risk by using less-hazardous chemicals (if possible) and by eliminating exposure possibilities by using eye protection and skin protection.
- *Prepare* for emergencies by learning locations of eyewashes, safety showers, sinks, and spill kits.

References

1. E. O. Wilson. BrainyQuote; available at http://www.brainyquote.com/quotes/quotes/e/eowilson198256.html (accessed June 16, 2009).
2. American Industrial Hygiene Association Laboratory Safety Committee. Hydrofluoric Acid Burn from Trifluoroacetic Acid; available at http://www2.umdnj.edu/eohssweb/aiha/accidents/chemicalexposure .htm#Trifluoracetic (accessed August 15, 2009).
3. Eileen B. Segal. First aid for a unique acid: HF. *Journal of Chemical Health and Safety* 5(5); 28–31 (1998).
4. Honeywell. *Recommended Medical Treatment for Hydrofluoric Acid Exposure*, Honeywell, Morristown, NJ, May 2000; available at http://membership.acs.org/F/FLUO/Links.htm (accessed August 15, 2009).
5. *Bretherick's Handbook of Reactive Chemical Hazards*, 7th edition (P. G. Urben, ed.), Elsevier, New York, 2007, Chlorine, #4041, pp. 1446–1448, 1457.

QUESTIONS

1. Corrosive chemicals usually involve what kind of reaction(s)?

   (a) Acid–base
   (b) Redox
   (c) Acid–base plus redox
   (d) Acid–base and/or redox

2. For some oxidizing acids, like nitric acid, the corrosive power of the chemical increases as concentration increases because of which of the following?

   I. There is more oxidizing agent available.
   II. The oxidizing agent is pH dependent and lower pH increases the oxidation half-cell potential.
   III. The oxidizing agent is pH dependent and higher pH increases the oxidation half-cell potential.

   (a) Only I
   (b) I and II
   (c) I and III
   (d) Only II

3. Aqua regia is a mixture of

   (a) HCl and H_2SO_4
   (b) HNO_3 and H_2SO_4
   (c) HNO_3 and HNO_2
   (d) HCl and HNO_3

4. Aqua regia can dissolve precious ("unreactive") metals such as Pt and Au since

   (a) It has a high concentration of a strong oxidizing agent

   (b) The oxidizing half-cell potential is very high due to the very acidic nature of the solution
   (c) It has complexing anions present
   (d) All of the above

5. NaOH destroys living tissue quite well since it reacts readily with

   (a) Proteins
   (b) Esters
   (c) Acids
   (d) Proteins and esters

6. Acetic acid can be corrosive

   (a) At high concentrations when inhaled
   (b) Since it is a good dehydrating agent
   (c) Since it causes burns when exothermically reacting with water in tissue
   (d) All of the above

7. Phosphoric acid

   (a) Is corrosive since it has an unusually large K_a
   (b) Is corrosive since it has an unusually small K_a
   (c) Can cause pulmonary edema
   (d) Is usually found as a solid

8. HF is corrosive at concentrations that are

   (a) ≥ 0.01 M
   (b) ≥ 0.1 M
   (c) ≥ 1 M
   (d) ≥ 10 M

9. HF should only be used

   (a) If there has been special training on using this chemical
   (b) If medical treatment for HF exposures are immediately available
   (c) If a response plan for HF exposure is in place
   (d) All of the above

10. Which halogen is relatively safe to handle in elemental form?

    (a) Cl_2
    (b) Br_2
    (c) I_2
    (d) None of the above

11. Reducing agents such as active metals and some metal hydrides

    (a) Are not corrosive since they are reducing, not oxidizing, agents

    (b) Are unreactive except in the presence of moderately strong oxidizing agents
    (c) Often produce hydroxide and flammable H_2
    (d) Generally do not react with water

12. Water-reactive, dehydrating agents

    (a) Are often solids
    (b) Should be flushed immediately with large amounts of water upon exposure
    (c) Should be scraped away before flushing with water
    (d) Both (a) and (c)

13. Phenol is

    (a) Used as a commercial throat spray at 1% concentration
    (b) A solid compound that reacts quickly upon contact with skin
    (c) Also called carbolic acid
    (d) All of the above

5.2.2

THE CHEMISTRY OF FIRE AND EXPLOSIONS

Preview Section 5.1.2 has already introduced the topic of flammable chemicals, how they are classified by the NPFA and the GHS, the fire triangle, and the basic terminology for characterizing flammable liquids. In this section we continue the discussion of the nature of fire and explosions as chemical reactions.

Among the notable things about fire is that it also requires oxygen to burn—exactly like its enemy, life. Thereby are life and flames so often compared.

Otto Weininger, Austrian philosopher[1]

INCIDENT 5.2.2.1   ETHER FIRE[2]

A student poured some diethyl ether ("ether") into a beaker from a 5-gallon can and began heating the beaker on a hot plate. He worked on an open lab bench using a hot plate that was not spark proof. As vapors rolled out of the beaker, down the hot plate, and over the bench, the hot plate thermostat clicked on, producing a spark. The ether vapors were ignited resulting in a flash fire that burned the hair off the student's arm and eyebrows. The student staggered backwards and fortunately the vapors were consumed instantly so that his clothing did not catch fire. However, the careless student had left the cap off the 5-gallon can of ether. Now the can looked like a large burner with a flame about 6 inches or higher coming out of the spout as the ether vapor burned. Another student saw what was happening, grabbed an extinguisher, and put out the fire. There were no injuries requiring medical treatment and no damage to the laboratory.

What lessons can be learned from this incident?

Fires Are Chemical Reactions!

Fires can be devastating to any laboratory. One of the principal causes of fires is the careless use of flammable chemicals. Reinforcing the need for prevention of fires is an important part of your safety education, and developing an understanding of the chemistry of fire can make it easier to prevent fires from occurring.

Fire is a chemical reaction that is an oxidation process resulting from a flammable material being ignited in the presence of an oxidizing agent, usually oxygen in our air. This generates products including energy in the form of heat and light. The release of energy from a reaction is termed an exothermic reaction.

Combustion can be complete in an atmosphere of excess oxygen, or incomplete if the fire stops before all the flammable is consumed. This reaction generates products in various states of oxidation. An example of complete combustion of elemental carbon is shown by Equation 5.2.2.1, where carbon dioxide is the only product. To start the fire, some source of energy is required. Usually a flame or spark can ignite the flammable material. That is, the carbon has to be ignited before the oxidation (burning) can proceed. This demonstrates the three components of the fire triangle: a flammable substance, an oxidizing agent, and a source of energy.

$$C(s) \quad + \quad O_2(g) \quad \rightarrow \quad CO_2(g) \qquad \Delta H = -393.5 \text{ kJ/mol } C(s)$$

(5.2.2.1)

Laboratory Safety for Chemistry Students, by Robert H. Hill, Jr. and David C. Finster
Copyright © 2010 John Wiley & Sons, Inc.

This can also be viewed in terms of the oxidation states of the reactants and products. Elemental carbon and elemental oxygen are both in the oxidation state of zero. Combustion is called "oxidation" because some of the atoms in the fuel increase in oxidation state. In this example, carbon increases from zero to $+4$ and the elemental oxygen is reduced from zero to -2 (per atom).

$$C(s) \quad + \quad O_2(g) \quad \rightarrow \quad CO_2(g) \quad \Delta H = -393.5 \text{ kJ/mol } C(s)$$
oxidation state : $\quad$ 0 $\qquad\qquad$ 0 $\qquad\qquad$ $+4-2$ $\hspace{4cm}$ (5.2.2.2)

Incomplete combustion arises when a substance is burned in the absence of adequate oxygen.

$$2C(s) \quad + \quad O_2(g) \quad \rightarrow \quad\quad 2CO(g) \quad\quad \Delta H = -221.0 \text{ kJ} \hspace{2cm} (5.2.5.3a)$$
$$C(s) \quad + \quad \tfrac{1}{2}O_2(g) \quad \rightarrow \quad\quad CO(g) \quad\quad \Delta H = -110.5 \text{ kJ/mol } C(s) \hspace{1cm} (5.2.2.3b)$$
$$\text{carbon} \qquad \text{oxygen carbon monoxide}$$
oxidation state : $\quad$ 0 $\qquad\quad$ 0 $\qquad\quad$ $+2-2$

You can see why this is called "incomplete combustion," since the carbon does not get fully oxidized to the $+4$ state. Also, as you would expect, less energy (-110 kJvs. -393 kJ) is released per mole of carbon burned (see *Chemical Connection 5.2.2.1* Using Bond Energies to Understand Heats of Reaction).

CHEMICAL CONNECTION *5.2.2.1*

USING BOND ENERGIES TO UNDERSTAND HEATS OF REACTION

A close look at Equations 5.2.2.1 and 5.2.2.3b reveals something that may seem nonintuitive at first. We can view the combustion of $C(s)$ to $CO(g)$ as "halfway" to the complete combustion of $CO_2(g)$ since the oxidation of carbon has gone only from zero to $+2$, "halfway" toward the $+4$ value in CO_2. We might first guess that the "halfway combustion" would produce about half of the -393.5 kJ of energy released, or about -197 kJ. But Equation 5.2.2.3b indicates only -110.5 kJ. Why?

The heat of reaction always depends on the total enthalpy of the products minus the total enthalpy of the reactants. However, this "halfway" point does not nearly produce "half" of the exothermic energy. To understand this, we can examine the reaction in terms of breaking and making bonds.

Let's look at the combustion that produces $CO(g)$ first. An approximate way to calculate the enthalpy of reaction is to subtract the sum of the bond energies of the products from the sum of the bond energies of the reactants. Since $C(s)$ is not a small, discrete molecule, identifying the bond energy is less obvious, but we can calculate the effective bond energy of $C(s)$ (graphite) using the bond energies of O_2 and CO and the enthalpy of the reaction.

$$C(s) + \tfrac{1}{2}O_2(g) \quad \rightarrow CO(g) \qquad\qquad \Delta H = -110.5 \text{ kJ/mol } C(s) \quad (5.2.2.3b)$$
Bond energies (BE) : ? $\quad$ $1/2 \times 498$ kJ/mol $\quad$ 1072 kJ/mol

$$\Delta H = -110.5 \text{ kJ} = [BE(C) + 1/2 \times 498 \text{ kJ}] - [1072 \text{ kJ}]$$

$$BE(C) = 713 \text{ kJ}$$

We can next use this to calculate the enthalpy of reaction for Equation 5.2.2.1 (even though we already know the experimental value):

$$C(s) \qquad\quad + \qquad O_2(g) \qquad \rightarrow \qquad CO_2(g)$$
BE : 713 kJ/mol $\qquad$ 498 kJ/mol $\qquad$ 2×799 kJ/mol

$$\Delta H \quad = [713 \text{ kJ} + 498 \text{ kJ}] - [2 \times 799 \text{ kJ}]$$
$$= -387 \text{ kJ}$$

This reasonably agrees with the experimental value of -393.5 kJ.

So, back to the original question: Why isn't the enthalpy of combustion that produces $CO(g)$ about one-half the enthalpy of combustion that produces $CO_2(g)$? The answer lies in the relative bond energies of the triple bond in CO and the double bonds in CO_2. While the triple bond in CO is one of the strongest bonds known, each of the double bonds in CO_2 is also very strong, and *much stronger* than the double bond in O_2. Thus, when CO_2 is formed, the very strong double bond energies (2×799 kJ/mol) help stabilize the products and this contributes to the exothermicity of the reaction much more than the single triple bond in CO (1×1072 kJ/mol). Requiring an extra half-mole of O_2 bond to be broken doesn't "cost much" energetically, and the overall reaction forming CO_2 becomes disproportionately exothermic.

Another way to view this is to look at the combustion of $CO(g)$:

$$CO(g) \quad + \quad \tfrac{1}{2}O_2(g) \quad \rightarrow \quad CO_2(g)$$

$BE:$ 1072 kJ/mol $\tfrac{1}{2} \times 498$ kJ/mol   799 kJ/mol

$$\Delta H = [1072 \text{ kJ} + \tfrac{1}{2} \times 498 \text{ kJ}] - [2 \times 799 \text{ kJ}]$$
$$= -277 \text{ kJ}$$

Thus, the "second half" of the combustion is quite exothermic; much more so than the "first half."

Why Are Fires So Dangerous?

There are three main effects from a fire that are dangerous: heat released, toxic by-products from the combustion, and oxygen consumed. These three dangers will be discussed in the context of assuming there is a fire in a confined room or set of rooms (as opposed to an outdoor fire).

The energy released is mostly in the form of heat energy and infrared and visible radiation. The heat that you "feel" when sitting several feet from a campfire is the infrared radiation and this can be quite considerable. In a confined fire, such as a bedroom or living room, the heat released can be so great that at some point even though the flames are not yet touching other contents of the room, those contents finally spontaneously burst into flames. This point is called the "flashover" in firefighter terminology and represents an extremely dangerous condition since a "small fire" is instantly a "large fire." The temperature at which flashover occurs is in the range of 480–650 °C (900–1200 °F). The ignition temperature of carbon monoxide is 609 °C and the combustion of this gas (produced in the fire) contributes to flashover.

A great many toxic by-products are generated in a fire. Since many flammable materials are various forms of hydrocarbon compounds, the chemically obvious products are carbon dioxide and carbon monoxide. Carbon dioxide is a simple asphyxiant since, as the concentration of O_2 necessarily decreases in air, the concentration of CO_2 increases. CO, on the other hand, is a chemical asphyxiant since it attaches to hemoglobin in the blood and prevents oxygen transport throughout the body.

Depending on what is burning in a fire, other toxic materials may be generated. Most obviously, smoke, which is a mixture of particulate matters and the gaseous products associated with fire, can be toxic since the particulate matter impedes lung function. Other gases in smoke can include HCN, HCl (especially from burning PVC pipes) and other hydrogen halides, nitrogen oxides (NO_x), and other organic irritants. The presence of plastics in various sorts in the fire contributes significantly to the load of toxic substances in smoke. The leading cause of death in fires is smoke inhalation.

The third danger from a fire is the consumption of oxygen. As Table 5.2.2.1 shows, adverse physiological effects begin when the O_2 drops from about 21% to 19% and more serious effects appear at 16% and lower. This is rarely a cause of death, however, since smoke generally causes death more rapidly than oxygen depletion. An additional contributing factor to oxygen depletion that is largely of

TABLE 5.2.2.1   **Effects of Oxygen Deficiency on the Human Body** [3]

Percent O_2	Possible results
20.9	Normal
19.0	Some unnoticeable adverse physiological effects
16.0	Increased pulse and breathing rate, impaired thinking and attention, reduced coordination
14.0	Abnormal fatigue upon exertion, emotional upset, faulty coordination, poor judgment
12.5	Very poor judgment and coordination, impaired respiration that may cause permanent heart damage, nausea, and vomiting
< 10	Inability to move, loss of consciousness, convulsions, death

concern only to firefighters is the presence of steam as a displacing agent. When water is applied to a fire, tremendous amounts of steam are generated and this necessarily lowers oxygen concentration (see *Chemical Connection 5.2.2.2* How Much Steam Is Produced from Liquid Water?). This is one, of several, reasons why firefighters wear supplemental air systems (SCBA, self-contained breathing apparatus) when fighting interior fires.

CHEMICAL CONNECTION *5.2.2.2*

HOW MUCH STEAM IS PRODUCED FROM LIQUID WATER?

When a firefighter is applying water from a fire hose to put out a fire, some of the water vaporizes to steam. How much? We can do a simple calculation to show the "expansion" factor of water to steam.

Let's assume that we start with 1 liter of liquid water and vaporize this. We'll change this volume of liquid to moles and then use the ideal gas law at a typical "fire" temperature of 1000 °C (= 1300 K).

$$1.00 \text{ L water} \left(\frac{1000 \text{ mL water}}{1.00 \text{ L water}} \right) \left(\frac{1.00 \text{ g water}}{1.00 \text{ mL water}} \right) \left(\frac{1 \text{ mol water}}{18.0 \text{ g water}} \right) = 55.5 \text{ mol water}$$

$$PV = nRT$$

$$V = \frac{nRT}{P} = \frac{(55.5 \text{ mol})(0.08206 \text{ L} \cdot \text{atm/K} \cdot \text{mol})(1300 \text{ K})}{1.00 \text{ atm}} = 5900 \text{ L steam}$$

Thus, the "expansion factor" under these conditions is about 5900, although the actual expansion factor is likely much lower since the steam will reduce in temperature considerably as it mixes in a room. This illustrates why the vaporization of water can lead to a significant decrease in the percentage of oxygen in a room and, in this regard, water is acting as a simple asphyxiant. (See Section 4.1.2.) The positive aspect of this, though, is that as the steam acts as a simple asphyxiant, it will also reduce the O_2 level to concentrations that might no longer support the fire. Thus, the steam "smothers" the fire.

A further danger for firefighters is that this superheated steam can easily burn unprotected skin. This is why firefighters wear protective gear that *completely* covers exposed skin when fighting fires.

The movie *Backdraft* (Universal Studios, 1991) also provides another example of understanding the chemistry of fire. An enclosed space or a room with windows and door closed is almost, although not completely, "sealed." When a fire burns, it will consume oxygen in the room and when the oxygen drops below about 16% the fire will self-extinguish, even with plenty of fuel (furniture, etc.) and heat in the room. Two-thirds of the fire triangle is present and if a door or window is opened there can be a quick rush of fresh air into the room. This can result in an explosion called a "backdraft." While many scenes in movies often depict fires quite inaccurately, the movie *Backdraft* actually showed a good example of a backdraft explosion when a firefighter inadvertently opened a door to a room in backdraft conditions.

How do firefighters detect and prevent backdrafts once the conditions are ripe? A tell-tale sign is "smoked-over" windows since the oxygen depleted fire tends to be very smoky. Also, smoke can be seen "puffing" in and out around doorways or leaks in windows. The only way to open the room without causing a backdraft is through "vertical ventilation," which means opening a large hole in the ceiling of the room. This allows heat to escape the room before enough air is admitted to cause a backdraft explosion. For this reason, and others, firefighters are often seen cutting large holes on roofs of buildings in the early stage of firefighting (often *before* entering a structure with a hose line) to allow dangerous heat and gases to safely escape.

Got Gas?

Natural gas (methane, CH_4) is piped into chemical laboratories. This highly flammable gas is used for Bunsen or other burners to supply heat for chemical reactions in beakers and flasks. Of course, methane is also used in homes for furnaces, hot water heaters, and stoves and ovens.

Methane has a GHS flammability rating of 1 (NFPA rating of 4), putting it in the "extremely flammable" category. A methane leak in a lab is extremely dangerous. The lower explosive level (LEL) is 4.5% and the upper explosive level (UEL) is 16.5%. A sulfur compound, often ethyl mercaptan or a mixture of mercaptans, is added to natural gas to give it an unpleasant odor and allow for easy detection.

The complete combustion of methane produces carbon dioxide and water:

$$CH_4(g) + 2O_2(g) \rightarrow CO_2(g) + 2H_2O(g) \quad \Delta H = -891 \text{ kJ}$$

The reaction is very exothermic. If a laboratory has a methane leak and the gas finds an ignition source, the result will be an explosion. It is very important to make sure stopcocks and valves are turned off completely when finished using this flammable gas.

Explosions from Fires — The Spectacular BLEVE

As we presented in Section 5.1.2, the most common liquids used in laboratories are flammable solvents. And most of these solvents have a reasonably high vapor pressure, which allows them to readily form flammable vapors. If these solvents are used on an open benchtop they can "pool" and travel and may find an ignition source such as an open flame or electrical spark. The possibility of igniting this flammable vapor is high, as illustrated in the incident described at the beginning of this section. Working in a chemical hood and avoiding ignition sources are the important preventative steps in avoiding lab fires started in this fashion.

If these precautionary steps are not taken, there may be not only a fire but perhaps an explosion. You have all seen explosions in the movies or perhaps even on television, but do you really know what is happening and why there was an explosion? An explosion is the sudden release of energy in the form of heat and light. In the discussion below we will be describing explosions caused by fires.

If flammable vapors collected within the confines of a room, equipment, or building, and the concentration is between the LEL and UEL, it only takes a spark to ignite this cloud (see Section 5.1.1 for more information about LEL and UEL). If the rate at which energy is released is faster than the ability of the local area to dissipate that energy through convection and heat capacity, a shock wave at supersonic velocity is formed. This is an explosion. This shock wave constitutes a wave of high-pressure gas that can do considerable damage. Explosions are more likely to occur near the stoichiometric point of the reacting species and less likely to occur the closer the reacting ratios are to just above the LEL or just below the UEL. The incident described at the beginning of this section more likely represents a "flash fire" than an explosion. Both events are dangerous, of course, but the flash fire doesn't have the associated shock wave.

There is another kind of explosion that can occur in labs and elsewhere, and it illustrates many chemical and physical principles. A *boiling liquid expanding vapor explosion* (BLEVE, pronounced "blev-ee") results when a container of flammable liquid is exposed to high heat and the internal pressure produced becomes so great that the integrity of the container eventually fails, resulting in a rapid release of vapors that are ignited instantaneously—a very violent explosion. The BLEVE is often depicted in movies, and it is a very real concern for firefighters and other personnel who can be injured from shrapnel from these containers and nearby materials destroyed by the explosion. Let's consider more how a BLEVE might occur in a laboratory situation.

Many flammable liquids are stored in containers of various sizes. Such containers could be 4-L bottles, 5-gallon cans, or even 55-gallon drums. Usually, these are sealed or capped. If any of these containers is in a lab fire the pressure of the vapor inside the container will start to increase for two reasons. First, we know that $PV = nRT$. The pressure in the sealed container will be about 1 atm, or slightly higher if the liquid has vaporized a bit since the container was sealed. If the temperature rises from about 300 K to about 1300 K (a typical fire temperature), the pressure will increase to about 4.3 atm. This alone may rupture the container.

Furthermore, the Clausius–Clapeyron equation shows the nonlinear relationship between vapor pressure and temperature:

$$P_{\text{vap}} = \beta e^{(-\Delta H/RT)}$$

where P_{vap} is the vapor pressure of the liquid, β is a constant characteristic for the gas, ΔH is the enthalpy of vaporization, R is the ideal gas law constant, and T is the temperature in kelvin units. As temperature increases, the exponential term becomes a smaller negative value and P_{vap} increases exponentially. As more molecules enter the gas phase, the number of moles (n) increases and this will increase the pressure above the 4.3 atm estimated above. For any container, there will be some pressure at which some part of the container will experience failure and rupture. When this happens a large cloud of flammable vapor will be produced.

If this flammable vapor cloud finds an ignition source, as it likely would in a fire, and the vapor concentration is between the LEL and UEL, then the rapidly expanding cloud will burn, probably explosively. This process will likely vaporize the rest of the liquid, which will also subsequently burn. The term *boiling liquid expanding vapor explosion* is easily understood. It is also possible that, even without the ignition source, the vapor cloud could explode if the temperature is above the autoignition temperature—although for most liquids the autoignition temperature is quite high. (See Section 5.1.2 and Table 5.1.2.3.)

The other scenario for a BLEVE is less likely in a lab but more commonly occurs in home or industrial settings. First, let's consider a liquefied petroleum gas (LPG) container that is used for home barbeques. (LPG is typically at least 90% propane with the remainder being butane and propylene.) This is typically a 20-pound (4.76 gallon) container. When filled, the container is about 80% liquid and 20% vapor over the surface of the liquid by volume. At ambient temperatures (about 70 °F or 21 °C) the vapor pressure inside the tank is about 8.4 atm. (Propane has a normal boiling point of −42 °C; this is the temperature at which the vapor pressure is 1.0 atm.) These containers have relief valves set at about 25–32 atm. The cylinders are designed to rupture at pressures not less than about 65 atm.[4] So what happens if such a cylinder is in a fire?

As in the laboratory scenario, the pressure inside the tank will increase as the temperature of the cylinder increases in the fire. However, when the pressure in the tank reaches about 25–32 atm, the relief valve will open and the gas is released. The jet of gas will likely start on fire since an ignition is available in a fire and the resulting gas becomes a torch. Since the liquid inside the tank has a reasonably good heat capacity, the temperature of the tank does not rise rapidly, and the pressure in the tank cannot go above the relief valve pressure. These two factors delay the rupture of the tank. However, as the gas burns the level of the liquid in the container eventually decreases and a larger surface area of the tank is exposed to fire conditions and temperatures without the benefit of the heat-absorbing liquid inside. At some point, the integrity of the tank will fail, and the BLEVE will occur.

Propane tanks located at rural homes may have hundreds of gallons of LPG and rail tank cars have from 4000 to 45,000 gallons. BLEVEs from rail tank cars are spectacular, sending shrapnel as far as 1/2 mile from the site of the explosion and producing fireballs several hundred feet in diameter. There are many spectacular videos of BLEVEs on the YouTube web site found at http://www.youtube.com/.

Working with Flammable Chemicals: RAMP

Given the obvious hazards of flammable chemicals, it makes sense to limit their use when possible. We can use RAMP to work safely with flammable chemicals:

- *Recognize* the appropriate flammability hazard associated with the solvents and containers holding flammable liquids.

- *Assess* the level of risk under the circumstances by considering means by which the vapor might be generated and released and determine if any sources of ignition are available.

- *Minimize* the risk by substituting a less-flammable solvent if possible, by using less solvent if possible, by working in a ventilated area to limit vapor concentrations, and eliminating sources of ignition.

- *Prepare* for emergencies by locating and knowing how to use the appropriate fire extinguisher. Maintain a current inventory of chemicals for use by emergency personnel.

References

1. O. WEININGER. BrainyQuote; available at http://www.brainyquote.com/quotes/authors/o/otto_weininger.html (accessed June 16, 2009).
2. ROBERT HILL. Personal account of an incident.
3. Compressed Gas Association, 2001. Found in U.S. Chemical Safety Hazard Investigation Board. Safety Bulletin. Hazards of Nitrogen Asphyxiation; available at http://www.csb.gov/assets/document/SB-Nitrogen-6-11-03.pdf (accessed September 14, 2009).
4. National Propane Gas Association. Report on Testing and Assessment of CG-7 Pressure Relief Valve and Propane Cylinder Performance. Volume 1: Results and Evaluation. Prepared by D. R. STEPHENS, M. T. GIFFORD, R. B. FRANCINI, D. D. MOONEY (Battelle). January 31, 2003; available at http://www.propanecouncil.org/files/10202_Battelle_Cylinder_Report_Final.pdf (accessed September 14, 2009).

QUESTIONS

1. In combustion reactions

   (a) The fuel is reduced and the oxygen is oxidized
   (b) The fuel and the oxygen are both oxidized
   (c) The fuel and the oxygen are both reduced
   (d) The fuel is oxidized and the oxygen is reduced

2. Incomplete combustion arises from

   (a) The absence of enough fuel and produces carbon dioxide in a reduced state
   (b) The absence of enough oxygen and produces carbon dioxide in a reduced state
   (c) The absence of enough fuel and produces carbon monoxide
   (d) The absence of enough oxygen and produces carbon monoxide

3. With regard to sustaining life, which of the following are the three main dangerous results of a fire?

   I. The heat released
   II. Toxic by-products
   III. Reduced fuel
   IV. Reduced oxygen levels

   (a) I, II, and III
   (b) I, II, and IV
   (c) I, III, and IV
   (d) II, III, and IV

4. The leading cause of death in fire is due to

   (a) Lack of oxygen
   (b) Heat
   (c) Inhalation of smoke
   (d) None of the above

5. At what oxygen level do humans start to show reduced physiological characteristics?

   (a) 12.5%
   (b) 14.0%
   (c) 16.0%
   (d) 19.0%

6. A backdraft explosion can occur when

   (a) There is a lack of fuel in a partially burned room
   (b) There is a lack of oxygen in a partially burned room
   (c) A door is opened *into* a room instead of opening into a hallway or outside
   (d) A fire is very smoky

7. The odorant in natural gas is usually

   (a) An amine
   (b) A phosphorus compound
   (c) A mercaptan
   (d) An ester

8. An explosion can occur when

   (a) The rate of energy released in a chemical reaction is faster than the ability of the area to dissipate the energy through convection and heat capacity
   (b) The rate of energy released in a chemical reaction is slower than the ability of the area to dissipate the energy through convection and heat capacity
   (c) The energy released in a chemical reaction is almost exactly equal to the ability of the area to dissipate the energy through convection and heat capacity
   (d) Any of the above

9. A BLEVE is an explosion where

   (a) A container of flammable liquid has ruptured in the presence of an ignition source
   (b) The flammable liquid in a container can no longer absorb enough heat to keep the container cool enough not to rupture
   (c) A container has reached an internal pressure above the limit of the pressure relief valve
   (d) All of the above

10. A BLEVE is not a likely lab event since

   (a) Relief valves on compressed gases in labs will always prevent a BLEVE
   (b) Special additives are present in high-pressure lab tanks that inhibit explosions
   (c) It requires that there be an active fire impinging upon a sealed vessel containing a flammable liquid
   (d) Vessels used in lab environments are constructed so that they cannot explode

5.2.3

INCOMPATIBLES—A CLASH OF VIOLENT PROPORTIONS

Preview   This section describes incompatibles, why they should not be mixed, and steps to prevent incompatibles from coming in contact with each other.

To every action there is always opposed an equal reaction.

Sir Isaac Newton, *Principia Mathematica. Laws of Motion, III*[1]

INCIDENT 5.2.3.1   EXPLODING HAZARDOUS WASTE[2]

A large glass capped bottle being used in a laboratory to collect nitric acid waste was located in a chemical hood when it spontaneously exploded, spraying nitric acid and glass pieces throughout the lab. A student was working at a computer in the lab when the explosion occurred, but he was not injured. Other waste containers in the hood were either destroyed or cracked. An investigation revealed that the nitric acid waste bottle had originally contained methanol. It was estimated that the explosion occurred about 12–16 hours after the nitric acid had been added to the bottle.

What lessons can be learned from this incident?

Incompatibles—A Chemical Overview

Incompatible chemicals are combinations of substances, usually in concentrated form, that react with each other to produce very exothermic reactions that can be violent and explosive and/or can release toxic substances, usually as gases. There have been many incidents involving incompatible chemicals because these kinds of hazards were not recognized. In these unintentional situations the risks are considerable since no one is expecting the reaction to occur and proper precautions are likely not in place. Sometimes there may be no one around, which presents other dangers, such as extended fire or property damage, since no quick emergency response takes place. In incidents where gases are generated in a closed system the additional hazard of an explosion due to the rupture of a container from high pressure is also possible, as happened in Incident 5.2.3.1. It is worth noting that sometimes in the laboratory we mix dilute solutions of incompatibles for some purpose, but because they are not concentrated they do not react violently and only release limited heat in their reaction.

There are hundreds of chemicals in common use and thousands of less frequently used chemicals that could be used as part of any particular lab experiment or process. Memorizing all the combinations and permutations of potentially incompatible chemicals is not possible. The goal in this section is to learn about some specific incompatible mixtures but more generally to learn about combinations of incompatible chemical groups. Some fundamental chemistry can help you think about the properties and structures of molecules that might predict them as chemicals having "incompatible partners."

Most chemical reactions that are exothermic will be either acid–base reactions or oxidation–reduction reactions. Thus, as you think about any single chemical as something that may have "incompatible partners," you can ask:

Laboratory Safety for Chemistry Students, by Robert H. Hill, Jr. and David C. Finster
Copyright © 2010 John Wiley & Sons, Inc.

Is this chemical a strong acid?

Is this chemical a strong base?

Is this chemical easily oxidized?

Is this chemical easily reduced?

Table 5.2.3.1 presents a useful overview of the most common incompatible classes of chemicals with examples. This table is probably too long to memorize, and it is *not* comprehensive, either. Understanding, rather than memorizing, this table is possible if the chemistry behind all of the entries is appreciated. Let's review these fundamental categories as a prelude to discussing the most significant incompatible combinations below.

Acid–Base Incompatibles—Overview

The most common incompatibles that you will likely encounter in the laboratory will be acids and bases. Mixing acids and bases can produce violent, even explosive, reactions that release a lot of heat (exothermic reactions). The six strong acids that you are likely to encounter are listed in Table 5.2.3.2. Each of the four acids in common usage has particular chemical features that make it desirable in particular reactions. The common strong bases are listed in Table 5.2.3.3. Clearly, a reaction between a strong acid and a strong base will be very exothermic, but as Table 5.2.3.1 shows, strong acids reacting with weak bases and strong bases reacting with weak acids can be dangerous, too. The exothermicity of the reaction will depend on the concentration of the solution. Weaker concentrations not only have fewer moles of reagents to react, but the relative amount of water (with its high heat capacity) is higher so the heat energy released is better absorbed and the temperature rise is moderated.

As an illustration of the danger of mixing strong acids and bases, let's calculate the heat released, and temperature increase, when mixing 100 mL of 12 M HCl and 100 mL of 12 M NaOH. The heat of reaction is -55.8 kJ/mol, which leads to 67 kJ of energy being released from 1.2 mol of reactants. Assuming the solution has a heat capacity of about 4 J/g·°C, we calculate a temperature change of about 84 °C. If the reactants start at room temperature, about 25 °C, the final temperature of about 109 °C suggests a very violent, uncontrolled reaction! In practice, even using less-concentrated solutions, but assuming that much of the exothermicity of the reaction will be in "spot heating," it is easy to see that such a mixture might cause dangerous spattering of a hot solution of a strong acid or base. And if this is inadvertently done in nonborosilicate glassware, such as an ordinary glass bottle, there is a good chance that the bottle might shatter from the rapid temperature change.

Often the exothermicity of the acid–base reaction is enough to produce a violent reaction, but there are some instances where it is not the exothermicity *per se* of the reaction that is so dangerous as much as the products. We look at a few common examples.

Any soluble cyanide salt plus any strong acid produces toxic HCN gas:

$$NaCN(aq) + HCl(aq) \rightarrow NaCl(aq) + HCN(g)$$

Any soluble azide salt plus any strong acid produces toxic and explosive HN_3 gas:

$$NaN_3(aq) + H_2SO_4(aq) \rightarrow Na_2SO_4(aq) + HN_3(g)$$

Any soluble sulfide salt plus any strong acid produces toxic and flammable H_2S gas:

$$Na_2S(aq) + 2HCl(aq) \rightarrow 2NaCl(aq) + H_2S(g)$$

The three reactions shown above involve strong acids, but these reactions can also occur using weak acids. However, acid–base equilibria always shift to the side of the reaction with the weaker acid and base. The K_a and K_b (and pK_a and pK_b) values indicate the reaction below will be shifted to the right.

$CN^-(aq)$	+	$CH_3CO_2H(aq)$	$\leftrightharpoons$	$HCN(g)$	+	$CH_3CO_2^-(aq)$
stronger base		stronger acid		weaker acid		weaker base
$K_b = 1.6 \times 10^{-5}$		$K_a = 1.7 \times 10^{-5}$		$K_a = 6.2 \times 10^{-10}$		$K_b = 5.8 \times 10^{-10}$
$pK_a = 4.79$		$pK_b = 4.76$		$pK_a = 9.21$		$pK_b = 9.24$

TABLE 5.2.3.1 Incompatibles Chemicals—A Limited Listing[3-5]

Class	Types of incompatibles	Examples of incompatibles
Acid *incompatibles:* *substances listed to the right react violently with acids*	Hydroxides	NaOH, KOH
	Inorganic azides	Sodium azide (produces toxic HN_3)
	Chlorates	Potassium chlorate
	Cyanides	Potassium cyanide (produces HCN gas)
	Carbides	Calcium carbide (produces flammable C_2H_2)
	Hydrides	Sodium hydride (produces flammable H_2)
	Oxides	Calcium oxide
	Perchlorates	Potassium perchlorate
	Sulfides	Sodium sulfide (produces H_2S)
	Organic peroxides	Benzoylperoxide, C_5H_5COO—$OOCC_6H_5$
Base *(strong) incompatibles:* *substances listed to the right react violently with bases*	Acids	HCl, H_2SO_4, CH_3COOH
	Inorganic cyanides	Sodium cyanide
	Organic acyl halides	Acetyl chloride
	Organic anhydrides	Acetic anhydride
	Organic nitro compounds	Nitrobenzene
Water-reactives: *substances listed to the right react with water*	Alkali/alkaline earth metals	Sodium, potassium
	Metal carbides	Calcium carbide
	Metal hydrides	Sodium hydride, lithium aluminum hydride
	Nonmetal hydrides	Boranes, silanes
	Alkali/alkaline earth metals oxides	Calcium oxide
Pyrophorics: *substances listed to the right react in air*	Some finely divided metals	Magnesium, zinc
	Alloys of reactive metals	Potassium–sodium alloy
	Alkylmetals	*t*-Butyllithium, trimethylaluminum
	Selected main group elements	White phosphorus
	Metal hydrides	Potassium hydride
	Nonmetal hydrides	Diborane, phosphine
	Iron sulfides	FeS (moist), FeS_2 (powdered)
	Alkylphosphines	Diethylphosphine
	Some organometallics	Bis(cylclopentadienyl)manganese
Oxidizing agents: *substances listed to the right are easily oxidized*	Organic compounds	Acetic acid, aniline
	Metals	Sodium, magnesium
	Metal hydrides	Sodium hydride
	Main group elements	Phosphorus, sulfur, carbon
	Main group compounds with hydrogen	Ammonia
Reducing agents: *substances listed to the right are easily reduced*	Chlorates, perchlorates	ClO_3^-, ClO_4^-
	Chromates	CrO_4^{2-}, CrO_3
	Halogens	F_2, Cl_2
	Nitrates	NO_3^-
	Peroxides	Na_2O_2, H_2O_2
	Persulfates	$S_2O_8^{2-}$
	Permanganates	MnO_4^-

TABLE 5.2.3.2   Six Common Strong Acids

Acid	Formula	Molarity of concentrated acid	Concentration (%)	Frequency of use
Hydrochloric acid	HCl	12	36%	High
Nitric acid	HNO_3	16	70%	High
Sulfuric acid	H_2SO_4	36	98%	High
Perchloric acid	$HClO_4$	12	70%	Occasional
Hydrobromic acid	HBr	8.9	48% (azeotrope)	Rare
		14	69%	
Hydroiodic acid	HI	7.1	48	Rare
		8.8	57% (azeotrope)	

TABLE 5.2.3.3   Strong Bases

Base	Formula	Molarity of concentrated base	Concentration (%)	Solubility in g/100 mL ($^\circ$C)	Frequency of use
Sodium hydroxide	NaOH	19.4	50.5	42 (0 $^\circ$C)	High
Potassium hydroxide	KOH	11.7	45.0	107 (15 $^\circ$C)	High
Calcium hydroxide	$Ca(OH)_2$	Partially soluble $K_{sp} = 8 \times 10^{-6}$	Partially soluble	0.185 (0 $^\circ$C)	Occasional
Lithium hydroxide	LiOH	Not readily available	Not readily available	13 (20 $^\circ$C)	Occasional
Rubidium hydroxide	RbOH	Not readily available	Not readily available	180 (15 $^\circ$C)	Rare
Cesium hydroxide	CsOH	Not readily available	Not readily available	395 (15 $^\circ$C)	Rare

Strong Oxidants and Reductants—An Overview of Redox Incompatibles

Identifying chemicals that are easily reduced (strong oxidants) or easily oxidized (strong reductants) is less simple, but there are some guides. We can look at reduction potentials for various chemicals. A very brief version of a reduction table is shown in Table 5.2.3.4. We will consider general patterns of oxidation states using the periodic table. Let's look at Table 5.2.3.4 and how this might help you identify strong oxidizing agents and strong reducing agents.

Chemicals that are *reactants* at the top left of the table (boldfaced) are very easily reduced and are therefore good *oxidizing agents*. Some trends to note are the elemental halogens and compounds that have atoms in relatively high oxidation states, such as N(+5) in nitrate and Mn(+7) in permanganate. Note that when oxygen is in an oxidation state that is greater than −2, it is in an "oxidized" state. So, oxygen in peroxide is at −1 and oxygen in O_2 is at zero.

Nitric acid and nitrate have been involved in the majority of incidents involving incompatible chemicals. The very fact that nitric acid is such a good relatively inexpensive oxidizing agent makes it a very common chemical in labs and in many reactions. The virtue of good oxidizing power is simultaneously its greatest hazard.

Another strong acid that is a very strong oxidizing agent is perchloric acid, $HClO_4$. In fact, it is quite likely that you have never used perchloric acid since it presents unusual hazards. While cold 70% perchloric acid is only a weak oxidizing agent, hot solutions are powerful oxidizers and pure perchloric acid explodes upon contact with organic compounds and materials (such a cloth, wood, and rubber). Perchlorate salts are shock-sensitive and may easily detonate. Section 5.3.3 will discuss this circumstance further.

Permanganate is also a strong oxidizing agent and frequently used in titrations in that role. However, permanganate solutions used in titrations are often relatively dilute, such as 0.1 M, and the solutions decompose over the course of days.

TABLE 5.2.3.4   A Short List of Reduction Potentials Showing Strong Oxidizing Agents and Strong Reducing Agents

Reduction half-reaction	E°(V)	Strong oxidizing agent on left	Strong reducing agent on right
$F_2(g) + 2e^- \rightarrow 2F^-(aq)$	2.87	X	
$H_2O_2(aq) + 2H^+(aq) + 2e^- \rightarrow 2H_2O(l)$	1.78	X	
$MnO_4^-(aq) + 8H^+(aq) + 5e^- \rightarrow Mn^{2+}(aq) + 2H_2O(l)$	1.507	X	
$ClO_4^-(aq) + 8H^+(aq) + 8e^- \rightarrow Cl^-(aq) + 2H_2O(l)$	1.389	X	
$Cl_2(g) + 2e^- \rightarrow 2Cl^-(aq)$	1.36	X	
$Cr_2O_7^{2-}(aq) + 14H^+(aq) + 6e^- \rightarrow 2Cr^{3+}(aq) + 7H_2O(l)$	1.232	X	
$O_2(g) + 4H^+(aq) + 4e^- \rightarrow 2H_2O(l)$	1.229	X	
$Br_2(l) + 2e^- \rightarrow 2Br^-(aq)$	1.09	X	
$NO_3^-(aq) + 4H^+(aq) + 3e^- \rightarrow NO(g) + 2H_2O(l)$	0.957	X	
$I_2(s) + 2e^- \rightarrow 2I^-(aq)$	0.54	X	
$H_2(aq) + 2e^- \rightarrow 2H^-(s)$	−2.23		X
$Al(OH)_4^-(aq) + 3e^- \rightarrow Al(s) + 4OH^-(aq)$	−2.328		X
$K^+(aq) + e^- \rightarrow K(s)$	−2.379		X
$Na^+(aq) + e^- \rightarrow Na(s)$	−2.71		X
$Ca^{2+}(aq) + 2e^- \rightarrow Ca(s)$	−2.868		X
$Li^+(aq) + e^- \rightarrow Li(s)$	−3.04		X
$Cs^+(aq) + e^- \rightarrow Cs(s)$	−3.06		X

Hydrogen peroxide is an excellent oxidizing agent and at concentrations greater than 30% present considerable fire and explosion hazards. It reacts with a large number of organic and inorganic compounds.

Chemicals that are *products* at the bottom of the table (boldfaced) are very easily oxidized (to the reactants shown in the half-reaction) and will be good *reducing agents*. Many metals, called the *active metals*, oxidize easily and, as you might guess, hydrogen in the −1 oxidation state (hydride) is also easily oxidized. The substances are strong reducing agents.

Fairly exhaustive lists of classes of molecules and ions that are good oxidizing agents and reducing agents exist. An excellent source for this kind of information is *Bretherick's Handbook of Reactive Chemical Hazards*.[5] Since most chemists and chemistry students will not memorize those lists, it is important to know some generalities about various chemicals and, probably more importantly, to be ready to frequently consult these lists whenever working with chemicals.

A few additional comments and precautions are helpful in using Table 5.2.3.4:

- All of these half-reaction potentials are shown for standard state conditions where all concentrations are 1 molar and all pressures are 1 atm. (These conditions are rare in the lab.) The Nernst equation must be used to determine reduction potentials at nonstandard state conditions.

- This is a *thermodynamic* table. Activation energy barriers might allow some thermodynamically spontaneous reactions to be very slow or nonreacting.

- Many inorganic and organic salts that have anions that are oxo anions with central atoms in high oxidations are good oxidizing agents. This includes chlorates, chromates, nitrates, nitrites, perchlorates, permanganates, peroxides, and persulfates. Other hazardous anions are azides, cyanides, fulminates, and hydrides.

- Organic compounds are generally easily oxidized, but since these are not generally reactions that occur in aqueous solution, they do not appear in reduction potential tables. The trends are as follows:

  - All organic compounds can be oxidized, some better than others.
  - Some organic functional groups are particularly reactive, including acyl halides, anhydrides, azides, diazonium salts, nitroso compounds, peroxides, fulminates, and nitriles.

You will see most of these chemicals somewhere in Table 5.2.3.1.

Sometimes chemicals react with each other that, at first, would seem compatible. A classic example of this is nitric acid and acetic acid. Can two acids react with each other? As you can see below, this is not an acid–base reaction but instead is a redox reaction—acetic acid and nitric acid produce toxic NO_2 gas:

$$CH_3CO_2H(aq) + 8HNO_3(aq) \rightarrow 2CO_2(g) + 8NO_2(g) + 6H_2O(l)$$

This reaction has an obvious lesson for chemical storage: don't assume that the most common use of a chemical (e.g., nitric acid as an acid) limits its ability to function otherwise. Storing these two acids together, which at first might seem reasonable, presents a considerable hazard. This is discussed more in Sections 8.1.1 and 8.3.4.

Chemicals That React With Water — Water-Reactives

We ordinarily consider water to be fairly nonreactive. Water-reactives are chemicals that react violently with water, releasing large amounts of heat and sometimes flammable gases or toxic gases, often resulting in fires or explosions.

As Table 5.2.3.5 shows, most notorious in this group are the alkali metals—lithium, sodium, and potassium. In Table 5.2.3.3 you can see that these are the metals where the cation has a large, negative reduction potential. (Thus, any chemical that is the product of a reduction with a large, negative reduction potential might be water reactive.) There have been laboratory fires caused by the reaction of water with these metals, particularly sodium and potassium. The National Research Council included these water reactives in their "Dirty Dozen" listing to highlight the risk of laboratory fires when using these metals.[3] Sodium is often used to "dry" organic solvents and fires have occurred because researchers assumed that all of the sodium metal has been consumed or reacted and have dumped the residue into a sink, where water came in contact with unreacted sodium metal which caught fire[6] (see Incident 5.1.2.1). The equation below shows the reaction between sodium and water that results in sodium hydroxide and the release of hydrogen:

$$2Na(s) + 2H_2O(l) \rightarrow 2NaOH(aq) + H_2(g)$$

When chemists or firefighters refer to a "metal fire" it is actually the hydrogen gas that is produced that is on fire.

Other classes of chemicals are also water-reactive. Table 5.2.3.4 lists the classes of chemicals and sample reactions. All of the reactions in the table are acid–base reactions (unlike the active metals above, which are redox reactions).

In Table 5.2.3.5 nonmetal oxides are also considered to be "acid anhydrides" in some categorization systems since addition of water produces acids. Note also that many nonmetal oxides, like CO_2 and SO_2, react with water but not in a fashion that would be considered "incompatible," since the reaction is not very exothermic or dangerous. For example, exhaling CO_2 through a straw into a glass of water does not produce a violent reaction.

Also in Table 5.2.3.5, the example under nonmetal oxides refers to P_4O_{10}. This compound, called tetraphosphorus decaoxide, is also commonly called "phosphorus pentoxide" and the formula is

TABLE 5.2.3.5 Examples of Water-Reactive Classes of Compounds

Chemical class	Example
Alkali metals	$M + 2 H_2O(l) \rightarrow 2 MOH\ (aq) + H_2\ (g)$
Organometallics	$RMgX + H_2O \rightarrow RH + Mg(OH)X$
Acyl chlorides	$RCOCl + H_2O \rightarrow RCOOH + HCl$
Main group metal halides	$TiCl_4 + 2H_2O \rightarrow TiO_2 \cdot (H_2O)_n + 4HCl$
Metal oxides	$CaO + H_2O \rightarrow Ca(OH)_2$
Acid anhydrides	$(RCO)_2O + H_2O \rightarrow 2\ RCOOH$
Nonmetal oxides	$P_4O_{10} + 6 H_2O \rightarrow 4 H_3PO_4$

frequently written as "P_2O_5." This is an issue of nomenclature, since for many years the formula of the compound was considered to be "P_2O_5" until the structure was eventually determined to be P_4O_{10}. To this day, many chemists still call this "P-two-O-five" or "phosphorus pentoxide."

When handling a water-reactive compound you should carefully plan how you will use it in a manner that prevents contact with water, including moisture in the air. In some cases, sloppy handling of the chemical, perhaps even in humid air, will cause it to react slowly and prevent further use of the chemical as a reagent. In other situations, reaction with water, particularly liquid water, can be disastrous and dangerous.

Finally, we note that many strong acids and bases with high concentrations react violently when diluted with water due to large energies of solvation. (This is explained in detail in *Chemical Connections* 5.1.1.2 and 5.1.1.3.) It may seem odd to consider an aqueous solution "water-reactive" but it is smart to be mindful of these situations. Particularly reactive in this regard are "syrupy," 90% orthophosphoric acid and 98% sulfuric acid.

Pyrophorics—Incompatibles with Air

Pyrophoric comes from the Greek word *purophoros* meaning "fire-bearing." Pyrophorics are chemicals that ignite spontaneously in the presence of air. These are compounds that are oxidized by oxygen in the air or react so quickly with moisture (water) in the air that they ignite in the presence of air. Finely divided metal powders can be pyrophoric, as well as reactive metal alloys, metal hydrides, and some metal salts. Calcium, zirconium, uranium, and magnesium powders are examples of pyrophoric metals. The equation below shows a representative reaction of a metal with oxygen:

$$Zn(s, dust) + 2H_2O(l) \rightarrow Zn(OH)_2(s) + H_2(g)$$

Lithium aluminum hydride, sodium borohydride, and sodium hydride are pyrophoric hydrides that react with moisture in the air. Lithium aluminum hydride is one of the NRC's "Dirty Dozen"—it has been the cause of many fires. The equation representing the reaction of a hydride with moisture in air is shown below:

$$LiAlH_4(s) + 4H_2O(l) \rightarrow LiOH(aq) + Al(OH)_3(s) + 4H_2(g)$$

While hydrocarbons are thermodynamically unstable in an oxygen-containing atmosphere, they are kinetically very slow to react. However, many boranes (B_xH_y) and silanes (Si_xH_y) are both thermodynamically unstable and they react quickly in air. Boranes were explored as possible rocket propellants during 1940–1960. Two examples are shown below, and the thermodynamic stability of the oxides that are products contributes significantly to the exothermicity of the reactions:

$$B_2H_6(g) + 3O_2(g) \rightarrow B_2O_3(s) + 3H_2O(l)$$

$$SiH_4(g) + 2O_2(g) \rightarrow SiO_2(s) + 2H_2O(l)$$

One other chemical bears some mention here since its reactivity is probably unexpected. Elemental phosphorus may look like an innocent-enough element in the main group block of the periodic table, and in some ways it is. In fact, it appears as three allotropes: red phosphorus, black phosphorus, and white phosphorus. None of these is naturally occurring—phosphorus occurs as phosphate in nature. White phosphorus is P_4, with the four atoms bonded in a tetrahedral shape, requiring $60°$ bond angles that are highly strained and reactive. It spontaneously ignites in air above $35\,°C$, while the other allotropes are air-stable. Since white phosphorus is pyrophoric, it is stored under water to prevent contact with air (oxygen). The reaction of white phosphorus with oxygen in the air is shown in the equation below. It is interesting to note that the product, P_4O_{10}, is very water-reactive as discussed above.

$$P_4(s) + 5O_2(g) \rightarrow P_4O_{10}(s)$$

Some pyrophoric reagents such as *tert*-butyllithium are purchased as solutions in flammable solvents such as hexane. Should the pyrophoric compound be exposed to air it can ignite and catch the available fuel (i.e., hexane) on fire. An incident that illustrates this possibility is found in Incident 7.3.5.1.

When handling a pyrophoric chemical, *and before you take the first steps to open the container*, plan how you will use it in a manner that prevents contact with air. Pyrophorics should only be handled by chemists with the knowledge and skills to work with these chemicals safely. If you are contemplating handling a pyrophoric for the first time, you should seek the advice and skill of an experienced chemist before handling pyrophorics.

Storing Incompatible Chemicals

Incompatibles must not come in contact with each other and should not be stored together. This is particularly true in geographic areas with significant seismic activity, where earthquakes are possible. While we typically think of California as earthquake-prone, in fact, many areas of the country are susceptible to earthquakes, so its good practice to keep this in mind no matter where you are.

But even those who do not live in these areas could experience some natural weather damage or event that could result in shelves collapsing and broken bottles of chemicals mixing. If you keep your incompatibles separated, then the risk of incompatibles coming in contact is reduced. As with all hazardous chemicals, you can reduce the risk by minimizing the quantities of chemicals being used and stored. Be prudent in ordering highly reactive chemicals, such as strong acids, strong bases, strong oxidizing agents, and strong reducing agents. Most highly reactive chemicals are incompatible with some other chemicals. Be sure that you learn all you can about potential incompatibles. The storage of chemicals is discussed in more detail in Sections 8.2.1 and 8.3.3.

Working with Incompatible Chemicals

Using RAMP, we can identify the prudent steps to take to avoid unexpected reactions between incompatible chemicals.

- *Recognize* chemicals that you are using (or storing) that are considered to be "incompatible" with other chemicals. If uncertain, do the necessary library or online research to learn more about unfamiliar chemicals.
- *Assess* the risk level when using or storing these chemicals. Review chemical procedures to *assess* the possibility that incompatibles might unintentionally mix.
- *Minimize* risks by using incompatible chemicals only when necessary and limiting quantities that are used. Design experimental procedures to *minimize* the chance for unintentional interaction between incompatibles. Store incompatibles in separate locations. *Minimize* storage volumes by ordering only as much as reasonably necessary.
- *Prepare* for emergencies by knowing how to respond to unexpected reactions between incompatibles and how to respond to spills.

References

1. *Bartlett's Familiar Quotations*, 17th edition, Little, Brown, and Co., Boston, 2002, p. 291–4.
2. American Industrial Hygiene Association. Laboratory Health and Safety Committee, Laboratory Safety Incidents, Glass Waste Bottle Ruptures; available at http://www2.umdnj.edu/eohssweb/aiha/accidents/glass.htm#Wastes (accessed October 16, 2009).
3. National Research Council. *Prudent Practices in the Laboratory: Handling and Disposal of Chemicals*, National Academy Press, Washington, DC, 1995, pp. 51–54.
4. L. BRETHERICK. Incompatibles. In: *Handbook of Chemical Health and Safety* (R. J. ALAIMO, ed.), American Chemical Society Oxford University Press, Washington, DC, 2001, pp. 338–343.
5. L. BRETHERICK. *Bretherick's Handbook of Reactive Chemical Hazards*, Vols. 1 and 2, 7th edition (P. G. URBEN, ed.), Elsevier, New York, 2007.
6. P. ZURER. Fire guts University of Texas chemistry lab. *Chemical & Engineering News*, pp. 10–11 (October 28, 1996); M. B. BRENNAN. Laboratory fire exacts costly toll. *Chemical & Engineering News*, pp. 29–34 (June 23, 1997).

QUESTIONS

1. What class of chemicals is incompatible with azides, cyanides, hydrides, perchlorates, and sulfides?

   (a) Acids
   (b) Bases

(c) Oxidizing agents

(d) Reducing agents

2. What class of chemicals is incompatible with anhydrides, organic nitro compounds, and acids?

(a) Acids

(b) Bases

(c) Oxidizing agents

(d) Reducing agents

3. What class of chemicals is incompatible with air?

(a) Acids

(b) Bases

(c) Pyrophorics

(d) Reducing agents

4. What class of chemicals is incompatible with chromates, peroxides, and permanganates?

(a) Acids

(b) Bases

(c) Oxidizing agents

(d) Reducing agents

5. What class of chemicals is incompatible with hydrides, carbides, and alkali metals?

(a) Pyrophorics

(b) Reducing agents

(c) Bases

(d) Water or aqueous solutions

6. Mixing a concentrated strong acid with a concentrated strong base is dangerous because

(a) The reaction between H^+ and OH^- is very exothermic

(b) Acid–base reactions are always dangerous

(c) When using concentrated solutions, the relative amount of water available to absorb heat is small

(d) Both (a) and (c)

7. Chemicals with large, positive reduction potentials will be which of the following?

I. Strong oxidizing agents

II. Strong reducing agents

III. Easily oxidized

IV. Easily reduced

  (a) I and III

  (b) II and III

  (c) I and IV

  (d) II and IV

8. Chemicals with large, negative reduction potentials will be which of the following?

I. Strong oxidizing agents

II. Strong reducing agents

III. Easily oxidized

IV. Easily reduced

  (a) I and III

  (b) II and III

  (c) I and IV

  (d) II and IV

9. What is the most common oxidizing agent used in chemistry laboratory experiments?

(a) Cl_2

(b) $HClO_4$

(c) H_2O_2

(d) HNO_3

10. Why are perchlorate salts unusually hazardous?

(a) They are toxic and volatile.

(b) Some are shock-sensitive.

(c) They are strong bases.

(d) They are water-reactive.

11. Which is *not* a good reducing agent?

(a) Na

(b) NaH

(c) Au

(d) H_2

12. When nitric acid is mixed with acetic acid, the acetic acid behaves as

(a) An acid

(b) A bas

(c) An oxidizing agent

(d) A reducing agent

13. Which class of chemicals is water-reactive?

(a) Alkali metals

(b) Acid anhydrides

(c) Nonmetal oxides

(d) All of the above

14. Which is *not* pyrophoric?

(a) Silane, SiH_4

(b) Methane, CH_4

(c) Borane, B_2H_6

(d) White phosphorus, P_4

5.3.1

GAS CYLINDERS AND CRYOGENIC LIQUID TANKS

Preview This section discusses the hazards of gas cylinders and cryogenic liquid tanks, and safety measures to take when handling or working with these containers.

Talking is a hydrant in the yard and writing is a faucet upstairs. Opening the first takes all the pressure off the second.

Robert Frost[1]

INCIDENT 5.3.1.1   LIQUID NITROGEN TANK EXPLOSION[2]

A postdoctoral researcher was using a tank of liquid nitrogen. During the early morning hours, the tank suddenly exploded, blowing out the bottom of the cylinder and projecting it upward through the concrete ceiling into a mechanical room above, leaving a 20-inch diameter hole in the 6-inch concrete floor of the mechanical room. The 6-inch concrete floor under the cylinder in the laboratory below was cracked and tile on the floor was blown around the room as shrapnel. There were no injuries in this explosion, but the damage to the entire floor of laboratories was so extensive they were closed for a month. The cost of the repairs approached $500,000. Investigation revealed that the pressure relief devices had been removed and replaced by brass plugs.

INCIDENT 5.3.1.2   INAPPROPRIATE GAS CYLINDER CAP REMOVAL INCIDENT[2]

A large gas cylinder containing carbon dioxide was delivered to a laboratory. When the lab personnel had difficulty removing the cylinder cap, they inserted a long screwdriver through the holes in the cap as a means of leverage to twist off the cap. Although the cap did not loosen, the twisting motion loosened the main valve and carbon dioxide discharged from the cylinder. The valve could not be easily reclosed and the scientists had to evacuate the room and call for help.

What lessons can be learned from these incidents?

Gas Cylinders, Liquid Tanks, and Their Hazards

Chemists use chemicals of all sorts, including chemicals in gaseous form. These gases are almost always provided to the laboratory in gas cylinders that are under high pressure. The gas in a cylinder can be used as a component in a synthesis of a needed compound, an inert gas for an air-sensitive reaction, a carrier gas in chromatographic separation, or a gas for a torch for sealing or molding glass.

The gases used in laboratories can be stored either under high pressure in the gaseous state or as cooled liquids. For example, both N_2 and He can be stored at high pressure or can be stored at very low temperatures in the liquid state. We will use the term "gas cylinder" to refer to containers that store chemicals in the gaseous state. For cooled, or "cryogenic," liquids, we will refer to the containers as "tanks." Both types of container have relief valves that prevent conditions of high pressure that could compromise the integrity of the container.

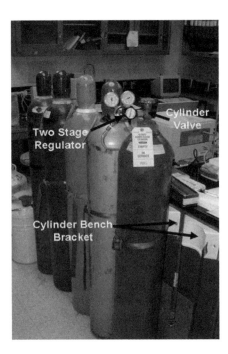

FIGURE 5.3.1.1   Gas Cylinders Showing Safety Brackets, Valves, and Regulators. Gas cylinders must always be secured to a laboratory bench or other immobile object so that they cannot fall over, snap off the regulator, and turn into missiles.

Gas cylinders in the laboratory come in a wide variety of sizes. The most common laboratory cylinder is just under 5 feet tall and 9 inches in diameter and is typically used for nitrogen, hydrogen, air, argon, helium, acetylene, oxygen, and many other gases (see Figure 5.3.1.1). For some "specialty" gases where less quantity is needed, smaller "lecture bottle" (about 15 inches tall and 2 inches in diameter) are used. Cryogenic liquids, most commonly N_2 and He, can be stored in tanks with volumes ranging from hundreds to thousands of liters.

There are many laboratory procedures that could not be accomplished without these gases, and the gas containers used to transport and store them. This section will mostly discuss gas cylinders. Gas cylinders present hazards beyond those encountered in handling liquid and solid chemicals. We will discuss each of these in a general sense, but it will be up to you to know the hazards of the particular chemical you will be using. Gases and their cylinders present a number of potential hazards associated with them. These are summarized in Table 5.3.1.1 and each entry will be discussed more below.

TABLE 5.3.1.1   Main Hazards and Preventative Measures for Gas Cylinder and Regulator Use and Storage

Hazard	Preventive measures
Cylinder valve snaps off	Keep valve cap on when not in use.
	Keep cylinder secured.
	Move cylinder only with a cart.
Asphyxiation (from leaking gas or massive discharge)	Check valves and regulators for leaks.
Escape of high-pressure, reactive, flammable, cryogenic, or toxic gas	Prevent circumstances where valve or regulator can break off.
	Have a response plan in the event of a large discharge of gas.
	Store in ventilated area.
Valve or regulator malfunction	Use only appropriate regulator.
	Replace washers periodically, if used.
	Do not grease, lubricate, or use Teflon tape.
	Inspect for damage or foreign material before use.

High Pressure, But Not From a Salesperson

High pressure is perhaps the most important hazard of a gas stored in the cylinder. A typical gas cylinder usually comes filled at a pressure of around 2200–2400 psi (pounds per square inch) (see *Special Topic 5.3.1.1* Pressure Units). Upon the sudden release of this pressure the gas can be very dangerous. Of course, as the temperature rises in the laboratory or place where the cylinder is stored, the cylinder pressure also rises, and the greater the pressure the more stress there is on the integrity of the cylinder. To take measures to prevent explosions, manufacturers install a relief valve that is designed to rupture at some prescribed cylinder pressure, temperature, or combination of temperature and pressure. Most of these relief devices open when a disk ruptures or melts. Less commonly, a spring-loaded relief valve is installed that can open at some high pressure and reclose when the cylinder pressure is reduced.

SPECIAL TOPIC 5.3.1.1

PRESSURE UNITS

Scientists have several units for measuring pressure and many fields of research use different units that have become common over time. The SI unit for pressure is the pascal (Pa) and it is a derived unit that is force/area. One pascal is equal to one newton per square meter: $1\,Pa = 1N/m^2$. Table 5.3.1.2 shows some useful comparisons of commonly used pressure units.

TABLE 5.3.1.2   Commonly Used Pressure Units

	Pa	bar	atm	mm Hg (torr)	psi
1 pascal, Pa (N/m²)	1	1×10^{-5}	9.87×10^{-6}	7.50×10^{-3}	1.45×10^{-4}
1 bar	1×10^5	1	0.987	7.50×10^2	14.5
1 atmosphere, atm	1.01×10^5	1.01	1	760	14.7
1 torr or mm Hg	1.33×10^2	1.33×10^{-3}	1.31×10^{-3}	1	1.93×10^{-2}
1 pound per square inch, psi	6.89×10^3	6.87×10^{-2}	6.80×10^{-2}	51.7	1

A gas cylinder in a lab that has about 2200 psi is also at 150 atm. Perhaps this gives a better intuitive feel for the pressure in the cylinder.

It is also worth noting that a pressure gauge reads the pressure *above local atmospheric pressure*. Thus, a pressure reading of "3 atm" indicates a total pressure of 4 atm. A pressure gauge reading zero indicates that the pressure inside the tank is the same as outside. This is the condition if the valve is completely open and the tank is "completely" discharged. The same is true for the pressure for car tires: a "flat" tire will give a reading of 0 psi, although the pressure in the tire is about 1 atm.

The most vulnerable part of a gas cylinder is its valve (see Figure 5.3.1.1). If a valve is damaged it could leak or malfunction. The worst scenario is... well do you remember how you can blow up a balloon, release it and it goes flying through the air? If the valve of a gas cylinder is suddenly snapped off, the cylinder becomes a "rocket" smashing everything in its path. This is a clear example of Newton's third law: "For every action there is an equal and opposite reaction." This can happen if a cylinder is not carefully secured and falls, hitting the valve on the floor and snapping it off. While this is a rare event, it can and has happened from careless handling of gas cylinder—we will talk about prudent transporting and handling of gas cylinders below. (A spectacular demonstration of this setup by the Mythbusters television show is available at http://www.youtube.com/watch?v = ejEJGNLTo84.)

Any Place Without Oxygen Is a Big Problem — You Can't Breathe!

The rapid release of a gas, such as nitrogen, helium, argon, or other gases, can also reduce or eliminate oxygen from the space where this happens. The air we breathe contains about 21% oxygen, and if that oxygen is displaced from the area where we are by the rapid release of another gas, and we do not immediately evacuate, we could asphyxiate (die from a lack of oxygen). At oxygen concentrations below 16% at sea level, decreased mental effectiveness, visual acuity, and muscular coordination occur. At 15% oxygen, we experience impaired thinking and at 12% we lose judgment skills. At oxygen concentrations below 10%, loss of consciousness may occur, and below 6%, death will result[3,4] (see Table 4.1.2.2). Furthermore, many have died trying to rescue people in oxygen-deficit atmospheres when they entered the space not knowing that the space could not support life. In fact, over 60% of the deaths in "confined spaces" are would-be rescuers. (See Section 4.1.2 for a more thorough discussion of asphyxiation hazards.)

This kind of event occurring in a laboratory is extremely rare because most laboratories use only limited quantities of gases, and there is usually adequate ventilation to remove gases and vapors. Nevertheless, the sudden release of an inert gas in a "confined" space (a space with limited or restricted entry and exit that has inadequate ventilation) can create a very dangerous environment that does not support life. You should remember this potential hazard if a tank suddenly releases its contents in a laboratory. Thinking in advance about this event and having some response plan is the prudent course of action.

The Usual Suspects—Flammable, Corrosive, Toxic, Reactive

Gases, the contents of gas cylinders, can have significant hazardous properties. Flammable, corrosive, toxic, or reactive gases are especially hazardous since if released they are totally airborne, can be inhaled, and can be ignited by local ignition sources. A flammable gas that is inadvertently released in a laboratory could be very dangerous if the air concentration reaches explosive limits and it encounters a source of ignition. The National Fire Protection Association standards and local fire and building codes establish limits on the number of flammable or oxygen cylinders allowed in a laboratory—you should check with your local safety officer about these limits.

The release of a reactive gas could result in a violent reaction or an explosion upon accidental release into the laboratory environment. A gas that is corrosive or toxic can readily be inhaled if a release happens in an open laboratory space. This could result in significant personal injury. You must always be aware of the hazardous properties of the gases you will be using in the laboratory.

Each of these groups must be stored separately and not be stored with other incompatible gases (see Section 5.2.3). When using these gases it is important to frequently check your gas lines for leaks. As with other chemicals, avoid contact with gases and use protective clothing and gloves to protect your skin from exposure. Wear your chemical goggles when handling compressed gases, and if you are using corrosive gases, use a face shield in addition to your goggles as well as other appropriate protective equipment.

Cryogenics—Now These Are Really Cold

Some chemicals that are gases at normal temperatures can be condensed under extremely high pressure to become cryogenic liquids. These liquids are often very useful in laboratory work, but they are cold—and baby, we mean really cold! The most common cryogenics are liquid nitrogen, liquid argon, and liquid helium, with boiling points of 77 K, 87 K, and 4 K, respectively. Any direct skin contact with cryogenic liquids can result in very serious frostbite including permanent damage at the point of contact. Furthermore, cryogenic liquids can rapidly change into gaseous forms—even explosively if entrapped. Cryogenics require special attention and training and we will discuss these in more detail in Section 5.3.9. You should not be handling cryogenics without some specific education and training about their hazards, use, and handling.

Gas Cylinder Regulators—Getting the Gas Out

Now that you know something about the hazards of gas cylinders, you need to know how you can safely work with these in the laboratory. Gas cylinders come with a valve on the top but if you open

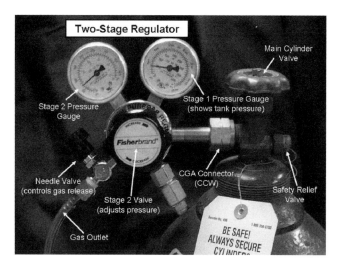

FIGURE 5.3.1.2 Two-Stage Regulator on a Gas Cylinder. The gas in the main cylinder is released when the main tank valve on the cylinder is opened. The safety relief valve will automatically open if excess pressure builds up in the cylinder. A two-stage regulator is connected to the tank with a CGA connector and as the gas passes through the main cylinder valve, it enters the regulator, where the pressure in the tank is registered on the first gauge. The pressure is controlled by the Stage 2 Valve and the second gauge shows the pressure in the second stage. The needle valve is opened to the desired place to allow gas to flow to the instrument being used.

the valve, you will be releasing the gas at the pressure within the tank—this is not useful and can be dangerous. To safely use a gas cylinder requires that you use it with a regulator (see Figure 5.3.1.1).

The two most common types of regulator are single-stage and two-stage. Regulators reduce the pressure supplied from the cylinder to a manageable level for the instruments and equipment used in the laboratory. The single-stage regulator reduces pressure in one step and the two-stage regulator does this in two steps. The selection depends on the need, but generally the two-stage is used by many since it gives constant flow until most of the gas has been used (see Figure 5.3.1.2). For smaller cylinders like lecture bottles, manual flow control valves are often used that do not have pressure gauges and do not control pressures—they are used only to dispense the gas.

The regulator and its associated connecting devices, such as meters, valves, hoses, tubing, or piping, all must be compatible with the gas you will be using. Standards for regulator materials and connections are set by the Compressed Gas Association (CGA). Specific regulators are designated for particular gases and are often stamped with a CGA number. Table 5.3.1.3 lists the appropriate regulator for some common gases. (More extensive versions of this list are in Resources #3, #10, and #11 in Table 5.3.1.4.) Make sure you select the right regulator. Your compressed gas supplier will know which regulator and connectors to select.

Why are there so many kinds of regulators? Different gases have different reactive properties that require different kinds of regulators. While brass is commonly used in the regulator component for noncorrosive gases (since it is just malleable enough to seal minor imperfections in the threads), corrosive gases require the use of more corrosion-resistant metals and alloys. And some alloys will react with flammable gases, like acetylene.

TABLE 5.3.1.3   Regulators for Common Lab Gases

Gas	Regulator, CGA number
CO_2, $Ni(CO)_4$, some halocarbons	320
BF_3, HCl, HBr, HI, H_2S, SiF_4	330
H_2, CO, CH_4, PH_3	350
Acetylene and most organic gases	510
O_2 (only)	540
N_2, Ar, He	580
Air, SF_6	590
BCl_3, Cl_2, NO, NO_2, SO_2, PF_5, O_3, some halocarbons, PH_3	660
NH_3, some organic amines	705

FIGURE 5.3.1.3   Notch in Connecting Nut Indicating CCW Thread. For most normal gases the CGA connector attaches clockwise (CW), but for flammable gases the CGA connector attaches counterclockwise (CCW).

A point of caution—some gases can be used with more than one regulator (such as PH_3 with CGA 350 or CGA 660). However, this does not imply that all of the gases for both regulators are interchangeable. Thus, CO can be used with CGA 350 but it cannot be used with CGA 660.

The most important safety consideration is to be sure that the regulator, connectors, and valves are designed for the gas and that they are made of material that is compatible with the gas being used. Clockwise (CW) is a right-handed thread. Counterclockwise (CCW) is a left-handed thread. CCW threads have notches cut into the closing nut for identification (see Figure 5.3.1.3). If the middle digit of the CGA valve number is even, the thread is clockwise. If the middle digit is odd, the thread is counterclockwise. Use Table 5.3.1.3 to make absolutely certain that you are using the appropriate regulator for the gas in the cylinder.

Gas regulators should never be lubricated with grease or oil, but this is especially important for regulators used with oxygen or other oxidizing gases since, even at room temperature, oxidizing gases at high pressure can react with the easily oxidized grease or oil. If they become unworkable, the regulator should be returned to the manufacturer for service.

The fittings on CGA connections should not be lubricated nor should Teflon tape be used. In fact, it is not the thread itself that is making the seal, but the male and female end of the two parts that are connected. Sealants should also not be used on compression fittings (such as Swagelock® fittings). In addition to other means of identification, CGA fittings have parallel threads. Some non-CGA fittings have tapered threads and it is appropriate to use sealants on these fittings for nonoxygen or nonoxidizing gases. If washers are used anywhere in a gas system, they should be regularly inspected and replaced when necessary.

When you are ready to connect the regulator to the cylinder, you will need to remove the cylinder's valve cap. Most of the time this is easy, but sometimes it may be stuck or hard to turn. There are commercially available tools (see Figure 5.3.1.4) that are designed to provide leverage in removing the cap. (These tools also help open a resistant cylinder valve wheel.) An adjustable wrench with a strap attachment can also be used. Do not try to remove the cap by inserting something in the holes on the side of cap, such as a large screwdriver. This could accidentally open or damage the valve inside and start a leak. (See Incident 5.3.1.2.) If you cannot get the cap off, return it to the supplier or ask the supplier to come out to remove the cap. Do not use oil or grease on the valve cap.

As an example of how to connect a regulator, consider the two-stage regulator in Figure 5.3.1.3. You will note two valves: one is the regulator outlet valve and the other is the regulator control valve. Make sure that both valves are closed. The regulator outlet valve screws in clockwise to the closed position. The regulator control valve is usually a large valve projecting out of the front of the regulator. Turn this valve counterclockwise to close the valve.

Connect the regulator to the cylinder using the proper wrench—don't use pliers. Open the cylinder tank valve and you will see the pressure register on the first stage of the regulator. Now adjust the

FIGURE 5.3.1.4 Tool to Remove Cylinder Cap. Do not try to force a stuck cylinder cap with a long screwdriver or other tool that goes through the cap. This may inadvertently open the tank by moving the main tank valve.

second stage to the desired setting by turning the regulator control valve clockwise. Finally, open the outlet valve to release the gas. You will want to be sure that all connections from the tank to the entrance to the instrument are not leaking. You can check for leaks using a soapy solution that detects leaks as bubbles. You can also purchase solutions made to check for leaks.

And remember that in addition to the regulator and associated fittings, the rest of the system must also be able to function safely with the gas in use. Resources #3 and #11 in Table 5.3.1.4 have extensive listings of the compatibility of many gases with metals, plastics, and elastomers.

Securing Your Cylinders — Falling Cylinders Are Dangerous!

One of the most important safety steps in handling cylinders is to ensure that they are secure wherever they are located. Never leave a gas tank unsecured. Unsecured, loosely secured, or improperly secured gas cylinders are dangerous, and detection of these should lead to immediate actions to properly secure these hazards.

In the laboratory, there are a large variety of tank restraints (straps or chains) available including floor stands, wall brackets, and laboratory bench brackets (see Figure 5.3.1.1). Whatever the choice, be sure that you use a device to secure the cylinder so that it cannot fall. This is true not only for large cylinders but for small lecture bottles. Cylinders must also be secured in their storage areas. This is often accomplished by putting a small group of gas cylinders in an area that can be chained. Incompatible gases should not be stored together. Using a chain allows users to expand the restraint as the number of tanks is increased or contract the restraint as the number decreases. Keep the chain tight around the upper third of these tanks and do not let it droop below this level since tanks could topple over.

Moving Your Cylinders — The No Roll-Slide-Drag Zone

You should only move gas cylinders using a cart designed for moving cylinders. Cylinders should be secured in the carts, usually with a chain (see Figure 5.3.1.5). You should not move any cylinder with a regulator. Remove the regulator before you move the cylinder.

You may see some people rolling cylinders by holding the top and leaning the tank slightly on its side. This is an unsafe practice. It is easy to lose control of a heavy cylinder and drop it, thereby presenting an opportunity to break or damage the cylinder valve—creating your very own rocket.

TABLE 5.3.1.4   Reference Materials: Useful Web Sites and Print Resources

Resources	*Printed materials*
#1	*Prudent Practices in the Laboratory: Handling and Disposal of Chemicals*, National Research Council, National Academic Press, Washington, DC, 1995 (also available online at http://www.nap.edu/ catalog.php?record_id=4911)
#2	*Handbook of Chemical Health and Safety* (R. J. Alaimo, ed.), Oxford University Press ACS, Washington, DC, 2005, pp. 377–382
#3	*The Laboratory Companion*, 2nd edition (Gary S. Coyne, ed.), Wiley-Interscience, Hoboken, NJ 2006, Chapter 5
Resources	*Internet resources*
	There are many good academic safety pages:
#4	http://www.ilpi.com/inorganic/glassware/regulators.html
#5	http://www.pp.okstate.edu/ehs/LINKS/gas.htm
#6	http://oregonstate.edu/ehs/bulletin/si22.htm
#7	http://www.uwm.edu/Dept/EHSRM/LAB/labgascyl.html
#8	http://safety.science.tamu.edu/cylinderhazard.html.
#9	Air Products has 28 "Safetygrams" available at: http://www.airproducts.com/Responsibility/EHS/ ProductSafety/ProductSafetyInformation/safetygrams.htm
#10	Scott Specialty Gases links for several topics: http://www.scottecatalog.com/ScottTec.nsf/All?ReadForm
#11	Scott Specialty Gases *Safety Design Handbook* contains many useful tables of data regarding gases and regulators, including: Gas Compatibility Guide, pp. 9–11 Gas Characteristics, pp. 34–35 Physical Properties of Gases, p. 36 Cylinder Valve Outlets and Connections, p. 37 CGA Fittings, p. 38 Cylinder Specifications, p. 40 Cylinder Markings, p. 41 http://www.scottecatalog.com/DSGuide.nsf/74923c9ec562a6fb85256825006eb87d/7 ce3b3615d27ce98852568f6004da2e1/$FILE/designandsafetyhb2006.pdf

Do NOT roll cylinders—use a cart. Similarly, do not slide or drag cylinders but instead use a cart. Anywhere a cylinder is stored is a No Roll-Slide-Drag Zone for cylinders.

Storing Gas Cylinders—Dry, Secured, and Ventilated

Gases must be clearly identified with legible labels and markings. Do not trust or use color coding for identifying the contents of a gas cylinder. Gas cylinders should be stored in a dry and well-ventilated area. These areas should also be away from mechanical or physically active areas, electrical circuits and panels, and sources of heat. Often these areas are outdoors but covered to keep them out of rain and direct sunlight—heat increases pressure and too much pressure could release gas through safety relief valves. Temperatures of gas cylinder storage areas should never be allowed to exceed 125 °F. Highly toxic gases should be stored inside a well-ventilated area that can be secured. Gas cylinder storage areas may need to meet additional requirements of local area building codes.

Cylinders should be stored in small, chemically compatible groups. Gas cylinder storage areas must be labeled with signs similar to this: "No Smoking or No Open Flames Allowed in this Area!" Signs should be clearly visible above each area indicating the appropriate storage groups—nonflammables, flammables, oxidizers and oxygen, and toxic gases. Strong oxidizers, including oxygen, must be kept some distance from flammable or organic gases and materials—25 feet minimal separation has been suggested.

FIGURE 5.3.1.5   Cylinder Secured on Gas Cylinder Cart. Gas cylinders must always be moved on a cart, and the cylinder must be secured to the cart.

Cylinders must be secured upright at all times using brackets with chains, straps, or cables that hold the cylinders fast around the upper third of the tank. This practice should also include empty or "MT" cylinders (also stored separately from full tanks), since many of these are likely to have residual gas under some pressure. Additional restraints should be used in earthquake zones.

Using Compressed Gases in the Lab: RAMP

Table 5.3.1.1 summarizes the main hazards and the preventive steps to avoid accidents with gas cylinders. Table 5.3.1.4 lists many resources that also have excellent advice about the uses of gases and handling of gas cylinders and regulators. However, you should not consider that reading this section of this book teaches you everything you need to know about handling gases and gas cylinders. Seek advice from experienced persons before using gas cylinders.

- *Recognize* which gases present various hazards, beyond just high pressure.
- *Assess* the level of risk based on the identity of the gas and the location where it is being used or stored.
- *Minimize* the risk by methodically using valves and regulators correctly, using appropriate personal protective equipment, and storing gases safely.
- *Prepare* for emergencies by anticipating slow or fast leaks and having a response plan prepared in advance.

References

1. *Bartlett's Familiar Quotations*, 17th edition (Justin Kaplin, ed.), Little, Brown, and Co., Boston, 2002, p. 671, line 16.
2. Laboratory Health and Safety Committee, American Industrial Hygiene Association. Laboratory Safety Incidents: Gas Cylinders; available at http://www2.umdnj.edu/eohssweb/aiha/accidents/cylinder.htm (accessed October 16, 2009).
3. U.S. Chemical Safety Hazard Investigation Board. Safety Bulletin, 2003-10-B, Hazards of Nitrogen Asphyxiation; available at http://www.csb.gov/assets/document/SB-Nitrogen-6-11-03.pdf (accessed October 16, 2009).
4. Centers for Disease Control and Prevention. National Institute for Occupational Safety and Health. NIOSH Respirator Selection Logic 2004; available at http://www.cdc.gov/niosh/docs/2005-100/chapter5.html (accessed September 15, 2009).

QUESTIONS

1. Which three of the following types of gas-containing cylinders are most common?

   I. Small lecture bottles
   II. Tanks about 5 feet tall and 9 inches in diameter
   III. Tanks about 4 feet tall and 18 inches in diameter
   IV. Large tanks that store liquid nitrogen or helium

   (a) I, II, and III
   (b) I, II, and IV
   (c) I, III and IV
   (d) II, III, and IV

2. What is the most common pressure found in "full" tanks of compressed gas?

   (a) 1200 psi
   (b) 1800 psi
   (c) 2200 psi
   (d) 3000 psi

3. What device is designed to prevent a gas cylinder from overpressurizing?

   (a) The regulator
   (b) The cylinder cap
   (c) The relief valve
   (d) The cylinder valve

4. What is the most vulnerable part of a gas cylinder without a regulator attached?

   (a) The cylinder cap
   (b) The relief valve
   (c) The valve
   (d) The body of the cylinder

5. "Inert gases" such as nitrogen or helium pose a health hazard because

   (a) Once inside the body they are not inert
   (b) They can displace enough oxygen from air to make the air unsafe
   (c) They react with oxygen in the air
   (d) They are very cold

6. There are regulations that limit the number of

   (a) Gas cylinders allowed in a lab, depending on the size of the lab
   (b) Total cubic feet of compressed gases in a lab
   (c) Cylinders that can have open valves at any single point in time
   (d) The number of oxygen and flammable gas cylinders in a lab

7. Using personal protective equipment is

   (a) Less important when using gases since they cannot splash
   (b) Less important when using gases since gases are usually unreactive
   (c) More important because gas cylinder explosions are fairly common

   (d) Equally important as when working with other chemicals in labs

8. Using liquid He, Ar, or N_2 poses a health hazard due to their

   (a) Potential for asphyxiation
   (b) Cryogenic properties
   (c) Reactivity.
   (d) Both (a) and (b)

9. The advantage of a two-stage regulator over a one-stage regulator is that it

   (a) Is less expensive
   (b) Allows for a more constant flow rate of the gas
   (c) Is less likely to malfunction
   (d) Has an emergency shut-off feature

10. Gas cylinders and gas regulators

    (a) Are interchangeable
    (b) Are each designed to work with particular gases
    (c) Have the same threads
    (d) Are usually made from the same metal

11. Brass is used in many regulators since it is

    (a) Very unreactive with almost all gases
    (b) Designed to fail before the cylinder explodes
    (c) Soft enough to allow for good seals
    (d) The cheapest metal alloy

12. When connecting the regulator to the cylinder it is best to

    (a) Use Teflon tape
    (b) Use a nonreactive grease
    (c) Use a drop of water
    (d) None of the above

13. What is the correct order in getting a gas into a system once the cylinder cap is removed?

    I. Open the main valve.
    II. Open and adjust the regulator valve.
    III. Attach the regulator.
    IV. Open the outlet valve.

    (a) I, III, II, IV
    (b) IV, III, I, II
    (c) III, II, I, IV
    (d) III, I, II, IV

14. When removing a "resistant" cylinder cap,

    (a) Never use any tool other than your hands
    (b) Use a long tool, such as a long screwdriver, that fits all the way through the cap to get the best leverage
    (c) Only use a tool designed specifically to remove gas cylinder caps
    (d) Use a hammer

15. Gas cylinders should be secured when which of the following conditions exists?

    I. The valve is open.

II. The tank is in an occupied lab.
III. The tank is in storage.
IV. The tank is empty.

   (a) I and II
   (b) I, II, and III
   (c) II, III, and IV
   (d) I, II, III, and IV

16. Gas cylinders should be stored

   (a) In areas where incompatible gases can be separated from each other
   (b) In areas protected from the weather if outside
   (c) In areas with good ventilation
   (d) All of the above

5.3.2

PEROXIDES—POTENTIALLY EXPLOSIVE HAZARDS

Preview This section discusses potentially explosive peroxides as hazardous contaminants in some chemicals, including their formation, detection, and care in handling.

Something deeply hidden had to be behind things.

Albert Einstein[1]

INCIDENT   ROTARY EVAPORATOR EXPLOSION[2]

A student was conducting undergraduate research that involved evaporating the solvent from a precipitate using a rotary evaporator. When she manipulated the round bottom flask containing the residue it exploded, spraying her with glass. The flying glass hit her safety goggles and also cut her forehead. An ambulance transported her to the hospital where she received treatment including stitches for the cuts above her eyes. Investigation found that tetrahydrofuran (THF) and diethyl ether were the solvents used. THF samples tested positive for high concentrations of peroxides.

What lessons can be learned from this incident?

Prelude

This section provides an introduction to the safety and reactivity concerns of peroxides and peroxide-forming compounds. If you are conducting a lab experiment that involves any of the categories of compounds discussed in this section, we strongly advise that further resources be consulted for more specific information about the particular chemical to be used. We urge this extra caution here due to the extraordinary danger of explosion under comparatively mild action of "agitation" or "shaking."

Some of the comments concern peroxide detection, ordering and storing chemicals, and dealing with peroxide-laden solvents. These are not likely to be actions that you ordinarily undertake as a student. Stockroom managers typically deal with these situations but it is still very important that you be aware of peroxide-related hazards as a student. However, since students are often employed as helpers in stockrooms, you may find yourself facing some of the hazards described in this section and knowing about these hazards in advance is extremely important. Of course, the odds are good that when you are not a student someday, you will be handling peroxide-forming solvents under less supervision than you experience as a student.

Peroxides—Capable of Supplying an Unexpected Explosive Surprise!

Peroxides are group of chemicals that contain an —O—O— peroxide linkage. The largest classes of peroxides are the carbon-based organic peroxides (R—O—O—R′) or hydroperoxides (R—O—O—H). Some inorganic chemicals and organometallics also form peroxides. Experience has demonstrated the need for careful handling of peroxides or peroxide-containing materials because most of these compounds are inherently unstable and are prone to decompose violently from shock, friction (such as opening a bottle cap), or heat resulting in an explosion. Part of the inherent reactivity of peroxides arises from the relatively weak oxygen–oxygen bond at about 207 kJ/mol. Compare this to C—C, 345

Laboratory Safety for Chemistry Students, by Robert H. Hill, Jr. and David C. Finster
Copyright © 2010 John Wiley & Sons, Inc.

TABLE 5.3.2.1 Types of Chemicals that May Form Dangerous Peroxides[3−10]

Ethers, acetals with α hydrogens
Alkenes, alkylacetylenes with allylic hydrogens
Vinyl halides, vinyl esters, vinyl ethers
Dienes, vinylacetylenes
Alkylarenes, alkanes, cycloalkanes with tertiary hydrogens
Acrylates, methacrylates
Secondary alcohols
Aldehydes
Ketones with α hydrogen on a secondary carbon
Ureas, amides, lactams with a hydrogen–carbon–nitrogen linkage
Alkali metals
Metal amides
Metal alkoxides
Organometallics with metal–carbon bonds

kJ/mol, C—O, 358 kJ/mol, and C—H, 411 kJ/mol. Furthermore, upon dissociation, the products are free radicals, which are highly reactive.

While peroxides themselves are used as reagents and handling these safely is critically important, there is also significant concern for peroxides as *contaminants* in other chemicals, especially organic solvents. Peroxides are often formed within solvents slowly and these solvents (solutions) become increasing hazardous as concentrations increase over time. Peroxide formation from a variety of organic compounds is a free-radical, autoxidation process involving molecular oxygen. These reactions are initiated by light (including sunlight), a radical generator, or peroxides themselves.

Many classes of chemicals can form peroxides as shown in Table 5.3.2.1. Ethers are the most recognized class of organics that can form peroxides that have resulted in explosions with serious injuries, as illustrated by Incident 5.3.2.1. Diisopropyl ether, cited in Incident 8.3.3.1, is notorious for forming peroxides—it has two α hydrogen atoms that can easily be reacted since the resulting radical is stabilized by its proximity to the ether oxygen. But there are other classes of compounds that also readily form dangerous peroxides, including, for example, acetals, alkenes, aldehydes, amides, lactams, and vinyl monomers. The formation of peroxides is generally favored by molecular structures that support stabilized free radicals.[3] (See *Chemical Connection 5.3.2.1* Free-Radical Autoxidation and Stability.)

CHEMICAL CONNECTION *5.3.2.1*

FREE-RADICAL AUTOXIDATION AND STABILITY

Some organic molecules have a propensity to lose a hydrogen atom due to the presence of some initiator (In) or due to reaction with a photon of light to form a free radical:

$$(1) \quad \text{R–H} + \text{In•} \rightarrow \text{R•} + \text{H–In}$$

In the presence of oxygen, a diradical, the organic molecule becomes a peroxide radical (continuing the radical propagation):

$$(2) \quad \text{R•} + \text{O=O} \rightarrow \text{R–O–O•}$$

This peroxide radical can react with another organic molecule to produce a hydroperoxide:

$$(3) \quad \text{R–O–O•} + \text{R–H} \rightarrow \text{R–O–O–H} + \text{R•}$$

Reactions 2 and 3 become a cycle with the net effect

$$(4) = (2) + (3) \quad \text{R–H} + \text{O=O} \rightarrow \quad \text{R–O–O–H}$$

As long as there is molecular oxygen present, the organic starting material can be converted to peroxides.

But, obviously, not all organic molecules get converted to peroxides even in the presence of molecular oxygen. Why do some molecules get converted more easily than others? The answer lies in the stability of the free radical initially formed.

The stability of the free radical depends on the ability of the radical electron to be delocalized beyond the confines of one atom. As with carbanions and carbocations, delocalization enhances stability. As an example, let's consider the free radical localized on the nonmethyl carbon atom in the ethyl radical. This electron is in the sp^3 hybrid orbital and, when this orbital is aligned (periplanar) with a *bonding pair* in a neighboring C—H bond, some of the electron density of the bonding pair can overlap with the "half-filled" orbital containing the single electron. This interaction is called hyperconjugation. (This argument is frequently used in the discussion of carbocation stability, too.) Note that a methyl radical cannot experience this stabilization (since there are no neighboring C—H bonds) and an electron on the central carbon of a *tert*-butyl group also cannot experience this stabilization since there are no neighboring C—H bonds. A hydrogen atom that is on the C atom *adjacent* to the site of stabilization is called the α-H or alpha hydrogen.

The effect of this delocalization can be seen in the energy required to homolytically break a C—H bond in various organic molecules. *The more stable the radical product, the more easily the bond is broken*. This is easily seen in the series methyl > primary > secondary > tertiary in Table 5.3.2.2.

TABLE 5.3.2.2   Relative Stabilities of Hydrocarbon Radicals

Molecule (C— H bond in boldface)	Description	Bond dissociation energy (kJ/mol)	Stability of radical
C_6H_5–**H**	Phenyl	460	Relatively unstable
CH_2=CH(–**H**)	Vinyl	452	
H_3C–**H**	Methyl	435	
H_3C–CH_2(–**H**)	Primary	410	
$(CH_3)_2CH$(–**H**)	Secondary	397	
$(CH_3)_3C$(–**H**)	Tertiary	385	
H_2C=CH–CH_2(–**H**)	Allylic	372	
H_5C6–CH_2(–**H**)	Benzylic	356	Relatively stable

The allylic and benzylic groups provide even more delocalization due to further resonance stabilization. At the top end of the table are the relatively unstable phenyl and vinyl groups. Why are they so unstable? In these two systems the unpaired electron is in an sp^2 orbital that is *perpendicular* to the C—H bonding orbitals in neighboring CH groups. This decreases the opportunity for effective orbital interaction and delocalization.

The comments above pertain to unsubstituted hydrocarbons. Tables 5.3.2.1 and 5.3.2.3 show many other examples of substituted hydrocarbons with the heteroatoms usually being oxygen or chlorine. It is generally possible to argue for the stability of the intermediate radical as playing an important role in the ease of peroxide formation. Alternately, the mechanism of the attack of molecular oxygen on some species can be facilitated by the heteroatoms.

Without doubt, the most infamous compounds in terms of ether-forming ability are ethers but this is due to the common use of ethyl ether and THF as solvents rather than inherent risk level. In fact, these two ethers are both in Group 2, the least hazardous of the three groups in Table 5.3.2.3. The argument for the stability of the intermediate radical is essentially the same as for the alkyl radicals above except that the lone electron undergoes hyperconjugation with the lone pair electron on the adjacent (ether) oxygen. In other words, α-H in an ether is susceptible to homolytic cleavage and the formation of a radical.

Finally, we note that methyl *tert*-butyl ether (MTBE) was for several years a common gasoline additive used to reduce pollution and enhance octane rating. (It has since been eliminated due to groundwater contamination from leaking storage tanks.) The use of this ether is fortuitous: since it has methyl and *tert*-butyl groups adjacent to the oxygen, there are no α-hydrogens that can engage in hyperconjugation with the lone pair electrons on the oxygen. How fortunate, since otherwise the fuel additive would have been prone to peroxide formation in storage tanks containing hundreds or thousands of gallons of gasoline!

TABLE 5.3.2.3   Peroxidizable Chemicals[3−11]

Group 1: Severe Peroxide Hazards—Chemicals forming peroxides that may explode without concentration.

Discard 3 months after opening.

Diisopropyl ether	Butadiene (liquid)	Potassium metal	Potassium amide
Vinylidene chloride	Chloroprene (liquid)		Sodium amide
Divinyl ether	Tetrafluoroethylene (liquid)		
Divinyl acetylene			

Group 2: Significant Peroxide Hazards—Chemicals forming peroxides that may explode on concentration.

Discard 12 months after opening.

Acetal	Cyclooctene	Dioxane	Methyl cyclopentane
Acetaldehyde	Cyclopentene	Furan	Methyl-*i*-butylketone
Benzyl alcohol	Decahydronaphthalene	Glyme	2-Propanol
2-Butanol	Diacetylene	4-Heptanol	Tetrahydofuran
Cumene	Dicyclopentadiene	2-Hexanol	Tetrahydronaphthalene
Cyclohexanol	Diethyl ether	Methylacetylene	Vinyl ethers
Cyclohexene	Diglyme	3-Methyl-1-butanol	Secondary alcohols

Group 3: Significant Peroxide Hazards—Chemicals that may violently polymerize due to peroxide formation.

Test inhibited chemicals 6 months after opening and discard 12 months after opening.

Discard uninhibited chemicals within 24 hours after opening.

Acrylic acid	Chlorotrifluroethylene	Vinyl acetate	Vinylidiene chloride
Acrylonitrile	Methyl methacrylate	Vinylacetylene	
Butadiene	Styrene	Vinyl chloride	
Chlorprene	Tetrafluoroethylene	Vinylpyridine	

Table 5.3.2.3 presents a compilation of examples of peroxidizable chemicals from several sources.[3−11] (In this instance, "peroxidizable" means "peroxide-forming," not "capable of reacting with peroxide.") This table does not list all the chemicals that could form dangerous concentrations of peroxides. Peroxide-forming compounds are often grouped based on their relative reactivity, as shown in Table 5.3.2.3. Group 1 chemicals are especially hazardous because they may form peroxides that can explode without concentration. Group 2 chemicals may form peroxides that explode upon concentration. Group 3 chemicals may spontaneously polymerize explosively with peroxide formation. There are other listings of chemicals that may form peroxides but do not fit into these groups. [3]

Most of the compounds in Tables 5.3.2.1 and 5.3.2.3 are organic compounds, but some inorganic compounds are peroxides and peroxide-formers. An easy rule of thumb is that most compounds that have "per-" in the name are chemicals that warrant close examination in terms of reactivity and stability. (The exceptions to this rule are compounds that use "per-" as a prefix indicating "fully substituted," as in "perchloroethylene.")

Preventing Dangerous Peroxide Situations—Practices That Can Protect You

Since the presence of peroxides in organic solvents represents a significant hazard, we can think about ways to avoid, minimize, and respond to these situations. We list 10 steps to consider below and organize these steps into the following categories: *avoiding peroxides, working with peroxides*, and *managing chemicals with high peroxide levels*.

Avoiding Peroxides

1. *Using inhibitors.* Inhibitors are chemicals that are added (by manufacturers) in very low concentration to peroxidizable chemicals to prevent or retard the formation of peroxides (see *Chemical Connection 5.3.2.2 Inhibiting Peroxide Formation*). If the potential peroxide-former that you need is available with an inhibitor, this should be your choice. There are various chemical inhibitors, such as

2,6-di-*tert*-butyl-4-methylphenol, also known as BHT (butylated hydroxy toluene). Peroxide formation in diethyl ether can be inhibited by iron. Purchasing diethyl ether in steel cans is a good option to help prevent the formation of unsafe peroxides. Storage under an inert gas can also help retard the formation of some peroxides. However, over time these inhibitors are degraded as they react to protect against peroxidation and once eliminated, the formation of peroxides in that chemical will proceed as it would without an inhibitor. Explosions with inhibited peroxide-formers have been reported.[4] Thus, even peroxide-formers with inhibitors need to be under surveillance.

CHEMICAL CONNECTION *5.3.2.2*

INHIBITING PEROXIDE FORMATION

Solvents that are predisposed to form peroxides frequently have trace amounts of peroxide inhibitors added to them by manufacturers. The two most common are butylated hydroxytoluene (BHT) and butylated hydroxyanisole (BHA)—BHA is a mixture of two isomers (see Figures 5.3.2.1 and 5.3.2.2).

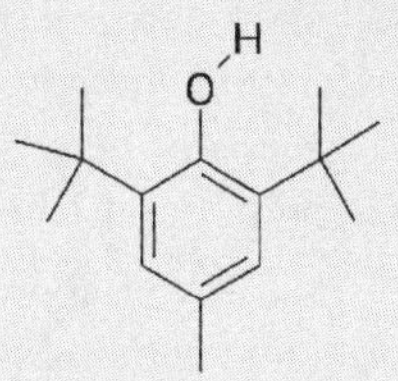

FIGURE 5.3.2.1 BHT [2,6-bis(*tert*-butyl)-4-methylphenol]. A common peroxide inhibitor.

(a)   (b)

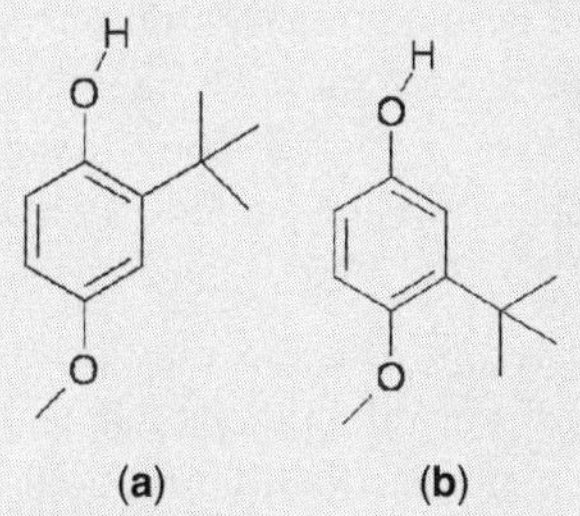

FIGURE 5.3.2.2 BHA [mixture of (a) 2-*tert*-butyl-4-hydroxyanisole and (b) 3-*tert*-butyl-4-hydroxyanisole]. Isomers that are common peroxide inhibitors

Why do these molecules inhibit peroxide formation? We see for BHT in Figure 5.3.2.3 below that the O—H in BHT is susceptible to cleavage and the hydrogen is transferred to the carbon radical to form R—H. Due to resonance stabilization and considerable steric hindrance from the adjacent *tert*-butyl groups, the BHT radical is exceptionally stable and unreactive. This terminates the chain reaction mechanism of radical propagation and each BHT molecule therefore inhibits the formation of thousands of other peroxides. Thus, only trace amounts of BHT are necessary.

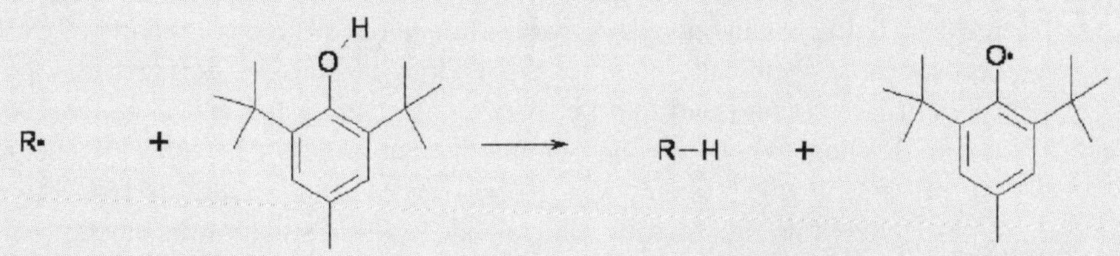

FIGURE 5.3.2.3 BHT Radical Reaction. BHT scavenges alkyl radicals, and the BHT radical is very stable and unreactive.

2. *Limiting use and storage.* It is always good practice to minimize and limit the amount of chemical that you order and use in your laboratory. The smaller the quantity of a chemical that you have on hand, the smaller the risk you will face from the hazards of that chemical. You should not purchase or have on hand more than you can use in a 3-month period. Make sure when ordering that the peroxide-forming chemical has a peroxide inhibitor. Any peroxide-forming chemical that does not contain an inhibitor should frequently (at least one a month) be monitored and tested for peroxides. Peroxide-forming chemicals should be stored in amber bottles in cool places away from light. (See Step #4, below, on labeling, too.)

3. *Alternate solvents.* Substituting a non-peroxide-forming solvent for a peroxide-forming solvent is always a good practice when this is chemically reasonably. For example, you could find a substitute for diisopropyl ether, such as *n*-propyl ether. While the latter compound can form peroxides, it does so to a much lesser extent than diisopropyl ether.

Working with Peroxides

4. *Labeling.* In general, all chemicals, but especially peroxide-formers, should be labeled with the following: (1) *date received*, (2) *date opened*, and (3) *date to discard*. It is preferable that this label be a brightly colored label that will attract attention; however, you should review your storage for chemicals with limited storage times on a periodic basis. If a peroxidizable chemical has remained unopened from the manufacturer, it may be keep up to 18 months.[3] Table 5.3.2.3 indicates various time frames in which the groups of peroxide-forming chemicals should be discarded. This can only be managed safely, of course, with good labeling.

5. *Suspecting peroxides.* If the appropriate schedule of disposal is not followed or if you come across a chemical that is old and perhaps suspect in quality, you should take great care in handling these peroxide-forming chemicals. It is possible to test for the presence of peroxides and decontaminate or remove these peroxides, but these operations are not without risk. Furthermore, a solvent can evaporate, resulting in increased peroxide concentrations that could be dangerous. Any old containers of chemicals that might contain peroxides must be handled with great care, particularly those in Groups 1, 2, or 3, since agitation, shaking, or friction could result in violent decomposition of the peroxides. This is not a situation that students should address; contact more experienced lab personnel to handle peroxide-containing solvents.

6. *Detecting peroxides.* There may be times when you need to know the peroxide content of a chemical and there are several methods that test for the presence of peroxides, including iodide methods, ferrous thiocyanate methods, titanium sulfate methods, and test strip methods.[3] These methods each have their limitations—some will not detect the presence of all peroxide forms. These methods should not be used to test alkali metals or amides since they react violently with water. Test strips offer some advantages in that they detect a wide group of different peroxides, can be used easily, and are convenient. However, they have limited shelf life and may be beyond the budget of some. For example, potassium iodide-starch test strips are available that can detect peroxides below 100 ppm. The presence of peroxides is detected by deep dark blue (virtually black) color on the test strip from the reaction of iodine (from potassium iodide reaction with peroxide) and starch. We will not discuss these peroxide test methods in detail here, but you should know that they are available.

7. *Limit distillations or evaporation of peroxide-forming chemicals.* Solvents or chemicals that can form peroxides should not be distilled to dryness or allowed to evaporate to dryness—the resulting residue could contain enough peroxides to explode. There are reports of explosions from distillations of peroxidizable chemicals. At times distillations of peroxide formers may be deemed essential. In these cases, it is recommended that peroxide content be tested and if above 100 ppm, the chemical should not be distilled. When carrying out any distillation of a peroxidizable chemical, you should ensure that the process is halted when about 20% of the starting volume is left. There have been suggestions to add a nonvolatile oil, such as mineral oil, to the distillation to ensure that it does not go to dryness. Furthermore, remember that the product of any distillation will not contain any inhibitor and is more prone to peroxide formation in the future.

8. *Avoid solution contamination.* It is essential to never return unused solvent (that may contain peroxides) to a stock container, since this increases the risk of peroxide formation in the stock container. Avoid using a metal spatula since metal atoms can participate in the catalytic formation of peroxides.

Managing Chemicals with High Peroxide Levels

9. *Emergency response actions.* What do you do if you find an old chemical or other solvent suspected of containing dangerous peroxides? First, remember not to panic—this bottle has been there some time without a problem, and as long as you do not touch the bottle, it will be safe to examine it. Next, seek to learn as much as possible from the label on the bottle, and examine the contents of the bottle by peering through the glass—remember do *not* touch the bottle or attempt to remove its cap. You may need a flashlight to peer into the bottle or use it as a backlight. Sometimes, particularly in old bottles of peroxidizable chemicals, peroxides may be observed as white crystals around the cap, a glass-like solid mass within the bottle, wisp-like crystals or precipitate in solution, cloudiness in a solution, or perhaps clear contamination where the liquid has evaporated leaving behind a solid mass. If any of these are detected or suspect, do not attempt to move or touch the bottle. Never attempt to open any container of a peroxider-former if you detect visible potential crystalline peroxides. This container will likely need to be handled by experts in disarming bombs—the bomb squad of your local law enforcement agency. Seek the immediate advice of your instructor, professor, or safety professional. They should be able to help decide the prudent course of action.

10. *Decontaminating peroxide-containing solvents.* There are published methods that can be used to remove and clean up peroxide-containing solvents. However, these are not without risk. Various methods have been used including passing the solvent through activated alumina, shaking with ferrous sulfate reagent solution, or allowing the solution to stand over molecular sieves.[3] These methods should be carried out by experienced personnel. Furthermore, any chemical known to contain high levels or visually observable amounts of peroxides should be handled only by professionals in handling potentially explosive materials.

Working with Peroxides and Peroxide-Forming Solvents: RAMP

Due to the uncommon shock sensitivity of peroxide-containing solutions, this represents one of the most hazardous situations in chemistry labs and stockrooms.

- *Recognize* what solvents are potential peroxide-formers.
- *Assess* the level of risk for a solvent based on age and the presence or absence of inhibitors.
- *Minimize* the risk by using substitute solvents, using solvents with inhibitors, minimizing storage and amounts of peroxide-forming solvents, testing for peroxides, and avoiding processes that concentrate peroxides. Do not handle solvents that are suspected to have high levels of peroxides.
- Since "peroxide emergencies" are likely to be explosions, the best step to *prepare* for the emergency is to avoid it. However, in the event that a potentially explosive peroxide is discovered, a trained disposal team should be called who can manage the event.

References

1. ALBERT EINSTEIN. The Einstein Letter That Started It All. *New York Times Magazine*, August 2, 1964. In: *Bartlett's Familiar Quotations*, 17th edition, Little, Brown, and Co., Boston, 2002, p. 684, #2.
2. University of California Industrial Hygiene and Safety Committee. Lessons Learned. Peroxide Explosion Injures Campus Researcher; available at http://ucih.ucdavis.edu/pages/lessons.cfm (accessed September 5, 2009).
3. R. J. KELLY. Review of safety guidelines for peroxidizable organic chemicals. *Journal of Chemical Health and Safety* 3(5):28–36 (1996).
4. H. L. JACKSON, W. B. McCORMACK, C. S. RONDESTVEDT, K. C. SMELTZ, and I. E. VIELE. *Journal of Chemical Education* 47(3):A175–A188 (1970).
5. R. J. KELLY. Peroxidizable Organic Chemicals. In: *Handbook of Chemical Health and Safety* (R. J. ALAIMO, ed.), American Chemical Society Oxford University Press, Washington, DC, 2001, pp. 361–370.
6. I. KRAUT. Explosive and Reactive Chemicals. In: *Handbook of Chemical Health and Safety* (R. J. ALAIMO, ed.), American Chemical Society Oxford University Press, Washington, DC, 2001, pp. 404–410.

7. *Bretherick's Handbook of Reactive Chemical Hazards*, Vol. 2, 7th edition (P. G. URBEN, ed.), Elsevier, New York, 2007, pp. 317–321.

8. L. BRETHERICK (editor). *Hazards in the Chemical Laboratory*, 4th edition, The Royal Society of Chemistry, London, Alden Press, Oxford, 1986, pp. 72–73.

9. World Health Organization (WHO)/International Union of Pure and Applied Chemistry (IUPAC). *Chemical Safety Matters*, Appendix B, Peroxide-Forming Chemicals, Cambridge University Press, Cambridge, UK, 1992, pp. 242–244.

10. National Research Council. *Prudent Practices in the Laboratory: Disposal and Handling of Chemicals*, National Academy Press, Washington, DC, 1995, pp. 54–56.

11. D. E. CLARK. Peroxides and peroxide-forming compounds. *Journal of Chemical Health and Safety* **8**(5):12–21 (2001).

QUESTIONS

1. It is safe to work with a particular peroxide-forming solvent as long as you

   (a) Have a general understanding of how peroxides form
   (b) Understand the general conditions under which peroxides can be hazardous
   (c) Avoid heating the solvent
   (d) Read in detail about the particular peroxide-forming capability of the solvent you are using

2. Which category of compounds has been associated with the dangers of peroxides?

   (a) Organic peroxides
   (b) Hydroperoxides
   (c) Some inorganic and organometallic peroxides
   (d) All of the above

3. What action might cause a peroxide-containing solvent to react violently?

   (a) Heat
   (b) Shock
   (c) Friction
   (d) All of the above

4. What chemical feature of peroxides makes them unusually reactive?

   (a) The very weak carbon–oxygen bond
   (b) The relatively strong oxygen–oxygen bond
   (c) Upon dissociation, radicals are formed
   (d) Two adjacent atoms, both with lone pair electrons

5. Approximately how many categories of chemicals are "peroxide formers?"

   (a) 3
   (b) 8
   (c) 12
   (d) 23

6. The chemicals that are "severe peroxide hazards" should be discarded how many months after opening?

   (a) 3
   (b) 6
   (c) 12
   (d) 24

7. Which method is *not* a desirable strategy for avoiding peroxides?

   (a) Using inhibitors
   (b) Using alternate, non-peroxide-forming solvents
   (c) Avoiding using solvents that contain any oxygen atoms
   (d) Limiting use and storage

8. What information should be included on a label for a peroxide-forming chemical?

   (a) Date received, date opened
   (b) Date received, date to discard
   (c) Date opened, date to discard
   (d) Date received, date opened, date to discard

9. As a student, if you encounter an old container of a peroxide-forming chemical, you should

   (a) Notify a more experienced chemist about the container after moving it to a safe location
   (b) Notify a more experienced chemist about the container without touching or moving it
   (c) Test it immediately
   (d) Have it tested only if there is visible evidence of peroxides

10. When distilling a peroxide-forming solvent, you should

    (a) Periodically test the distillate for peroxides
    (b) Perform a low-pressure distillation with no heat
    (c) Never distill the solvent pot to dryness
    (d) Distill to dryness only if you are certain an inhibitor is present

5.3.3

REACTIVE AND UNSTABLE LABORATORY CHEMICALS

Preview   This section discusses the hazards of unstable chemicals that are encountered in research and other laboratories.

The first requirement of an explosive is stability—but only metastability. This can mean that what you performed without accident today may still blow your silly head off if repeated tomorrow.

Bretherick's Handbook of Reactive Chemical Hazards[1]

INCIDENT 5.3.3.1   THE FLAMING GLOVEBOX[2]

A chemist conducted research on organometallic compounds within a glovebox in a nitrogen atmosphere since these compounds react with oxygen or moisture. With his hands in the glovebox gloves, he unscrewed the cap from the vial. As he twisted the cap, a sudden detonation blew off three fingers on the chemist's right hand and he was taken to the hospital. The hospital called to ask if someone from the laboratory could try to retrieve the fingers for reattachment to his hand. As two other chemists began procedures to do this, air entered the glovebox through the damaged glove, igniting something and flames came shooting out of the damaged glove port. The two chemists fearing an imminent explosion ran for cover. However, as the oxygen was consumed, the flames went out spontaneously. The two chemists returned and inserted a dry chemical extinguisher to cool and quench any existing hot spots and prevent further fires. The large mess in the glove box now made it impossible to quickly and safely recover the lost fingers in a timely manner. It took several days of meticulous and careful work to safely remove the remaining hazardous chemicals and debris from the glovebox before the fingers were located.

What lessons can be learned from this incident?

Hazards of Unstable Laboratory Chemicals

This section is not about explosives, but rather laboratory chemicals that may under certain conditions be unstable and may, if not handled properly, violently decompose resulting in an explosive event—that is, they could "blow your silly head off." Reports of explosions in research laboratories are not common, but they do occur. Most of these were preventable and perhaps even predictable if the researchers had carefully considered information about these or related compounds.

Chemists use a large variety of chemicals including some that are reactive and in some cases unstable. Nevertheless, these compounds are often very useful because they *are* reactive and they are used to carry out varied chemical transformations and syntheses that could not be accomplished in any other way—they are frequently the reagents of choice for a particular reaction. It is essential that every practicing chemist have some understanding of and the ability to recognize these hazardous compounds, and learn how they can be safely handled.

An Explosion—It's All About Energy

Chemical reactions involve energy changes and many of these reactions are exothermic in nature, releasing energy usually in the form of heat or light. An unstable or reactive chemical presents a hazard

---

Laboratory Safety for Chemistry Students, by Robert H. Hill, Jr. and David C. Finster
Copyright © 2010 John Wiley & Sons, Inc.

because the rate of release of energy from its decomposition is so fast that the excess heat cannot be absorbed by the surrounding environment. As a result this excess energy is released through an explosion. Explosions are rapid chemical reactions that, upon initiation, burn, deflagrate, or detonate. Upon detonation explosive compounds may release large volumes of heat and gases and may also generate hypersonic blast waves with velocities up to 28,000 feet per second. In a confined space pressures may reach 4 million pounds per square inch and any weakness in a vessel may result in fragmentation that generates shrapnel that can cause severe injury and damage to nearby people or property.

In a typical explosive event an unstable compound instantaneously releases energy and gases, such as CO, CO_2, and H_2O (as steam). Depending on the elements in the compound, other gases could also be released. The high energy and heat combined with the large volume of gases released are a formula for potential disaster. For example, 1 mole-volume of an unstable solid or liquid chemical explosive, such as nitroglycerin, upon detonation could release 1000 times that volume in gases at ambient temperatures, but since the temperature in the explosive products could reach perhaps 3000 °C, the volume of gas could be more than 10,000 times the volume of the original solid or liquid. See *Chemical Connection 5.3.3.1* Gas Expansion Factor.

CHEMICAL CONNECTION 5.3.3.1

GAS EXPANSION FACTOR

Let's start with 1 mol of liquid nitroglycerin. The reaction is

$$CH_2(NO_3)CH(NO_3)CH_2(NO_3)(l) \rightarrow 3CO_2(g) + 2\tfrac{1}{2}H_2O(g) + 1\tfrac{1}{2}N_2(g) + \tfrac{1}{4}O_2(g)$$

What is the volume of 1 mol of *liquid* nitroglycerin, using the density 1.6 g/mL?

$$(1 \text{ mol } C_3H_5N_3O_9(l)) \left(\frac{227 \text{ g}}{1 \text{ mol}} \right) \left(\frac{1 \text{ mL}}{1.6 \text{ g}} \right) \left(\frac{1 \text{ L}}{1000 \text{ mL}} \right) = 0.14 \text{ L}$$

When 1 mol of nitroglycerin detonates, it produces 7.25 mol of gaseous products. We can calculate the volume of this quantity of gas at a temperature of 3000 °C (a typical explosion temperature):

$$V = nRT/P = (7.25 \text{ mol})(0.08206 \text{ L·atm / K·mol})(3273\text{K})/(1 \text{ atm}) = 1900 \text{ L}$$

So, the original 0.14 L of $C_3H_5N_3O_9$ (l) produces 1900 L of gases, an expansion factor of 14,000!

Let's assume that you have a modest quantity of nitroglycerin, 10 mL, in a capped vial that has a volume of 50 mL. If this were to spontaneously detonate, the 10 mL would expand (at 1 atm) to 10 mL × 14,000 = 140,000 mL or 140 L. When this gas is confined to a volume of 50 mL, the pressure inside the vial will be (1 atm)(140 L)/(0.50 L) = 280 atm. This pressure would surely exceed the integrity of the vial and a violent explosion would occur.

Recognizing Chemicals with Explosive Properties

Some reactive laboratory chemicals are unstable with potential explosive properties requiring special consideration and handling. Table 5.3.3.1 provides a list of common classes of compounds that often have unstable members and have potential to explode under certain conditions—this is not a comprehensive listing and not all chemicals in these classes have explosive properties. Our knowledge of these compounds comes from experience—that is, probably someone experienced an explosion with these compounds.

When handling a chemical that is reactive and may be unstable, it is imperative that you evaluate any reported adverse events from this compound or its structural relatives. We highly recommend that you learn to effectively use one particular outstanding reference—*Bretherick's Handbook of Reactive Chemical Hazards*—it is in its 7th edition at the time of the writing of this book.[1] In this two-volume set, individual compounds are listed by their molecular formula with references to hazardous properties and reported incident. The second volume identifies reactive classes of chemicals. It is a valuable tool for researchers and chemists who use reactive or unstable chemicals.

5.3.3 REACTIVE AND UNSTABLE LABORATORY CHEMICALS

TABLE 5.3.3.1 Reactive Laboratory Chemicals with Explosive Properties[1,3]

Chemical class	General groups in a class	Examples
Acetylenic compounds: —C≡C—	Alkynes; haloacetylenes; metal acetylides; acetylenic peroxides; other acetylides	Acetylene; chloroacetylene; cupric acetylide; acetylenedicarboxaldehyde
Azides: —N$_3$	Acyl azides; metal azides; nonmetal azides; organic azides; other compounds containing azide moieties	Acetyl azide; lead azide; ammonium azide; benzyl azide; azido-2-propanone
Azo compounds: C—N=N—C	Selected azo compounds	Dimethyl azoformate; methyldiazene; azoisobutyronitrile
Diazo compounds: —N$_2$	Organic diazo compounds; metal diazo compounds	Diazomethane; diazoacetonitrile; lithium diazomethanide
Fulminates: —C≡N → O	Metal fuminates	Mercury fulminate; sodium fulminate
Nitrides: —N^{3-}	Metal and nonmetal nitrides	Lead nitride; silver nitride; disulfur dinitride; pentasulfur hexanitride
Aci-nitro: —C=N(O)O$^-$	*Aci*-nitro salts	Ammonium *aci*-nitromethanide; potassium phenyldinitromethanide
Organic nitro compounds: C—NO$_2$	Nitroalkyls; polynitro-aromatic or alkyl compounds	Nitromethane; tetranitromethane; trinitroresorcinol; 1,3,6,8-tetranitronaphthalene
Nitroso compounds: C—NO;N—NO	Organic nitroso compounds; N-nitroso compounds; inorganic nitrosyl compounds	2-Nitrosophenol; *N*-nitrosoacetanilide; nitrosylcyanide
Organic nitrites: C—O—NO	Acyl or alkyl nitrites	Acetyl nitrite; trifluoroacetyl nitrite; methyl nitrite
Organic nitrates: C—O—NO$_2$	Acyl or alkyl nitrates	Acetyl nitrate; benzoyl nitrate; methyl nitrate; glycerol trinitrate
Some nitrogen-containing compounds	Oximes; isoxazoles; triazenes; nitrogen halides; N-metallics	2-butanone oxime; 3-methyl-5-aminoisoxazole; 1,2-diphenyltriazene; nitrogen trichloride; hexamminechromium (III) nitrate
Organic peroxides: C—OO—C	Acyl or alkyl peroxides; peroxyacids; peroxyesters	Diacetyl peroxide; bis(trifluoroacetyl) peroxide; dimethyl peroxide; peracetic acid; *tert*-butylperoxybenzoate
Organic hydroperoxides: —C—OOH	Alkyl hydroperoxides	Allyl hydroperoxide; bis(2-hydroperoxy 1 methyl-2-pentyl) peroxide
Chlorites: —ClO$_2$	Chlorite salts	Lead chlorite; silver chlorite; tetramethylammonium chlorite
Chlorates: —ClO$_3$	Metal chlorates	Silver chlorate; potassium chlorate; sodium chlorate
Perchlorates: —ClO$_4$	Alkyl perchlorates; aminemetal perchlorate salts; diazonium perchlorates; metal perchlorates; nonmetal perchlorates; perchlorates of nitrogenous bases; perchloryl compounds	Methyl perchlorate; hexaamminenickel perchlorate; benzenediazonium perchlorate; mercuric perchlorate; nitronium perchlorate

Unstable chemicals or chemicals with explosive properties are termed to be "self-reactive." This self-reactive property is a result, essentially, of having relatively weak bonds within the compound itself so that its decomposition is driven to produce compounds with relatively strong bonds. (See *Chemical Connection 5.3.3.2* Why Are Explosives So Explosive?) To initiate decomposition only requires a small amount of energy such as slightly elevated temperature, a sudden shock, friction, or contact with an ignition source. A self-reactive chemical has a high heat of decomposition—a lot of energy is released when it decomposes to its resulting products.

CHEMICAL CONNECTION *5.3.3.2*

WHY ARE EXPLOSIVES SO EXPLOSIVE?

The explosion of nitroglycerin is very exothermic at -1500 kJ/mol $C_3H_5N_3O_9$ (*l*) and the role of this exothermicity is to release considerable heat, which contributes to the expansion of the gases. Why is this so exothermic? It is the *stability of the products* (more so than the "instability" of the reactant) that leads to this highly exothermic reaction. We can calculate the heat of reaction by using heats of formation:

$$C_3H_5N_3O_9(l) \rightarrow 3CO_2(g) + 2\tfrac{1}{2}H_2O(g) + 1\tfrac{1}{2}N_2(g) + \tfrac{1}{4}O_2(g)$$

Heat of formation (kJ/mol) : -364 -393.5 -241.8 0 0

Heat of reaction $= [3 \text{ mol } (-393.5 \text{ kJ/mol}) + 2.5 \text{ mol } (-241.8 \text{ kJ/mol})] - [1 \text{ mol } (-364 \text{ kJ/mol})]$

$= (-1180.5 \text{ kJ} + -604.5 \text{ kJ}) - (-364 \text{ kJ})$

$= (-1785 \text{ kJ})\quad - (-364 \text{ kJ})$

$= -1421 \text{ kJ}$

Alternately, using Table 5.3.3.2, we can calculate an approximate value of the reaction by using average bond energies. Subtracting the total bond energy of the reactant from the total bond energy of the products gives us an approximate heat of reaction: 8325.5 kJ -6848 kJ $= 1477.5$ kJ. This gives an approximate value since the bond energy data are only average values. The agreement with the value calculated using heats of formation is adequate.

TABLE 5.3.3.2   Average Bond Energies in Nitroglycerin and Its Detonation Products

	Bond energy (kJ)	Number of bonds	Total bond energy (kJ)
Reactant			
C—C	346	2	692
C—H	411	5	2055
C—O	358	3	1074
N—O	201	6	1206
N=O	607	3	1821
Sum			6848
Products			
C=O	749	6	4494
O—H	459	5	2295
N ≡N	942	1.5	1413
O=O	494	0.25	123.5
Sum			8325

Explosives are good examples of this kind of self-reactivity. These are often organic molecules with high nitrogen and oxygen content that decompose to produce N_2, CO_2, and H_2O (as steam)—products with strong bonds. These compounds can also be generally viewed as molecules that have both the oxidizing and reducing agent contained within its own formula, either as a molecule or a salt. Good examples of this are nitrated organic molecules, peroxyacids, and organic cations (usually amines) with nitrate or perchlorate anions. This peroxy-, nitro-, nitrate, perchlorate species are excellent oxidizing agents and the rest of the organic molecule is readily oxidized. Such molecules can be heat-sensitive, shock-sensitive, or both. (See also Section 5.2.3 on incompatibles.)

The other class of molecules that self-reacts consists of monomers that readily form polymers—often these are alkenes or dienes. The polymerization process is usually free-radical generated and with polymeric products sometimes involving thousands of monomer linkages that produce viscous materials. The very large amount of heat cannot be dissipated quickly and runaway reactions can occur, leading to fires and explosions. (This is also briefly discussed in Section 5.3.2 on peroxides.)

Finely divided laboratory chemicals, such as catalysts on carbon or metals, can cause fires and perhaps even explosions. The high reactivity of these substances is due largely to the high surface area associated with very small particles. This is therefore a kinetic, rather than thermodynamic, effect. Similarly, clouds of dust particles have caused explosions in industrial settings, although we know of no examples of such explosions in laboratories. You should always be wary of suspensions of combustible or oxidizable materials in air and make sure that they do not come in contact with an ignition source.

Many of these compounds are sensitive to mechanical manipulation and may explode from friction, as in the screw cap explosion described in Incident 5.3.3.1, or by the action of using a metal spatula on a solid. Diazomethane is a very useful reagent produced to make methyl esters, but experience has demonstrated that the production of this compound should never be carried out with ground glass joints, since it has the propensity to crystallize out on the ground glass and explode spontaneously. Some organic azides have the same properties. Peroxides are among the most hazardous reactive compounds used in the laboratory and are sensitive to shock, particularly in dry or concentrated forms (see Section 5.3.2 on peroxides). Most of these compounds are also very sensitive to heating, sparks, or sources of ignition.

Generally, the drier or more desiccated a compound with potential explosive properties becomes, the greater the risk for an explosion. Picric acid, aka trinitrophenol, is related very closely to TNT (trinitrotoluene) and it is used in some laboratory operations. Normally, picric acid comes in a moist state from the supplier, and in this state it is safe to handle and store. However, when picric acid reaches a dry state it has the same potential as TNT to be explosive. Metal salts of picric acid are also known to be shock-sensitive. Picric acid needs to be on a timed disposal protocol to ensure that it is discarded before it can dry out and become hazardous.

There are other compounds that, when used in the laboratory, can also produce serious hazards for maintenance workers if not disposed of properly. Sodium azide is often used as a preservative for laboratory water to prevent formation of bacteria. However, if poured down the sink, the azide can react with brass, copper, or lead in the piping to produce heavy metal azides—highly explosive compounds. Maintenance workers and plumbers have been injured while working on laboratory sinks from inappropriate disposal of azides into laboratory sinks.

Perchloric acid is used in digestion of samples for varied analyses. Most laboratories that carry out perchloric acid digestions do these on a regular basis and they must use hoods designed especially for the use of perchloric acid. In these digestion processes perchloric acid can be released into the air, and perchloric acid and perchlorate residues trapped within the ventilation system are removed using a special wash down apparatus. Perchloric acid digestions should only be carried out in these specially designed hoods and should not be used in other laboratory hoods, especially not those using organic chemicals. Explosions have occurred when maintenance people have worked on hoods used for perchloric acid digestions without proper wash down accessories. Perchloric acid digestions carried out at room temperature do not generate fumes and do not require special perchlorate fume hoods. Occasional small scale digestions can alternatively be safely carried out using special commercially available laboratory glass apparatus designed for these purposes.

Predicting Explosive Properties

There are many published works that seek to predict the explosive properties of chemicals. Some of these are useful but there are almost always exceptions in their predictive powers—thus, none are fully predictive. Probably the most widely known method used to predict or identify compounds or reaction mixtures with explosive properties is called oxygen balance.[4] The potential to be explosive increases as a compound's composition approaches a "zero oxygen balance." This means that when the stoichiometric composition of a compound is such that it provides all of the oxygen needed to oxidize all of the other elements to their preferred state, then it has a zero oxygen balance.

If a compound has a zero oxygen balance, this compound would be predicted to be very unstable and explosive—and it is. For example, ethylene glycol dinitrate (ethanediol dinitrate) has a zero oxygen balance and it is explosive and is used with nitroglycerin in commercial explosives. This compound has within its own composition enough oxygen to produce fully oxidized products and is thus self-reactive. You can write its decomposition equation as a perfectly balanced equation with the addition or subtraction of oxygen—thus, it has a "zero oxygen balance."

$$CH_2(NO_3)CH_2(NO_3)(l) \rightarrow 2CO_2(g) + 2H_2O(g) + N_2(g)$$

Nitroglycerin, a high explosive, has a positive oxygen balance (expressed in %) of $+3.5\%$, meaning that it has more oxygen than required to oxidize its elements. TNT (trinitrotoluene) has a negative oxygen balance of -74% and it requires additional oxygen to completely oxidize its elements. This method can also be used for predicting the explosive properties of mixtures. However, as with many predictive methods, exceptions do exist so it is important to always view these with some reservation. See *Chemical Connection 5.3.3.3* Calculating Oxygen Balance.

CHEMICAL CONNECTION 5.3.3.3

CALCULATING OXYGEN BALANCE

Calculating oxygen balance is a method that can be used to predict the chemical instability or explosive properties of some compounds. The closer a chemical is to a "zero oxygen balance," the more unstable or explosive it will be. A compound with zero oxygen balance has all the oxygen needed to produce its fully oxygenated products upon decomposition. We can calculate the oxygen balance easily using the formula below and the results expressed in %. Nitrogen in a compound is not counted and assumed to become N_2. If there are other species that react with oxygen, then you would account for these, too. For example, sulfur could form SO_2 (counts as two oxygens) and metals could form oxides (M_2O counts as $\frac{1}{2}$, MO counts as 1, MO_2 counts as 2, M_2O_3 counts as 3, etc.). Thus, you would add another term for sulfur or metal oxides in the equation (adding this to the total of carbon and hydrogen atoms).

$$\text{Oxygen balance}(\%) = \frac{-16[(2 \times C) + (H/2) - O]}{MW} \times 100$$

where 16 is the atomic weight of oxygen, C is the number of carbon atoms, H is the number of hydrogen atoms, O is the number of oxygen atoms, and MW is the molecular weight of the compound.

Lets look at the compound ethylene glycol dinitrate, whose formula is $CH_2(NO_3)CH_2(NO_3)$. If this compound decomposes to its fully oxygenated products, CO_2 and H_2O, we can write the following balanced equation:

$$C_2H_4N_2O_6(l) \rightarrow 2CO_2(g) + 2H_2O(g) + N_2(g)$$

The oxygen balance (OB%) is

$$OB\% = \frac{-16[(2 \times 2) + (4/2) - 6]}{152} \times 100 = 0$$

Another example is nitroglycerin, whose formula is $CH_2(NO_3)CH(NO_3)CH_2(NO_3)$. A balanced decomposition equation is

$$C_3H_5N_3O_9(l) \rightarrow 3CO_2(g) + 2\tfrac{1}{2}H_2O(g) + 1\tfrac{1}{2}N_2(g) + \tfrac{1}{4}O_2(g)$$

Note that the balanced equation shows an excess of oxygen from the decomposition. You can calculate that this compound has an excess of oxygen ($OB\% = +3.5\%$)—more than it needs to fully oxidize carbon and hydrogen to CO_2 and H_2O. This compound is said to have a positive oxygen balance. Compounds that have a negative oxygen balance need more oxygen to completely decompose. For example, to balance a decomposition equation for trinitrotoluene, aka TNT, you would need to subtract 5.25 oxygen molecules. Thus, TNT has a negative oxygen balance:

$$C_6H_2(NO_2)_3CH_3(s) \rightarrow 7CO_2(g) + 2\tfrac{1}{2}H_2O(g) + 1\tfrac{1}{2}N_2(g) - 5\tfrac{1}{4}O_2(g)$$

$$OB\% = -74\%$$

Mixing a compound with a positive oxygen balance with a compound with a negative oxygen balance in the right proportions could lead to a reaction with a near zero oxygen balance—it will have good explosive properties. Thus, compounds such as ammonium nitrate ($OB\% = +20\%$) or ammonium perchlorate ($OB\% = +27\%$) can be mixed with compounds such as trinitrotoluene, aka TNT ($OB\% = -74\%$), to make a more effective explosive mixture.

Most organic compounds that you are familiar with are relatively oxygen-poor in terms of this analysis. Ethanol has an OB of -128% and hexane has an OB of -362%. These substances are quite flammable, but not at all self-explosive.

There are other methods that predict reactivity more reliably than the oxygen balance method. These methods, however, require the use of data generated by measurements. Stull devised a relatively simple "Reaction Hazard Index" or RHI that used both kinetic and thermodynamic data measurements. This index is a graphic model (nomograph) that uses the *Arrhenius activation energy* and the *decomposition temperature*.[5] The latter term is the maximum adiabatic temperature reached by the products of a decomposition reaction. If data are available or can be measured, the RHI may be a very useful method to predict reactivity. Coffee described a method that predicts the explosive potential of a compound using thermal stability (measured), impact sensitivity (measured), and the heat of reaction (calculated).[6] Compounds found to be thermally unstable and sensitive to impact were explosive.

Minimizing Adverse Events When Working with Unstable Chemicals

The key to minimizing adverse events is to recognize when compounds may be unstable. The preceding paragraphs gave you some information about this. You should always conduct a search to find out if the compounds you will be using have been reported to exhibit unstable properties or adverse events, such as explosions. If you know the molecular formula, it is easy to find information in *Bretherick's Handbook of Reactive Chemical Hazards*.[1] If you have a colleague who has used this compound, you should seek to learn about that person's experience, keeping in mind that experience with no events does not mean that adverse events are unlikely. If you can find a less hazardous chemical that can effect the same functional change needed for your experiment, consider substituting this chemical for the more hazardous one. If you cannot find information about your compound(s), but you need to use this compound or compounds of related structures, it is best to be conservative and handle these compounds with care and respect, using methods to protect yourself and others. It is likely that, if you work in chemical research, you will need to use unstable, highly reactive, and perhaps even potentially explosive materials to accomplish your goals.

To minimize adverse events, such as explosions, with unstable compounds, you must be able to control the rate of the reaction. The two factors that are most important in this are the concentration of the reactants and the rise in the temperature of the reaction. You should seek to minimize the concentration of the reaction by using diluted solutions of these compounds—in other words, do not use concentrated solutions. Bretherick's suggests concentrations of 10% or less, perhaps 5% or 2% to carry out an initial reaction.[1] If you have not performed this reaction previously or used this reagent previously, it is prudent to keep concentrations low. Even after you have carried out this reaction several times, be conservative in increasing the concentrations. You must be careful to avoid mechanical friction or shock of the unstable materials. Avoid concentrating the solution or heating solutions to dryness. The *DOE Explosives Safety Manual*[7] provides guidance for using explosive chemicals in the laboratory.

The Arrhenius equation predicts that the rate of reaction increases exponentially with an increase in temperature. Bretherick[7] noted that an increase of $10\,^\circ C$ in reaction temperature can increase the reaction rate by a factor of 2. (See *Chemical Connection 5.3.10.1* in Section 5.3.10 for more discussion of the Arrhenius equation and reaction rates.) Thus, it is critical that temperature be adequately controlled to prevent the reaction from accelerating to a dangerous rate. Be prepared to provide adequate control of the temperature—you will need to measure the temperature and have cooling means readily available. If you are unable to control the temperature, an explosion may occur. You should try to avoid systems that hold in heat—adiabatic systems. You can often control a reaction by controlling the rate of addition of a reagent. You should consider the best way to provide adequate mixing for the reaction.

Scaling up reactions can bring difficulties in controlling temperatures, and any changes in scale need to be done carefully. Be prepared to deal with reactions where you cannot control the temperature—remove the source of heat, any insulating materials, stop adding reagents, maintain shielding between you and the reaction vessel, know where the fire extinguisher is located and how to use it, and alert others that something adverse may be happening. Even a small explosion can be dangerous, so you must take steps to minimize the chances of these occurrences. *Prudent Practices in the Laboratory*[9] provides a good discussion about working with reactive or explosive chemicals.

Some suppliers of reagents may provide guidance or instructions for the safe use of these reactive chemicals, and you would be prudent to follow their guidelines. There may be specialized safety equipment or apparatus available for carrying out this type of reaction. Some companies such as Aldrich

Chemical Co. sell these kinds of laboratory equipment. It is prudent to use shielding between you and reaction vessels. These operations should be carried out in a hood that is free of other experiments or bottles of chemicals or chemical waste. As with all operations with hazardous chemicals, use and keep on hand only the minimum quantities needed to carry out your research.

Working with Potentially Explosive Laboratory Chemicals: RAMP

Reactive chemicals, including those with explosive properties, are often important tools for chemical research. Most of these chemicals can be used safely if you are armed with the knowledge and skill and have clear plans to handle these in your laboratory.

- *Recognize* the types of chemicals that may have explosive potential and learn all that you can about previous incidents as you embark on using these compounds.

- *Assess* the level of risk of the reagents you will be using for your experiments.

- *Minimize* the risk by knowing what you are handling and how to safely handle these. When using reactive chemicals with potential explosive properties, keep the concentration of chemicals low and maintain close control of temperatures. Use appropriate protective measures to minimize exposures.

- *Prepare* to handle emergencies such as explosions, should they occur. Talk with others who have had experience with these compounds. Alert other laboratory partners of your work with potentially explosive materials. Know how and what to do in advance, should an explosion occur.

References

1. *Bretherick's Handbook of Reactive Chemical Hazards*, Vol. 2, 7th edition (P. G. URBEN, ed.), Elsevier, New York, 2007, p. 131.
2. R. HILL. Observed incident in late 1960s.
3. National Research Council. *Prudent Practices in the Laboratory: Handling and Disposal of Chemicals*. Chapter 3. Evaluating Hazards and Assessing Risks in the Laboratory. 3.D.3 Explosive Hazards. National Academy Press, Washington, DC, 1995, pp. 54–57.
4. W. C. LOTHROP and G. R. HANDRICK. The relationship between performance and constitution of pure organic explosive compounds. *Chemical Reviews*, **44**: 419–445 (1949).
5. D. R. STULL. CX. Linking thermodynamics and kinetics to predict real chemical hazards. *Journal of Chemical Education*, **51**(*1*): A21–A25 (1974).
6. R. D. COFFEE. XCII. Evaluation of chemical stability. *Journal of Chemical Education*, **49**: A343–A349 (1972).
7. Office of Environment, Health, and Safety, U.S. Department of Energy. *DOE Explosives Safety Manual, DOE M440.1-1A*. January 2006; available at http://www.directives.doe.gov/pdfs/doe/doetext/neword/440/m4401-1a.pdf (accessed July 26, 2009).
8. L. BRETHERICK. Reactive Chemical Hazards. In: Hazards in the Chemical Laboratory, 5th edition (S. G. LUXON, ed.), Royal Society of Chemistry, Cambridge, UK, 1992, pp. 50–65.
9. National Research Council. *Prudent Practices in the Laboratory: Handling and Disposal of Chemicals*. Chapter 5. Working with Chemicals. Section 5.G Working with Highly Reactive or Explosive Chemicals. National Academy Press, Washington, DC, 1995, pp. 96–104.

QUESTIONS

1. Why do chemists sometimes use chemicals that are inherently unstable or unusually reactive?

   (a) These are generally inexpensive chemicals.
   (b) Their reactivity usually guarantees very little generated waste that needs to be disposed of.
   (c) It is the very reactivity of the chemical that performs the necessary reaction.
   (d) These chemicals react at low temperatures, requiring little or no heat, which saves energy.

2. An explosion is characterized by a reaction that

   (a) Generates energy faster than it can be absorbed by the surrounding environment
   (b) Often generates large amounts of heat and gases
   (c) May generate a hypersonic blast
   (d) All of the above

3. In an explosion, the volume of gases generated is how many times greater than the volume of reactants?

   (a) 10–100 times
   (b) 100–1000 times
   (c) 1000–10,000 times
   (d) 10,000–100,000 times

4. About how many classes of compounds are considered to have explosive potential?

   (a) 3
   (b) 7
   (c) 12
   (d) 17

5. What book resource is considered the most useful reference when researching the reactivity of chemicals?

   (a) *Laboratory Safety for Chemistry Students*

(b) *Prudent Practices in the Laboratory*

(c) *Bretherick's Handbook of Reactive Chemical Hazards*

(d) U.S. Department of Energy, *DOE Explosives Safety Manual*

6. Explosive compounds generally have

   (a) Weak bonds and generate products that have strong bonds

   (b) Strong bonds and generate products that have weak bonds

   (c) Weak bonds and generate products that have even weaker bonds

   (d) Strong bonds and generate products that have even stronger bonds

7. Many explosive compounds are molecules that have

   (a) Both oxidizing and reducing groups

   (b) Both an acidic functional group and a basic functional group

   (c) Both oxidizing and acidic functional groups

   (d) Both reducing and basic functional groups

8. Polymer-forming compounds are considered hazardous because

   (a) The polymerization process is usually free-radical generated

   (b) The polymerization process can involve thousands of molecules

   (c) The heat generated often cannot be dissipated quickly, which accelerates the reaction

   (d) All of the above

9. What compound is usually stored wet or in solution, but becomes explosively dangerous when forming dry crystals?

   (a) Phenol

   (b) Sodium azide

   (c) Picric acid

   (d) Nitric acid

10. What circumstances are important to control to minimize the chance of explosions or out-of-control reactions?

    (a) Keep concentrations low and control the rate of rise of temperature.

    (b) Avoid the use of strong acids if weak acids can be substituted.

    (c) Avoid using catalysts, unless absolutely necessary.

    (d) Make sure that the reaction is adiabatic.

5.3.4

HAZARDS FROM LOW- OR HIGH-PRESSURE SYSTEMS

Preview   This section discusses the hazards associated with laboratory equipment at either low or high pressures.

There is no short cut to achievement. Life requires thorough preparation—veneer isn't worth anything.

George Washington Carver, American scientist[1]

INCIDENT 5.3.4.1   FILTRATION IMPLOSION[2]

A laboratory worker was conducting a routine filtration using a standard Erlenmeyer flask with a sidearm connected to a house vacuum system. The flask imploded and dispersed a spray of shards of glass and the solution being filtered. Fortunately, the filtration was in a chemical hood with the shield lowered and there were no injuries.

What lessons can be learned from this incident?

Applying Pressure in the Laboratory

Chemists often use either high pressure or low pressure in various laboratory operations—it is an essential part of our work and often the preferred method of accomplishing our tasks. It is common to use glass vessels in many pressurized or vacuum systems because they are relatively inexpensive, relatively chemically inert, and easy to work with. Nevertheless, there is always the possibility of failure of these glass vessels from scratches, stress fractures, or accidental bumping. The result can be an explosion or implosion that puts you and others in harm's way.

Chemical operations under high pressure are often regarded as high-risk hazards—there is always some risk of explosion with subsequent shards, shrapnel, fires, and toxic products. Always get assistance from an experienced researcher or a mentor who can teach you how to safely manage these high-pressure hazards.

The Simplest High-Pressure Reactors—Sealed Tubes

Before you consider using laboratory pressure reactors, learn all that you can about the chemistry of the reactants and products to assess the potential hazards, particularly explosive properties. *Bretherick's Handbook of Reactive Chemical Hazards* is highly recommended for evaluation of potential explosive properties of your reactants and related incidents.[3] Material Safety Data Sheets (MSDSs) may also have information about incompatibles and adverse reactions associated with the reactants.

We will discuss the use of laboratory pressure reactors—their hazards and safety precautions. This discussion is not designed to be comprehensive but rather to raise your awareness about the need for safety in managing high-pressure reactions—to reemphasize here you need to find someone with experience before tackling this kind of operation. Two examples of laboratory operations involving use of high-pressure reactors are sealed tube reactions and hydrogenations. Some institutions govern these operations and require permits before these are allowed.

Laboratory Safety for Chemistry Students, by Robert H. Hill, Jr. and David C. Finster
Copyright © 2010 John Wiley & Sons, Inc.

Sealed tube reactions are sometimes used when safer reactions under atmospheric conditions failed, and as you can guess from the name, basically a set of reactants are sealed in a tube and heated to drive the reaction to the desired products. However, any chemical heated in a closed system will build up pressure that can make the tube the source of a possible explosion (see Incident 7.3.7.1).

Preparing and handling sealed tubes should be regarded as high-risk operations due to potential explosive events. The glass reactor (tube) must be appropriate for sealed tube reactions—thick walled and stable to rapidly changing temperatures. Tubes made especially for this purpose are available commercially and are usually Pyrex® or quartz. See *Chemical Connection 5.3.4.1* Types of Laboratory Glass. Some glasses fracture under thermal shock so the right glass must be selected. Most glassblowers can provide these kinds of tubes and are a good source of information about the selection of the glass—thickness and temperature properties. The glassblower may be willing to help you seal the tube after charging or teach you how to seal the tube yourself. Sealing a tube requires experience and skill—you may have to practice this technique extensively. A poorly sealed tube could be dangerously susceptible to failure under pressure.

CHEMICAL CONNECTION *5.3.4.1*

TYPES OF LABORATORY GLASS

Glass is the term applied to amorphous solids, including those not only of silica but also of plastics, resins, and other silica-free amorphous solids. In the case of inorganic products such as silica, it is the product of fusion derived from being cooled but not crystallized. It refers to hard, brittle, and transparent solids that are commonly used for bottles, windows, mirrors, lenses, prisms, or eyeglasses. There are many types of silica-based glasses, such as soda-lime glass (most common), borosilicate glass, fused quartz and fused silica glasses, leaded glass, glass wool, and Muscovy glass. Ordinary glasses are transparent—they do not absorb visible light, unless a pigment such as iron oxide has been added. Very pure silica glasses (fused quartz or fused silica) do not absorb in the ultraviolet range as does ordinary glass.

The most common glass, soda-lime glass, is used to make about 90% of all glasses such as window panes, bottles, jars, and drinking glasses. Silica (SiO_2) is combined with varying amounts of sodium carbonate (Na_2CO_3) and lime (CaO), and smaller amounts of magnesium oxide (MgO) and aluminum oxide (Al_2O_3) to form soda-lime glass that is just over 70% silica. Combinations of silica with other oxides can produce other glasses, such as lead crystal glass (made with lead oxides). For laboratory use, a very important property of glass is its ability to undergo thermal shock—a sudden change in temperature. Ordinary soda-lime glass does not have good thermal shock properties—it easily shatters with sudden changes in temperature.

Borosilicate glasses are made from silica and boron oxide and they are well known to have good thermal shock properties—useful for laboratories where heating and cooling operations are common. The most commonly recognized name for borosilicate glasses in the laboratory is Pyrex®, but other names are also known, in particular, Kimax®. Pyrex is also sold for cooking ware in today's kitchens but it may not be borosilicate glass but rather a tempered soda-lime glass that has good thermal shock properties. Fused quartz and fused silica glasses are made primarily of silica and also have good thermal shock properties. Fiberoptic cables are made from a synthetic fused silica.

These borosilicate and fused quartz and silica glasses have very low coefficients of thermal expansion, α:

$$\alpha = \frac{1}{L_0}\frac{\partial L}{\partial T}$$

where L_0 is original length, ∂L is change in length, and ∂T is the change in temperature. The expansion coefficient can be expressed in linear terms (as above), area terms, or volumetric terms. For glasses the linear term is often used. Rearranging the equation above, we get

$$\frac{\Delta L}{L_0} = \alpha_{LL}\Delta T$$

The thermal expansion coefficients (ppm/K) for selected glasses are as follows:

Soda-lime glass	9
Lead crystal glass	7
Glass wool	10

Borosilicate glass	3.5
Fused quartz	0.59
Fused silica	0.55

How much will a 10.00-cm borosilicate glass rod elongate when heated from 300 K (room temperature) to 373 K (boiling water)?

$$\Delta L = (L_0)(\alpha)(\Delta T) = (10.00 \text{ cm})(3.5 \text{ ppm/K})(73 \text{K}) = 0.0025 \text{ cm} = 0.025 \text{ mm} = 25 \text{ } \mu\text{m} = 25 \text{ microns}$$

Of course, the other advantage to glass is that is it virtually insoluble in all common organic and inorganic solvents and solutions, which makes it *very* convenient for lab use! The important exceptions to this are concentrated strong bases and HF solutions. These solutions are typically stored in plastic bottles.

The tube is usually charged about half-full with the reactants and then the tube is flame-sealed. The sealed tube is usually placed in a special tube oven or some other heating device that raises the temperature, and the pressure increases as the reactants become volatilized. This reaction may be carried out a few minutes to hours or days, depending on the transformation sought. After the reaction, the tube is cooled by turning off the source of heating and allowing it to cool down, or by removing the tube to the air to allow a cool down process, or perhaps by a more aggressive cool down procedure. After cooling, the top of the tube is etched and broken off by gentle pressure, so the contents can be retrieved. Alternative reactors will be discussed below. Where might the hazards to this kind of high-pressure operation arise?

Tubes must not be filled more than half-full because of potential overpressurization. Flame-sealing is a potential hazard due to the potential for ignition of the reactants. If your reactants are flammable, the tube is cooled (often cryogenically) so the reactants do not ignite during the flame-sealing process. The sealed tube can now be loaded into a specially designed oven with a thick insulated door to contain the heat and contents in case of an explosion. This apparatus should be placed in a ventilated chemical hood if possible. A safety shield should always be between the researcher and tube at all times. When handling the tube after the reaction, use chemical splash goggles, a face shield, heavy leather gloves, and a towel to wrap around the tube. After cooling, the tube should be etched and the seal broken within the hood behind a shield, in case there may be pressure and/or toxic products.

You may want to consider alternatives to the use of sealed tubes. Glass reactors with screw-in Teflon® plugs and O-rings are commercially available. These plugs can be blown off if the pressure becomes too high and you also will need to ensure that the O-rings are compatible with the reactants and products. These plugs may also blow out under high pressure so you must ensure that you have some way to control such an event. Another alternative is the microreactor made for high-pressure operations using small quantities of reactants. Both of these alternatives offer some advantages over the classic sealed tube reactor, but are more expensive, cannot achieve the high pressures of the sealed tube, and might not produce the quantity of material that could result from sealed tubes. Nevertheless, based on risk considerations, alternatives may be more desirable for your work.

Laboratory Hydrogenations—A Common High-Pressure Operation

Hydrogenations are often high-pressure operations that utilize catalysts with hydrogen, and often flammable reactants—these can be risky operations but with the right equipment, education, training, and assistance from experienced researchers these operations can be carried out safely. Most researchers use commercially available reactors for hydrogenations (see Figure 5.3.4.1).

In academic labs, hydrogenation experiments are usually done on small quantities of reagents, usually less than 1 gram of an alkene or alkyne. In these situations, it is not necessary to worry about the inherent exothermicity of the reaction. However, on larger scale reactions, it is wise to consider how much heat will be liberated and if there are circumstances that might result in a reaction that could go out of control. Some of the catalysts used in hydrogenations are air-sensitive and should be added to a flask first under a blanket of nitrogen gas, before any flammable solvent is added. It is best to

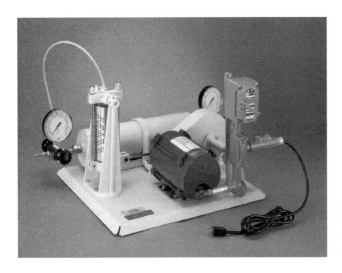

FIGURE 5.3.4.1   Hydrogenation Apparatus. This apparatus allows the hydrogenation reaction to be conducted at a pressure of up to 5 atm. (Courtesy of Parr Instruments, Moline, Illinois.)

conduct this experiment for the first time under the guidance of someone who has experience with these reactions.

Classic laboratory reactors especially designed for laboratory hydrogenations are available in a variety of sizes and designs. They are designed to ensure safety of operation when used within limitations defined by manufacturers—each reactor has limits based on capacity, pressure, vessel material, and design uses. Reactors are made of glass, glass in metal jackets, or metal.

It is important to select the right size and equipment for the job, using the manufacturer's guidelines. It is also important that you or others are taught how to use this equipment properly and safely—this is often accomplished by following the manufacturer's instructions for installing, operating, and maintaining this equipment. Not carefully following the manufacturer's instructions can lead to increased risks, especially if the limitations of the equipment are exceeded. Clearly, this is an operation that requires careful attention to detail and to recognition of the limitations of your equipment—a mistake here can be dangerous.

It is always prudent to take additional safety precautions to protect yourself from any potential explosion. Prevent any line-of-sight path to exposure from hot or burning chemicals, as well as missiles (glass and metal fragments if an explosion occurs) by using an appropriate shield. You can provide this added protection using chemical splash goggles, face shields, and portable shields made of materials that can withstand an explosive event. Placing the equipment in a hood will capture potential toxic or flammable releases should an explosion occur. Chemical hoods can also provide shielding if the doors are kept down all the way so that there is no line-of-sight path to the user. Since some hoods have vertical sliding doors rather than horizontal doors, it is not always feasible to protect from line-of-sight exposure, so portable shields will also likely be needed.

If the apparatus is too large for a hood, consider building a permanent fixed shield around this equipment to provide line-of-sight protection from all sides if there is a possibility of an explosion. You may want to consider using local exhaust ventilation to capture and remove as much of the volatile products as possible.

There are alternatives to classic high-pressure hydrogenations, including microreactors and compact flow reactors. Microreactors still use hydrogen gas but the scale is greatly reduced, thus reducing the risks. Flow reactors generate hydrogen in situ from water, thus eliminating hydrogen gas cylinders.

Low-Pressure Operations—Potential for Implosions and Toxic Exposures in the Lab and Elsewhere in the Building

Low-pressure operations are common in many laboratories. Laboratories are often equipped with a house vacuum system, vacuum pumps, rotary evaporators, water aspirators, vacuum concentrators, vacuum ovens, and other apparatus that operate under reduced pressure. Some labs build elaborate vacuum racks

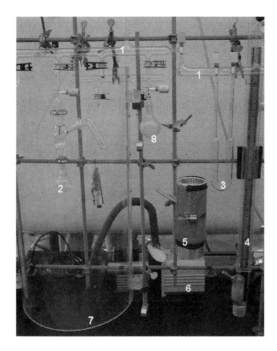

FIGURE 5.3.4.2 Laboratory Vacuum System. Low-pressure systems can have many configurations and components. This system shows a (1) manifold, (2) reaction flask, (3) trap, (4) manometer, (5) Dewar flask, (6) vacuum pump, (7) safety shield, and (8) storage bulb. (Photo courtesy of T. Leon Venable, Agnes Scott College, Decatur, Georgia.)

for special purposes (see Figure 5.3.4.2). All glass vessels under vacuum should be considered potential risks for implosions with the result being flying glass shards, chemical splashes, and fires. The violence of implosions is limited due to the fact that the pressure differential between inside and outside the vessel cannot be more than 1 atmosphere (unlike pressure vessels, where the pressure differential can be much greater).

A common use of vacuums involves their utility for removing volatiles from reactions or filtering materials. It is common to use house vacuum systems, water aspirators, or vacuum pumps for these purposes. Besides the risks of implosions, there can be potential hazards from exposures to toxic products released from the vacuum exhaust. Maintenance workers have been exposed to toxic chemicals that were allowed to enter into a house vacuum system. Others have been exposed to toxic products from unventilated vacuum pumps used to evaporate volatile organics. It is important to use techniques to set up traps to prevent these kinds of releases of hazardous materials.

There are a variety of trapping techniques for these systems. In line filters can easily trap particulates from vacuum systems. Specific sorbents can be used in in-line tubes to trap reactive, corrosive, or toxic products—however, you must make sure that the sorbent is compatible with *all* of your products. To prevent water or other nonvolatile materials from entering the vacuum, you can use a filter flask in-line to prevent these from getting into the vacuum system. When trapping volatile solvents, such as you might when using a rotary evaporator, you should use a cold trap with sufficient size and low enough temperature to prevent the vapors from entering the vacuum system. The most common cold trap utilizes dry ice in ethanol or isopropanol—this will provide cooling to −78 °C. (See Section 5.3.9 for handling cryogenics.) You should not use liquid nitrogen for this purpose since it could condense oxygen and can react violently with organics or explode when cooling is removed (see Figure 5.3.4.3).

You should use many of the same kinds of protection for vacuum systems as you use for high-pressure systems—chemical splash goggles, face shield, and protective shielding. The glass equipment must be designed for vacuum conditions; for example, some flasks are made with flat bottoms and these should not be used on rotary evaporators since flat surfaces are inherently subject to failure when the two sides are at different pressures. Always check equipment for cracks, scratches, or etches that could weaken the vessels and cause collapse under vacuum. Vacuum desiccators should be handled carefully after evacuation—you should wrap these with fiber or electrical tape to prevent flying glass in the event of implosion.

Vacuum pumps present their own hazards. Vacuum pumps are heavy and are often placed on floors—be sure that they are out of traffic paths to prevent trips and falls. Unguarded pulleys are

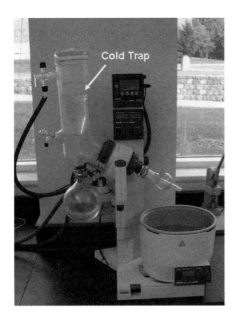

FIGURE 5.3.4.3   Cold Trap on Rotary Evaporator. In order to condense the vapors being removed from the reaction flask at the right (not shown in this image), the cold trap is often filled with dry ice in ethanol or isopropanol. The condensed solvent is collected in the flask under the cold trap and relatively little of the solvent vapor gets into the pump at the other end of the tubing.

dangerous and you should not use a vacuum pump that is not guarded. Body parts, clothing, and jewelry can be caught in this moving part and cause serious injury. Inadequate trapping of volatiles on the front end of a vacuum system can result in the release of those volatiles or toxic products back into the room, where the vacuum pump is located (if not otherwise vented), producing a hazard to those working there. Proper trapping can prevent this; however, it is good practice to also use local exhaust ventilation to remove any products of outgassing from the vacuum pump.

Working with Low- or High-Pressure Systems: RAMP

Some laboratories commonly use systems at high or low pressure. It is important to select the correct glassware or other equipment for the expected pressure. Predict, as best as you possibly can, the generation of gases for reactions in closed systems and have a plan to respond to unexpected explosions (or implosions), including an assessment of the toxicity of any chemicals released.

- *Recognize* equipment and experiments that involve high- or low-pressure conditions.
- *Assess* the risk of these hazards by considering the quality of the glassware used.
- *Minimize* the risk of implosions and explosions by the use of appropriate glassware, doing preexperiment calculations that predict expected pressures, and the use of appropriate personal protective equipment and other safety measures.
- *Prepare* to deal effectively with unexpected release of chemicals.

References

1. G. W. Carver. BrainyQuote; available at http://www.brainyquote.com/quotes/authors/g/george_washington_carver.html (accessed June 18, 2009).
2. American Industrial Hygiene Association, Laboratory Health and Safety Committee. Laboratory Incidents; available at http://www2.umdnj.edu/eohssweb/aiha/accidents/glass.htm (accessed September 20, 2009).
3. *Bretherick's Handbook of Reactive Chemical Hazards*, Vols. 1 and 2, 7th edition (P. G. Urben, ed.), Elsevier, New York, 2007.

QUESTIONS

1. Why is glass used in many high- or low-pressure reaction systems?

   (a) It is relatively inexpensive.

   (b) It is relatively inert.

   (c) It is easy to work with.

   (d) All of the above.

2. What is a disadvantage of using glass in high- or low-pressure systems?

   (a) Many gases tend to discolor the glass.

   (b) It is very hard to adequately evacuate glass systems due to leaks at ground glass joints.

   (c) The integrity of the glass can fail due to scratches, stress fractures, and accidental bumping.

   (d) Glass often fractures unexpectedly when immersed in liquid nitrogen.

3. Using a sealed glass tube requires

   (a) That the correct kind of glass be selected

   (b) Careful sealing of the glass to avoid failure under pressure

   (c) The use of relatively thick-walled glass

   (d) All of the above

4. After a sealed tube, high-temperature reaction has been run, the glass must be

   (a) Broken with a hammer

   (b) Immersed in liquid nitrogen and them allowed to crack due to thermal shock

   (c) Etched and broken in a chemical hood using adequate protective equipment

   (d) Cooled very slowly

5. An alternative to a sealed glass reactor is using

   (a) Standard taper glassware with plugs that are secured with wires

   (b) Standard taper glassware with plugs that are secured with rubber bands

   (c) Standard taper glassware with a rubber septum that provides a safety valve for high pressures

   (d) Glass reactors with screw-in Teflon® plugs and O-rings

6. Hydrogenation reactions done in academic labs are typically done on what scale of alkene or alkyne?

   (a) nanogram (ng)

   (b) milligram (mg)

   (c) gram (g)

   (d) kilogram (kg)

7. What safety equipment is commonly used with hydrogenation apparatus?

   (a) Safety shield

   (b) Chemical hood

   (c) Metal shields

   (d) Both (a) and (b)

8. Vaccum line systems

   (a) Can implode, but this is far less dangerous than an explosion

   (b) Can implode, but the attached pump usually prevents any chemicals from escaping

   (c) Can implode, which can send glass and chemicals flying in all directions in a lab

   (d) Cannot implode

9. When using vacuum systems, it is common to use traps

   (a) Since they limit the size of any implosion

   (b) Since they should prevent the release of any toxic materials into the lab atmosphere

   (c) To prevent air from entering the pump

   (d) That act as a ballast for the volume of the enclosed system

10. Vacuum pumps sometimes

    (a) Have unguarded pulleys that present a "moving part" hazard

    (b) Are vented into the lab atmosphere

    (c) Are placed on floors and present a trip hazard

    (d) All of the above

5.3.5

ELECTRICAL HAZARDS

Preview This section discusses common electrical hazards encountered in the laboratory.

Bring in the bottled lightning, a clean tumbler, and a corkscrew.

Charles Dickens from *Nickolas Nickleby* [1]

INCIDENT 5.3.5.1   ELECTROPHORESIS SHOCK[2]

A researcher was carrying out an electrophoresis operation when he received a powerful electrical shock. The electrophoresis unit was connected to a high-voltage source using a stackable banana plug that had no protective insulation or guarding. The shock occurred when his right elbow and right knee simultaneously touched the plug and another contact point. Investigations believe the shock could have been fatal if the contact points had been on opposite sides of his body.

INCIDENT 5.3.5.2   FATAL ELECTROCUTION FROM FAULTY CIRCUIT DESIGN[3]

A Ph.D. scientist was constructing a rack of lights with an electrical timer to study plant growth. Using an adapter that converts a three-prong plug into a two-prong (also known as a "cheater"), he connected the three-prong light rack to the two-prong timer, thus removing the grounding line in the circuit. The light fixture was rated at 800 mA but the lamps used were 1500 mA. The excess current drawn in the system likely melted some insulation in the transformer coil of the light ballast, which energized the ballast cover and the light fixture with almost 400 V. The system was not plugged into a GFCI outlet. The scientist grabbed the light fixture and probably brushed against a metal sink and was fatally electrocuted.

What lessons can be learned from these incidents?

Electrical Hazards are Common

Electrical hazards are frequently encountered in the laboratory, where there are many kinds of apparatus, equipment, and instruments that utilize electricity. The primary hazards of direct exposure to electricity are electrocution, electrical shocks, burns, or falls that are secondary to electrical shocks. Incident 5.3.5.1 illustrates the very real possibility of electrical shock from exposures in the laboratory. Although rare, there have been fatalities in the laboratory from unsafe practices with electrical apparatus and equipment.

Review of Basic Electrical Concepts

An electric current is the flow of electrons through wires, electrical devices, solutions, or other conducting materials. Current (I) is the number of electrons flowing past a point in a unit of time. A current of 1 ampere (A), known as the amp, is equal to 6×10^{18} electrons/second.

A current requires an electric potential, in volts (V), to drive or push the electrons through a circuit. The greater the potential or voltage, the greater is the flow of current. Electric potential is analogous to water pressure and the current is analogous to water flow. The greater the water pressure the faster and more powerful is the flow of water.

The flow of current depends on the resistance of the materials that it encounters as well as the shapes of these materials. Resistance (R) is measured in ohms (Ω). The relationship between voltage, current, and resistance is

$$V = IR$$

$$\text{volt} = \text{amp} \cdot \text{ohm}$$

Conductors are materials that have low resistance to current and electrons will flow freely through them if there is a circuit. Good conductors are copper, aluminum, iron, and ionic solutions, such water with salts and minerals. *Insulators* have very high resistance and it is difficult or virtually impossible for electrons to flow through these materials. Good insulators are rubber, plastics, glass, and ceramics.

Current cannot flow if it does not have a circuit. A circuit is a "loop" that allows the electrons to flow through until they get back where they started. Switches are often used to interrupt circuits. When you flip on a light switch you complete the circuit. Because components and connections in circuits sometimes fail, they are often connected to ground wires as a safety measure. These wires are attached to many other ground wires that ultimately are attached to a rod that is driven into the ground. So if a component fails, the flow of electrons can be diverted into the "ground" circuit.

There are two kinds of current, direct current (DC) and alternating current (AC). DC flows in one direction at a set potential, while AC flows in both directions at a defined frequency. Ordinary household and lab outlets are AC at 60 Hz (frequency) and about 120 volts.

Hazards from Electrical Contact

A current can injure or kill you if you become part of the circuit and as little as 0.1 A can be fatal. Your body offers low resistance to the flow of electrons, and if you give the current a path to a good ground connection you will become a conductor in completing the needed circuit. It is critically important that you use insulation as a means to prevent yourself from coming in contact with an electric current.

Tap water conducts electricity very well due to various ionic substances dissolved in it. (Very pure deionized or distilled water will have low conductivity, and the purity of water is often measured in this fashion.) Touching a live circuit with wet hands (and lower resistance) allows a greater current to flow. Otherwise stated, if V is constant, when R is smaller, I is larger: $I = V/R$.

Table 5.3.5.1 shows the effect of various currents on the body. Of particular note is the range of current where muscle contractions make is difficult to let go of the current source. You should know that the smallest fuses and circuit breakers are generally 15 amps so an electrical shock from a live electrical AC circuit has easily enough to cause the most serious effects, including death without tripping the circuit breaker.

Preventing Electrical Shock

Clearly, your efforts in dealing with electrical safety in the laboratory must focus on prevention. You must always seek to keep out of electricity's path. So let's talk about how you can get into the path from laboratory equipment or apparatus.

You probably know that electrical circuits have devices to protect against overcurrents (too many amps for the circuit) or from a ground fault condition (where the circuit inadvertently connects to

TABLE 5.3.5.1   Effects of Electrical Current on the Human Body[4]

Current in amps	Current in milliamps	Effect
0.001	1	Slight tingling
0.005	5	Slight shock; possible to let go
0.006–0.03	6–30	Painful shock; hard to let go
0.050–0.15	50–150	Extreme pain; cannot let go; death is possible
1–10	1000–10,000	Death is probable

FIGURE 5.3.5.1 Ground Fault Circuit Interrupter (GFCI). These outlets are required at any location near water sources or sinks, although older laboratories may not have these if they have not been upgraded to meet current electrical codes.

electrical "ground"). However, circuit breakers and fuses do not protect you from an electrical shock: they protect equipment and wiring and are designed to prevent fires due to sparking and overheating. Ground fault circuit interrupters (GFCIs) are designed to protect people because they instantaneously detect the wrong (false) path to ground (the "ground fault") and break the circuit before injury can occur (see Figure 5.3.5.1). Modern electrical codes require GFCIs near all sinks and sources of water since the likelihood of ground faults near water is much higher (since water conducts electricity so well).

You can get into the "path" to become part of the circuit if you ignore or are subjected to equipment or apparatus that (1) does not protect you from an existing electrical circuit, (2) is improperly wired, (3) is improperly grounded, or (4) is unprotected from ground faults.

Damaged electrical cords or improperly modified electrical cords can expose you to a circuit. This means when you bypass protections by using the wrong cord (e.g., a nongrounded two-prong plug versus a three-prong plug with a ground), you are putting yourself and others in jeopardy. This should help you understand why "lab-made" electrical apparatus can be dangerous if they do not provide the proper protections. Incident 5.3.5.2 provides a tragic example of a combination of several electrical "mistakes" that led to a fatality.

Using Electrical Equipment

Underwriter's Laboratory (UL®) certified equipment has met criteria for electrical safety and as long as you follow instructions for equipment and don't defeat or disable the protections, you should be safe with UL equipment. You should always be concerned about non-UL equipment, ungrounded equipment, or unprotected equipment or outlets that don't have ground fault protection. This is especially true about equipment and instruments near water and plumbing.

In the laboratory you may encounter equipment that does not always meet UL requirements or does not have adequate protection against circuits. Electrophoresis units may have uncovered, unguarded plugs, wires, or electrical clips that can provide electrical shocks, as in Incident 5.3.5.1 above. Lasers may have similar unguarded wiring. Other equipment such as electroporators, oil baths, electrolysis units, and other "homemade" laboratory apparatus that use unguarded electrical circuits are potential sources of electrical shock.

Look at your equipment to ensure that it is UL certified and it has not been modified in some way that could defeat that rating. If you have non-UL equipment or apparatus, you should seek the review of the electrical components by a licensed electrician. You should look to ensure that all equipment that operates around water or plumbing is plugged into GFCI outlets. Regularly inspect electrical cords for damage and ensure that these are the right cords for the job—they are not overloaded or compromised in some way. You should always turn off electrical power before touching wires or connecting electrical leads to an apparatus. Use one hand at a time to connect leads. Keep apparatus or equipment away from sinks or other water sources. You should never override safety devices or interlocks—they are there to protect you.

Electrical cords and electrical power strips can pose hazards if they are used in inappropriate places. Figure 5.3.5.2 shows a power strip lying on its side in a hood. What possible incidents could result from this situation? What if a flammable liquid were being used in the hood? What if a solution spilled in the hood? Figure 5.3.5.3 shows a commonly encountered situation where multiple power strips are plugged into a circuit and multiple cords are plugged into those power strips. This could result in a circuit overload or overheating in the cords.

Avoiding Electrical Contact in the Lab: RAMP

Electrical hazards of one kind or another are present in all labs. Using UL-certified equipment, not overriding safety measures, and using extra vigilance around sources of water usually prevents electrical accidents. Extremely high-voltage sources must be used with extra care.

If someone is in contact with an electrical circuit, do *not* touch them since you will be exposed to the same circuit. Seek to *safely* eliminate the source of the current before touching the patient or

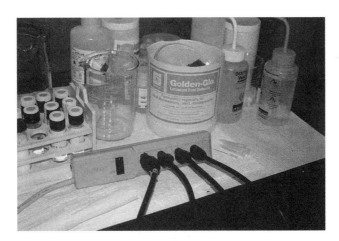

FIGURE 5.3.5.2   Outlet in Chemical Hood. A strip outlet in a chemical hood represents a fire hazard since the outlets and switch are not spark-proof and flammable chemicals are often used in hoods. The dangling cords might also present a trip hazard. A liquid spill could also present a shock hazard. (From *Journal of Chemical Health and Safety* **7**(5): 6 (2000). Used with permission. Copyright © 2000 Division of Chemical Safety and Health of the American Chemical Society.)

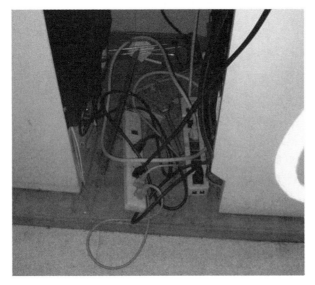

FIGURE 5.3.5.3   Outlets Clustered on Floor. It is common to use electrical strips since there are often multiple modules of an instrument(s) that need to have electricity. It is easy to forget that plugging in lots of equipment into these electrical strips could present a fire hazard. If the numerous cords are not neatly contained they can present other hazards such as trip hazards or shock hazards if a water spill enters into the strip.

rendering any first aid. See Sections 2.1.1 and 2.1.2 for more information about responding to electrical emergencies.

- *Recognize* unsafe electrical wiring or equipment and sources of high voltage or current.

- *Assess* very carefully the risks of home-built apparatus and/or sources of high voltage or current.

- *Minimize* the risk of exposure by *eliminating* the means by which you might contact electrical circuits.

- *Prepare* to respond to electricity accidents by knowing what to do and what *not* to do if someone gets shocked by an electric circuit.

References

1. *Bartlett's Familiar Quotations*, 17th edition, (Justin Kaplin, ed.), Little, Brown, and Co., Boston, 2002, p. 496, line 24.
2. American Industrial Hygiene Association, Laboratory Health and Safety Committee. Laboratory Safety Incidents: Electrical Shock from Electrophoresis Unit; available at http://www2.umdnj.edu/eohssweb/aiha/accidents/electrical.htm#Electrophoresis (accessed September 20, 2009).
3. American Industrial Hygiene Association, Laboratory Health and Safety Committee. Laboratory Safety Incidents: Electrocution Due to Improper Use of Equipment; available at http://www2.umdnj.edu/eohssweb/aiha/accidents/electrical.htm (accessed September 20, 2009).
4. National Institute for Occupational Safety and Health. Electrical Safety. Safety and Health for Electrical Trades Student Manual; available at http://www.cdc.gov/niosh/docs/2002-123/2002-123b.html#2 (accessed September 27, 2009).

QUESTIONS

1. The purpose of the "ground wire" in an electric circuit is
   - (a) To provide for the flow of electrons to the negative terminal of a battery
   - (b) To take the place of a circuit breaker in electronic configurations where breakers cannot be used
   - (c) To provide for the safe flow of electrons to a grounded line if a live wire touches part of an instrument or device that is not supposed to be energized
   - (d) All of the above

2. What value is enough current to cause tingling at the exposed site?
   - (a) 1 milliamp
   - (b) 10 milliamps
   - (c) 100 milliamps
   - (d) 1000 milliamps

3. The smallest fuses and circuit breakers are typically
   - (a) 15 milliamps
   - (b) 150 milliamps
   - (c) 1500 milliamps
   - (d) 15,000 milliamps

4. What current is high enough to cause death in some instances?
   - (a) 1 milliamp
   - (b) 10 milliamps
   - (c) 100 milliamps
   - (d) 1000 milliamps

5. Circuit breakers and fuses are designed to
   - (a) Protect humans from mild shocks
   - (b) Protect humans from severe shocks
   - (c) Protect humans from fatal shocks
   - (d) Protect circuits and equipment from overheating

6. Ground fault circuit interrupters
   - (a) Act as a "backup" system if a fuse fails
   - (b) Activate if a current to ground is detected
   - (c) Are not located near water sources since they would activate prematurely in many situations
   - (d) Activate in electrical systems that do not have fuses or circuit breakers

7. What condition can cause a scientist to become part of an electric circuit?
   - (a) Improper wiring
   - (b) Improper grounding
   - (c) Exposed electrical wires
   - (d) All of the above

8. What kind of lab equipment might have exposed circuits?
   - (a) Electrophoresis units
   - (b) Lasers
   - (c) Electrolysis units
   - (d) All of the above

9. Safety devices such as interlocks should
   - (a) Be used with electrical circuits that cannot have GFCIs or fuses
   - (b) Never be overridden
   - (c) Be used most of the time, except when very high voltages are involved
   - (d) Be used only when using devices near sinks and other water sources

10. If someone is in contact with an energized electrical circuit, it is best to
    - (a) Immediately grab the wire that is causing the exposure
    - (b) Immediately grab the person to remove them from the exposure
    - (c) Call 911 before doing anything
    - (d) Turn off all power to the circuit before touching the person

5.3.6

HOUSEKEEPING IN THE RESEARCH LABORATORY—THE DANGERS OF MESSY LABS

Preview This section discusses hazards arising in the lab from poor housekeeping.

If I had looked around the lab systemically, spotting hazards, these bottles would have been deemed inappropriately stored.

Jeremy White,[1] University College, London

INCIDENT 5.3.6.1   ACIDS STORED ON THE FLOOR[2]

A researcher and a colleague were working together in a laboratory when the colleague accidentally knocked over and broke a large glass bottle sitting on the floor. She slipped and fell into a liquid that was glacial acetic acid. She received some chemical burns but was released later from the hospital without long-term health effects.

What lessons can be learned from this incident?

Prelude

This section is about the topic of the best ways to maintain a safe working area in research laboratories. We will use the term "housekeeping" in a general way to describe the behaviors of laboratory workers with regard to keeping labs clean and orderly enough so as not to pose hazards for the occupants.

When Is "Messy" the Same as "Not Safe"?

Chemistry research labs in academic institutions are notoriously messy. The question to be considered here is: When does "messy" become "unsafe"? Let's look at some pictures to get some ideas.

We will start with Figure 5.3.6.1. What do you see that might be cause(s) for concern? Would you think this is messy or unsafe? What if you had to reach over this mess to get something? It seems likely you might knock something off onto the floor—maybe a bottle or perhaps a piece of glassware? While parts of the lab are very orderly (shelving to the right), clearly the primary bench in front is so cluttered it is difficult to find a place to work. Do you want to clean up after someone else, or even yourself, before you start your own work?

Now take a look at Figure 5.3.6.2. What kinds of unsafe things might lurk here? Broken glassware? Broken glassware in water is difficult to see. Exactly how would one make use of this sink without taking the glassware out? This also makes one wonder what was in all that dirty lab ware! Might contaminants simply get flushed down the sink instead of being disposed of properly?

Figure 5.3.6.3 shows another common housekeeping problem—the laboratory hood. Is there equipment that does not need to be there? Are chemicals being stored inside this that should be elsewhere? When the hood is so cluttered would you be tempted to do work in some other place? Because it is going to take effort to clean this up, would you perhaps even work outside the hood (for work that should be *in* the hood)?

Laboratory Safety for Chemistry Students, by Robert H. Hill, Jr. and David C. Finster
Copyright © 2010 John Wiley & Sons, Inc.

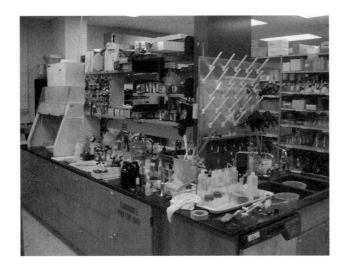

FIGURE 5.3.6.1   Messy Lab Bench. The bench is so cluttered as to make it unusable in its present state. Bottles might be bumped off the edge. Where would you be able to work?

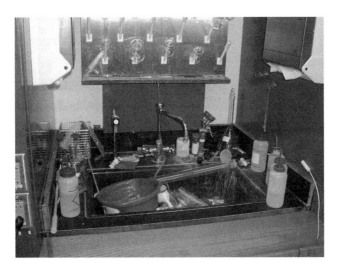

FIGURE 5.3.6.2   Cluttered Lab Sink. This sink is so cluttered with glassware that just trying to move something might result in breakage, and broken glass under water is hard to see. There might be odors coming from the sink. Obviously, the sink cannot be easily used without removing what is already in the sink.

FIGURE 5.3.6.3   Cluttered Lab Hood. Lab hoods are well-known sources of clutter. Unused flasks, bottle beakers, and other lab ware are often left behind from previous experiments so that the user has to move things about to make room for his/her new work. What will you do if you were in a hurry and did not want to take time to clean this up?

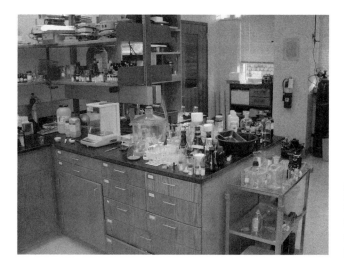

FIGURE 5.3.6.4 Cluttered Laboratory. Clearly, there are many bottles, vials, and containers at the edge waiting to be easily bumped onto the floor. There are other hazards in this lab. Can you spot them?

FIGURE 5.3.6.5 Blocked Laboratory Exit. How rapidly could you exit through this door in an emergency? In the daily routine of working in a lab it is easy to move something to a space that is not being used. Make sure that you don't block exits and that you have a clear access for emergencies.

Figure 5.3.6.4 shows another cluttered lab bench. The bottles on the edge of the bench are just waiting for someone to knock them off. Also, note the materials high up on the top shelf. Some of that is heavy and it might be difficult to retrieve without presenting an unnecessary hazard and risk. Is the gas lecture bottle stored properly? See anything on the floor that might be a tripping hazard?

The last two photos, Figures 5.3.6.5 and 5.3.6.6, have clearer safety hazards—a blocked exit and a blocked safety shower/eyewash station.

Why are some labs not in an orderly condition? There are several reasons for this, some of which may offer some reasonable justification for the condition of the lab and some of which do not. First, academic labs are occupied primarily by students who have little training in safety and may not recognize the hazards associated with messy labs. In fact, most of their "experience" in labs prior to working in a research lab has been in labs associated with courses. These labs are usually well organized with relatively few chemicals on hand in order to minimize hazards and optimize safety. It is not surprising, then, that students entering a research lab would not naturally and readily take it upon themselves to keep a lab organized and clean.

Second, labs are generally supervised by chemistry faculty who may or may not actually spend much time in the lab. Chemistry professors, like students, have fairly erratic or weak education in safety and (unfortunately, in our judgment) some of them place modest value on issues of safety, particularly in comparison to the goal of producing results. The degree, and quality, of supervision varies with the result that many academic research labs actually have little supervision or guidance with regard to the issue of safety. Labs under faculty supervision might take on the same state of "orderliness" as

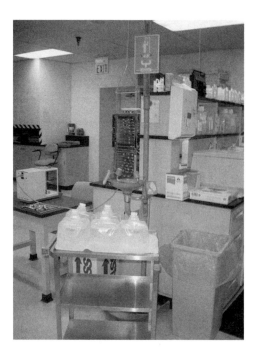

FIGURE 5.3.6.6   Cluttered Laboratory. In addition to the carts blocking the safety eyewash and safety shower, there are several other unsafe conditions in this laboratory. Can you spot them? (Courtesy of Harry J. Elston, Ph.D., CIH, Midwest Chemical Safety LLC, Buffalo, IL.)

the professor's office (and a casual survey of faculty offices usually reveals an extraordinary spectrum from "neat and clean" to "fire hazard").

Third, academic labs are often shared by multiple students with multiple sets of chemicals and multiple sets of lab apparatus. A lack of awareness about the experiments of others can lead to unsafe conditions since dangerous reagents might be left on benches or in hoods without the consent or knowledge of everyone working in the lab. And the old saying applies here—"When a committee is in charge, nobody is in charge." Or in this case, when a group of students are left to work in common areas no one feels the responsibility to clean up.

Finally, the nature of academic research is such that it is usually a sporadic activity that is far from the daily 9 to 5 activity that is more associated with chemistry labs in industry. Research projects are attended to in the context of other labs and courses and it is easy to find oneself "squeezing in" research activities into hourly or half-day sessions. This schedule makes it easy to focus on productivity instead of safety and leaving experimental setups in "mid-experiment" at times, which leaves labs with chemicals and equipment "out on the bench" and not always in a safe fashion.

These four factors—limited safety education, limited supervision, multiple lab workers, and sporadic activity—all make academic labs prone to "messiness." While perhaps understandable, this nonetheless often leads to unsafe environments where chemicals are not stored properly and experiments-in-progress are left in unsafe conditions.

How should labs look? Labs with active research projects in progress likely will not have empty hoods and clean benchtops, with every piece of equipment and glassware not in use neatly stored in cupboards and drawers. An active lab probably *should* look like "something is going on in here." Perhaps not all glassware is cleaned immediately after use and perhaps there are a few solutions or solvents being temporarily stored in chemical hoods. Lab notebooks might well be left on benchtops. The question becomes "Is this a safe lab?" not "Is this a pristine lab?"

Possible Effects of Messy Labs

Let's consider some conditions of messy labs that create hazards.

Sometimes, perhaps frequently in labs where students have only a few hours at a time to work on research projects, it is not possible to completely "finish" a given experiment or completely clean up after an experiment is run. Leaving a small amount of a solvent in a capped bottle that is appropriately labeled in a fume hood for one day is probably not unsafe. Leaving small amounts of nonvolatile reagents in capped or covered containers appropriately labeled on a benchtop is probably not unsafe. Not putting away an unused hot plate is not unsafe. What conditions start to create unsafe environments? Here are some examples:

- On a benchtop, 200 mL of unused solvent is left in an uncovered beaker. If this is volatile (as most solvents are) then it will slowly vaporize and create an unsafe breathing environment. If this is flammable (as many solvents are) then the vapors might travel and seek out an ignition source. It is also susceptible to just being knocked over and spilled.

- A 5-L jug of concentrated nitric acid might conveniently be placed on the floor (particularly if there isn't *room* on the benchtop or in a chemical hood for temporary storage) and the hazard of this is obvious if someone inadvertently kicks the jug and causes a spill.

- A solution of weak acid (or base, or anything) is temporarily left in a beaker *without a label*. Someone else might spill this or assume that the contents are known and use it or dispose of it improperly (perhaps just to use the container!). Using a solution that is the wrong solution or pouring something down the sink that shouldn't be poured down the sink can have many unpleasant and dangerous effects.

- Used glassware that has not been cleaned both takes the glassware out of service for another's use and may create unpleasant and unsafe odors in a lab.

- Larger pieces of equipment, such as a 5-L Dewar flask or a safety shield, might temporarily be placed on the floor, perhaps even near an exit door (Figure 5.3.6.5). These can become trip hazards and might impede rapid exit during an emergency.

- Spills that are not cleaned up create hazards that may or may not be obvious to other lab personnel. Solutions that evaporate may leave unsafe residues. Solids dropped around the common analytical balance can present a hazard to the next users of the balance.

- A hood is so cluttered or obstructed by equipment and materials that experiments with volatile or toxic materials that should be conducted within hoods are carried out on an open bench.

- Shelves above shoulder height are loaded with heavy materials, books, or equipment such that they could be overloaded and collapse or with vibration fall on people below.

Finally, in labs with multiple users, all of whom leave multiple solutions, reagent bottles, and clean or dirty glassware on benches or in hoods collectively reduce the "operating space" in the lab. A cluttered hood used for temporary storage ultimately becomes a space where the intended function of the hood—a location to conduct experiments—is compromised. Benchtops that are so cluttered that experiments cannot be set up properly become spaces where incompatible chemicals come in close contact or flammable solvents find ignition sources. Capped and covered solvents are more likely to be knocked over and create hazardous situations. Unused stirplate/hot plates or other electrical equipment usually have dangling electrical cords that take up space in an unwieldy fashion.

Authorities such as the National Research Council have long concluded that there is a correlation between housekeeping (orderliness) and safety.[2] Incidents, such as the one cited at the beginning of this section, where housekeeping is a cause or contributing factor, are largely anecdotal. Over time, though, many organizations have come to assume that good housekeeping is a good indication of a safe workplace.

Nevertheless, there have been studies in industrial areas that have established a clear relationship between housekeeping and incidents. One study found that increased housekeeping resulted in fewer lost workdays from injuries, and based on this study an objective improvement in housekeeping of 20% resulted in almost 15% fewer injuries.[3] Also noted was that productivity, quality, and morale were improved with better housekeeping.

An even more striking study over a 3-year period found that almost 46% of injuries were associated with poor housekeeping.[4] The author concluded that poor housekeeping was a frequent source of incidents and safety monitoring should be implemented to maintain improved housekeeping. He goes on to suggest that as part of "process safety" (see Section 7.3.7) all incidents, however minor, should be monitored since these are potential indicators of a larger problem.

Falling Is Not Graceful

Slips, trips, and falls are often the leading cause of significant injuries in a workplace, and this can be true in laboratories—particularly those with poor housekeeping and storage practices. Falls can result in serious injuries, including broken bones, and lost time from the laboratory. Chemists are a frugal bunch and they don't like to throw away things. When coupled with the fact that most laboratories have limited storage space, the result is that containers, boxes, bottles, cans, and even books seem to find their way to the floor, where they can become trip hazards. Lab equipment not being used sometimes finds its way to the floor. Electrical cords and computer cables can become trip hazards if they are not taped down and kept from traffic areas. Sometimes the clutter ends up blocking access to work areas, hoods, emergency equipment, or even second doorways. Recognition of hazards and persistence in preventing them are required to keep your laboratory safe.

Keeping Labs Safe

First, and most importantly, you are responsible for your own safety in any lab. If the behavior of supervisors (or professors) and co-workers (other students) creates a hazardous lab situation, that is something that needs to be addressed. Don't let the poor habits of others or an "unsafe culture" of a particular lab jeopardize your safety. Conducting your own periodic "scene survey" or inspection (see Section 8.3.1) can be useful in maintaining a safe lab and protecting yourself.

The rules are pretty easy to construct (and perhaps this should be a periodic "training session" for any group of lab workers):

- Lab benchtops and hoods are not storage areas. Some temporary storage in ongoing research projects is likely inevitable but this must be managed carefully.
- Lab benchtops and hoods should not become so cluttered with chemicals and equipment that conducting experiments becomes hazardous.
- Keep floors and exit pathways clear.
- Clean glassware and equipment immediately (or as soon as reasonably possible).
- Dispose of lab waste in an appropriate manner.
- All reagents and solutions should be clearly labeled.

Good housekeeping will also maintain ready availability to safety equipment and first aid materials. Cluttered lab spaces may impede their access and use.

RAMP

- *Recognize* hazards in a laboratory arising from poor housekeeping.
- *Assess* the risks of these hazards and how you might be exposed.
- *Minimize* the risk of lab accidents and exposure with good housekeeping practices.
- *Prepare* to deal with emergencies by making sure that safety and first aid equipment are readily accessible and exit pathways are not blocked.

References

1. L. DICKS. Keeping Safe: Some Cautionary Tales. ScienceCareers.org, August 4, 2006; available at http://sciencecareers.sciencemag.org/career_magazine/previous_issues/articles/2006_08_04/noDOI.928066 2141629900561 (accessed August 26, 2006); or alternatively go to http://sciencecareers.sciencemag.org/career_magazine/previous_issues, click on 2006, then click on August 4, then click on "Keeping Safe: Some Cautionary Tales."

2. National Research Council. *Prudent Practices in the Laboratory: Handling and Disposal of Chemicals*, National Academy Press, Washington, DC, 1995, p. 84.

3. P. McCON. Housekeeping and injury rate: a correlation study. *Professional Safety* **42**(*12*):29–33 (1997).

4. K. SWAT. Monitoring accidents and risk events in industrial plants. *Journal of Occupational Health* **39**: 100–104 (1997).

QUESTIONS

1. What factors contribute to "messy labs" academic environments?

   (a) Many of the occupants are students with limited training in safety.

   (b) Faculty mentors offer varying degrees of supervision in labs.

   (c) "Productivity" can sometimes override "safety and cleanliness."

   (d) All of the above.

2. Why is "messy" not automatically "dangerous"?

   (a) Because most labs are, in fact, somewhat messy yet perfectly safe.

   (b) Because it is not practical to keep labs clean.

   (c) Because a "working lab" is likely to have experiments in preparation, in progress, or in a cleanup stage that doesn't necessarily present significant hazards.

   (d) Because some "clean" labs are dangerous, too.

3. Which situation is *not* dangerous?

   (a) A bottle of strong base placed on the floor

   (b) An unlabeled beaker of a clear liquid

   (c) A spill that is contained inside a chemical hood

   (d) None of the above

4. Using a chemical hood for storage space is acceptable when

   (a) The chemicals being stored are nonvolatile

   (b) There is no other safe place to store chemicals

   (c) No more than one-half of the surface area is occupied

   (d) None of the above

5. Which has been associated with "orderliness" in the work environment?

   (a) Fewer "lost work" days

   (b) Fewer injuries

   (c) Improved morale

   (d) All of the above

6. When is it appropriate to store boxes of chemicals on the floor in a laboratory?

   (a) When they are too heavy to lift by only one person

   (b) When they will be stored there for no more than 24 hours

   (c) When a sign is posted just above the location indicating the contents of boxes

   (d) Never

5.3.7

NONIONIZING RADIATION AND ELECTRIC AND MAGNETIC FIELDS

Preview   This section presents an overview of regions of nonionizing electromagnetic radiation and electric and magnetic fields and their hazards in chemistry labs.

In the right light, at the right time, everything is extraordinary.

Aaron Rose, film director and art curator[1]

INCIDENT 5.3.7.1   UV LIGHT EXPOSURE[2]

A chemist was working in a new laboratory using gels for purifying DNA on a light box with an ultraviolet (UV) light source. The new researcher decided to follow the example of the other researchers in the lab who did not wear safety goggles or any other kind of specialized eye protection. She continued to cut and remove the DNA bands throughout the day, but in the latter part of the day her eyes started to itch. After leaving work her eyes got worse and she felt like something was in her eyes. She was taken to the emergency room when the pain had become so great that she could not open her eyes and was diagnosed with burns from UV light exposure. Fortunately, she recovered without permanent damage. When she returned to the laboratory she learned that the other researchers who had done this work had worn their own personal eyeglasses, which had at least partially protected them from the UV radiation.

What lessons can be learned from this incident?

Prelude

The topic of radiation is generally divided into the two main categories of ionizing radiation and nonionizing radiation. The section discusses only nonionizing radiation or regions of the electromagnetic spectrum with relatively low energy that have little or only modest interaction with living cells. Section 5.3.8 discusses ionizing radiation.

Since radiofrequency radiation is also associated with the effects of electric and magnetic fields, we also discuss the presence and hazards of these fields in this section.

Low Photon Energy = Low Hazard

Electromagnetic radiation appears as a stream of photons, each of which has an energy related to the wavelength and frequency using

$$E = h\nu = hc/\lambda$$

where E is the energy (J) of the photon, h is Planck's constant (6.626×10^{-34} J $\cdot$ s), c is the speed of light (3.00×10^8 m/s), ν is the frequency (s^{-1} or Hz, hertz), and λ is the wavelength (m). Photons with relatively long wavelengths (and low frequency) will have relatively low energy. In this section we briefly discuss photons with wavelengths in the visible region and longer. (Section 5.3.8 discusses higher-energy photons and radiation.) See Figure 5.3.7.1 and Table 5.3.7.1.

The exact boundaries of these regions are not consistently reported in various sources, although in some instances (such as in regulations promulgated by the Federal Communication Commission) very

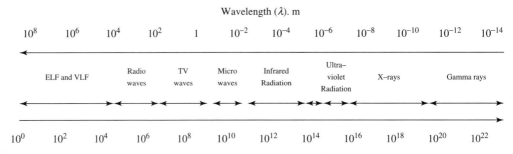

Frequency (ν), Hertz = Hz = s^{-1}

The visible spectrum is the narrow band between Infrared and Ultraviolet radiation

FIGURE 5.3.7.1   Electromagnetic Spectrum. The wavelengths/frequencies of various kinds of radiation are shown in the diagram.

TABLE 5.3.7.1   Features of Nonionizing Radiation in the Electromagnetic Spectrum

Region	Frequency	Frequency	Wavelength
	(Hz)	(MHz, 10^6 Hz; GHz, 10^9 Hz; THz, 10^{12} Hz; PHz, 10^{15} Hz)	
Low frequencies	1–30,000		3×10^8–1×10^4 m
RF			
Radio, AM	3×10^5–1.5×10^6	0.3–1.5 MHz	1000–200 m
Shortwave	1.65×10^6–5.4×10^7	1.6–54 MHz	187–5.55 m
TV, FM radio	5.4×10^7–8.9×10^8	54–890 MHz	5.55–0.187 m
Cell phones		(824–848 MHz)	
Radar	1.3×10^9–1.6×10^9	1.3–1.6 GHz	
Microwaves (some sources list microwaves as a subset of RF, allowing RF to go up to 300 GHz)	1.6×10^9–3×10^{11}	1.6–300 GHz	187–1 mm
IR	3×10^{11}–4×10^{14}	0.3–400 THz	1 mm–750 nm
Vis	4×10^{14}–7.5×10^{14}	400–750 THz	750–400 nm
UV	7.5×10^{14}–3×10^{15}	0.75–30 PHz	400–100 nm
UV-A (near UV)	7.5×10^{14}–9.4×10^{14}	0.75–0.94 PHz	400–320 nm
UV-B (middle UV)	9.4×10^{14}–1.1×10^{15}	0.94–1.1 PHz	320–280 nm
UV-C (far UV)	1.1×10^{15}–3.0×10^{15}	1.1–3.0 PHz	280–100 nm

specific ranges of frequency are designated. It is helpful to remember that these definitions are human inventions; it is the behavior of the radiation (not the label assigned to it) that is important. We will use the common terms shown in Figure 5.3.7.1 with the frequency and wavelength limits indicated in Table 5.3.7.1.

We will discuss these regions in the order of increasing energy.

Extremely Low Frequency and Very Low Frequency Radiation

Extremely low frequency (ELF) and very low frequency (VLF) photons have frequencies in the range of approximately 3–30,000 Hz. This radiation is rarely used in an experiment in a chemistry lab, except for the ubiquitous presence of 60-Hz fields associated with all common AC electricity. The World Health Organization concluded that there is "limited" evidence of an association of ELF and childhood leukemia—meaning there is at least a single study that suggests an association.[3] Health and safety studies about ELFs and VLFs are also commonly referred to as "EMF" (electromagnetic field) studies. Static magnetic and electric fields are discussed below. See *Special Topic 5.3.7.1* Effects of Nonionizing Radiation on Humans.

SPECIAL TOPIC 5.3.7.1

EFFECTS OF NONIONIZING RADIATION ON HUMANS

One of the main points of this section on nonionizing radiation is that *in lab settings* (with the exception of light from lasers) the various regions of the electromagnetic spectrum from radiowaves to visible light pose no special hazards. From a toxicological point of view, we would say that in the lab you'll likely get exposed only to fairly low doses on an acute time scale. A larger question that both OSHA (for occupational settings) and our modern society faces is the effect of low doses of these low-energy forms of radiation given over a longer period of time; this is *chronic* exposure.

The opportunities for concern are considerable. We are all exposed to 60-Hz AC fields due to line voltage that surrounds us in many situations, but perhaps with the exception of sleeping under an electric blanket (with close proximity and a "continuous" dose) the electric and magnetic fields from electric lines are miniscule. Concerns have been raised, however, about living in the near vicinity of very-high-voltage power lines. There has also been concern raised about the hazard level for police officers who use radar guns. And nearly everyone now uses a cell phone where 800 MHz radiation is being generated in close proximity to the brain, in some cases perhaps for hours each day. And, of course, we are all familiar with the "sunburn" that occurs when skin is exposed to excess UV radiation (either from the sun or in a "tanning booth").

With the exception of the known correlation between "tanning" and subsequent increase in risk for skin melanoma (cancer), epidemiological studies have not found convincing evidence that nonionizing radiation is linked to various human ailments. But, as with chronic toxicity studies for chemicals, the absence of convincing evidence is not definitive proof of safety (just as "acquittal" is not the same "proof of innocence" in our system of jurisprudence). OSHA and others continue to study these possible relationships between chronic exposure and nonionizing radiation. Animal studies and epidemiological studies attempt to examine the doses of radiation and possible carcinogenic, reproductive, neurological, cardiovascular, brain, and behavior, hormonal, and immune system effects.

Radiowaves or Radiofrequency Radiation

Radiofrequency (RF) radiation occupies the region of approximately 30 kHz to 300 MHz in the electromagnetic spectrum. The most common use of RF is in nuclear magnetic resonance (NMR) spectrometers.

The frequency of radiation used in NMR depends on the field strength of the magnet used in the instrument. Frequencies range from 60 to 600 MHz for commercial instruments. Modern NMR equipment uses RF pulses in Fourier transform systems. There are no safety concerns associated with these frequencies of radiation.

Microwave Radiation

Microwave (MW) radiation occupies the region of approximately 1.6–30 GHz in the electromagnetic spectrum. EPR instruments use radiation in several bands in the gigahertz range: 1–2 GHz (L-band), 2–4 GHz (S-band), 8–10 GHz (X-Band), 35 GHz (Q-band), and 95 GHz (W-band), although the 8–10 GHz range is the most common. There are no safety hazards associated with this radiation.

Microwave spectroscopy is less commonly used but can gather information about the rotational states of gas-phase molecules. These are often "home-built" spectrometers found in research labs.

Microwave ovens are now being used in chemistry labs as a source of heat. These ovens produce radiation at 2.54 GHz. Federal standards limit the amount of radiation that can leak from an oven to levels far below those known to harm people. The only hazards associated with these ovens are the problems of arcing if a metal object is placed inside or the oven door interlock switch is damaged or overridden, which would allow the oven to generate radiation while the door is open.

Infrared Radiation and Visible Light

Infrared (IR) radiation and visible light are used frequently in chemistry labs in spectrometers to analyze molecular vibrations and electronic transitions in atoms, molecules, and salts. These instruments generally pose no safety hazards.

We began this section with the idea that low-energy photons present small hazards. While generally true, this should be modified to include the photon flux as the energy of the radiation is increased. Thus, while ordinary visible light is generally considered harmless, very high intensity light (with high photon flux) can be hazardous. This is certainly true for light from lasers. See Section 7.3.3 for more specific information about working with lasers.

Ultraviolet Radiation

Ultraviolet (UV) radiation is generally considered to be "nonionizing" but at the high-energy end of the UV spectrum, photons will have enough energy to ionize valence electrons in atoms and molecules. Ultraviolet photoelectron spectroscopy (UPS) is an example where a helium discharge lamp emits radiation at 21.2 eV (58.5 nm, 5.12×10^{15} Hz) and this radiation examines the valence electrons in molecular orbitals.

Like IR and visible radiation, UV light is used in spectrometers (sometimes in conjunction with visible light). In these instruments, there is little hazard from the UV source. However, there are many other sources of UV light in labs and sometimes these are fairly intense sources. Common sources are "UV light boxes" or "transilluminators" that are used to observe substances that fluoresce under UV light. These are also sometimes called "black light" boxes since they produce little or no visible radiation. UV lamps can also be used to kill bacteria (germicidal lamps) or as a source of energy in chemical reactions and these reactions must be carried out in quartz flasks since ordinary lab glass absorbs UV light.

UV light boxes, transilluminators, and UV lamps can emit intense radiation and it is necessary to use PPE or other shields to avoid excess skin and eye exposure to UV radiation. UV light boxes typically project UV light onto a TLC plate or other fluorescing item and the resulting visible light is viewed through a window that filters UV light. The light sources for light boxes are typically not affixed permanently to the boxes. It is easy to remove them and, of course, the protective shielding of the box is absent. These devices can be used without any special eye protection. Transilluminators and other UV sources are unenclosed systems and should only be used while wearing eye protection specially designed to filter UV light. UV-filtering shields are also available. Clothing generally protects the skin adequately, but the use of body shields and face shields may be necessary to protect exposed skin on hands, arms, and the face and neck.

UV light, of course, is invisible, but some sources emit enough visible radiation that it is possible to "see where the UV is." This is not reliable, however, since sometimes the visible radiation is too weak to be readily observed. It is important, then, to know when the source is "on" and exactly how much of the UV light is dispersed.

The damage from UV light exposure is dependent on wavelength and dose (which depends on intensity and distance from the source). The most damaging region is 200–315 nm, in the UV-B and UV-C regions (see Table 5.3.7.1); many UV sources emit bands with peaks at either 254 nm ("shortwave UV") or 365 nm ("longwave UV"). UV will cause damage to the skin in the form of "sunburn." Damage to the cornea occurs most acutely at 270 nm with the effect of aversion to bright light, itching, and pain. Corneal eye cells are replaced about every 48 hours so the effect in usually nonpermanent.

Extremely intense beams of IR, visible, and UV light are possible using lasers and this situation is quite common in advanced and research labs. Extra precautions must be taken to avoid eye contact with laser beams and this is discussed more fully in Section 7.3.3.

UV light is capable of generating ozone from the oxygen in air. A sufficiently bright source for at sufficiently long time could generate trace amounts of ozone, which is a toxic gas. The odor threshold (0.005–2 ppm) spans the TLV range (0.05–0.2 ppm) so odor detection is not a reliable method to prevent overexposure. Along with many other good reasons for lab ventilation, this is a reason to use UV lamps only in adequately ventilated areas.

Electric and Magnetic Fields

The presence of static magnetic fields of any consequence in labs is uncommon, except for modern NMR instruments. Magnetic fields are measured in either gauss (G) or tesla (T, 1 T = 10,000 G). The

Earth's magnetic field ranges from 0.3 to 0.7 G. The field strength inside the probes of superconducting magnets used with NMR instruments ranges from 7.1 to 14.2 T (for instruments with proton frequencies of 300–600 MHz, respectively). Newer NMR instruments have very good shielding capabilities; older instruments have poorer shielding. While there is no evidence that strong magnetic fields have any negative biological effect (and MRI patients are in fields ranging from 1.0 to 3.0 T) there are secondary hazards. Persons with heart pacemakers or other medical implants that could be affected by magnetic fields should not get close to NMR magnets. It is common to define, for example, a "5-gauss" line around the magnet in all directions, beyond which these persons should not enter. (Although not safety-related, magnetic bank or ID cards or ferromagnetic substances in general should stay outside this line.)

As noted above with regard to ELF and VLF radiation, there is no evidence that electric fields have an effect on humans (except in cases where pacemakers are used). In fact, there are no common lab situations where laboratorians would be exposed to strong static electric fields.

Working with Nonionizing Radiation: RAMP

With the exception of UV light, there are few hazards that are associated with microwave, RF, and IR radiation, visible light, and electric and magnetic fields. UV light should only be used with appropriate forms of shielding for the eyes and skin.

- *Recognize* sources of UV radiation.
- *Assess* the level of risk for exposure based on intensity, exposure duration, wavelength, and shielding.
- *Minimize* the risk for exposure by using shielding.
- There is no particular medical response to excess UV exposure to the cornea other than rest. Thus, the best *preparation* for an emergency is to make sure it cannot happen. Skin exposure can be treated as one treats a sunburn.

There are many web sites that present information about nonionizing radiation. Here are two from OSHA and the National Institutes of Health (NIH): http://www.osha.gov/SLTC/radiation_nonionizing/index.html; http://www.niehs.nih.gov/health/topics/agents/emf/.

References

1. Aaron Rose. The Quotations Page; available at http://www.quotationspage.com/quotes/Aaron_Rose/ (accessed September 20, 2009).
2. J. KING. Wear Your Safety Goggles. ScienceCareers.org, August 4, 2006; available at http://sciencecareers.sciencemag.org/career_magazine/previous_issues/articles/2006_08_04/noDOI.12595683603361266879 (accessed August 26, 2006); or alternatively go to http://sciencecareers.sciencemag.org/career_magazine/previous_issues, click on 2006, then click on August 4, then click on "Wear Your Safety Goggles."
3. World Health Organization. Extremely Low Frequency Fields: Environmental Health Criteria, No. 238, WHO 2007; available at http://www.who.int/peh-emf/publications/elf_ehc/en/index.html (accessed January 27, 2009).

QUESTIONS

1. Which part of the electromagnetic spectrum is *not* considered to be nonionizing radiation?

   (a) Microwaves
   (b) Short wave radiofrequencies
   (c) X rays
   (d) Radar

2. The 60-Hz frequency associated with standard electrical current is categorized as what band of radiation?

   (a) ELF
   (b) VLF
   (c) Shortwave

   (d) None of the above

3. What instrument used in chemistry laboratories uses RF radiation?

   (a) EPR instrument
   (b) Microwave spectrometer
   (c) NMR instrument
   (d) Lasers

4. Which devices emit potentially dangerous levels of UV light?

   (a) Light boxes for TLC
   (b) Transilluminators

(c) UV lamps for use in chemical reactions in quartz flasks

(d) All of the above

5. Damage to the cornea occurs mostly from which band of UV light?

(a) A

(b) B

(c) C

(d) None of the above

6. Generally, static magnetic fields

(a) Have biological effects only at field strengths greater than 1 G

(b) Have biological effects only at field strengths greater than 1000 G

(c) Have biological effects only at field strengths greater than 1 T

(d) Have no biological effects

5.3.8

AN ARRAY OF RAYS—IONIZING RADIATION HAZARDS IN THE LABORATORY

Preview   This section presents an overview of radiation properties and hazards in chemistry labs.

This incident was the culmination of multiple failures of the NIST safety procedures and policies and their application, which could have prevented or mitigated the incident. . . . It is apparent that no one seriously considered the possibility of a spill, and no response to such an event was planned.

J. Michael Rowe,[1] nuclear safety consultant, referring to Incident 5.3.8.1

INCIDENT 5.3.8.1   PLUTONIUM SULFATE SPILL AND CONTAMINATION[2]

A researcher unintentionally cracked a glass vial containing 530 mg of plutonium sulfate with a total radioactivity of 44.1 millicuries. After placing the vial in a can and putting it in a locked cabinet, he moved the contaminated notebook to another location, washed his hands in a sink, used the men's room, and went to his office. Later, a second researcher examined the can using ungloved hands to confirm the vial was broken. The hands of the two researchers and the lab sink were found to be contaminated with plutonium. Thirteen other lab researchers used the lab before they learned about the possible radioactive contamination. Fearing contaminated shoes, the lab workers were instructed to remove shoes and socks in a nearby hallway and subsequently walked on the contaminated floor. Later, 29 employees were tested for contamination. The lab was sealed until further investigation could take place by qualified personnel. It was determined that about 60 mg of the plutonium was missing.

What lessons can be learned from this incident?

Prelude

The topic of radiation is generally divided into the two main categories of ionizing radiation and nonionizing radiation. The section discussed only ionizing radiation. As you might guess, ionizing radiation is a form of radiation capable of producing ions when interacting with atoms and molecules. This radiation is more energetic than nonionizing radiation, and therefore more capable of causing damage to living systems. Nonionizing radiation is represented by the "lower energy" regions of the electromagnetic spectrum: visible, infrared, microwave, and radiowave regions.

Types of Radiation—Waves and Particles, and Both

While electromagnetic radiation, in particular, visible light, was understood as a wave phenomenon in the mid-1800s through Maxwell's equations, it was not until 1896 when Becquerel discovered radioactivity and in 1899 when Rutherford identified alpha and beta particles that the larger picture of wave-based and particle-based radiation started to be understood. However, early on it was only understood that "something" was being emitted from radioactive substances and whether the emissions were particles or electromagnetic radiation was unclear. The early names, alpha, beta, and gamma "particles" (still in

TABLE 5.3.8.1   Forms of Ionizing Radiation and Their Sources

Radiation type	Mass (amu)	Charge	Range of energy	Radioactive sources	Devices
Alpha, α	4.0001	+2	4–6 MeV	^{210}Po, ^{222}Rn, ^{238}U, ^{241}Am	
Beta, β	0.000549	−1	eV–4 MeV	^{14}C, ^{32}P, ^{35}P, ^{228}Ra	Cathode ray tube
Gamma, γ	0	0	eV–4 MeV	^{99m}Tc, ^{60}Co	
X ray	0	0	eV–0.1 MeV		X-ray diffraction, X-ray fluorescence, electron microscope
Neutron, n	1.0009	0	MeV–MeV	Fission reactor (^{235}U), ^{252}Cf, ^{241}Am/Be	

use today), reflect the uncertainty about the nature of these emissions at the time of discovery and the assumption that they were in fact particles. "X rays" were named with an "X" since no one knew what they were when first discovered. As you likely now know, alpha and beta particles are, in fact, particles with mass and charge. Gamma "particles" (and X rays) are photons of high-energy electromagnetic radiation. Neutron "radiation" is, of course, a stream of particles.

The work of Planck, Einstein, and deBroglie in the early 1900s brought more clarity to the understanding of the wave nature of particles and the particle nature of waves. Even today wave–particle duality is certainly an unintuitive concept since we don't experience the dual nature of matter and energy in our macroscopic world. But, even with this modern understanding, we still use the terms "beta radiation" and "gamma rays," for example, even though the simplest understanding makes one think of beta *particles* as electrons (with mass and charge) and gamma *radiation* (without mass or charge).

Table 5.3.8.1 helps us organize the array of kinds of ionizing particles and forms of electromagnetic radiation. In this table the charge of an electron is −1 or 1.602×10^{-19} coulombs.

Radioactive decay processes involve the emission of a particle and/or photon (a gamma ray) from the nucleus of an atom. (See *Chemical Connection 5.3.8.1: Radioactive Decay*—A First-Order Reaction). Alpha decay is the ejection of an alpha particle from the nucleus of the atom (Equation 5.3.8.1) and produces a "daughter" nucleus that has two fewer protons and a decrease of four mass units. The velocity of the alpha particle accounts for the energy range of 4–6 MeV shown in Table 5.3.8.1. While alpha radiation can cause damage to tissues, it can only do so if the source is ingested or inhaled because the energy of alpha emitters is usually very weak and can readily be stopped by a sheet of paper.

$$^{224}_{88}\text{Ra} \rightarrow {}^{220}_{86}\text{Rn} + {}^{4}_{2}\text{He} \tag{5.3.8.1}$$

Beta decay (Equation 5.3.8.2) occurs when a neutron in the nucleus spontaneously changes to a proton and an electron and the electron is ejected from the atom. Thus, the atomic number increases by one, but the overall mass change is zero. (Strictly speaking, the mass will change by a very slight amount in any nuclear process and can be calculated using $\Delta E = \Delta mc^2$. These mass changes are neither easily measured nor important in chemistry labs.) The energy of the particles will be dependent on the applied accelerating voltage.

$$^{228}_{88}\text{Ra} \rightarrow {}^{2208}_{89}\text{Ac} + {}^{0}_{-1}\text{e} \tag{5.3.8.2}$$

In both alpha and beta decay the resulting nucleus is frequently in an excited state, which, upon relaxation, sometimes releases a photon of gamma radiation. So, often there is a gamma photon (or photons) associated with these particulate decays.

When dealing with radioactive materials we are usually concerned about alpha, beta, and gamma radiation. However, we also mention here that some instruments in chemistry labs produce X rays and, less commonly, neutron radiation.

CHEMICAL CONNECTION *5.3.8.1*

RADIOACTIVE DECAY—A FIRST-ORDER REACTION

Radioactive decay is perhaps the most commonly used example of a first-order process. Simply put, the rate of decay depends directly on the amount of radioactive substance, and this can be expressed with two simple equations:

$$\text{Rate} = kN \tag{5.3.8.3}$$

$$\ln(N_0) - \ln(N_t) = \ln(N_0/N_t) = kt \tag{5.3.8.4}$$

$$N = N_0 e^{-t/\tau} \tag{5.3.8.5}$$

where N is the quantity of radioisotope (as grams or moles), k is the first-order rate constant, t is the time of decay, and $\tau = t_{1/2}$, the half-life of the radioactive isotope. Since $\ln 2 = 0.693$, for all first-order processes, $k = 0.693/\tau$ or $\tau = 0.693/k$. Tables of data about radioisotopes almost always list the half-lives. The values of τ can range from less than a nanosecond to billions of years. In general chemistry, the first-order rate equation is usually in the form shown in Equation 5.3.8.4, while nuclear physicists render the same equation as Equation 5.3.8.5.

The magnitude of the half-life determines some practical limits to using radioisotopes in chemical reactions. For example, the introduction of ^{32}P into a molecule is practical since the half-life of 14.3 days allows a reasonable window for conducting reactions and following the fate of the ^{32}P atom. If ^{34}P were introduced, with a half-life of 12.4 seconds, most of the ^{34}P would have disappeared by the time any practical reaction chemistry could be performed. As Figure 5.3.8.1 and Table 5.3.8.2 show, in a little as 10 half-lives (2 minutes) about 99.9% of the ^{34}P will have disappeared.

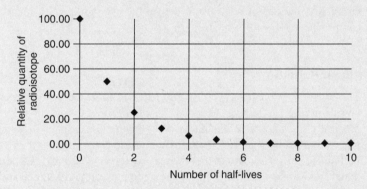

FIGURE 5.3.8.1 First-Order Decay of Radioisotope ^{34}P Through 10 Half-Lives. After 10 half-lives, only 0.1% of the starting amount of isotope remains.

TABLE 5.3.8.2   Data for Figure 5.3.8.1

Number of half-lives	Relative quantity of radioisotope
0	100.00
1	50.00
2	25.00
3	12.50
4	6.25
5	3.13
6	1.56
7	0.78
8	0.39
9	0.20
10	0.10

^{3}H, known as tritium, is an isotope commonly used in radioactive work. It has a half-life of 12.3 years, which makes it stable enough to use in various laboratory procedures. Another radioactive material sometimes used in laboratory experiments is ^{14}C that has a long half-life of 5730 years, which makes it convenient for dating materials of historical interest up to many tens of thousands of years old.

^{239}Pu, however, has a half-life of 24,110 years. This is a component of nuclear waste from light water reactors that provide electricity in the United States (and elsewhere). Storing or even burying this waste, with the expectation that it will "cool down" from a radioactive point of view, is clearly impractical since even one half-life greatly exceeds the history of human civilization. Storing these wastes in geologically stable underground depositories is the current plan for the disposal of high-level radioactive waste but it is easy to see that guaranteeing their safety for many half-lives (or many tens of thousands of years) is uncertain.

How might we encounter ionizing radiation in chemistry labs? Radioactive atoms are often inserted into molecules and then "followed" in a reaction scheme or a biological medium to determine the pathway and/or fate of the molecule. Some radioisotopes commonly used for this are ^{3}H ($t_{1/2} = 12.3$ yr) and ^{32}P ($t_{1/2} = 14.3$ days). The hazard level associated with these procedures is related to the quantity of the isotope, the decay products and their energy, and the radioactive half-life.

X rays are encountered in chemistry labs where X-ray diffraction analysis is performed or in the technique of X-ray fluorescence (XRF). There are devices associated with these instruments that protect users from exposure to beams of X rays and it is important to be mindful of these devices. Neutron radiation is encountered in instruments that measure neutron scattering or diffraction or in neutron activation analysis. The level of hazard is associated with the energy (velocity) of the neutrons. Ordinarily, the use of neutrons is tied to the availability of a nuclear reactor, where controlled fission can produce a stream of neutrons.

Hazards of Radiation

The cells in our bodies are constantly exposed to low doses of ionizing radiation, but the repair mechanisms are very effective. However, at high doses, the repair mechanisms become overwhelmed and medical effects are then inevitable. Very dramatic examples of large doses of radiation have been the nuclear bombs used in WWII at Hiroshima and Nagasaki and the fire and steam explosion, leading to a partial nuclear meltdown, at the nuclear reactor in Chernobyl, Ukraine, in 1986. Thousands died immediately and shortly thereafter in Japan due to radiation poisoning. The immediate death toll at Chernobyl due to radiation exposure was 28, and the long-term effect of dispersed radioactive substances has been estimated to be 1800 cases of thyroid cancer.[3] (America's most famous nuclear incident at the Three Mile Island nuclear power plant in Pennsylvania in 1979 involved no significant release of radiation or radioactive isotopes.)

The effects of exposure to significant doses of radiation can be both immediate and/or delayed. *Stochastic* effects are those where the probability of the effect (but not the degree) is related to the dose. *Nonstochastic* effects are those where the severity of the effect is related to the dose. In both cases, however, *less exposure is safer*. In species other than humans (but not in humans), it has been demonstrated that abnormalities of offspring are related to radiation exposure in parents. Radiation is also known to have a teratogenic effect on fetuses and embryos.

The effect of radiation at the cellular level is dependent, of course, on the ability of the photon or particle to penetrate protective layers of clothing (in humans) and the skin. Alpha particles have very low penetrating ability and are considered to be only an "internal" hazard—they are only hazardous if the radioactive parents get inside your body. For example, inhaling radon gas with subsequent alpha decay inside the lungs is very hazardous and is the source of concern about chronic exposure to radon in homes. See *Special Topic 5.3.8.1* Radon—A Significant Public Health Concern and Table 5.3.8.3. Beta particles are more penetrating but can still be stopped by a thin sheet of plastic or metal, depending on the beta particle energy, the exact plastic or metal, and its thickness.

SPECIAL TOPIC *5.3.8.1*

RADON—A SIGNIFICANT PUBLIC HEALTH CONCERN

Radon, a radioactive gas, is a product of the decay of uranium (^{238}U) and thorium (^{232}Th). It is a public health concern because its two isotopes, ^{222}Rn and ^{220}Rn, constitute more than half of the Earth's background radiation and is the second leading cause of lung cancer in the United States (smoking being first). Radon is an α emitter with a half-life of 3.8 days but since it is noble gas and is not reactive, it does not stay in the lungs long enough to do damage. However, its daughters, ^{218}Po and ^{216}Po, are solid α emitters and can be trapped in the lungs and cause tissue damage through continued decay as illustrated below:

$$^{222}_{86}\text{Rn} \rightarrow \alpha + {}^{218}_{84}\text{Po} \rightarrow \alpha + {}^{214}_{82}\text{Pb} \rightarrow \beta + {}^{214}_{83}\text{Bi} \rightarrow \beta + {}^{214}_{84}\text{Po} \rightarrow \alpha + {}^{210}_{82}\text{Pb} \rightarrow$$

$$\beta + {}^{210}_{83}\text{Bi} \rightarrow \beta + {}^{210}_{84}\text{Po} \rightarrow \alpha + {}^{206}_{82}\text{Pb}$$

$$^{220}_{86}\text{Rn} \rightarrow \alpha + {}^{216}_{84}\text{Po} \rightarrow \alpha + {}^{212}_{82}\text{Pb} \rightarrow \beta + {}^{212}_{83}\text{Bi} \rightarrow (\beta + {}^{212}_{84}\text{Po}) + (\alpha + {}^{208}_{81}\text{Tl}) \rightarrow$$

$$\alpha + \beta + {}^{208}_{82}\text{Pb}$$

Some regions of the country have soils, rocks, and mineral deposits rich in uranium or thorium. Radon gas can seep through cracks and openings in the lower parts of some buildings or houses and can accumulate in significant concentrations—exceeding the level of ~0.4 pCi/L found in the air outside of enclosed structures.

How much lung cancer is caused by radon? This is a complicated question and we have no *direct* epidemiological studies that compare lung cancer rates between large groups of humans exposed to either no, low, or high levels of radon. The best studies that we have are from the analysis of lung cancer rates in uranium miners since radon levels are high in uranium mines. These studies are complicated by (1) the assumption that there is a linear relationship between dose and response, (2) the assumption of no-threshold level, and (3) the recognition that uranium miners usually are cigarette smokers and that they inhale excessive amount of dust (both of which are synergistic effects with lung cancer). Based on these studies, however, a cancer rate model for exposure produces the expected cancer rates shown in Table 5.3.8.3. (A nonlinear model would likely reduce these risks considerably. We don't know, with certainty, which model is correct. The linear model is the "accepted model" at this time.)

TABLE 5.3.8.3   Predicted Lung Cancer Rates Based[a] on Radon Level Exposure [5]

Radon level (pCi/L)	Lifetime risk, never smoker, per 10,000	Lifetime risk, current smoker, per 10,000	Lifetime risk, general population, per 10,000
8	150	1200	450
4	77	620	230
2	37	320	120
0.4	7	64	24

[a] Assumes a constant 70-yr exposure.

The EPA selected 4 pCi/L as the "action level" above which it recommends that a home undergo some radon mitigation. The average level inside homes in the United States is about 1.3 pCi/L. Current technology allows radon levels to be reduced to ~4 pCi/L for a cost of $800 to $1200. Sometimes federal regulations are based on "what can be done," not "what is ideal." This is reasonable and practical. This is called a *technology-based* regulation (see Section 7.1.1). The official acceptable cancer rate that is used for many exposure regulations set by the EPA is "1 in a million." You can see in Table 5.3.8.3 that the predicted rate of cancer at 4 pCi/L is 23,000 per million. The EPA does not call 4 pCi/L a "safe" or "acceptable" level, but rather an "action level." This is an example of using ALARA (see text) as the basis for regulation.

X rays have generally very good penetrating ability and the degree of penetration depends on the energy of the photon. Medical X rays readily demonstrate that the X rays thoroughly penetrate most tissue but are impeded by bones, teeth, and mercury amalgams in teeth. Gamma rays are extremely penetrating and their high energy predicts their excellent ionizing ability. Neutrons also have good penetrating ability and their degree of hazard will be related to their velocity.

Two theories exist regarding the relationship between dose and effect. The *linear hypothesis* argues that the only safe level of exposure is zero and there is a linear relationship between the dose and the effect or probability of the effect. (There are also variations of this model where the relationship is quadratic or some other nonlinear relationship, but still maintaining a no-threshold baseline.) The *nonlinear* model uses the concept of the threshold level (which is generally believed to be the model most appropriate for chemical toxins; see Sections 4.1.1 and 4.2.1). The National Research Council (a nonprofit organization that functions under the auspices of the National Academy of Sciences, the National Academy of Engineering, and the Institute of Medicine) periodically issues reports related to the current understanding of the epidemiology of radiation damage. The most recent of these reports[4] is referred to as BEIR VII (BEIR = *b*iological *e*ffects of *i*onizing *r*adiation) and uses the linear, no-threshold model.

As with chemical toxins, the ability to confidently predict the dose–response relationship is complicated by the ability to reliably detect low-dose responses that can be statistically linked to a possible causative agent. Furthermore, on average, U.S. citizens absorb about 360 millirem (mrem) of radiation annually from a variety of natural (80%) and anthropogenic (produced by human activity, 20%) sources. See *Special Topic 5.3.8.2* Measuring Radiation. Also, as with chemical toxins, the response induced by radiation is affected by the general health of an individual, exposure to other hazardous materials, and age. Given all these uncertainties, it is not surprising that different persons might come to different conclusions about the relative risk, for example, of absorbing an additional 5 mrem per year if annual exposure is already 360 mrem per year. Most regulatory agencies assume the linear hypothesis. The posture taken by most health and safety professionals is therefore to keep doses *as low as reasonably achievable*. This is commonly referred to by the acronym ALARA. Thus, safety regulations focus on minimizing exposure to radiation, but recognize that setting a zero exposure level may not be practical.

SPECIAL TOPIC *5.3.8.2*

MEASURING RADIATION

Measuring a quantity of radiation is both simple and complicated since we may be interested in either the quantity of radioactive substance or the relative effects of the radiation on a human. Let's look at both kinds of measurements.

A becquerel is one nuclear transformation (disintegration) per second, but given the size of the mole you can guess that measurements using becquerel units lead to very large numbers. A more commonly used measure is the curie, which is the amount of radiation produced by 1 gram of ^{226}Ra. One curie (Ci) is equal to 3.7×10^{10} transformations per second. The Nuclear Regulatory Commission (NRC) establishes certain maximum levels of quantities of radioactive substances for various kinds of licensing and uses microcuries (μCi) as the common dimension for the quantities of radioisotopes on hand.

However, scientists (and the public) are often more interested in the relative hazard posed by the dose of a radioactive substance than the absolute amount (in grams or curies). The dose of radiation received has commonly been expressed in a couple of ways. The dose can be expressed as a rad (*r*adiation *a*bsorbed *d*ose) that is equal to 0.01 J of energy absorbed per kg of absorber and it is not limited to the type of radiation. This is a reasonable dimension since the amount of energy absorbed is related to the subsequent damage. The rad has been replaced by the International System of Units (*Système International [SI]*) measurement unit called the gray that is equivalent to 100 rad.

Another common way to express dose is to take into account the type of particles since different particles have different energies and different capabilities to cause damage to living material. The rem (*r*oentgen *e*quivalent *m*an) is derived from the product of the dose in rad units times a factor called the RBE (*r*elative *b*iological *e*ffectiveness). RBE is sometimes referred to as the *quality factor* (QF). Thus, the rem represents a dose from a specific kind of radiation—rem is derived by multiplying QF times rad (1 for β, γ, and X rays; 20 for α; 2 for slow neutrons; 10 for fast neutrons and protons). The most common dimension to measure the radiation damage from various sources used in the United States is the millrem (mrem); other countries use the SI measurement unit—the sievert (Sv), which is equal to 100 rem.

The use of these various units is rational, but confusing. As in other areas of science, "historical" units often stay in use for a long time even after SI units have been officially adopted. Table 5.3.8.4 shows the various units of measurement for radiation and some useful conversion factors.

TABLE 5.3.8.4 Units for Measuring Radiation

	Common unit	SI unit	Conversions
Radioactivity	curie (Ci)	becquerel (Bq)	1 Bq = 1 disintegration/s
			1 Ci = 3.7×10^{10} disintegrations/s
			1 Ci = 3.7×10^{10} Bq
Absorbed dose	rad	gray (Gy)	1 Gy = 100 rad
		(1 Gy = 1 joule/kg)	
Dose equivalent	rem	sievert (Sv)	1 Sv = 100 rem
			For β, γ, and X ray:
			1 rad = 1 rem; 1 Gy = 1 Sv
			For thermal neutrons:
			1 rad = 2 rem; 1 Gy = 2 Sv
			For α and fast neutrons:
			1 rad = 20 rem; 1 Gy = 20 Sv
Exposure (as measured by radiation monitors)	roentgen (R)	coulomb/kilogram (C/kg)	1 C/kg = 3880 R

There is no simple conversion between radioactivity (e.g., Ci) and the other dimensions regarding dose since the quantity of radiation absorbed will depend upon the amount of radiation that actually interacts with tissue (which could be assumed to be 100% for an internal absorption but would be much less for an external exposure), the duration of the exposure, the half-life of radioisotope, and the type of radiation (α, β, γ, X ray, or neutron).

The multiple conversions shown in Table 5.3.8.4 between rad and rem (and Gy and Sv) are due to the different relative biological effectiveness factors for different kinds of radiation.

Finally, we recognize that, as with chemical toxicity, we should consider the effects of both acute and chronic exposure. High doses from radioisotopes seem an unlikely event. The nonlaboratory, intentional, and fatal poisoning in 2006 of Alexander Litvinenko, a former member of the Russian Federal Security Service, is an exceptional case study, however, where it appears that a tiny dose of ^{210}Po was used to murder him.[11] Also unlikely, but possible, is a high-level exposure of radiation due to an accident at a nuclear reactor. Table 5.3.8.5 shows the effects of large radiation doses (measured in rem). In comparison, Table 5.3.8.6 shows dose levels (in mrem) for a variety of events, both chronic and acute.

TABLE 5.3.8.5 Effects of Acute Radiation Doses on a 70-kg Human

Dose[5] (rem)	Symptoms
< 5	None
50	Temporary reduction in white blood cell count
100–200	Nausea and fatigue
300–500	Nausea, vomiting, diarrhea; loss of hair and appetite; death possible with no medical attention
500–1200	Probable death within a few days
>10,000	Death within hours

TABLE 5.3.8.6   Estimated Absorbed Doses of Radation from Varied Sources

Radiation source	Type	Estimated dose
Cosmic radiation at sea level (from space)	Chronic exposure	26 mrem/yr (Ref. 5)
Terrestial radiation (from ground)	Chronic exposure	30 mrem/yr (Ref. 5)
Radon gas	Chronic exposure	200 mrem/yr (Ref. 5) (dependent on location and building materials)
Total exposure for U.S. citizens living at sea level	Chronic exposure	Average of 360 mrem/yr (Ref. 5) (could be higher with medical procedures)
Air travel at 39,000 ft for 5 hours	Single event	2.5 mrem (Ref. 5)
Chest X ray	Single event	6 mrem (Ref. 5)
Dental X ray	Single event	2–9 mrem (Ref. 6)
Mammogram	Single event	70 mrem (Ref. 6)
Smoking cigarettes	Annual	16,000 mrem/yr (Ref. 7)
Single ^{60}Co radiation treatment for cancer	Single event	200,000 mrem (Ref. 8)

Working with Radioisotopes and Other Sources of Ionizing Radiation: RAMP

All procedures involving radioisotopes and other sources of radiation must be undertaken with utmost care. The very fact that radiation is invisible and otherwise not immediately sensed requires that safeguards must be in place and completely understood by laboratory workers to avoid exposure to anything but the most trivial levels. Labs using radioisotopes must be identified. Licensure, usually by state agencies operating on behalf of the Nuclear Regulatory Commission, is required. Special training about the use of radioisotopes is required. In general, the following strategies will minimize exposure:

1. Minimize the amount of radioisotope used.
2. Maximize the distance between you and the radioisotope. Radiation from a point source decreases in relation to the square of the inverse of the distance.
3. Minimize the time that radioisotopes are handled.
4. Use shielding when possible to do so. A piece of plastic that is 0.5-inch thick will protect against most beta particles (depending on energy). Lead shields are the best for gamma and X-ray radiation.

TABLE 5.3.8.7   Printed and Internet Resources About Radiation

Printed Resources

Handbook of Chemical Health and Safety (R. J. Alaimo, ed.), American Chemical Society Oxford University Press, Washington, DC, 2001

CRC Handbook of Laboratory Safety, 5th edition, A. Keith Furr, CRC Press, Boca Raton, FL, 2000

Internet Resources

http://www.osha.gov/SLTC/radiation/index.html
http://www.uvm.edu/~radsafe/
http://www.rss.usda.gov/
http://www.drs.uiuc.edu/rss/
http://www.nrc.gov/materials/medical.html
http://www.radiationanswers.org
http://www.hps.org/
http://www.epa.gov/radon/index.html
http://www.ans.org/

Not surprisingly, there are special rules about securing and handling radioisotopes, the disposal of radioactive wastes, and how to respond to a spill of radioactive materials. Everyone working with these substances must be trained in their use and be completely familiar with appropriate protocols. Devices and instruments that generate ionizing radiation must be used only by trained operators. Safeguarding devices should never be overridden.

- *Recognize* sources of ionizing radiation.

- *Assess* the level of risk for exposure based on lab procedures.

- *Minimize* the risk for exposure by using the four strategies listed above.

- *Prepare* for an unexpected exposure by knowing the correct response procedure.

This has been only a brief introduction to radiation safety. For more detailed discussions see Table 5.3.8.7.

References

1. National Institute for Standards and Technology. Memorandum from NIST Director. Pu Contamination Incident at Boulder; available at http://www.nist.gov/public_affairs/releases/rowe.pdf (accessed September 20, 2009).
2. National Institute for Standards and Technology. NIST News Release. Information and Updates Relating to June 8, 2008 Plutonium Incident at the NIST—Boulder Laboratory; available at http://www.nist.gov/public_affairs/releases/boulder-incident.html (accessed September 20, 2009).
3. International Atomic Energy Agency. Feature Stories. Frequently Asked Chernobyl Questions; available at http://www.iaea.or.at/NewsCenter/Features/Chernobyl-15/cherno-faq.shtml (accessed September 20, 2009).
4. National Research Council. *Health Risks from Exposure to Low Levels of Ionizing Radiation: BEIR VII Phase 2*. The National Academies Press, Washington, DC, 2005; a summary of this report is available at http://dels.nas.edu/dels/rpt_briefs/beir_vii_final.pdf (accessed February 11, 2009).
5. American Nuclear Society. Common Sources of Radiation; available at http://www.ans.org/pi/resources/dosechart/ (accessed February 11, 2009).
6. Health Physics Society. Radiation Exposure from Medical Diagnostic Imaging Procedures; available at http://hps.org/documents/meddiagimaging.pdf (accessed February 11, 2009).
7. Health Physics Society. How Much Radiation Dose Is Received by Cigarette Smoker? Available at http://www.hps.org/publicinformation/ate/q3137.html (accessed February 22, 2009).
8. G. J. SCHREIBER. Radiation Therapy, General Principles; available at http://emedicine.medscape.com/article/846797-overview (accessed February 22, 2009).
9. Radiation Answers. Answers to Questions About Radiation and You; available at http://www.radiationanswers.org/radiation-and-me/effects-of-radiation.html (accessed September 20, 2009).
10. Environmental Protection Agency. Radon. Health Effects; available at http://www.epa.gov/radon/healthrisks.html (accessed September 20, 2009).
11. JOHN EMSLEY. Polonium and the Poisoning of Alexander Litvenenko. In: *Molecules of Murder: Criminal Molecules and Classic Cases*, RSC Publishing, Cambridge, UK, 2008, Chapter 10, p. 204.

QUESTIONS

1. Which feature of an alpha particle can vary and, in doing so, affect the hazard level of the radiation?

   (a) Mass
   (b) Charge
   (c) Velocity
   (d) None of the above

2. Alpha radiation is generally considered not to be hazardous since the particles

   (a) Decay almost immediately after being emitted from a nucleus
   (b) Have very low penetrating power
   (c) Are absorbed in the air
   (d) Have such low velocities

3. Which is *true*?

   (a) Alpha decays usually also produce beta emissions.
   (b) Beta decays usually also produce alpha emissions.
   (c) Alpha and beta decays usually also produce gamma emissions.

   (d) All of the above.

4. X-ray radiation from diffraction and fluorescence instruments is

   (a) Not very dangerous because the X rays are absorbed in the air
   (b) Very dangerous but poses little real hazard since instruments are designed to completely enclose all beams
   (c) Very dangerous and users must not override devices designed to protect from exposures
   (d) Dangerous only when at "full power"

5. For radiation, stochastic effects are those where

   (a) The probability of the effect is related to the dose but the degree of the effect is not related to the dose
   (b) The probability of the effect and the degree of the effect are related to the dose
   (c) The probability of the effect is not related to the dose but the degree of the effect is related to the dose
   (d) Neither the probability of the effect nor the degree is related to the dose

6. What is the order of penetrating ability?

    (a) Alpha > beta > gamma
    (b) Beta > alpha > gamma
    (c) Gamma > alpha > beta
    (d) Gamma > beta > alpha

7. Regulatory agencies use which dose–effect model and recommend what maximum exposure level?

    (a) Linear, ALARA
    (b) Nonlinear, ALARA
    (c) Linear, zero
    (d) Nonlinear, zero

8. What is the average radiation dose received per year by U.S. citizens?

    (a) 20 mrem
    (b) 360 mrem
    (c) 20 rem
    (d) 360 rem

9. A dental X ray typically involves what dose of radiation?

    (a) 0.5 mrem
    (b) 5 mrem
    (c) 360 mrem
    (d) 5 rem

10. What shielding material is best for X-ray and gamma radiation?

    (a) None is needed since these types of radiation are absorbed by air
    (b) Plastic
    (c) Lead
    (d) No shielding is effective

5.3.9

CRYOGENIC HAZARDS—A CHILLING EXPERIENCE

Preview This section describes the hazards of cryogens, and procedures to minimize the risk in handling these chemicals.

> *There's simply never an adequate excuse for not wearing safety glasses in the laboratory at all times.*
>
> Dr. K. Barry Sharpless, Nobel Prize Laureate, after losing an eye in a cryogenic explosion[1]

INCIDENT 5.3.9.1 EXPLODING NUCLEAR MAGNETIC RESONANCE (NMR) TUBE[1]

As a professor was preparing to leave for home he took off his safety glasses and put on his coat. On his way out he saw a graduate student carrying out a procedure to prepare sealed NMR tubes (long thin-walled tubes about 25 cm × 0.3 cm OD) using a liquid nitrogen bath. Since he and the student had discussed this procedure, but neither had done this before and he stopped to enquire how it was going. He picked up one of sealed tubes to examine it and held it up to the light. When the tube frosted over, he wiped away the frost and saw that the tube seemed to be overfilled with solvent. However, as he watched, suddenly the solvent level dropped to the normal level. At that moment he realized that some oxygen gas had been cooled to a liquid state in the nitrogen bath and had been sealed in the tube. And now as the temperature of the tube rose, the liquefied gas had revaporized, likely creating enormous pressure within the tube. The tube suddenly exploded and glass shards were sprayed into his eye and into his face. He lost vision in one eye but only suffered superficial cuts to the rest of his face.

What lessons can be learned from this incident?

What Are Cryogens?

Kryos is the Greek word for icy cold and cryogens represent a class of chemicals that are extremely cold. Cryogens are gases that have been cooled to the liquid or solid states at temperatures $\leq$ 200 K, (or $-73\,°C$, or $-100\,°F$). The extremely cold temperatures of cryogens present special hazards that deserve very careful and thoughtful consideration before they are used. Anyone working with cryogens should learn about their properties in general and specifically about the properties of the cryogen that they will be using. You should seek out someone who has direct experience with cryogens before attempting to use them. The most common laboratory cryogens and their boiling points and common uses are listed in Table 5.3.9.1.

In the laboratory these cryogens are used for cooling in various operations, including for cooling solutions in glass vials and tubes before they are sealed as described in Incident 5.3.9.1. Cryogens are also used commonly as cooling traps to capture volatiles traveling through a system. Using liquids that freeze below 195 K, "cold baths" at 195 K can be prepared using dry ice. Ethanol dry ice baths are commonly used for this purpose. Using liquids that freeze above 195 K, the cold bath can be prepared down to the freezing point of the liquid (or at the freezing point the liquid may be a "slushy" mixture).

Some instruments, such as nuclear magnetic resonance (NMR) spectrometers that use superconducting magnets, require cooling by cryogens such as liquid nitrogen and liquid helium. Some freezers use liquid nitrogen to maintain samples at low temperatures.

Laboratory Safety for Chemistry Students, by Robert H. Hill, Jr. and David C. Finster
Copyright © 2010 John Wiley & Sons, Inc.

TABLE 5.3.9.1   Common Cryogens

Cryogen	Boiling point (°C)	Boiling point (K)	Uses of solid or liquid
CO_2 (solid, "dry ice")	−78 (sublimes)	195 (sublimes)	Cold baths
N_2 (l)	−196	77	Coolant, superconducting magnet coolant
He (l)	−269	4	Superconducting magnet coolant

Hazards of Operations Using Cryogens

The main hazards specifically associated with the use of cryogens are severe or dangerous frostbite, explosions, asphyxiation, and other specific hazards associated with particular chemicals such as being flammable, corrosive, reactive, or toxic.

Any contact of your skin or eyes with cryogens or materials cooled by cryogens will cause very serious injuries from freezing or frostbite. (Chapter 2 describes what to do in the event of a medical emergency.) If your skin comes in contact with material that has been cryogenically cooled, it will stick to that material and removal could result in tearing away the skin in contact with this material. These types of injuries are extremely painful as the tissue thaws (frozen tissue will be painless at first), and they compromise the skin so that infections, edema, and dangerous blood clots may become serious problems. Prevention of these injuries is essential when using cryogens. Below we will discuss the appropriate personal protective equipment to use.

Another kind of injury can occur from the explosions caused by cryogens. Any cryogen is capable of producing huge amounts of gas if the cryogen itself is warmed above its boiling point. Vessels or containers must not be sealed, plugged, or covered to allow buildup of pressure that could lead to a pressure explosion. (If a container is capable of withstanding extremely high pressures, then it can be sealed but this would be a very rare situation. A common example, that is not really a "lab" situation, is a carbon dioxide fire extinguisher where the liquid carbon dioxide has a vapor pressure of about 57 atm at room temperature.)

Explosions can also occur due to the ability of the cryogen to liquefy other gases. When using cryogens, one must always be concerned about a gas being cooled below its condensation temperature and becoming trapped in some line or vessel. When this system is rewarmed it will revert to its gaseous state. Especially be aware that oxygen has a higher boiling point than nitrogen (the most common cryogen used), so it can be condensed out from the air in an "open" system or inadvertently in a system where there is an "air leak." This is exactly what happened in Incident 5.3.9.1 at the beginning of this section. The liquid nitrogen bath (at 77 K) caused gaseous oxygen from the air to condense to its liquid form (BP = condensation temperature = 90 K) in the tube through a leak in the system during cooling. Then, as the temperature of the tube rose above oxygen's boiling point, the oxygen revaporized, the pressure in the tube increased dramatically, the thin glass NMR tube could not sustain the pressure, and it exploded. (Perhaps just a pinhole leak first existed, allowing the oxygen to enter the tube and condense. Even if the leak continued to exist, it could probably not allow the rapidly expanding gas to leak out fast enough to prevent the explosion.) The result was an explosion known as a BLEVE—boiling liquid expanding vapor explosion (although unlike most BLEVEs, the expanding vapor does not burn and produce a large fireball, as discussed in Section 5.2.2).

Another potential hazard is asphyxiation from the release of inert cryogens into a room or confined space. If a liquid cryogen is spilled and subsequently vaporized, the gas expands considerably. For example, if 1 liter of liquid would expand to as much as 600–1200 volumes of gas at ambient conditions, this is enough to easily displace oxygen in a confined space and make it an area that cannot sustain life (see *Chemical Connection 5.3.9.1* Oxygen Concentrations in a Laboratory with a Spilled Cryogen). Since these gases are odorless and colorless, they present an unsuspected hazard. Thus, cryogens should not be used in confined spaces, such as basements, tunnels, or other depressed areas. Areas where large quantities of these types of cryogens are used are often equipped with oxygen detectors so if the oxygen level goes below a safe level, an alarm will sound and alert the occupants to leave the area.

CHEMICAL CONNECTION *5.3.9.1*

OXYGEN CONCENTRATIONS IN A LABORATORY WITH A SPILLED CRYOGEN

Let's assume that you are working in a small research lab. The size of the lab is 8ft × 16ft × 10ft. There is a single fume hood that is in "quiet mode" that is causing the volume of the lab air to "turn over" every 30 minutes. If you knock over a Dewar flask of liquid nitrogen that has 2.0 L of liquid in it, this will vaporize in less than 1 minute. How safe is it to be in this room? Liquid nitrogen has a density of about 0.808 g/mL. What if this spill involved liquid helium or liquid argon?

Liquid Nitrogen Calculation

The volume of the room is 1280 ft^3, which can be converted to liters:

$$1280 \text{ ft}^3 \times 1728 \text{ in.}^3/\text{ft}^3 \times 16.39 \text{ cm}^3/\text{in.}^3 \times 1 \text{ L}/1000 \text{ cm}^3 = 36,000 \text{ L}$$

The liquid nitrogen becomes gaseous to add the following volume of nitrogen in the air:

$$2.0 \text{ L liq } N_2(1000 \text{ mL liq } N_2/\text{L liq } N_2)(0.808 \text{ g liq } N_2/\text{mL liq } N_2)(1 \text{ mol } N_2/28 \text{ g } N_2) = 58 \text{ mol } N_2$$

$$V = nRT/P = 58 \text{ mol}(0.08206 \text{ L} \cdot \text{atm/K} \cdot \text{mol})(300 \text{ K})/(1 \text{ atm}) = 1400 \text{ L}$$

Note that the expansion ratio of liquid nitrogen becoming a gas is 1:700.

$$\text{"New" atmosphere volume} = 36,000 \text{ L} + 1400 \text{ L} = 37,400 \text{ L}$$

(The actual volume would still be 36,000 L, of course, but this fictitious volume makes the next calculation easier.)

$$\text{Original volume of oxygen}: \qquad 36,000 \text{ L}(0.21 \text{ } O_2) = 7600 \text{ } O_2 \text{ L}$$

$$\text{New percent } O_2: \qquad (7600 \text{ L}/37,400 \text{ L}) \times 100\% = 20.3\% \text{ } O_2$$

Therefore, O_2 is now 20.3%, down from about 21%. Any atmosphere with less than 19.5% O_2 is considered to be hazardous. At 16% there will be "increased pulse and breathing rate, impaired thinking and attention, reduced coordination." (See Table 4.1.2.2.)

Liquid Argon

$$2.0 \text{ L liq Ar}(1000 \text{ mL liq Ar/L liq Ar})(1.4 \text{ g liq Ar/mL liq Ar})(1 \text{ mol Ar}/39.9 \text{ g Ar}) = 70 \text{ mol Ar}$$

$$V = nRT/P = (70 \text{ mol})(0.08206 \text{ L} \cdot \text{atm/K} \cdot \text{mol})(300 \text{ K})/(1 \text{ atm}) = 1700 \text{ L}$$

The expansion ratio is 1:850.

$$\text{"New" atmosphere volume} = 36,000 \text{ L} + 1700 \text{ L} = 37,700 \text{ L}$$

$$\text{New percent } O_2: \qquad (7600 \text{ L}/37,700 \text{ L}) \times 100\% = 20.1\% \text{ } O_2$$

Liquid Helium Calculation

If this incident involved liquid helium, the method of calculation would be the same (He liquid density = 0.18 g liq He/mL).

$$2 \text{ L liq He}(1000 \text{ mL liq He/L liq He})(0.18 \text{ g liq He/mL liq He})(1 \text{ mol He}/4\text{g He}) = 90 \text{ mol He}$$

$$V = nRT/P = (90 \text{ mol})(0.08206 \text{ L} \cdot \text{atm/K} \cdot \text{mol})(300 \text{ K})/(1 \text{ atm}) = 2200 \text{ L}$$

The expansion ratio is 1:1100.

$$\text{"New" atmosphere volume} = 36,000 \text{ L} + 2200 \text{ L} = 38,200 \text{ L}$$

$$\text{New percent } O_2: \qquad (7600 \text{ L}/38,200 \text{ L}) \times 100\% = 19.8\% \text{ } O_2$$

In all three of these instances, the atmosphere is not compromised with regard to human health, although the He vaporization gets close. Obviously, larger volumes of any of these liquid cryogens could easily reduce the percent O_2 to less than 19.5%, the lower limit for a "safe" atmosphere.

It is important not only to consider the hazards associated with the low temperatures of cryogens, but to remember that each cryogen itself may have other hazardous properties specific to that particular chemical. Liquid nitrogen and liquid helium present only asphyxiation hazards as gases while carbon

FIGURE 5.3.9.1   Assorted Dewar Flasks. Dewar flasks come in a variety of shapes and sizes.

dioxide has an IDLH of 40,000 ppm (or 4% by volume). Most other possible cryogens are also only simple asphyxiation gases: H_2, Ne, air, O_2, and freons. While not typically used as cryogens, other condensed hydrocarbons such as liquid propane or liquefied natural gas (LNG, mostly CH_4) are highly flammable, as is liquid H_2. Liquid oxygen can react violently with organic materials, so it is essential to keep these materials away from liquid oxygen. In fact, liquid oxygen spilled on asphalt can ignite if a tool is dropped on the surface or even by pedestrian traffic.[2] Cloth that is soaked in liquid oxygen becomes extremely flammable.

Special Equipment to Handle Cryogens

Dewar flasks are especially designed to be used with cryogens (Figure 5.3.9.1). They are double-walled flasks and the space between the two walls has been evacuated and sealed under high vacuum. Many Dewar flasks are made of glass and the walls have been coated inside with silver as a reflective coating to minimize radiative heat transfer. This high-vacuum system is subject to failure and collapse under sudden thermal or mechanical shock. Stainless steel Dewar flasks are available but are generally more expensive than the glass Dewar flasks and have higher evaporation rates.

Never fill a cryogenic container more than 80% full so there is room for expansion of the cryogen. Use tongs to remove anything that is submerged in the cryogenic solution and be sure that you do not touch that object or the tongs after removal until they have warmed to ambient temperature.

Glass Dewar flasks should be wrapped with electrical, plastic, or fiber tape to minimize the risk of shards and flying glass. Some Dewar flasks come with metal jackets to minimize this risk. If metal Dewar flasks are available and suitable for your use, they are preferred over glass Dewar flasks. Dewar flasks often come with cork tops that are not meant to seal the flask but to keep dirt, debris, and moisture out of the flask and prevent splashes from reaching the carrier during transport. Dewar flasks containing cryogens should never be sealed or plugged. Dewar flasks come in various sizes and shapes and you should select the Dewar flask that is best for the operation at hand.

Small Dewar flasks can be hand-carried, providing appropriate personal protective equipment is used. As Dewar flasks increase in size, a rolling dolly or laboratory cart should be used to move them to the needed location. Use care in transporting any cryogenic to avoid splashing and bumping. Always keep cryogenic vessels upright and vertical.

Large commercial Dewar flasks used in the transport of cryogens, such as liquid nitrogen, are made of stainless steel (Figure 5.3.9.2). Never modify or change the safety features of these flasks (see

FIGURE 5.3.9.2 Large Commercial Dewar Tank. This tank holds about 180 L of liquid nitrogen. It is possible to withdraw the liquid directly from this tank and transfer the liquid to a smaller Dewar flask using the tube at the left in the picture.

FIGURE 5.3.9.3 Phase Separator. This device allows a cryogenic liquid to be transferred with minimal splashing. (Courtesy of Office of Environmental Health and Radiation Safety, University of Pennsylvania, Philadelphia, PA.)

Incident 5.3.1.1 in Section 5.3.1 on gas cylinders). When transferring cryogens from a storage tank to an open flask, you should use a phase separator (Figure 5.3.9.3). This is a tube with a device at the end that separates the gas and the liquid and allows the liquid to be transferred with less splashing.[3] Transfer of cryogens should only be carried out in well-ventilated areas to prevent hazard from asphyxiation.

Liquid nitrogen and liquid helium are used in cooling NMR superconducting magnets. The rapid loss of liquid nitrogen or liquid helium is known as "quenching" and it results in the rapid release of a large amount of gas that could displace oxygen from the air, if this occurs in a small room or confined space. This could pose an asphyxiation hazard and you should be aware of this hazard if you use cryogenically cooled instrumentation.

First-time users of large commercial Dewar flasks should develop an understanding of what is normal—lines that are pressure relief lines may frost over and it is normal for these tanks to "hiss" as pressure is relieved. These large tanks are very heavy and are susceptible to tip over if tilted at too steep an angle. Hand rolling these tanks is an unsafe practice and can damage the internal structure of the tank, resulting in rupture of the inner flask. Furthermore, because of the heavy contents tip over is a real possibility. If a tank tips over, immediately evacuate the area and alert others not to enter the area until the gas has escaped from the room and the oxygen level is safe. Make sure that you have the proper equipment to safely and gently move these tanks.

Safety Measures in Handling of Cryogens

You should not attempt working with cryogenics unless you have specific education and practical training using these materials. First and foremost you must protect your body from exposure to the cryogen. The most common cryogenics are CO_2, N_2, and He so we focus our discussions of protection on these materials. Comments about liquid nitrogen will also apply to liquid helium.

First, it is important to prevent any direct contact between a cryogen and your skin. You should use insulated gloves (Figure 5.3.9.4) and/or tongs to pick up small blocks of dry ice (solid CO_2). Similarly, when using small volumes of liquid nitrogen, wearing gloves is a prudent measure. Insulated gloves should be "loose enough" so that you will be able to remove or shake them off quickly if the cryogen gets into the gloves. You should consider taking steps to protect any other parts of your skin (arms, legs, feet) that might be exposed by wearing long sleeve shirts, long pants, and shoes that protect your feet. Pants without cuffs and that cover your shoes would protect your feet from splashes that might get into your shoes. It is prudent to remove metal jewelry, such as rings, watches, bracelets, necklaces, and earrings, since they might become extremely cold if the cryogen were to come in contact them—thus creating an additional hazard for you.

Chemical splash goggles should always be worn since there is some risk that a cryogen could unexpectedly explode or a Dewar flask could implode due to thermal or mechanical shock. Additionally, using a face shield would be a prudent practice to protect your face from splashes of cryogens. Acrylic or glass shields can also be used between you and the cryogenic vessel to provide a greater degree of protection.

When dealing with larger quantities of liquid nitrogen, such as during transfers into larger Dewar flasks (>2 L), you should take additional steps such as using protective boots or a heavy rubber or leather apron. You must always be aware that transfer lines, uninsulated pipes, vessels, or containers might be at or near the temperature of the cryogen, and you must avoid skin contact with these.

A very common use of cryogenic liquids is in association with the superconducting magnets for high-field NMR instruments. It is important to follow the manufacturer's instructions and local protocols for these procedures and work with others who have experience in this process when new to working with these instruments.

Working with Cryogens: RAMP

- *Recognize* what substances are cryogens and the general and specific hazards of the cryogen.
- *Assess* the level of risk based on the cryogen, its intended use, and the potential for exposure.

FIGURE 5.3.9.4 Insulated Gloves for Handling Cryogenics. The gloves on the left are best for holding cold objects since they will have better dexterity than the glove on the right (which can also be used for hot objects).

- *Minimize* the risk by using appropriate personal protective equipment, appropriate ventilation, appropriate equipment for a given procedure, and safe procedures.

- *Prepare* for an emergency by having a response procedure planned for a spill, quench, skin exposure, or equipment failure. (Medical responses are discussed in Chapter 6.)

References

1. Used with permission from Dr. K. Barry Sharpless.
2. THOMAS M. FLYNN. *Cryogenic Engineering*, 2nd edition, CRC Press, Boca Raton, FL, 2004, p. 799.
3. E. WARZYNIEC. Safe handling of compressed and cryogenic liquids. *Journal of Chemical Health and Safety* **7**(*3*): 34–36 (2000).

QUESTIONS

1. Which substance is *not* a commonly used cryogen?
   - (a) $H_2O(s)$
   - (b) $CO_2(s)$
   - (c) $He(l)$
   - (d) $N_2(l)$

2. Exposure of the skin to a cryogen
   - (a) Poses little risk since the skin quickly vaporizes the cryogen
   - (b) Usually leads to dead skin that heals with surprising quickness
   - (c) Causes frostbite
   - (d) Freezes blood below the skin, which causes a temperature drop throughout the body

3. Common cryogens pose an explosion hazard since
   - (a) They are very flammable
   - (b) If they vaporize or sublime in a constant-volume container a very high pressure can rapidly be produced
   - (c) They are usually stored in sealed containers
   - (d) When in contact with glass, the glass usually shatters

4. Cryogens can cause asphyxiation when
   - (a) They are consumed in the liquid state
   - (b) Rapid vaporization displaces oxygen from the atmosphere
   - (c) They cool down the air in a room enough to freeze the lungs upon inhalation
   - (d) They become colorless as gases

5. Containers designed to hold cryogens are called
   - (a) Cryogenites
   - (b) High-vacuum flasks
   - (c) Double-reflective beakers
   - (d) Dewar flasks

6. Dewar flasks are often wrapped with tape in order to
   - (a) Protect the user's hands from the cold temperature
   - (b) Provide a better grip
   - (c) Minimize flying glass in the event of an implosion
   - (d) Provide more insulation to help keep the cryogen cold

7. Dewar flasks should be
   - (a) Sealed tightly when holding a cryogen to minimize cryogen loss
   - (b) Sealed tightly when holding a cryogen to minimize the danger of asphyxiation
   - (c) Covered loosely to keep out contaminants or prevent spillage during transport
   - (d) Covered loosely to minimize the danger of asphyxiation

8. When transferring a liquid cryogen from one container to another,
   - (a) The systems should be connected to each other but sealed from the atmosphere
   - (b) A phase separator should be used
   - (c) The process should be done in a small, unventilated room to prevent cryogen escape throughout a building
   - (d) The fastest, and safest, way is to simply pour from one vessel into another

9. The process of "quenching"
   - (a) Occurs when there is a rapid release of vaporized cryogen
   - (b) Is both inevitable and desirable when using liquid cryogens
   - (c) Poses little hazard other than hearing damage
   - (d) Involves the destruction of the cryogen by using a very hot solid object

10. What practice is risky when using a cryogen?
    - (a) Wearing "loose" gloves
    - (b) Wearing jewelry
    - (c) Handling dry ice with nitrile gloves
    - (d) All of the above

5.3.10

RUNAWAY REACTIONS

Preview This section describes runaway reactions, their impact, their causes, and steps to prevent these incidents.

Those who cannot remember the past are condemned to repeat it.

George Santayana, American philosopher[1]

INCIDENT 5.3.10.1   RUNAWAY REFLUX LEADS TO EXPLOSION[2]

A technician was carrying out a reaction with ethylene glycol and phosphorus pentasulfide in hexane. The reaction flask was heated by an electric mantle in a hood. The procedure called for maintaining the temperature at or below $60\,°C$ by controlling the power to the mantle. As the reaction proceeded he realized that the temperature was rising rapidly, and at $177\,°C$ he turned off the power to the mantle. At this point he opened the hood to remove the flask from the heating mantle. As he did this the flask exploded, causing burns to his face and eyes. Pieces of the flask were scattered over the laboratory.

INCIDENT 5.3.10.2   RUNAWAY GRIGNARD REACTION[3]

A researcher prepared a Grignard reagent using tetrahydrofuran as the solvent. Magnesium turnings were added to the solvent followed by iodine to initiate the reaction. A bromobenzene derivative was added to the reaction mixture incrementally, and the reaction was left unattended in a chemical hood. After about 20–30% of the bromobenzene derivative had been added, a sudden exothermic reaction caused the solvent to boil, blowing most of the reaction mixture out of the flask.

What lessons can be learned from these incidents?

Runaway Reactions—A Cause of Significant Incidents

As an advanced undergraduate in chemistry you are likely to be involved in one or more research projects with one of your professors. It is also likely that you have heard about explosions in a laboratory as well as explosions in facilities that handle chemicals. While it is unlikely that you will be involved in any project where this could happen, you should have at least some understanding of one of the major causes of many explosions—runaway reactions.

A runaway reaction is an unexpected event in which the rate of reaction increases significantly, resulting in a significant increase in temperature from the failure to control or maintain the temperature of the reaction.[4-8] In a runaway reaction the excess heat produced exceeds the capacity of the surrounding environment to absorb this heat and an explosion is a likely result. Any reaction that is exothermic is capable of getting out of control if the heat generated by this reaction cannot be absorbed by the surrounding environment. This adverse event could result from the chemical reaction itself, from the decomposition of reactants or products, from a side reaction, or from storage or purification.

When the heat generated by an exothermic reaction surpasses the ability of the system to remove or control the excess heat, the additional heat increases the temperature of the reaction mixture linearly.

However, this excess heat causes the rate of reaction, and correspondingly the rate of heat generation, to increase exponentially (see *Chemical Connection 5.3.10.1* Using the Arrhenius Equation to Examine Reaction Rate Increases). The resulting reaction may begin slowly, but it accelerates rapidly until it has become "runaway" or out of control, feeding itself with its own heat. A general approximation is that the reaction rate doubles with every $10\,°C$ rise in temperature and the rate of heat generation rises in the same way.[4] The temperature escalation occurs so rapidly that reactants or products may be vaporized or decomposed into other products, which leads to a higher pressure than can be contained by the reaction vessel. The result almost certainly will be an explosion, possibly with fire and/or release of toxic or flammable gases.

CHEMICAL CONNECTION *5.3.10.1*

USING THE ARRHENIUS EQUATION TO EXAMINE REACTION RATE INCREASES

A general form of a rate equation is

$$\text{Rate} = k[A]^a[B]^b$$

where A and B are reactants, a and b are the orders of the reactions with respect to A and B, and k is the rate constant. The value of the rate constant is given by the Arrhenius equation:

$$k = Ae^{-E_a/RT}$$

where A is the frequency factor (a product of a collision frequency factor and orientation probability factor), E_a is the activation energy, R is the ideal gas law constant, and T is the temperature (in kelvin). Relative increases in k (and rate) can be assessed by

$$\text{Rate}_{high}/\text{Rate}_{low} = k_{high}/k_{low} = e^{-E_a/RT_{high}}/e^{-E_a/RT_{low}}$$

This is often rewritten as

$$\ln(\text{Rate}_{high}/\text{Rate}_{low}) = (E_a/R)(1/T_{low} - 1/T_{high})$$

Table 5.3.10.1 illustrates how the k increases, which is represented by the "increase in rate" factor.

TABLE 5.3.10.1   Effect of Temperature and Activation Energy on Reaction Rate

Line	T_{low} (K)	T_{high} (K)	E_a (kJ/mol)	Increase in rate
1	298	308	10	1.14
2	298	348	10	1.79
3	298	398	10	2.76
4	298	308	50	1.93
5	298	348	50	18.2
6	298	398	50	159
7	298	308	100	3.71
8	298	348	100	330
9	298	398	100	25400

Let's imagine that you have a reaction with an $E_a = 50$ kJ/mol. You heat this reaction to a modest temperature of $75\,°C$ (438 K) and the reaction undergoes a nice enhancement factor of 18.2, compared to room temperature ($25\,°C$, 298 K) (line 5 in Table 5.3.10.1). However, if this reaction is exothermic and the combined heat capacities and heat transfer rate of the flask and solvent (particularly when nestled in a heating mantle for a round-bottom flask) cause the released heat to increase the temperature of the system, line 6 shows that this reaction can quickly accelerate even more, releasing more heat, causing a greater rise in temperature. This accelerating reaction becomes an out-of-control reaction, even to the point of exceeding the boiling point of the solvent and likely causing an explosion.

In addition, the collision frequency (represented by A above) increases with temperature, crudely as a function of $T^{1/2}$, but this varies depending on the nature of the reaction. Nonetheless, the higher collision frequency will increase the reaction rate *even more* than just the rate increase due to overcoming the E_a more easily as temperature increases.

Table 5.3.10.1 also illustrates that the rule of thumb about "doubling rate for every $10\,°C$ increase in temperature" applies only to reactions with E_a values around 50 kJ/mol. (The exact "doubling value" is 52 kJ/mol, but only for the $10\,°C$ increase from 298 K to 308 K.) Other $10\,°C$ changes and other E_a values will quickly violate this "rule of thumb."

Most laboratory reactions are carried out on a small scale, where adequate cooling and agitation can prevent a runaway reaction. Nevertheless, runaway reactions have resulted in significant laboratory incidents.[9] Incident 5.3.10.1 is an example of a runaway reaction that resulted in an explosion with injuries. Incident 5.3.10.2 illustrates how common reactions such as those involving Grignard reagents can result in runaway reactions with adverse results in the laboratory.

As the scale of a reaction goes up, the challenges for maintaining adequate control of an exothermic reaction increase. Thus, if you were to take a reaction in the laboratory and decide to scale it up to make more of your desired product (a common practice), you should be aware of the importance of providing adequate cooling and mixing for the reaction. Scaling up reactions can provide challenges in controlling the heat of an exothermic reaction because the heat increases within the volume of the reaction mixture, but the transfer of heat takes place only at the surfaces that are in contact with the reaction mixture—the sides of the flask. Thus, the ratio of the reaction mixture volume to the surface area available for heat transfer is used as a measure to evaluate the risk of controlling an exothermic reaction. This is one reason that, when a reaction has been deemed worthy of commercial interest, the pilot plant scale is used next. In the pilot plant stage, the reaction is carried out under a larger scale to ensure that this can be done safely. Only after this is successfully done can the reaction be scaled up to an industrial plant operation.

Predicting Runaway Reactions

To predict the capability of a reaction to runaway requires that someone carries out a hazard evaluation and a risk assessment of the reaction including its reactants, its possible products, and side reactions/products. The hazards of the chemical reactants and products can often be found in the literature. *Bretherick's Handbook of Reactive Chemical Hazards* provides several examples of reactions that can runaway.[9] These examples include:

- Butene isomers catalyzed by aluminum chloride to form polymers
- Polymerization of phenol and formaldehyde to form resins
- Nitrobenzene production using nitric acid and sulfuric acid
- 2,4,5-Trichlorophenol preparation from hydrolysis of 1,2,4,5-tetrachlorobenzene with sodium hydroxide

This should also include looking at the thermal decomposition of these materials, whether an exothermic reaction is possible, and how much heat and gas could be generated. Thermodynamic and kinetic calculations can also be part of the assessment. Actual testing and measurements employing calorimetric methods may be used for reactions. Overall, there is no set procedure for evaluating all reactions, and not all reactions warrant a thorough investigation. Carefully reviewing the literature can also provide suggestions for carrying out safer reactions. See *Chemical Connection 5.3.10.2* Grignard Reagents and Improved Measures for Safety. As we have always recommended, seek experience in doing a risk assessment of a reaction. This includes your professor, mentor, instructors, and graduate students. We can learn a lot from past experiences and looking at industrial experiences can also teach us something.

CHEMICAL CONNECTION 5.3.10.2

GRIGNARD REAGENTS AND IMPROVED MEASURES FOR SAFETY

Grignard reagents are exceptionally useful tools in synthetic organic chemistry. Victor Grignard won a Nobel Prize in 1912 for the discovery of this reaction. These reagents are not only useful on the laboratory scale but they are also used in the chemical industry at pilot and plant scales. As Incident 5.3.10.2 illustrates, this reaction can present a safety hazard that can potentially lead to a runaway reaction if not carried out under optimal conditions.

Bretherick's Handbook of Reactive Chemical Hazards notes that preparing these reagents can present practical difficulties and lists some methods to bring better control in these preparations.[12]

- While Grignard reagents are typically prepared using highly flammable ethers, such as diethyl ether or tetrahydrofuran, others have suggested that glycol ethers, such as diglyme, can improve safety since they have much higher flash points and higher boiling points.[12,13]

- It is not unusual for the initiation of the reaction to be difficult, delayed, or slow—this can happen if the magnesium turnings are not relatively fresh. The reaction can be started by adding a Grignard reagent from a previous reaction. Iodine has often been used to initiate the reaction but it can have an induction period that can make it difficult to control. Methyl iodide or dibromoethane have also been used to initiate the reaction.[11,14] An "activated" magnesium, prepared using potassium metal, anhydrous magnesium chloride, and potassium iodide, reacts more effectively than other methods.[14]

- Active monitoring of the reaction mixture using near infrared (NIR) has been effective in following the formation of the Grignard reagent.[15]

Learning from Industrial Experience

Several sources have identified runaway reactions as major causes in serious industrial incidents.[10,11] These incidents occurred principally because the reaction was not controlled for any number of reasons, such as changes in standard operating procedures or bypassing of safety measures. The chemical industry is a safe one that actually carries out many of these kinds of operations safely everyday. It is usually only when a series of errors occur that these incidents occur.

Studies of industrial incidents have shown that certain types of reactions have resulted in runaway incidents and we can learn from their experience. Incidents involving the following processes (listed in decreasing order of incidents) have resulted in runaway reactions:[5]

- Polymerization and condensation
- Nitration
- Sulfonation
- Hydrolysis
- Salt formation
- Halogenation
- Alkylation (Friedel–Crafts)
- Amination
- Diazotization
- Oxidation
- Esterification

However, the incidents themselves were not due to the processes but rather errors in using the processes.[5] The errors can be summarized as failures in:

- Adequately understanding the chemistry and thermodynamics of the reaction
- Controlling the temperature of the reaction

- Adding reaction materials properly
- Maintaining systems properly
- Adequately agitating the reaction mixture
- Controlling the quality of raw materials
- Following established procedures

As a student working in the laboratory, you can learn from these experiences by ensuring that you understand the reaction, assess its hazards and risks, and learn how to carry out the reaction properly by taking steps to follow procedures as closely as possible. Remember that scaling up a reaction is not just a matter of multiplying everything by a factor; it is also giving clear thought and planning for managing a larger scale reaction.

One last point here is that major incidents are almost always preceded by a series of other seemingly isolated minor incidents or near misses. Maintain awareness of these smaller incidents, learn from them, and take steps to prevent them from happening again. This will go a long way in preventing a major incident.

Runaway Reactions: RAMP

As noted above, runaway reactions are always preventable.

- *Recognize* the hazard posed by the type of reaction and its scale.
- *Assess* the level of risk by estimating the exothermicity of the reaction and the ability of the system to absorb the heat that is released.
- *Minimize* the risk by keeping the reaction at a reasonable scale and/or providing for a method of cooling through temperature control.
- *Prepare* for emergencies by having a plan to respond to a runaway reaction.

References

1. G. SANTAYANA. BrainyQuote; available at http://www.brainy quote.com/quotes/quotes/g/georgesant101521.html (accessed March 19, 2009).
2. Manufacturing Chemists Association. *Case Histories of Accidents in the Chemical Industry* 1:52 (1962).
3. UK Chemical Reaction Hazards Forum. Grignard Erupts from Flask; available at http://www.crhf.org.uk/incident144.html (accessed April 14, 2009).
4. Health and Safety Executive, United Kingdom. Chemical Reaction Hazards and the Risk of Thermal Runaway; available at http://www.hse.gov.uk/pubns/indg254.htm (accessed March 19, 2009).
5. J. BARTON and R. ROGERS (editors). Chemical Reaction Hazards: A Guide to Safety, 2nd edition, Gulf Publishing Company, Houston, TX, 1997.
6. Chemical Emergency Preparedness and Prevention Office, Environmental Protection Agency. How to Prevent Runaway Reactions—Case Study: Phenol-Formaldehyde Reaction Hazards; available at http://www.epa.gov/emergencies/docs/chem/gpcasstd.pdf (accessed March 19, 2009).
7. L. VAN ROEKEL. Thermal Runaway Reactions: Hazard Evaluation. In: *Chemical Process Hazard Review* (J. M. HOFFMANN and D. C. MASER, eds.), ACS Symposium Series 274, American Chemical Society, Washington, DC, 1985, Chapter 8.
8. J. L. GUSTIN. Runaway reactions, their courses, and the methods to establish safe process conditions. *Risk Analysis* 22(4):475–481 (1992).
9. *Bretherick's Handbook of Reactive Chemical Hazards*, Vols. 1 and 2, Elsevier, New York, 2007.
10. U.S. Chemical Safety and Hazard Investigation Board. Hazard Investigation: Improving Reactive Hazard Management; available at http://www.chemsafety.gov/index.cfm?folder=completed_investigations&page=info&INV_ID=21 (accessed March 21, 2009).
11. TA-CHENG HO, YIH-SHING DUH, and J. R. CHEN. Case studies of incidents in runaway reactions and emergency relief. *Process Safety Progress* 17(4):259–262 (1998).
12. *Bretherick's Handbook of Reactive Chemical Hazards*, 7th edition (P. G. URBEN, ed.), Vol. 2: Grignard Reagents; Elsevier, New York, 2007, pp. 157–158.
13. F. CAREY and R. J. SUNDBERG. *Advanced Organic Chemistry*, Springer, New York, 2007, p. 621.
14. R. D. RIEKE, S. E. BALES, P. M. HUDNALL, T. P. BURNS, and G. S. POINDEXTER. Highly reactive magnesium for the preparation of Grignard reagents: 1-norboranecarboxylic acid. Organic Syntheses Collective Volume 6, p. 845 (1988); Vol. 59, p. 85 (1979); available at http://www.orgsyn.org/orgsyn/prepContent.asp?prep=CV6P0845 (accessed April 15, 2009).
15. J. WISS, M. LÄNZLINGER, and M. WERMUTH. Safety improvement of a Grignard reaction using on-line NIR monitoring. *Organic Process Research and Development Journal* 9(3):365–371 (2005).

QUESTIONS

1. A runaway reaction will occur when

   (a) The rate at which heat produced by a reaction exceeds the rate at which it can be absorbed by the surrounding environment
   (b) A side reaction is exothermic
   (c) Reactions are run in closed vessels
   (d) Impurities in solvents reduce the heat capacity of the solvent

2. Runaway reactions

   (a) Are rare because excess heat usually slows down exothermic reactions
   (b) Would be common if condensers were not routinely used in most reactions
   (c) Get out of control since as heat is released, the reaction rate increases, which further increases the rate at which heat is released
   (d) Usually self-extinguish when all of the solvent has been vaporized

3. "Scaling up" a reaction usually

   (a) Reduces the chance for a runaway reaction because the heat capacity of the system is larger
   (b) Reduces the chance for a runaway reaction because it is harder to add heat to larger systems
   (c) Increases the chance for a runaway reaction because heat transfer occurs at surfaces and the volume-to-surface area value increases

   (d) Increases the chance for a runaway reaction because heat transfer occurs at surfaces and the volume-to-surface area value decreases

4. Predicting that a reaction will be a "runaway risk" is

   (a) Relatively easy if you know the heat of reaction and the order of the reaction
   (b) Relatively easy if you know the heat of reaction and the heat capacity of the solvent
   (c) Not very easy since it is usually hard to determine if a reaction will be exothermic or endothermic
   (d) Usually not very easy since many variables contribute to the rates of heat production and absorption in a reaction

5. What category(ies) of reactions tend to have the most "runaway reaction" events?

   (a) Hydrolysis
   (b) Alkylation (Friedel–Crafts)
   (c) Oxidation
   (d) Polymerization and condensation

6. Major incidents tend to occur when

   (a) There have been a series of minor incidents that precede a major incident
   (b) Reaction materials have not been added together properly
   (c) There is an inadequate understanding of the chemistry and thermodynamics
   (d) All of the above

5.3.11

HAZARDS OF CATALYSTS

Preview   This section discusses the special hazards of catalysts, where you might encounter these, and how you can safely handle these materials in the laboratory.

It has pleased no less than surprised me that of the many studies whereby I have sought to extend the field of general chemistry, the highest scientific distinction that there is today has been awarded for those on catalysis.

Wilhelm Ostwald, German scientist[1]

INCIDENT 5.3.11.1   CATALYST FIRE[2]

The researcher was working alone, having completed a hydrogenation with a palladium charcoal catalyst. Since this catalyst was known to be pyrophoric, he filtered the hydrogenation solution to remove the catalyst using a nitrogen gas stream from an inverted funnel that was placed about 15 inches above the filter funnel. As the solvent was removed and the filter became dry, the catalyst caught on fire. Fortunately, he was working in a chemical hood and was able to extinguish the fire.

What lessons can be learned from this incident?

Catalysts in the Laboratory

Catalysts are important tools in the laboratory chemists' repertoire to transform reactants into desired products. They are also commonly used in industrial chemical processes, particularly in the petroleum industry. One report estimated that 60% of manufactured chemical products used chemical syntheses with catalysts and 90% of petroleum processing involved catalysts.[3] Enzymes are naturally occurring proteins and catalysts for biological processes. They are also widely used in industrial operations, particularly in the detergent and food industries.

Catalysts are often used in the laboratory for hydrogenations. These catalysts are often pyrophoric—they catch on fire in the air. If they are loaded with hydrogen, they can explode. The two most common kinds of hydrogenation catalysts encountered in the laboratory are finely divided metal catalysts, such as Raney® nickel, and precious metal catalysts, including platinum, palladium, rhodium, and ruthenium, especially on carbon powder. Both kinds of catalysts have been involved in fires and explosions. Palladium on carbon, a very common hydrogenation catalyst, is quite pyrophoric and has earned such a notorious reputation that it was listed as one of the "Dirty Dozen" by the National Research Council.[4]

Metal Catalysts

Raney nickel is produced when nickel–aluminum alloy is treated with sodium hydroxide that dissolves the aluminum, leaving many small pores. This newly "activated" porous structure has a very high surface area that is responsible for its catalytic behavior. (The surface area of activated aluminum is in the range of 300 m^2/g. The surface area of activated charcoal can be 500–2000 m^2/g.) Because it

Laboratory Safety for Chemistry Students, by Robert H. Hill, Jr. and David C. Finster
Copyright © 2010 John Wiley & Sons, Inc.

has a high surface area, it can store a high volume of hydrogen that makes it pyrophoric and requires handling in an inert atmosphere. Raney nickel should never be exposed to air. It is normally shipped as a slurry in water. Any exposures to Raney nickel should be considered hazardous since it can cause respiratory injury as well as eye and skin irritation.

Precious metal catalysts are often prepared using highly porous activated carbon that provides a high surface area. See *Chemical Connection 5.3.11.1* Automotive Catalytic Converters. A metal, such as palladium, is bound to the carbon to produce the catalyst, "palladium on carbon." These catalysts are also pyrophoric and pose significant hazard during the recovery phase after hydrogenations. Prior to hydrogen addition they can be used initially in a dry form, but it is safer to use the wet form and the wet form behaves like a dry, free-flowing powder. Using the wet form of this catalyst is a good safety measure. When charging vessels with these catalysts, you should first purge the reaction vessel with nitrogen or argon, and ensure that all vessels are grounded to prevent static electricity. (These apparatus are often made of metal, not glass, since the reactions are often run at high pressure.) All fine combustible powders should be considered at risk for dust explosions. Once these catalysts have been used for hydrogenations, they become saturated with hydrogen, are pyrophoric, and are fire and explosion hazards. Consider all spent (used) catalysts as pyrophoric.

CHEMICAL CONNECTION 5.3.11.1

AUTOMOTIVE CATALYTIC CONVERTERS

Most of today's motor vehicles contain a catalytic converter to reduce the toxicity of emissions from their internal combustion engines. These catalytic converters have been used for many years as a means to comply with EPA emission standards. Since 1981 most vehicles have used three-way converters. They convert emissions to safer products in three ways. Exhausts, before they reach the catalytic converter, contain nitrogen oxides, carbon monoxide, and some amount of noncombusted hydrocarbons from gasoline. In the catalytic converter they are converted to less toxic products, as shown in the following examples:

$$2\,NO_2 \rightarrow 2\,O_2 + N_2 \qquad \text{(reduction of nitrogen oxides)}$$
$$2\,CO + O_2 \rightarrow 2\,CO_2 \qquad \text{(oxidation of carbon monoxide)}$$
$$C_6H_{12} + 9\,O_2 \rightarrow 6\,CO_2 + 6\,H_2O \qquad \text{(oxidation of noncombusted hydrocarbons)}$$

These catalytic converters contain a high surface area, a honeycombed ceramic or stainless steel core that is coated with silica and alumina, called a washcoat. Precious metal catalysts, such as platinum, palladium, and rhodium, are added as a suspension to the washcoat. As the hot gases pass through the catalytic converter, they are converted by the catalysts to the reduced or oxidized products.

As you might guess, these catalytic converters are expensive, so it is important to take care of them. As with many catalysts they can be "poisoned" (rendered ineffective) by substances from the exhaust that can coat the working surface of the catalyst. Lead is the most common poison of these catalytic converters. From 1923 to 1986 gasoline contained tetraethyl lead as an antiknock additive in gasoline despite the fact that its health effects were known since the 1920s. It was removed from gasoline largely due to its poisoning effect on the catalytic converter. Other common poisons include manganese, silicon, and phosphorus.

Recovery of these metal catalysts after hydrogenations is a potentially hazardous operation. It is important to keep the filter cake containing the catalyst wet since it is more hazardous when dry. Inert gases such as nitrogen or argon should be used to purge the catalyst as it is being filtered. In Incident 5.3.11.1, the funnel providing the inert gas was too far above the filter apparatus and air was allowed to enter the mixture as it was filtered. Once it became dry, it ignited.

Enzyme Catalysts

Another type of catalyst is the enzyme and the nature of its hazards is different. Enzymes are naturally occurring proteins that carry out essential processes in our bodies and other living systems. Enzymes may be isolated from living organisms and these have been found to be useful tools for transformations in the laboratory. Laboratory exposure to these purified or powdered forms of enzyme can lead to sensitization, an allergic reaction causing rhinitis or asthma (or both). These allergic reactions can be very serious, resulting in sensitizations that make working in that laboratory untenable.

As with other chemicals in the laboratory, you must eliminate or minimize exposures to enzymes. When using an enzyme, carry out operations in a chemical laboratory hood to prevent exposure to aerosols that can result from handling enzymes or any solid. Protect your skin with protective gloves, long sleeves, and a lab coat. (We assume that you will always be wearing chemical splash goggles!) Wash any areas on benchtops that have been potentially contaminated with enzymes. These measures are particularly important with sensitizers, such as enzymes, to prevent an immunological response (allergy) from inadvertent exposures. After an allergic response has developed, even minute amounts of these materials can evoke a strong and even serious reaction.

RAMP

- *Recognize* that catalysts, even in small amounts, can be hazardous.

- *Assess* the areas of potential risk of exposures to catalysts.

- *Minimize* exposure to the hazards of catalysts by carefully planning experiments and taking steps to ensure you are protected from exposures and their consequences, such as fires, explosions, and allergic reactions.

- *Prepare* for emergencies that might result from handling catalysts—know what to do in case of fires and explosions. Know how to clean up spills of potentially hazardous materials.

References

1. W. Ostwald. BrainyQuote; available at http://www.brainyquote.com/quotes/quotes/w/wilhelmost311682.html (accessed June 20, 2009).
2. UK Chemical Reaction Forum: Catalyst Fire with Lone Working; available at http://www.crhf.org.uk/incident60.html (accessed November 5, 2008).
3. F. P. Daly. Hazardous Catalysts in the Laboratory. In: Handbook of Chemical Health and Safety (R. J. Alaimo, ed.), American Chemical Society Oxford University Press, Washington, DC, 2001, pp. 345–347.
4. National Research Council. *Prudent Practices in the Laboratory: Handling and Disposal of Chemicals*, National Academy Press, Washington, DC, 1995, p. 58.

QUESTIONS

1. What percentage of manufactured chemical products involves catalytic reactions?

   (a) 10%
   (b) 37%
   (c) 60%
   (d) 95%

2. What is a common laboratory procedure that uses a pyrophoric catalyst?

   (a) Esterification
   (b) Acid neutralization
   (c) Titrations using oxidants
   (d) Hydrogenation

3. Raney nickel

   (a) Is actually an aluminum compound
   (b) Has a high surface area that stores sodium hydroxide solution
   (c) Should be handled only in an inert atmosphere
   (d) Is normally shipped as a slurry in acetonitrile

4. When using precious metal catalysts,

   (a) They should be kept wet after use
   (b) The reaction vessels should first be purged with nitrogen or argon
   (c) Reaction vessels should be grounded to prevent static electricity sparks
   (d) All of the above

5. When using enzymes as catalysts,

   (a) They must be used in an inert atmosphere
   (b) Precautions must be taken to avoid exposure to powdered forms that can cause sensitization reactions
   (c) Protective equipment is generally not required since the enzymes are naturally occurring
   (d) Working in a chemical hood is necessary only when the enzyme is volatile

CHAPTER 6

RISK ASSESSMENT

WHO DECIDES what's "safe"? And how do they make that determination? Another principle of safety is *assessing the risks of hazards*. Understanding how we each view risk affects how we behave in the laboratory, and affects how we interpret safety information. Not everyone makes the same choice given a set of dangers and options. This chapter is about the process of risk assessment and the means by which scientists share information about hazard levels posed by chemicals. Students involved in research projects have to learn to develop their sense of judgment about risk as they plan and conduct experiments. Will you know how to decide what's "safe" when presented with the design of a new experiment?

INTRODUCTORY

6.1.1 Risk Assessment—Living Safely with Hazards Presents a broad view of risk assessment of laboratory hazards, and how to apply it to your laboratory operations.

INTERMEDIATE

6.2.1 Using the GHS to Evaluate Chemical Toxic Hazards Explains how you use the Globally Harmonized System of Classification and Labelling of Chemicals (GHS) to evaluate toxicity data to determine the relative risk associated with potential exposure to a toxic chemical.

6.2.2 Understanding Occupational Exposure Limits Explains the common systems for occupational exposure limits (OELs)—threshold limit values (TLVs) and permissible exposure limits (PELs)—and how to use these data to judge the relative hazard of a chemical.

ADVANCED

6.3.1 Assessing Chemical Exposure Provides a brief overview of some methods to measure and assess exposures to chemicals using air monitoring techniques and biological monitoring techniques.

6.3.2 Working or Visiting in a New Laboratory Presents questions and ideas to consider when starting work in a new laboratory or visiting a new laboratory.

6.3.3 Safety Planning for New Experiments Discusses how planning can help you prepare to conduct an experiment safely.

6.1.1

RISK ASSESSMENT—LIVING SAFELY WITH HAZARDS

Preview This section presents a broad view of risk assessment of laboratory hazards, and how to apply it to your laboratory operations.

If you want to become a chemist, so Liebig told me, when I worked in his laboratory, you have to ruin your health. Who does not ruin his health by his studies nowadays will not get anywhere in chemistry.

August Kekulé, discoverer of the benzene structure[1]

INCIDENT 6.1.1.1 FATAL LAB EXPOSURES[2]

Three researchers were synthesizing a series of new chemicals called bicycloheptadiene dibromides. These compounds had never been made or previously been reported. During their work to prepare these compounds, these chemists were exposed to these compounds but the extent of exposure was not reported and was probably unknown. Nevertheless, two of the three researchers developed similar pulmonary disorders that were fatal. The third scientist became sensitized to these compounds. Further research with these compounds was discontinued.

What lessons can be learned from this incident?

Risk in Labs

Many activities that we encounter every day have some risk associated with them. Driving a car is a risky endeavor, taking medicines with side effects has risks associated with it, working at a computer has ergonomic risks, and even walking down a flight of stairs can be risky. These are individual activities, but we can easily consider activities of our culture like operating a nuclear power plant, manufacturing cars, or substituting ethanol for gasoline as an automobile fuel, that also have risk–benefit trade-offs. Simply put: all activities have some level of risk associated with them despite our occasional desire for a "no risk" activity. We are all at risk of something every day in all that we do and there is nothing that we can do to avoid this.

Chemistry labs are no different. There are risks associated with using chemicals and equipment in labs and you have already read several sections in Chapters 3, 4, and 5 about recognizing a wide range of hazards in chemistry labs. One of the most important goals of this book is to help you develop a keen sense of how to recognize the risks associated with working in chemistry labs and how to reasonably minimize those risks (Chapters 7 and 8). Additionally, just by reading various sections in this book periodically, you will increase your awareness and vigilance for hazards. It is not realistic to have the goal of completely eliminating risk since the only way to do this is to avoid doing anything, either in the lab or elsewhere. This section is about how to assess risks in a more systematic fashion so that they can be better understood and minimized.

In the early days of chemistry over a hundred years ago, labs were not safe places. There were no fume hoods and no safety equipment, and the hazards of the chemicals were not always appreciated. The opening quote by Kekulé reflects the attitude that the risks of labs simply had to be endured and one's health necessarily suffered. This was true at the time, but today we have the knowledge and tools to drastically reduce the risk of harm when working in chemistry labs.

Hazard and Risk—They Are Different!

We previously defined hazard and risk in Section 1.1.1. Let's take another look at these definitions since it is important that you clearly understand the difference between these two terms. They are often confused and sometimes regarded as one in the same, particularly in the public's view.[3]

> A *hazard* is a potential source of danger or harm, where potential implies that something is capable of being dangerous or harmful. A chemical or piece of equipment may have one or more inherently hazardous properties. A hazard is always going to be with this material and its does not change. A hazardous material will be just as hazardous next week or next year as it is now.

> *Risk* is the probability of suffering harm from being exposed to an unsafe situation or a hazardous material. Risk can almost always be reduced to acceptable levels if the nature of the risk is recognized and understood since preventive measures can be implemented. For example, in the laboratory we are able to use a flammable chemical such as acetone because we can take precautions to prevent a fire by keeping the flammable material away from ignition sources and using ventilation to remove vapors so that flammable air concentrations do not form.

The *risk equation* shows a useful way to think about risk as a combination of the nature (severity) of the hazard and the likelihood of the hazardous event occurring.

$$\text{Risk level} = \text{Severity of the hazard} \times \text{Probability of exposure to the hazard}$$

This is more of a conceptual equation for us now, although it would be possible to assign numerical values for these variables. A risk can be tiny if the probability of the event is nearly zero (no matter how significant the hazard). For example, a sealed vial of a very toxic chemical presents virtually no risk if it is just sitting in a chemical hood and not being handled. The hazard level is high but there is virtually no probability of exposure, so the risk is very low.

Alternately, a lab procedure like the titration of an acid with a base to the endpoint can have a high probability of overshooting the endpoint (particularly by someone new to titrations) but the hazard is essentially zero since overshooting an endpoint is not a dangerous event. Thus, this is a very low-risk event.

Understanding risk as the combination of these two factors can greatly improve the analysis of risk. Sometimes it is better to reduce the hazard level (perhaps by substituting a flammable solvent with a nonflammable solvent) or reduce the amount of hazardous material handled, and sometimes it is better to reduce the probability of an event occurring (by carefully eliminating ignition sources when using flammable solvents). This leads us to consider "risk assessment" versus "risk management."

Risk Assessment Versus Risk Management

It can be helpful to distinguish between risk assessment and risk management. These are two procedures that work in sequence: one should first identify (assess) all of the risks associated with a chemical or procedure. This can often be relatively straightforward if one is thorough, is knowledgeable, and maintains a constant vigilance for hazards. But, for example, as you know from earlier chapters on hazard recognition, we sometimes don't fully know the level of health risk associated with some chemicals. A faulty piece of equipment can be very hazardous, too, and sometimes it would be nearly impossible to know about this until the equipment fails. The key here is to do the best risk assessment possible.

The process of risk management is sometimes easy and sometimes difficult and problematic. It is easy to wear personal protective equipment like safety goggles and to take steps like not using flammable solvents near ignition sources. The severity of a hazard is dependent not only on its nature but also on the amount of the hazard to which you are exposed. Reducing the amount of hazard is a key risk management strategy. Thus, if you are handling a chemical, the severity is greater if you are handling a container with 500 g than if you are handling a container with 1 g. This is also true if you are using a chemical or reaction that has explosive potential; if you reduce the amount then you reduce the severity of the hazard. If you are close to an unshielded explosion, the event is likely to be very severe, but if you are some distance away or you have shielded the explosive vessel from you and others the effect of the event will be less severe.

In some labs and in many industrial environments, however, there can be significant costs (in time and money) in reducing the risk level. Those who manage risk, which may be laboratory workers or their supervisors (who may or may not have technical training), have to make decisions about how to minimize risks and these are sometimes difficult judgment calls. Many incidents have been described throughout this book where the risk assessment and risk management processes simply did not work effectively.

The process of risk management is also guided by various federal laws related to the use of chemicals in labs and in industry. This is discussed more in Section 7.1.1.

Assessing Probability of Exposure to a Hazard

Sections 6.2.1 and 6.2.2 will describe in more detail how to assess the *hazard level* of chemicals. In many instances we'll find that we can apply hazard classes to chemicals that will give a good general indication of the danger from a chemical.

We can identify three factors that affect the *probability* of a lab accident: how we use chemicals, how we control the physical environment, and the behavior of other people. Let's look at each one.

For exposures to chemicals with regard to health effects, we have already presented the routes of exposure in Section 3.1.1: skin, eyes, inhalation, or ingestion. The probability of exposure will be determined by the degree to which the chemical is "released" and the probability of entering the body through the routes of exposure. Containing chemicals in closed or mostly closed systems and using fume hoods can avoid contact. Beyond that, laboratory workers use personal protective equipment as the final layer of defense. The combination of these two strategies can greatly minimize the possibility of exposure.

Other events in the lab such as fires, explosions, physical harm (such as slipping on a wet floor or having a gas cylinder fall over on your foot) can all be minimized by recognizing potentially hazardous situations, which requires a knowledge of hazards and a constant vigilance for these hazards. It is often quite easy to reduce the probability of some incident occurring by simply taking the time to step back and examine the situation and procedure. Even a tiny spill of water on a floor can be slippery and preventing a fall is easy by removing the spill. Gas cylinders can easily be secured. Virtually all fires can be prevented (and could have been prevented) by removing ignition sources from areas where flammables chemicals are used. Electrical shocks can be prevented by following the electrical code and using ground fault interrupters. Only rarely do lab accidents occur that were not preventable.

Human error is always preventable and you can have considerable control over the probability that *you* make some mistake in a lab. However, since you cannot control other people in the lab, someone else may cause an accident that injures you and is "beyond your control." Assessing the probability of other people making mistakes depends on understanding the experience and attitudes of other students or co-workers. It is simply prudent to remember that the overall risk assessment process must involve all sources of risk in any environment.

Risk Assessment

Table 6.1.1.1 presents the combination of "severity of the hazard" and "probability of exposure to the hazard" in the form of a matrix.[3-5] The risk is expressed in numbers 1 to 4, where the lower number represents higher risk—similar to the risk categories established by the Globally Harmonized System[6] (see Sections 3.2.1 and 6.2.1).

Severity can broadly be classified in terms of outcome: catastrophic, significant/serious, moderate, and minor.

TABLE 6.1.1.1   Risk Assessment Matrix

Severity of hazard	Probability of exposure to the hazard				
	Very likely	Likely	Possible	Unlikely	Very unlikely
Catastrophic	1	1	1	2	2
	Danger	Danger	Danger	Warning	Warning
Significant/serious	1	1	2	2	3
	Danger	Danger	Warning	Warning	Caution
Moderate	1	2	3	3	3
	Danger	Warning	Caution	Caution	Caution
Minor	3	3	4	4	4
	Caution	Caution	Care	Care	Care

Source: Adapted from Refs. 3–5.

- *Catastrophic* means that you and others could suffer fatal injuries, and/or damages to property could be huge.

- *Significant/serious* means that you and others could receive serious, perhaps life-threatening, injuries, and damages to property could be significant and costly.

- *Moderate* means that you or others could be injured and these injuries while perhaps painful are treatable, allowing you to survive, and only moderate damage to your laboratory is likely.

- *Minor* means that you could suffer minor injuries that require some first aid or self-treatment, with damage likely only to the immediate area.

Of course, this is mostly a conceptual table. While in some instances we may have some numerical rating of the severity of hazard, the rating of "probability of occurrence" is a judgment call based on the analysis of the situation. However, when using good judgment and perhaps seeking the judgment of experienced mentors and other lab workers, you can make a reasonable assessment of the overall risk level. Whenever the hazard–probability combination puts you in situations characterized as "Danger" or "Warning," it is time to reconsider the plan of action. There should not be a time when taking on these levels of risk is necessary. Figure out a way to reduce the hazard and/or reduce the probability. See *Special Topic 6.1.1.1* Perception of Risk.

SPECIAL TOPIC *6.1.1.1*

PERCEPTION OF RISK

Although there are countless studies of actual risks in terms of "percent fatality," the occurrence of diseases, or accident rates in industry, people often make subjective judgments about the relative hazard of a behavior or substance and the likelihood of its occurrence. For example, we know that about 40,000 people will die in highway accidents each year but nearly everyone engages in this activity on a daily basis. Despite the known risks of smoking cigarettes or illicit drug use, many people engage in these activities. To many people then, there are actions or behaviors that they have decided are "acceptable risks." What factors affect the perception of risk?

Why do some people fear flying more than driving in a car even though the statistics indicate that flying is a safer mode of transportation? Many factors affect this perception of risk. Table 6.1.1.2 shows a variety of factors that affect how people perceive risk.[7] When driving, a person has a reasonably high degree of *control* of the event; but when flying, passengers have *no control* over the plane. Driving is a *familiar* event; flying is *unfamiliar* (to some). It may seem *necessary* to drive (as a part of our lifestyle in the United States) but flying is *unnecessary*. The severity of a car accident can vary but many are *ordinary*; the severity of plane crashes almost always is *catastrophic*. All of these factors can lead someone to hold a statistically unjustified fear of flying.

Similarly, using a tanning booth is a choice many people (particularly college students!) make, even though the risks of skin cancer are very significant. Tanning booths are *natural* in the sense that they emit the same UV radiation as the sun. This is a *voluntary* activity, the effects are *immediate*, the *benefits* are *clear*, and the risk of skin cancer seems remote or *ordinary*. Skin cancer is not perceived as *catastrophic*. The use of these booths is

TABLE 6.1.1.2   Risk Perception Criteria

Criteria	Characteristics perceived as low risk	Characteristics perceived as high risk
Origin	Natural	Nonnatural
Volition	Voluntary	Involuntary
Effect manifestation	Immediate	Delayed
Severity	Ordinary	Catastrophic
Controllability	Controllable	Uncontrollable
Benefit	Clear	Unclear
Familiarity	Familiar	Unfamiliar
Exposure	Continuous	Occasional
Necessity	Necessary	Luxury

Source: Reproduced with permission from *Chemical Risk: A Primer*, Copyright © 1984 American Chemical Society, ACS Office of Legislative and Government Affairs.

controllable, and a nice tan is a *familiar* and desirable attribute. Since sun exposure produces tanning, this booth tanning is just a part of the *continuous* process that occurs in today's world. For some, "this tan is something I want and I view it as *necessary* for being more attractive to others."

How do people view chemicals? Studies have shown that people will accept a risk level of 1 in 10,000 from aflatoxin in order to eat peanut butter and reject a risk level of 1 in 1,000,000 for some synthetic chemical that has some benefit also associated with it. *Natural* is seen as less risky than *synthetic* or *nonnatural*.

How does this affect your behavior in the lab? Since you are becoming *familiar* with chemicals you now know that there is no connection between hazard level and whether the chemical is *natural* or *synthetic*. Long chemical names that are *unfamiliar* to nonchemists do not scare you (except perhaps on organic exams). Thus, you are less susceptible to misperceiving risks for some of the variables in the table.

However, as you work more in labs, you work with more chemicals and they become *familiar* to you. You realize that you can work with chemicals safely, even though they exhibit a wide range of hazard level. But perhaps you can become "too familiar." Most chemicals that you use will be safe to work with if you take routine cautions and use PPE. However, some chemicals really *are* quite hazardous and you must not let your "familiarity" allow you to "underperceive" or underestimate a hazard and drop your guard in these cases. For many chemicals, too, we simply don't know the toxicity hazard and Incident 6.1.1.1 illustrates how a false assumption can be deadly. There can also be pressures (both in academia and industry) to "get experiments done quickly" and this can lead to shortcuts that compromise safety. Once you have done your 25th distillation, without mishap, you may get sloppy and take an unnecessary risk. You will become very familiar with working with flammable solvents and probably never have a fire and this may let you get less concerned about always scrutinizing the area for ignition sources.

As scientists, we can be somewhat less susceptible to making errors in assessing risk level based on the factors in Table 6.1.1.1, but we are not immune from them. Many lab incidents occur to experienced scientists who simply misjudged a risk level or knowingly took an unnecessary risk—they exhibited at-risk behavior.

1. *Extreme risk.* Fatalities are very likely to result; damage to facilities and surrounding areas is likely to be devastating. You must not undertake this task without taking strong risk reduction steps.

2. *High risk.* Serious injuries and fatalities may occur; damage to facilities is likely to be major. You must not undertake this task without taking strong risk reduction steps.

3. *Moderate risk.* Some serious injuries could occur; damage is likely to be restricted to only part of a building or only one laboratory. After you consider further risk reduction steps, you can undertake this task.

4. *Low risk.* Only minor injuries may occur; any damage will be only in the immediate area. You can undertake this task but should continually review the task to ensure that risk levels do not increase. Always look for other ways to reduce risks.

Carrying Out Risk Assessment and Risk Management—An Example

To use Table 6.1.1.1 you will need to know what hazard you will encounter in doing a given operation. Let's look at an example to illustrate how this might be used.

Task: *Prepare 1 liter of 1 M sulfuric acid from concentrated sulfuric acid.*

Concentrated sulfuric acid is very hazardous. It is a strong acid of very high concentration, an excellent dehydrating agent, and a corrosive and can cause severe burns to the skin and eyes. It will react with many other chemicals. Vapors are very hazardous to the lungs. It is often purchased in 4-L bottles. (See Table 6.1.1.3.)

Without taking any precautions, this activity would surely generate an *extreme risk* ("Danger") due to the number of "serious" events that could occur and the possibility that the events could occur. Surely, breathing the vapors is virtually guaranteed if a hood is not used. However, if all of the risk management strategies are used, one can reduce the probability assessments to "very unlikely." Overall, the procedure can be recategorized as "care," which suggests that it can be performed safely as long as the proper precautions are taken.

RAMP

All of the steps in RAMP are part of the risk assessment and management process.

- *Recognize* the hazards associated with chemicals and procedures. Consider *all* of the hazards of chemicals and *all* of the things that can go wrong, not just the obvious and likely event.
- *Assess* the probability of the hazardous event occurring.
- *Minimize* the hazards by eliminating them and/or taking steps to reduce the probability of the event occurring.
- Since probabilities can rarely be reduced to zero, *prepare* for emergencies that might result if *anything* goes wrong. Have an action plan in mind for spills and exposures.

TABLE 6.1.1.3   Example of Risk Assessment and Risk Management from Preparing Dilute Sulfuric Acid

Event (assumes no steps taken to prevent event or wear PPE)	Severity	Probability	Risk management strategy
Sulfuric acid spills on skin while pouring and measuring volume of acid	Moderate	Possible	Wear arm-length butyl gloves; wear lab coat, appropriate clothes, and shoes; use face shield; work carefully; know what to do if a spill occurs
Sulfuric acid spills with other reactive chemicals around	Moderate	Possible	Work in a clear area without other reactive chemicals
A bottle of sulfuric acid in transit from a storage cabinet to a hood drops and breaks	Moderate	Possible	Use a rubber bucket to carry the sulfuric acid bottle; purchase acid in plastic-coated bottle; know what to do if a spill occurs
Sulfuric acid vapors are breathed in	Serious	Very likely	Work in a chemical hood; keep heat out of fume hood; know what to do if some vapor is inhaled
Spattering occurs from mixing water into acid	Moderate	Possible	Add acid to water (not water to acid); know what to do if acid spatters due to incorrect mixing procedure
Sulfuric acid splashes in eyes	Serious	Possible	Wear face shield and splash goggles; work carefully; know what to do if acid gets in eyes

References

1. R. Purchase (editor). *The Laboratory Environment*, Special Publication No. 136, Royal Chemical Society, Cambridge, UK, 1994, as cited in *Prudent Practices in the Laboratory: Handling and Disposal of Chemicals*, National Academy Press, Washington, DC, 1995, p. 14.
2. S. Winstein. Bicycloheptadiene dibromides. *Journal of the American Chemical Society* **83**:1516–1517 (1961).
3. John A. Singley. Hazard versus risk. *Journal of Chemical Health and Safety* **11**(*1*):14–16 (2004).
4. J. Stephenson. *Systems Safety 2000*, John Wiley & Sons, Hoboken, NJ, 1991.
5. E. I. DuPont de Nemours and Company. *Safety Management Leadership Workshop for Battelle Executives and Managers, Participant's Workbook*, 2006.
6. United Nations. *Globally Harmonized System for Classification and Labelling of Chemicals (GHS)*, United Nations, New York, 2005.
7. *Chemical Risk: A Primer*. American Chemical Society information pamphlet, 1984, Department of Governmental Relations and Science Policy, American Chemical Society, 1155 Sixteenth Street, N.W., Washington, DC, 20036.

QUESTIONS

1. Which combination of parameters generates the highest risk level?

   (a) Low hazard and low probability of exposure
   (b) Low hazard and high probability of exposure
   (c) High hazard and low probability of exposure
   (d) High hazard and high probability of exposure

2. Which statement is *true*?

   (a) Risk assessment and risk management are essentially the same process
   (b) Risk assessment precedes risk management
   (c) Risk management precedes risk assessment
   (d) Risk assessment is usually easy and risk management is usually difficult

3. What is a common way to minimize exposure to chemicals, as a first measure of reducing the probability of exposure?

   (a) Using chemicals in a closed system
   (b) Using chemicals in a chemical hood
   (c) Using personal protective equipment
   (d) c, plus a or b

4. What word is used to represent the highest level of severity in risk assessment?

   (a) Catastrophic
   (b) Significant
   (c) Serious
   (d) Deadly

5. What level of risk is associated with using concentrated sulfuric acid if no precautions are taken?

   (a) Care; low risk
   (b) Caution; moderate risk
   (c) Warning; high risk
   (d) Danger; extreme risk

USING THE GHS TO EVALUATE CHEMICAL TOXIC HAZARDS

Preview This section explains how you can use the Globally Harmonized System of Classification and Labelling of Chemicals (GHS) to evaluate toxicity data to determine the relative risk associated with potential exposure to a toxic chemical.

Something is rotten in the state of Denmark.

William Shakespeare, *Hamlet*[1]

INCIDENT 6.2.1.1   HYDROGEN SULFIDE LEAK[2]

Hydrogen sulfide, a very toxic gas, was being used in an experiment within a chemical hood. The hydrogen sulfide gas cylinder was too large to fit into the hood and it was attached adjacent to, but outside, the hood. The laboratory worker detected the characteristic "rotten egg" odor of hydrogen sulfide and tried unsuccessfully to find the source. He evacuated the laboratory and called the health and safety office for help. The health and safety team entered the laboratory using self-contained breathing apparatus and finally found the leak at the valve of the cylinder. When they could not stop the leak, they moved it to the outdoors and contacted the supplier. After receiving further instructions, they used a wrench and hammer to close the valve and stop the leak. The malfunctioning cylinder was returned to the supplier.

What lessons can be learned from this incident?

Globally Harmonized System of Classification and Labelling of Chemicals (GHS)[3]

In Section 3.2.1 we introduced the GHS as a system that could be used to help you evaluate data pertaining to hazardous properties. GHS has been derived from a consensus of international participants and establishes the criteria that define hazard classes (or hazard types) of chemicals such as flammable, corrosive, or acute toxicity. Then each hazard class is subdivided into relative hazard categories (relative ratings or rankings). By using this kind of framework you can separate the hazards of chemicals into hazard classes, such as flammable, corrosive, acutely toxic, and other hazard classes. Once you have identified hazard classes for a chemical, you can then determine the relative category of the hazard using the described GHS criteria and this in turn will give you information to judge how hazardous it is to handle this compound in your experiments. Useful information will be provided by the GHS label on the chemical container and may also be found in Safety Data Sheets (SDSs).

You may recall that the GHS identifies sixteen classes of physical hazards, eleven classes of health hazards, and one class (two subclasses) of environmental hazards. This system may seem quite complicated, and perhaps unnecessarily so, but the simple truth is that there are many classes of hazards posed by chemicals and the GHS is an attempt to use a labeling system for users and handlers that can quickly identify the hazardous properties of chemicals. A total of 28 classes with several different rating systems is not simple. Our goal is to present the GHS without making it seem more, or less, complicated than it really is.

Laboratory Safety for Chemistry Students, by Robert H. Hill, Jr. and David C. Finster
Copyright © 2010 John Wiley & Sons, Inc.

TABLE 6.2.1.1   Health Hazards in the GHS

Health hazards	GHS rating system[3] (hazard categories)	See Table
Acute toxicity (oral, dermal, inhalation)	1, 2, 3, 4, 5	6.2.1.2, 6.2.1.3, 6.2.1.4
Skin corrosion/irritation	1A, 1B, 1C, 2, 3	6.2.1.5
Serious eye damage and eye irritation	1, 2A, 2B	6.2.1.5
Respiratory sensitization	1	6.2.1.5
Skin sensitization	1	6.2.1.5
Germ cell mutagenicity	1A, 1B, 2	6.2.1.5
Carcinogenicity	1A, 1B, 2	6.2.1.5
Reproductive toxicity	1A, 1B, 2	6.2.1.5
Target organ systemic toxicity (TOST): single exposure	1, 2, 3	6.2.1.5
Target organ systemic toxicity (TOST): repeated exposure	1, 2	6.2.1.5
Aspiration hazard	1, 2	6.2.1.5

The eleven health hazard classes and their category rating systems are reproduced in Table 6.2.1.1. As a reminder, we note that the hazard categories (HCs) are designed so that most hazardous substances have low HC values and the least hazardous substances have the higher HC values.

As we discuss the GHS ratings below, the chemical manufacturers or suppliers of these chemicals are responsible for classifying and categorizing the chemicals they have based on scientific data. We explain the category ratings to you so that you will have some understanding of the basis for the ratings. While the GHS is complex and detailed, you will not have to know about the process of classifying and categorizing chemicals in order to be able to make use of the ratings. Rather, your job will be to use the category ratings in assessing your hazards. Knowing that *the lower the number the more hazardous the chemical*, will help you in recognizing and dealing with your chemical hazards.

Although not obvious from Table 6.2.1.1, it is convenient to separate the first hazard class of "acute toxicity" from all of the others since it is the only class for which it is fairly easy to find *data* about the effect of exposures. It is possible to locate LD_{50} (ingestion) and LC_{50} (inhalation) values. (See Section 3.2.2.) Somewhat less common, but still available, are LD_{50} values for dermal exposure. In order that you can examine these kinds of data using the GHS system, we include three separate tables (Tables 6.2.1.2, 6.2.1.3, and 6.2.1.4) for these three modes of exposure regarding acute toxicity. We discuss these tables more below.

Using Acute Toxicity Data to Judge the Importance of Chemical Toxic Hazards

Data used in assessing the toxicity of chemicals comes from past experience, in part from actual human exposures in workplaces, from unintentional and intentional poisonings, but principally from animal testing. These data are sometimes found in MSDSs (or SDSs) and labels, but we have previously cautioned (in Sections 3.1.3 and 3.2.3) that the data in MSDSs may not always be reliable and we suggest that the best and most authoritative sources be used for toxicity data as described in Section 3.2.2.

When you do locate data about toxicity, such as an LD_{50}, what do you do with these data? How do you determine the relative hazard of the chemicals you are using? The GHS allows sorting by hazard class and category that was previously unavailable.

Generally, the first step in using the GHS will be assessing acute chemical toxicity information reported for chemicals by oral, dermal, or inhalation routes. Using Table 6.2.1.2, you can assess the relative acute oral toxicity of chemicals with reported information about oral LD_{50} values. The table provides a hazard statement and GHS "signal words" such as *Danger*. For acute oral toxicity, there are five hazard categories (HC 1–5) based on oral LD_{50} values. The GHS also uses pictograms as a form of labeling and identification of hazard category. In Figure 6.2.1.1 some of these pictograms are shown along with the health hazard classes they represent (see Figure 3.2.1.1 for other pictograms).

As you know, the route of toxicity is not only oral, but exposure through the skin and by inhalation can also occur. GHS hazard categories for dermal and inhalation hazards are shown in

TABLE 6.2.1.2   Relative Acute Toxicity Categories of Chemicals Using LD$_{50}$ Values by Oral (Ingestion)

Hazard statement	GHS hazard category (HC)	Signal word	Reported LD$_{50}$ (mg/kg)	Examples (toxicity other than oral may place these chemicals in another hazard category)
Fatal if swallowed	1	Danger	≤5	Botulinum toxin, phosphorus, sarin, potassium cyanide, sodium fluoroacetate, tetrodotoxin
Fatal if swallowed	2	Danger	>5 to ≤50	Acrolein, hydrogen cyanide, osmium tetroxide, sodium azide, sodium cyanide, trimethyltin chloride
Toxic if swallowed	3	Danger	>50 to ≤300	Aniline, acrylonitrile, chromium trioxide, diethylnitrosoamine, hydrazine, methyl iodide, sodium hydroxide, trifluoroacetic acid
Harmful if swallowed	4	Warning	>300 to ≤2000	Acetaldehyde, benzene, chloroform, chloromethyl methyl ether, dichloromethane, diethyl ether, formaldehyde, phenol, pyridine
May be harmful if swallowed	5	Warning	>2000 to ≤5000	Acetic acid, carbon disulfide, carbon tetrachloride, dimethylformamide, methyl ethyl ketone, sulfuric acid, tetrahydrofuran, toluene diisocyanate

Source: Parts of the information in this table were provided courtesy of the United Nations Economic Commission of Europe from the *Globally Harmonized System of Classification and Labelling of Chemicals*. Copyright © 2007 United Nations, New York and Geneva.

TABLE 6.2.1.3   Relative Acute Toxicity Categories of Chemicals Using LD$_{50}$ Values by Dermal Exposure

Hazard statement	GHS hazard category (HC)	Signal word	Reported LD$_{50}$(mg/kg)	Examples (toxicity other than dermal may place these chemicals in another hazard category)
Fatal in contact with skin	1	Danger	≤50	Sodium azide, crotonaldehyde
Fatal in contact with skin	2	Danger	>50 to ≤200	Hydrazine, allyl alcohol, benzenemercaptan, chloroacetonitrile
Toxic in contact with skin	3	Danger	> 200 to ≤1000	Acrolein, acrylamide, acrylonitrile, aniline, *t*-butylhydroperoxide, ethylene dibromide, formaldehyde, hydrogen peroxide, methyl iodide, phenol
Harmful in contact with skin	4	Warning	> 1000 to ≤2000	Acetic acid, acetonitrile, hydrogen cyanide, pyridine, potassium hydroxide
May be harmful in contact with skin	5	Warning	> 2000 to ≤5000	Aluminum trichloride, dimethylformamide, hexamethylphosphosamide

Source: Parts of the information in this table were provided courtesy of the United Nations Economic Commission of Europe from the *Globally Harmonized System of Classification and Labelling of Chemicals*. Copyright © 2007 United Nations, New York and Geneva.

TABLE 6.2.1.4   Relative Acute Toxicity Categories of Chemicals Using LC$_{50}$ Values by Inhalation

Hazard statement	GHS hazard category (HC)	Signal word	Reported LC$_{50}$	Examples (toxicity other than inhalation may place these chemicals in another hazard category)
Fatal if inhaled	1	Danger	≤100 ppm (gas);	Acrolein, arsine, chloromethyl methyl ether, diborane, hydrogen cyanide, hydrogen fluoride, iodine, mercury, nitrogen dioxide, ozone, toluene diisocyanate
			≤0.5 mg/L (vapor); ≤0.05 mg/L (dust, mist)	
Fatal if inhaled	2	Danger	> 100 to ≤500 ppm (gas);	Acrylonitrile, aniline, boron trifluoride, chlorine, diazomethane, fluorine, formaldehyde, hydrogen sulfide, phosgene
			> 0.5 to ≤2.0 mg/L (vapor); > 0.05 to ≤0.5 mg/L (dust, mist)	
Toxic if inhaled	3	Danger	> 500 to ≤2500 ppm (gas); > 2 to ≤10 mg/L (vapor); > 0.5 to ≤1 mg/L (dust, mist)	Ammonia, carbon monoxide, ethyl acetate, sulfur dioxide
Harmful if inhaled	4	Warning	> 2500 to ≤5000 ppm (gas); > 10 to ≤20 mg/L (vapor); > 1 to ≤5 mg/L (dust, mist)	bromine; dimethylformamide
May be harmful if inhaled	5	Warning	No airborne concentrations given, substances with LD$_{50}$ values in range of 2000–5000 mg/kg	

Source: Parts of the information in this table were provided courtesy of the United Nations Economic Commission of Europe from the *Globally Harmonized System of Classification and Labelling of Chemicals*. Copyright © 2007 United Nations, New York and Geneva.

Acute Toxicant (severe)

Carcinogen

Respirator Sensitizer

Reproductive Toxicant

Target Organ Toxicant

Germ Cell Mutagen

Acute Toxicant

Skin Irritant

Eye Irritant

Skin Sensitizer

FIGURE 6.2.1.1   Selected Pictograms Associated with the GHS Health Hazards. Some pictograms more clearly suggest a particular hazard than others. (Courtesy of the United Nations Economic Commission for Europe. Copyright © 2007 United Nations, New York and Geneva.)

Tables 6.2.1.3 and 6.2.1.4, respectively. Tables 6.2.1.2, 6.2.1.3, and 6.2.1.4 all show well-defined categories of toxicity based on LD_{50} and LC_{50} measurements. This makes it relatively easy to categorize a chemical with regard to toxicity. See *Chemical Connection 6.2.1.1* Dimensions for Airborne Concentrations.

CHEMICAL CONNECTION *6.2.1.1*

DIMENSIONS FOR AIRBORNE CONCENTRATIONS

Table 6.2.1.4 shows two different dimensions for gases and vapors: ppm and mg/L. Gases and vapors are the same thing: isolated molecules in the gas phase. The term "vapor" is often used when referring to a gas that has volatized from a liquid or solid. Why both "ppm" and "mg/L"?

It is often the case that researchers in various fields develop and use dimensions that are convenient at the time an experiment is done or measurement is made. When measuring pressure, for example, many labs still use atm, mm Hg, and torr even though the SI dimension is the pascal (Pa). When looking at LC_{50} data, it is common to find five different ways of representing the concentration of a chemical species in air. Four of these are based on "mass per volume" and are easy to interconvert:

$$1000 \; \mu g/L = 1 \; mg/L = 1 \; g/m^3 = 1000 \; mg/m^3$$

Also quite common is "parts per million" (ppm). Since this is a volume ratio and does not involve mass, the conversion between ppm and other dimensions requires the use of the molar mass of the chemical. Two ways of expressing this conversion are

$$ppm = (mg/m^3)(24.45)/(molar\ mass)$$

$$mg/m^3 = (ppm)(molar\ mass)/(24.45)$$

Why 24.45 as a conversion factor? Using the ideal gas law and the ideal gas law constant, this value can be calculated. (Your instructor may wish to have you prove this!)

When assessing the hazards of a chemical, all forms of toxicity (and other health hazards) must be considered. Let's take a look at a few examples.

Example #1. Let's imagine that you wish to use osmium tetroxide as a staining agent in electron microscopy. Using TOXNET,[4] we find an LD_{50} (oral, mice) of 162 mg/kg. If we use Table 6.2.1.2, this places OsO_4 in the HC 3, "Danger" rating, which suggests that it should be handled with great care with regard to ingestion. (You would also find that OsO_4 is a severe irritant to eyes and skin as a vapor, and as a solid it has a vapor pressure of 11 mm Hg at 27 °C. It has an odor threshold of 0.0019 ppm, a NIOSH REL 15-minute STEL of 0.0006 ppm, and an IDLH of 0.1 ppm. These data about occupational exposures will be discussed more in Section 6.2.2.) OsO_4 should be handled in a chemical hood, with impermeable gloves.

Example #2. Methyl iodide is a common methylating agent. It is a liquid with a boiling point of 42 °C. The LD_{50} is 76 mg/kg (oral, rat) and the LC_{50} data are 5 mg/L (inhalation, mouse) and 1.3 mg/L (inhalation, rat). This puts methyl iodide in HC 3 (danger) for oral toxicity and HC 2 (danger) or 3 (danger) for inhalation toxicity (See *Special Topic 6.2.1.1* Complications with LC_{50} Values and the Temptation to Overinterpret Data). Methyl iodide should be handled in a chemical hood, with impermeable gloves.

Example #3. Allyl alcohol is a liquid with a boiling point of 97 °C. Its LD_{50} value (85 mg/kg, oral, mouse) places it in HC 3 and the LC_{50} (165–520 ppm) puts it in HC 2. It is easy to avoid ingestion and inhalation using appropriate precautions. What if you spill this on your hand while working in a chemical hood? The dermal LD_{50} (rabbit) is 45 mg/kg, which puts in it HC 1! This is an instance where wearing the appropriate impermeable gloves is critically important. See Sections 7.1.3 and 7.2.2.

SPECIAL TOPIC *6.2.1.1*

COMPLICATIONS WITH LC$_{50}$ VALUES AND THE TEMPTATION TO OVERINTERPRET DATA

In Table 6.2.1.4 the LC$_{50}$ criteria are noted as 4-hour exposure times. It is also common to find LC$_{50}$ values based on 1-hour, 2-hour, or 8-hour exposures. Converting between "time frames" is not a "linear" calculation, and the conversion factors are not the same for gases/vapors versus dusts/mists. The GHS[5] indicates that 4-hour data can be converted to the 1-hour criteria by multiplying by 2 for gases and vapors and by multiplying by 4 for dusts and mists. There is no guidance about conversions between other time frames. How problematic is this?

Let's think about LC$_{50}$ data with the goal of using such data to determine a hazard level for a chemical. Imagine that you have a chemical with an LC$_{50}$ = "10 mg/L, 1-hour, rat." Using the conversion above, we would change this to "5 mg/L, 1-hour, rat" and using Table 6.2.1.4 we place this in HC 3. Section 4.1.1 describes the process by which LD and LC data are measured and the variety of uncertainties that need to be considered when extrapolating this to humans. Depending on the size of the animal study, we should perhaps not really use more than one significant figure in any LC value. Furthermore, different animals are used and Table 6.2.1.4 treats them all the same with a questionable assumption about how this transfers to human toxicity. In fact, any LC value might be viewed as an "order of magnitude" data point, with considerable uncertainty. Given this large uncertainty, we can imagine that the chemical above could easily be HC 2 or HC 4, and now the issue of "time conversion factors" seems less important.

Furthermore, there are no differences in lab procedures when it comes to handling HC 2 versus HC 3, so the distinction becomes less critical. In fact, the assumption is that any volatile liquid or solid or any gas should be handled in a chemical hood whether it is HC 2 or HC 5. One might take even greater precautions with regard to procedure, and planning for emergency response, with an HC 2 chemical but there are no specific guidelines in this regard.

Scientists like numbers and like to use quantitative criteria to sort and categorize things. In this process, however, it is important to critically analyze the quality of the data and the assumptions built into criteria and their application. Is an HC 3 chemical really more hazardous than an HC 4 chemical? Maybe, but we're likely not sure of that.

Assessing Other Acute and Chronic Health Hazards

Table 6.2.1.5 shows information about the highest hazard category for the other health categories in Table 6.2.1.1 apart from acute toxicity. These are corrosives, eye hazards, sensitizers, mutagens, carcinogens, reproductive toxicants, target organ toxicants, and aspiration hazards. You will see below that all of these classes seem less well defined with regard to placing chemicals in various hazard categories since they all lack the kind of quantitative criteria that LD$_{50}$ and LC$_{50}$ values afford. We have identified examples of chemicals that seem to fit in these hazard classes and categories in these hazard classes that you might encounter in the laboratory.

The GHS has criteria for sorting chemicals into hazard categories but these criteria are more complicated, more difficult to understand, and in some cases more subjective. Often data are missing or unavailable to classify chemicals in these hazard categories. Table 6.2.1.5 shows the GHS hazard statements and signal words used for these hazard classes and categories. The GHS labels require the hazard pictogram, the hazard signal words, and the hazard statement.

The GHS has not yet been implemented and we cannot provide an example of a "real" GHS label. The next best thing to this is Figure 6.2.1.2 taken from the OSHA web site.[5] Figure 6.2.1.2 lists elements that should be included on a GHS label. In Figure 6.2.1.2 a "mock" label illustrates the use of various elements for a substance called "ToxiFlam (Contains: XYZ)." This label illustrates a chemical that is both flammable and toxic; thus, it includes the flammable pictogram and the toxic pictogram. It includes the signal word "Danger." There are two hazard statements—"Toxic if swallowed" and "Flammable Liquid and Vapor." The label could include supplementary information from the supplier but none is included here. Precautionary measures are provided on this mock label, starting with "Do not eat, ..." A first aid statement is provided here as well as the name of the company. So while we don't have a real label, this label should be similar to a GHS label.

TABLE 6.2.1.5 GHS Classification for Other Selected Toxic Effects—Highest Hazard Class

Hazard class	Hazard statement	GHS hazard category (HC)	Signal word	GHS criteria	Examples
Skin corrosive	Causes severe skin burns and eye damage	1A, 1B, 1C	Danger	Produces destruction of skin tissue—visible necrosis of epidermis and dermis	Sodium hydroxide, potassium hydroxide, sulfuric acid, hydrofluoric acid, nitric acid
Eye hazard	Causes serious eye damage	1	Danger	Causes irreversible damage to cornea, iris, or conjunctiva	Sodium hydroxide, potassium hydroxide, sulfuric acid, hydrofluoric acid, nitric acid
Respiratory sensitizer	May cause allergy or asthma symptoms or breathing difficulties if inhaled	1	Danger	Human evidence of respiratory hypersensitivity, and/or positive from animal testing	Polyisocyanates, such as toluene diisocyanate; acid anhydrides, such as trimellitic anhydride, formaldehyde
Skin sensitizer	May cause an allergic skin reaction	1	Warning	Human evidence of skin sensitivity in a substantial number of persons or positive animal testing	Pentadecylcatechol (poison ivy allergen), formaldehyde, epichlorohydrin, 2,4-dinitrochloro-benzene, nickel
Mutagen	May cause genetic defects	1A, 1B	Danger	Known to cause heritable mutations in human germ cells or presumed to cause heritable mutations in human germ cells	There are no known human germ cell mutagens; identification of presumed human germ cell mutagens is unclear
Carcinogen	May cause cancer	1A, 1B	Danger	Known or presumed human carcinogen	Benzidine, 2-naphthylamine, dimethylni-trosamine, bis(chloromethyl) ether, ethylenimine, nickel carbonyl, β-propiolactone, vinyl chloride
Reproductive toxicant	May damage fertility or the unborn child	1A, 1B	Danger	Known or presumed human reproductive toxicant	Diethylstilbestrol, thalidomide, ethanol, organomercury compounds, lead, cocaine, 1,2-dibromo-3-chloropropane

(continued)

TABLE 6.2.1.5   (*Continued*)

Hazard class	Hazard statement	GHS hazard category (HC)	Signal word	GHS criteria	Examples
Target organ toxicant— single exposure	Causes damage to organs	1	Danger	Reliable evidence of human or animal toxicity or adverse effects on specific organs or systems upon a single exposure	Dimethylmercury (nervous system), organophosphates (nervous system), chlorine (lung), beryllium (lung), phosgene (lung), sodium hydroxide (skin, eyes)
Target organ toxicant— repeated exposures	Causes damage to organs	1	Danger	Reliable evidence of human or animal toxicity or adverse effects on specific organs or systems upon repeated exposures	Benzene (blood), acid anhydrides (immune system), 2,4-dinitrochloro-benzene (skin), carbon tetrachloride (liver), inorganic mercury salts (kidney), asbestos (lung), carbon disulfide (nervous system)
Aspiration hazard	May be fatal if swallowed and/or enters airways	1	Danger	Known human aspiration hazard or hydrocarbon with viscosity of ≤ 20.5 mm^2/s at 40 deg C	Gasoline, petroleum distillates

Source: Parts of the information in this table were provided courtesy of the United Nations Economic Commission of Europe from the *Globally Harmonized System of Classification and Labelling of Chemicals*. Copyright © 2007 United Nations, New York and Geneva.

These health hazards include:

- *Corrosives* that produce destruction of skin with damage to the epidermis and dermis. The highest hazard category is divided into three subcategories (HC 1A, 1B, 1C), depending on the time it takes to cause the skin damage.

- *Eye hazards* that cause irreversible damage to the eye's cornea, iris, or conjunctiva upon prolonged contact (HC 1).

- *Sensitizers* that affect either the lungs or the skin, causing "hypersensitivity" to the chemical. Respiratory sensitizers are HC 1 with a "Danger" signal word, and skin sensitizers are HC 1 with a "Warning" signal word.

- *Mutagens* that may produce genetic defects in humans (HC 1A) or in animals that meet testing criteria (HC 1B). There are no known human germ cell mutagens. The criteria for identifying potential germ cell mutagens of concern to humans seem to be evolving. Data from germ cell mutagenic testing require expert evaluation. This is a complicated, developing, and sometimes controversial area of scientific investigation.

- *Carcinogens* that are known to cause cancer in humans (HC 1A), and those presumed to cause cancer in humans (HC 1B). There are two hazard categories here because there is more evidence for HC 1A than HC 1B; however, the weight of evidence for both of these essentially means these are human carcinogens.

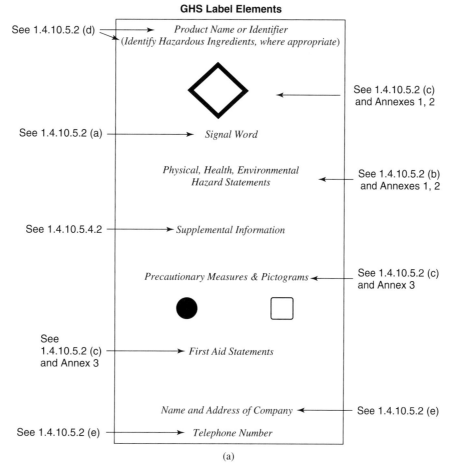

(a)

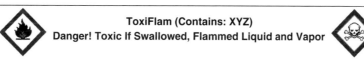

(b)

FIGURE 6.2.1.2 (a) Diagram of GHS Label Elements and (b) Mock GHS Label. The GHS has not yet been formally adopted by the United States, but the diagram of a GHS label and a mock GHS label illustrate how the GHS might be used in the United States. (From the Occupational Safety and Health Administration's Internet site; available at http://www.osha.gov/dsg/hazcom/ghs.html.)

- *Reproductive toxicants* that cause damage to the developing fetus resulting in birth defects and abnormalities in the baby (teratogens) or cause effects on the developing fetus in other ways (feto-toxicants). HC 1A compounds are known to cause and HC 1B are presumed to cause reproductive damage. (See also Section 4.1.2.)

- *Target organ toxicants* that specifically attack certain organs causing them damage that affects their ability to carry out normal functions. This can be from a single dose or repeated exposures.

- *Aspiration hazards* that can be fatal if swallowed and then enter the airways.

Note that some of the compounds listed in Table 6.2.1.5 are chronic toxicants where it takes small doses over a long period of time for the effects to become apparent. As also discussed in Section 4.2.1, studying chronic toxicity and other long-term effects is much more difficult than acute effects since the studies are inherently longer and more complicated. Methods for identifying chemicals' chronic effects, including cancer, are very poor, and our information about chronic toxicity is very scant. Chronic health effects do not appear immediately after a single dose of a chemical, but rather they usually appear after receiving small, continuous, or frequent doses over an extended period of time. These chronic effects may include, for example, neurological effects, skin effects, organ damage, or cancer. Lung cancer, emphysema, and oral cancer, for example, are types of chronic toxicity exhibited by cigarette smoking; the small doses over years of time accumulate until the effects finally begin to appear. For many chemicals we simply do not know the chronic effects.

Most of the knowledge we have about chemicals that cause human cancer or other chronic diseases has been learned through experience, that is, after a number of people using that chemical regularly over many years developed some sort of cancer. For example, vinyl chloride was recognized as a human carcinogen because it caused a rare type of cancer called angiosarcoma.[6] Benzene, on the other hand, was finally recognized as a human carcinogen through years of studying various exposed populations and establishing links between benzene-induced aplastic anemia and leukemia.[7]

Long-term effects cannot be predicted from acute toxicity measures. Chemicals that can cause delayed but severe effects such as allergic reactions are not detected by LD_{50} or LC_{50} values. For example, while hexane is not acutely toxic by the oral route (LD_{50} 28,700 mg/kg),[8] it is chronically toxic, causing significant neurological toxic effects with small doses over a long period of time.[9] Ethanol is acutely toxic only in very high doses (LD_{50} 7060 mg/kg).[8] However, ethanol ingested in lesser doses severely affects judgment and muscular coordination, and continuous doses of ethanol over a long period of time can cause cirrhosis of the liver, a condition frequently observed among alcoholics. Furthermore, children born to expectant women who consume ethanol can suffer a lifetime condition from fetal alcohol toxicity.

The hazard categorization of a chemical depends largely on the strength of the evidence that it causes harm over a period of time. This degree of uncertainty is often frustrating when deciding "how dangerous" a chemical might be, but for regulatory purposes we treat some chemicals as more dangerous than others because the evidence is more compelling. Future research and testing may, and probably will, change the hazard category labeling for some chemicals. For this reason it is best to always try to determine the most recent understanding of the toxicity of a chemical. TOXNET is an excellent resource for this.[4]

> *Example #4.* Let's imagine that you wish to run a reaction in the solvent carbon disulfide. A quick look in *Prudent Practices in the Laboratory*[10] indicates an LD_{50} (oral, rat) = 3188 mg/kg. Table 6.2.1.2 puts this in HC 5, "slightly toxic." However, this is by the route of ingestion. With a boiling point of 46 °C and vapor pressure of 300 mm Hg at 20 °C, a look at the inhalation toxicity seems wise. The LC_{50} values for animals range from 10 to 25 g/m^3. Using the appropriate conversion factor, 20 g/m^3 = 20 mg/L = 6400 ppm. This places CS_2 in the "moderately toxic" hazard category, HC 4, for acute toxicity by inhalation. A review of CS_2 using TOXNET[4] also suggests that this solvent is a severe skin and eye irritant, but "it has not been possible to verify these data." It is very easily absorbed through the skin. It is not a carcinogen nor is it genotoxic. There are other system damages in humans. Overall, the health hazards seem moderate. However, with a flash point of −30 °C this is obviously an extremely flammable liquid!

Example #5. Toluene diisocyanate (TDI) has an LD_{50} (rat, oral) = 2060 mg/kg, which places it in HC 5 with regard to ingested toxicity. However, the LC_{50} (various animals) = 0.06–0.35 mg/L (4 h), which places it in HC 1 for inhaled toxicity. Furthermore, it is described as a "powerful irritant to eyes, skin, and the respiratory tract," which places it in HC 1 for skin corrosive, respiratory sensitizer, and eye damage. Clearly, many precautions must be taken when working with TDI.

Example #6. Chloromethyl methyl ether (CMME) has an LD_{50} (oral, rat) = 500 mg/kg, which places it in HC 4 for ingested toxicity. However, the LC_{50} (rat) = 0.18 mg/L (7 h), which places it in HC 1 for inhaled toxicity. Furthermore, it is classified as a potent human carcinogen based on animal studies (HC 1 for carcinogens) and it causes second degree burns to the skin after a few minutes (HC 1 for skin corrosive). (See Incident 4.3.1.1.)

These examples illustrate that you must not forget that there are other hazardous properties that you should not overlook in your assessment of a chemical's risk, including chronic toxic effects, allergic effects, corrosive effects, flammability, or reactive properties. Furthermore, you should remember that you might not always find adequate hazard information about the chemicals you are using. Only a few thousand out of the millions of recognized chemicals have been evaluated for toxicity, and this is usually acute toxicity. Even fewer chemicals have been tested for chronic toxicity.

RAMP

- *Recognize* that classifying chemicals into relative hazard categories is a useful tool.

- *Assess* the exposure routes and the relative hazards of chemicals, a critical step in laboratory safety.

- *Minimize* exposures to chemicals. This is always a prudent path no matter what the relative hazard of the chemical might be. Remember that chronic toxicities *may not have yet been recognized*.

- *Prepare* for experiments by identifying steps that are to be taken in the event of an emergency such as an unexpected exposure.

References

1. W. SHAKESPEARE. *Hamlet*, Act I, iv, 90. *Bartlett's Familiar Quotations*, 17th edition (J. KAPLAN, ed.), Little, Brown, and Co., Boston, 2002, p. 202, line 11.
2. University of Arizona, Risk Management & Safety Department. Chemical Safety Bulletins, Incident: Toxic Gas Leak/Exposure; available at http://risk.arizona.edu/healthandsafety/chemicalsafetybulletins/toxicgasleak112707.shtml (accessed December 9, 2008).
3. United Nations. *Globally Harmonized System of Classification and Labelling of Chemicals (GHS)*, 2nd revised edition, United Nations, New York, 2007; available at http://www.unece.org/trans/danger/publi/ghs/ghs_rev02/02files_e.html (accessed November 19, 2008).
4. TOXNET. Available at http://toxnet.nlm.nih.gov/ (accessed December 9, 2008).
5. Occupational Safety and Health Administration. *A Guide to The Globally Harmonized System of Classification and Labelling of Chemicals (GHS)*; available at http://www.osha.gov/dsg/hazcom/ghs.html (accessed October 5, 2009).

6. W. C. COOPER. Epidemiologic study of vinyl chloride workers: mortality through December 31, 1972. *Environmental Health Perspectives* **41**:101–106 (1982).
7. U. RANGAN and R. SNYDER. An update on benzene. *Annals of the New York Academy of Sciences* **837**:105–113 (1997).
8. National Research Council. *Prudent Practices in the Laboratory: Handling and Disposal of Chemicals*, National Academy Press, Washington, DC, 1995; see Laboratory Chemical Safety Summary for hexane and ethanol.
9. Centers for Disease Control and Prevention. *n*-Hexane—related peripheral neuropathy among automotive technicians, California, 1999–2000. *Morbidity and Mortality Weekly Report* **50**(45):1011–1013 (2001); available at http://www.cdc.gov/mmwr/preview/mmwrhtml/mm5045a3.htm (accessed November 19, 2008).
10. National Research Council. *Prudent Practices in the Laboratory: Handling and Disposal of Chemicals*, National Academy Press, Washington, DC, 1995.

QUESTIONS

1. In the GHS, what are the number of classes of physical, health, and environmental hazards, respectively?

   (a) 1, 11, 16

   (b) 1, 16, 11

   (c) 11, 16, 1

   (d) 16, 11, 1

2. For which category of health hazard do we have relatively good data for many chemicals?

   (a) Germ cell mutagenicity
   (b) Eye damage and eye irritation
   (c) Skin corrosion
   (d) Acute toxicity

3. For the class of acute toxicity, the signal word "Danger" is used for what hazard category(ies)?

   (a) 1
   (b) 1, 2
   (c) 1, 2, 3
   (d) 1, 2, 3, 4

4. What range of LD_{50} values if considered "fatal if swallowed"?

   (a) $\leq$5 mg/kg
   (b) $\leq$50 mg/kg
   (c) $\leq$300 mg/kg
   (d) $\leq$2000 mg/kg

5. With regard to dermal exposure, what chemical is considered "fatal in contact with skin"?

   (a) Sodium azide
   (b) Allyl alcohol
   (c) Hydrazine
   (d) All of the above

6. What range of LC_{50} values is considered "fatal if inhaled" for gases?

   (a) $\leq$100 ppm
   (b) $\leq$500 ppm
   (c) $\leq$2500 ppm
   (d) $\leq$5000 ppm

7. Carbon disulfide is ranked as

   (a) Slightly toxic by ingestion, moderately toxic by inhalation, a carcinogen, but extremely flammable
   (b) Moderately toxic by ingestion, slightly toxic by inhalation, a carcinogen, but extremely flammable
   (c) Slightly toxic by ingestion, moderately toxic by dermal exposure, but extremely flammable
   (d) Slightly toxic by ingestion, moderately toxic by inhalation, genotoxic, and mildly flammable

8. Toluene diisocyanate is ranked as

   (a) Mildly toxic by ingestion, very toxic by inhalation, and a powerful sensitizer
   (b) Mildly toxic by ingestion, mildly toxic by inhalation, and a powerful sensitizer
   (c) Very toxic by ingestion, mildly toxic by inhalation, and a powerful sensitizer
   (d) Mildly toxic by ingestion, very toxic by inhalation, and very flammable

6.2.2

UNDERSTANDING OCCUPATIONAL EXPOSURE LIMITS

Preview This section explains the common systems for occupational exposure limits (OELs)—threshold limit values (TLVs) and permissible exposure limits (PELs)—and how to use these data to judge the relative hazard of a chemical.

Sometimes we see a cloud that's dragonish; a vapor sometime like a bear or a lion. . . .

William Shakespeare, *Antony and Cleopatra*[1]

INCIDENT 6.2.2.1   MERCURY POISONING IN AN ACADEMIC LAB[2]

A chemistry student went to the university clinic complaining of chronic fatigue, headaches, and mental lethargy. After no significant findings from the usual examinations and testing, the physician learned that the student had been in contact with mercury vapors. Urinary testing revealed that the student had chronic mercury poisoning. Further investigation within the chemistry department revealed widespread mercury contamination, probably from spills with mercury droplets being visible in some laboratories. Air concentrations ranged as high as 1 mg/m^3. [The recommended exposure limit, known as the ACGIH TLV, for mercury was at that time 0.1 mg/m^3; today's TLV is 0.025 mg/m^3.] Twenty-six of twenty-eight patients had abnormal elevated levels of mercury in their urine. Two other students, besides the initial case, also exhibited clinical chronic mercury poisoning. Decontamination with a mercury vacuum decreased mercury levels to acceptable limits.

What lessons can be learned from this incident?

Prelude

As a warning: this may be one of the most initially confusing sections in this book. You will find below more than a small handful of acronyms and references to some governmental and nongovernmental agencies that publish occupational exposure limits (OELs) that are mandatory or nonmandatory, respectively. The result is often a dizzying array of abbreviations. Sometimes, different agencies use the same data and language and sometimes they do not. This is simply the way this field has evolved over time, with registered trademarks and some unfortunate lawsuits along the way to complicate matters.

Ultimately, you should have two goals in mind as you read this: first, to understand the concept of OELs and their variations as tools to protect workers (including those in laboratories) from adverse chemical exposures and to understand that some OELs are voluntary and some are legal requirements; and second, to understand how you can use this information to assess the relative risk of a hazard.

One last comment about OELs before you learn more. Most uses of chemicals in the laboratory will not result in airborne concentrations anywhere near OELs because OELs were developed for industrial exposures that are usually higher than the laboratory. Exposures approaching or exceeding those limits are likely only to occur if you fail to use proper protections (such as a chemical hood or PPE) or if you have a spill in a laboratory.

Laboratory Safety for Chemistry Students, by Robert H. Hill, Jr. and David C. Finster
Copyright © 2010 John Wiley & Sons, Inc.

ACGIH Threshold Limit Values

In 1938 the American Conference of Governmental Industrial Hygienists (ACGIH) was established.[3] Industrial hygienists are safety professionals who work in various workplaces to help ensure the safety of the workforce by making measurements of chemical contaminants and providing recommendations for safe practices to avoid exposures to hazardous materials and operations.

The ACGIH is a nongovernmental organization, but as you will see, their data are sometimes, but not always, used by OSHA to set legal occupational exposure limits. The ACGIH is most well known for establishing threshold limit values (TLVs®).[3-7] These TLVs are *nonmandatory recommendations*. However, OSHA sometimes adopts the TLVs as permissible exposure limits (PELs, see below) and in such cases the PELs are *mandatory* regulations.

There are three ways that TLVs can be set. These are based on the "time frames" that employees in labs and industrial settings might encounter chemicals.

A very common TLV is the *time-weighted average* (TWA) value. Since workers might be exposed to airborne chemicals over longer periods of time, the TWA is an estimate of the "average" exposure to which an employee can be exposed for 5 days/week at 8 h/day over the course of a person's working career without experiencing any adverse effects. The TWA value would allow for some exposure *over* the TWA occasionally, as long as the *time-weighted average* is not exceeded over the course of a day or week. The model on which this calculation is built assumes a 150-pound male between the ages of 25 and 44. As noted elsewhere in Chapter 4 in the discussion of toxicology, there are individual variations but the model had to be built based on some particular set of assumptions. At the outset, however, we should recognize that TLV-TWA values are estimates. Compared to the established TLV, a value of 10% lower is not inherently safe and a value of 10% higher is not necessarily dangerous.

A second TLV commonly seen is the TLV-STEL, the *short-term exposure limit*. This is the maximum concentration to which workers can be continuously exposed for 15 minutes. It is recommended that no more than four such episodes occur in a given day and that at least 60 minutes separate each episode. Finally, the ACGIH also establishes a TLV-C, a *ceiling level*. This is a level that is not to be exceeded at any time. These values are the concentrations that are designed to protect one who has short-term exposures to chemicals such as solvents as you might have in a laboratory operation.

These voluntary standards are consensus values derived by experts from careful review of information from animal studies, accidental or intentional exposures, or workplace experiences with chemical exposures.[7] Each specific TLV is based on expert evaluation, and each is designed to protect most people over a lifetime of exposure, if exposures are kept below the recommended level.[4,7] The expert panel reached an agreement setting the TLV until such time as additional new information is available that would warrant consideration of further changes.

The ACGIH publishes an updated TLV listing each year,[6] and also publishes periodically a volume that summarizes the rationale for establishing the TLV, the *Documentation of TLVs® and BEIs®*.[7] Some TLVs are shown in Table 6.2.2.1. For example, you will find mercury had a recommended TLV of 0.025 mg/m^3 in 2000 and this limit is 4 times lower than the 1968 TLV cited in Incident 6.2.2.1,[2] demonstrating how TLVs are updated with new information. (See *Chemical Connection 6.2.2.1* What Is a ppm?)

CHEMICAL CONNECTION *6.2.2.1*

WHAT IS A PPM?

A ppm is a one part per million, that is, 1/1,000,000. This is a common way to express concentrations of chemicals in air or water. Another way to imagine 1 ppm is that this is the concentration of 1.5 teaspoons of oil added to a typical "tanker" of gasoline on the highway which has a volume of about 2000 gallons.

TABLE 6.2.2.1   American Conference of Governmental Industrial Hygienists' Threshold Limit Values (TLVs) for Selected Chemicals, 1999–2000[10] and Occupational Safety and Health Adminstration's Permissive Exposure Limits[9] (PELs) Expressed as Time-Weighed Average (TWA), Short-Term Exposure Limit (STEL), and/or Ceiling (C)[a]

Chemical name	Threshold limit value (ACGIH) (*Marks TLVs that are more restrictive than PELs)	Permissible exposure limit (OSHA)
Acetic acid	TWA 10 ppm; STEL 15 ppm	TWA 10 ppm (25 mg/m^3)
Acetone	TWA 500 ppm*; STEL 750 ppm	TWA 1,000 ppm (2,400 mg/m^3)
Acetonitrile	TWA 40 ppm; STEL 60 ppm	TWA 40 ppm (70 mg/m^3)
Ammonia	TWA 25 ppm* (17 mg/mg/m^3); STEL 35 ppm (27 mg/mg/m^3)	TWA 50 ppm (35 mg/m^3)
Benzene	TWA 0.5 ppm* (1.6 mg/m^3)—skin (human carcinogen); STEL 5 ppm (16 mg/mg/m^3)	TWA 1 ppm; STEL 5 ppm
Bromine	TWA 0.1 ppm; STEL 0.2 ppm	TWA 0.1 ppm (0.7 mg/m^3)
Carbon disulfide	TWA 10 ppm* (31 mg/m^3)—skin	TWA 20 ppm
Carbon monoxide	TWA 25 ppm*	TWA 50 ppm (55 mg/m^3)
Dichloromethane	TWA 50 ppm* (animal carcinogen)	TWA 25 ppm; STEL 125 ppm
Diethyl ether	TWA 400 ppm; STEL 500 ppm;	TWA 400 ppm (1,200 mg/m^3)
Formaldehyde	TWA C 0.3 ppm* (0.37 mg/m^3) (suspected human carcinogen)	TWA 0.75 ppm; STEL 2 ppm
Hexane (*n*-hexane)	TWA 50 ppm*	TWA 500 ppm (1,800 mg/m^3)
Hydrogen cyanide	STEL C 4.7 ppm*—skin	TWA 10 ppm—skin
Mercury (vapor)	TWA 0.025 mg/m^{3*}—skin	TWA 0.1 mg/m^3
Phenol	TWA 5 ppm—skin	TWA 5 ppm
Toluene diisocyanate	TWA 0.005 ppm* (0.036 mg/mg/m^3); STEL 0.02 ppm (0.14 mg/mg/m^3)	C 0.02 ppm (0.14 mg/m^3)

[a]Units are parts per million (ppm) and/or milligrams per cubic meter (mg/m^3).

For airborne concentrations a part per million can be converted to another common airborne measurement, mg/m^3, using the formula below for concentrations at 25 °C and 1 atmosphere pressure:

$$1\,ppm = 1part/10^6\;parts = 1\,\mu L/1liter = 1liter/10^6\;liters$$

$$Concentration(mg/m^3) = (1liter/10^6 liters)(10^3 liters/m^3)(1mole/24.45liters)(gram\;MW/mole)(10^3\;mg/gram)$$

$$Concentration(mg/m^3) = ppm \times MW/24.45$$

Using this formula, you could calculate the LC_{50} for methanol in rabbits in mg/m^3. The LC_{50} for methanol is 81,000 ppm—thus, the equivalent concentration in mg/m^3 would be $81,000 \times (32/24.45)$ or 106,000 mg/m^3.

Airborne concentrations are also expressed in µg/L—this is equivalent to mg/m^3—$(10^3\;\mu g/mg)/(10^3\;liters/m^3)$.

Analytical chemists use this information in calculating their results in measuring airborne concentrations. Suppose that a chemist measured the amount of the chemical acrylonitrile ($CH_2{=}CH{-}C{\equiv}N$) in a sampling device as being 30 µg. He knew from other measurements that this came from a 15-liter sample of air. He could next calculate the airborne concentration as 30 µg/15 liters or 2 µg/L or 0.9 ppm. To evaluate the significance of this exposure, the chemist could use the OEL for acrylonitrile of 2 ppm. Thus, the exposure is below the OEL. Nevertheless, it is not that much below the limit and it is prudent to minimize exposure to chemicals, particularly since the determination of OELs sometime changes and the values almost always become lower.

There are a few special cases of TLVs.

- TLVs are airborne concentration limits that usually assume inhalation as the route of exposure, but other designations are sometimes noted in the TLV list.[6] Some TLVs are labeled as "skin" or "dermal," meaning that exposure of the gas or vapor to the skin, mucous membranes, or eyes can cause a toxic effect (not simply dermatitis). The values for skin TLVs are generally higher than inhalation TLVs so if some skin exposure is significant it is likely that one is already inhaling air well above the inhalation TLV. Relatively few skin TLVs have been measured. One category of compounds for which they have been measured is organophosphate pesticides.

- The TLV list classifies some chemicals as carcinogens. "A1" are known human carcinogens, "A2" are suspected human carcinogens, and "A3" are animal carcinogens. (Note that the GHS does *not* include animal carcinogens in its hazard recognition system. Because GHS and TLV are two independent hazard rating systems, chemicals rated as animal carcinogens in the TLV listing will likely not be considered carcinogens under GHS.)

- Sensitizers are a designated special hazard in the TLV list. All sensitizers warrant special care to avoid exposures. Some compounds are listed in the TLV listing as asphyxiants, but they are not assigned numerical TLVs.

- Simple asphyxiants are usually relatively nontoxic, but present a special hazard if used in a confined space since they can displace oxygen and make the atmosphere unsuitable to sustain life. (Chemical asphyxiants, such as carbon monoxide, function by the chemical preventing oxygen uptake.)

You can use TLVs in a manner similar to that explained for evaluation of toxicity from the *Globally Harmonized System for Classification and Labelling of Chemicals* (*GHS*).[8] (This is also discussed in Sections 3.2.1 and 6.2.1.) Chemicals that are more hazardous have lower TLVs; chemicals that are less hazardous have higher TLVs. Table 6.2.2.2 suggests ranges of TLVs that can help you decide the hazard category of chemicals with TLVs.[9]

If available, TLVs provide an excellent guide to assist you in assessing the hazard category of a chemical since experts have already considered all pertinent and available information in deriving the

TABLE 6.2.2.2 Suggested Classification of Chemicals into Hazard Categories Using TLVs, PELs, STELs, or Other Occupational Limits[6,8–11]

Hazard category	TLVs, PELs, STELs	Examples
HC 1	≤1 ppm or ≤0.1 mg/m^3 or A1 carcinogens or sensitizers	Acrylamide, arsine, benzene, benzo[a]pyrene, bis(chloromethyl) ether, bromine, chlorine, diborane, lithium hydride, methylisocyanate, toluene diisocyanate, 2,4,6-trinitrotoluene (TNT), tri-o-cresylphosphate, vinyl chloride
HC 2	> 1 to ≤10 ppm or > 0.1 to ≤1 mg/m^3 or A2 carcinogens	Acetic acid, 1,3-butadiene, carbon tetrachloride, chlorobenzene, fluorine, formic acid, nitrogen dioxide, picric acid
HC 3	> 10 to ≤50 ppm or > 1 to ≤5 mg/m^3 or A3 carcinogens	Acetonitrile, adipic acid, carbon disulfide, cumene, dibutylphthalate, dichloromethane, n-hexane, nitromethane
HC 4	> 50 to ≤500 ppm or > 5 to ≤15 mg/m^3	Acetone, 2-butanone, calcium sulfate, cyclohexane, ethyl acetate, isopropyl ether, 1,1,1-trichloroethane
HC 5	> 500 ppm or > 15 mg/m^3	Ethanol, carbon dioxide, propyne, sulfur hexafluoride, trifluorobromomethane

TLV. Over the years since their development, TLVs have played a major role in improving the health of workplaces, including laboratories, by ensuring that airborne concentrations are kept to a safe level for those organizations that voluntarily adopted and used TLVs.

OSHA Permissible Exposure Limits: Regulatory (Nonvoluntary) Standards

Another type of exposure limit is called the permissible exposure limit (PEL), which is the language used by OSHA. OSHA generally uses ACGIH TLVs in their regulations, which makes these levels subject to regulatory enforcement rather than being voluntary. PELs were established by OSHA in 1972 when it adopted TLVs for 400 chemicals from the 1968 ACGIH's TLV list.[5] PELs are found in OSHA's regulations, which are contained in "29 Code of Federal Regulations, Part 1910 Subpart Z" within its tables and individual standards.[9] (Learn more about the Code of Federal Regulations in *Special Topic 1.3.3.1*.) Even though ACGIH TLVs get updated annually, OSHA continues to use the original PEL Table Z-1 from 1972 unless there are more recent established individual chemical standards. These regulations are used to cite employers who do not comply with PELs, which may lead to fines or even imprisonment for individuals for intentional or willful violations.

Table 6.2.2.1 contains the OSHA PEL values from 29 CFR Parts 1910.1000 Table Z-1, 1910.1000 Table Z-2, 1910.1028, 1910.1048, and 1910.1052.[9] If you compare OSHA's PELs in "Table Z-1" with a current TLV listing, you will find many TLVs to be more stringent because they are updated with more current information. Table 6.2.2.1 indicates examples of TLVs (marked with asterisk) that are more stringent than the PELs. You can use Table 6.2.2.2 in judging the relative hazard of chemicals having PELs, but we recommend that you use the most current TLV if available since the TLV was based on more current information.

Finding TLV, PEL, and Other OEL Data

It is worth mentioning that compared to other kinds of safety data, TLVs have been established by the ACGIH only for compounds that are industrially common enough to warrant the expense of undertaking the animal studies to set TLVs. There are about 500 TLV data, compared to the many, many thousands of chemicals in use and the many millions of chemicals known to exist. The ACGIH is a private organization that is funded in part by the sale of the TLV information, so this is not often as readily available in a single resource as other information except from an ACGIH member or a technical library. Many TLV and PEL values are listed in TOXNET. The *NIOSH Pocket Guide to Chemical Hazards* lists PELs and also lists another kind of nonmandatory OEL developed by NIOSH called *recommended exposure levels* (RELs) that were derived by NIOSH through its research program. Another NIOSH publication called the *Registry for the Toxic Effect of Chemical Substances (RTECS)* has more toxicological and OEL data. RTECS is only available through subscription; however, there are links to RTECS for the chemicals listed in the *NIOSH Pocket Guide to Chemical Hazards*, and you can find a lot of information about toxicity and OELs there. See *Special Topic 6.2.2.1* How to Use the OEL Information in an MSDS.

- At the TOXNET web site,[13] enter the name of the chemical and select "Search," select HSDB (Hazardous Substances Data Bank), and select the best match (usually the first item in the list). If TLV and PEL data exist, it will be in the category "Occupational Exposure Standards."
- At the web site for the *NIOSH Pocket Guide to Chemical Hazards*,[14] select "Index of Chemical Names"; next, go to the alphabetical listing at the time to find your chemicals, then select the name of the chemical. There will be a lot of information about the chemical. You can get even more information by clicking on the "RTECS" line, which gets you into that database for that particular chemical.

SPECIAL TOPIC *6.2.2.1*

HOW TO USE THE OEL INFORMATION IN AN MSDS

An MSDS for NO_2 has the following data listed:[12]

1991 OSHA PEL	15 min STEL: 1 ppm (1.8 mg/m^3)
1990 IDLH	50 ppm* (2008: 20 ppm)
1990 NIOSH REL	15 min STEL: 1 ppm (1.8 mg/m^3)
1992–1993 ACGHI TLV	8-h TWA: 3 ppm (5.6 mg/m^3)
	STEL: 5 ppm (9.4 mg/m^3)

Note: This particular MSDS shows a value of 50 ppm based on the 1990 NIOSH listing. However, this was subsequently lowered to 20 ppm. This illustrates a useful example of relying on a "dated" MSDS. It is always best to search for the best current information about chemicals.

What do all of these numbers really mean in terms of how dangerous the stuff is to breathe? Using Table 6.2.2.2 and the ACGIH TLV values, we can place NO_2 in Hazard Class 1 or 2. This requires that extreme care should be taken to only use NO_2 in a chemical hood

Another kind of OEL that you may encounter is IDLH—*i*mmediately *d*angerous to *l*ife or *h*ealth.[15] This term was developed by NIOSH for its respiratory program. The IDLH value is "the maximum concentration from which one could escape within 30 minutes without any escape-impairing symptoms or irreversible health effects" in the event that a respirator fails. However, you make every effort to IMMEDIATELY exit an area having an IDLH; 30 minutes is only a margin of safety if a respirator fails. An IDLH value is a concentration of an airborne toxicant that poses a threat likely to cause death or health effects or prevent escape from that environment.

Using TLVs and PELs in the Lab: RAMP

- *Recognize* the hazards associated with chemicals by locating TLVs and PELs using online or printed materials. Consider both inhalation and dermal exposure routes.
- *Assess* the probability that a chemical might exceed an OEL in a lab.
- *Minimize* the hazards posed by inhalation by using PPE and chemical hoods.
- *Prepare* for emergencies that might result if *anything* goes wrong. Have an action plan in mind for spills and exposures.

References

1. WILLIAM SHAKESPEARE. *Anthony and Cleopatra*, Act IV, xii, 2. *Bartlett's is Familiar Quotations*, 17th edition, Little, Brown, and Co., Boston, 2002, p. 223, line 33.
2. K. D. ROSE, E. W. SIMPSON, and D. WEED. Contamination by mercury in chemical laboratories. *Journal of the American College Health Association*, **20**:197–199 (1972).
3. ELIZABETH K. WEISBURGER. History and background of the Threshold Limit Value Committee of the American Conference of Governmental Industrial Hygienists. *Journal of Chemical Health and Safety* **8**(*4*):10–12 (2001).
4. GERALD L. KENNEDY, Jr. Setting a threshold limit value (TLV): the process. *Journal of Chemical Health and Safety*, **8**(*4*):13–15 (2001).
5. PHILIP L. BIGELOW. Application of Threshold Limit Values for chemical substances to employees in laboratories. *Journal of Chemical Health and Safety*, **8**(*4*):16–21 (2001).
6. American Conference of Governmental Industrial Hygienists. *TLVs*® *BEIs*® *Threshold Limit Values for Chemical Substances and Physical Agents and Biological Exposure Indices*, published yearly by the American Conference of Governmental Industrial Hygienists, Cincinnati, OH; available at http://www.acgih.org/home.htm (accessed February 21, 2008).

7. American Conference of Governmental Industrial Hygienists. *Documentation of TLVs® and BEIs®*, 7th edition, American Conference of Governmental Industrial Hygienists, Cincinnati, OH, 2007.

8. United Nations. *Globally Harmonized System of Classification and Labelling of Chemicals (GHS), 1st revised edition, United Nations, New York, 2005; available at* http://www.unece.org/trans/danger/publi/ghs/ghs_rev01/01files_e.html *(accessed February 11, 2008).*

9. Occupational Safety and Health Administration. 29 CFR 1910 Standards; available at www.osha.gov, under Standards (accessed February 1, 2008).

10. C. D. KLAASSEN. *Casarett & Doull's Toxicology: The Basic Science of Poisons*, 6th edition, McGraw-Hill, New York, 2001, Appendix, pp.1155–1176.

11. R. H. HILL, J. A. GAUNCE, and P. WHITEHEAD. Chemical safety levels (CSLs); a proposal for chemical safety practices in microbiological and biomedical laboratories. *Journal of Chemical Health and Safety* **6**(*4*):6–14 (1999).

12. Material Safety Data Sheet #7. Genium Publishing Company, One Genium Plaza, Schenectady, NY, 12304-4690.

13. National Institutes of Health. National Library of Medicine. TOXNET Toxicology Data Network; available at http://toxnet.nlm.nih.gov/ (accessed September 20, 2009).

14. National Institute for Occupational Safety and Health. *NIOSH Pocket Guide to Chemical Hazards*; available at http://www.cdc.gov/niosh/npg/ (accessed September 20, 2009).

15. National Institute for Occupational Safety and Health. NIOSH Documentation for Immediately Dangerous to Life or Health Concentrations. May 1994; available at http://www.cdc.gov/Niosh/idlh/idlhintr.html (accessed September 20, 2009).

QUESTIONS

1. Which is *true*?

   (a) TLVs and PELs are nonmandatory.
   (b) TLVs and PELs are mandatory.
   (c) TLVs are mandatory and PELs are nonmandatory.
   (d) TLVs are nonmandatory and PELs are mandatory.

2. What assumption is built into the TWA value with regard to years of exposure?

   (a) A working career
   (b) 10 years
   (c) 20 years
   (d) 30 years

3. A STEL is based on what duration of exposure?

   (a) 15 minutes
   (b) 30 minutes
   (c) 60 minutes
   (d) One 8-hour work day

4. A ceiling level is

   (a) A maximum concentration that is allowed at the ceiling of a laboratory
   (b) A minimum concentration that is allowed at the ceiling of a laboratory
   (c) A level that is not to be exceeded at any time
   (d) The level that may not be exceeded for more than 15 minutes

5. When OSHA uses a TLV in regulations,

   (a) The TLV becomes a mandatory PEL
   (b) The PEL is nonmandatory
   (c) It is required that the TLV be updated annually
   (d) Updated TLVs automatically become updated PELs

6. The easiest and best source for TLV data is

   (a) The manufacturer's MSDS
   (b) Wikipedia toxicity data
   (c) TOXNET
   (d) The OSHA web site

7. The IDHL is an OEL that uses what time frame as the criterion for escape?

   (a) 2 minutes
   (b) 5 minutes
   (c) 15 minutes
   (d) 30 minutes

6.3.1

ASSESSING CHEMICAL EXPOSURE

Preview This section provides a brief overview of some methods to measure and assess exposures to chemicals using air monitoring techniques and biological monitoring techniques.

My father is a chemist, my mother was a homemaker. My parents instilled in us the feeling that learning was the most exciting thing that could happen to you and it never ends.

Rita Dove, poet laureate[1]

INCIDENT 6.3.1.1 LABORATORY EXPOSURE TO SOLVENTS[2]

Laboratory technicians preparing samples for histological examination were using a surgical biopsy hood. During this procedure they were exposed to formaldehyde. Air sampling revealed average concentrations between 3 ppm and 7 ppm with highest air concentrations around 11 ppm. These concentrations exceeded a 1-ppm NIOSH ceiling value for a 30-minute sampling period. Modifications were made to the hood so that the face velocities in varied sash positions exceeded 100 ft/min and effectively captured vapors in the hood as demonstrated by smoke tubes. Follow-up sampling of airborne concentrations were now below the detection limit of 0.5 ppm.

What lessons can be learned from this incident?

Why, and When, to Sample?

We have learned methods to minimize exposures to chemicals during laboratory operations. But how can we know whether our methods are effective to prevent or minimize exposures? This is especially important to your safety if the chemicals being used are very toxic. Remember that the greater the toxicity the greater the hazard, and the greater the protection that is needed. Perhaps the hood in Incident 6.3.1.1 would have been sufficient for a much less toxic compound, but for a very toxic compound having a very low occupational limit, it was not adequate.

There are few circumstances where "constant" monitoring of an environment is required or recommended.

- Oxygen levels are often monitored in confined spaces that should not be entered unless it is known that the atmosphere is safe or one is wearing an SCBA.

- Carbon monoxide concentrations are often monitored in our own homes using commercially available detectors.

- Radioactive materials being used in laboratories require monitoring using meters to detect radioactive contamination and lab personnel wear "radiation detection badges" to detect personal exposures to radiation (see Section 4.3.2).

Generally, exposure to laboratory chemicals is very low, often nondetectable. Additionally, concerns about the cost and inconvenience of monitoring "chemical levels" in a lab and the wide range of chemicals used, usually limit the need for air sampling (air sampling could cost $100 to several hundred dollars per test). Furthermore, the prudent use of engineering controls (like hoods) and personal protective equipment lead chemists to conclude that a working environment is "safe" unless there is some specific cause for concern.

OSHA requires as part of a lab's Chemical Hygiene Plan that monitoring occur only when there is "reason to believe that exposure levels for that substance routinely exceed the action level (or in the absence of an action level, the PEL)" [see 29 CFR 1920.1450 (d)(1)[3]]. A "reason to believe" usually takes the form of a laboratory worker experiencing some physical symptom or a concern about handling something very toxic, especially among pregnant women.

Air Sampling in Assessing Chemical Exposures

Occupational health and safety specialists use air sampling to evaluate potential exposures to chemicals. There are many types of air sampling and the selection of the proper air sampling method is often a decision for an occupational safety and health specialist. Many years ago canaries were used as monitors in mining operations. If they died while in the mine, the miners knew the air was not safe. Fortunately, our methods today are a little more advanced. This section will provide a brief overview of some common methods used to measure air concentrations of chemicals in workplaces, including laboratories. [4,5] This review cannot be comprehensive and it is meant only to illustrate how air sampling is done so that you may have an understanding of its role in assessing chemical exposures.

There are two principal types of air samplers: active and passive. Active air samplers draw air into the sampler using a pump; while passive air samplers rely on diffusion from the air without a pump. There are many kinds of active air samplers. Sorbent tubes are tubes that are filled with a sampling medium that adsorbs the chemical when it is drawn into the tube. Choices of sampling media depend on the analyte and include activated charcoal, silica gel, XAD–2®, Tenax®, and Chromosorbs®, as well as specially coated sorbents. Impingers are active samplers that use liquid collection media. Filters of various types are used to collect particulates and aerosols. After collection of samples using these active samplers, they are returned to a laboratory for analysis. Figure 6.3.1.1 shows some typical sampling tubes filled with activated charcoal and Figure 6.3.1.2 shows an impinger that is used to trap air contaminants in a liquid solution.

In recent years passive samplers have become popular. They come with various kinds of analyte-specific sorbents, and the rate of diffusion for that particular device has been determined by the manufacturer. Some of these passive devices are color detector tubes and can be read directly after the sampling period. Other passive samples must be sent back to the supplier's laboratory for analyses. Figure 6.3.1.3 shows a passive sampling badge.

One of the most common active air sampling methods involves collection of an air sample that has been drawn through a tube containing activated charcoal. Because solvents are commonly used in laboratories, we will describe the common method of using charcoal tubes for measuring solvents.

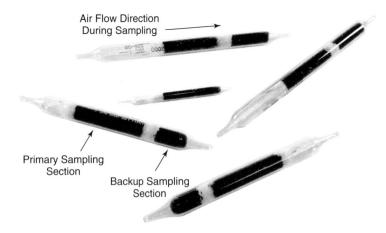

Air Flow Direction
During Sampling

Primary Sampling
Section

Backup Sampling
Section

FIGURE 6.3.1.1 Air Sampling Tubes with Activated Charcoal. Charcoal tubes have two sections, the primary section and the backup section. The primary section is extracted to determine the amount of the air contaminant from the charcoal. Air is actively drawn through the tube during sampling and the air volume is measured so that the amount of contaminant divided by the air volume gives air concentration. The backup section is used to detect if the primary section was overloaded. (Courtesy of SKC, Inc., Eighty Four, PA.)

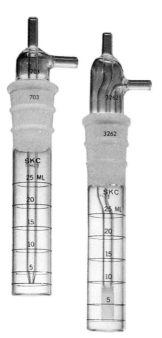

FIGURE 6.3.1.2 An Impinger for Air Sampling. The bottom portion of the impinger is filled with a measured quantity of a liquid. The air sample is drawn into the liquid from the opening at the top by using reduced pressure at the side port. The impinger on the right has a sintered glass attachment at the bottom, which creates many small bubbles in the air and optimizes the air–liquid interface. (Courtesy of SKC, Inc., Eighty Four, PA.)

FIGURE 6.3.1.3 Passive Sampling Badge. The passive sampling badge contains a sorbent or a filter coated with a specific reactant that traps the contaminant(s) in the air. Passive monitors work by diffusion of the contaminant from the surrounding environment into the sorbent and the determination of the average air concentration is determined by validation testing. (Courtesy of SKC, Inc., Eighty Four, PA.)

(See *Chemical Connection 6.3.1.1* Activated Charcoal and Adsorption.) Charcoal tubes come in various sizes but the most common is a 6 mm OD × 70 mm long glass tube that is filled with two sections (100 mg and 50 mg, respectively) of activated coconut-shell charcoal separated by glass wool or foam. The ends of the tube are flame sealed. The 100-mg section is for primary collection and the 50-mg section is to detect breakthrough or overloading in the front section (see Figure 6.3.1.1). The *NIOSH Manual of Analytical Methods* describes many procedures using charcoal tubes.[4]

CHEMICAL CONNECTION *6.3.1.1*

ACTIVATED CHARCOAL AND ADSORPTION

Activated charcoal is frequently used as the sampling medium in air sampling methods, especially charcoal tube methods. Activated charcoal can be produced from many materials but coconut-shell charcoal has the characteristics that make it very suitable for air sampling. Coconut shells are carbonized at temperatures of 600–900 °C in the absence of oxygen. The resulting carbon residue is activated by treatment with steam or carbon dioxide at high temperatures (600–1200 °C), which yields an ideal sampling medium for nonpolar organic molecules because it is highly porous and has a very high surface area of > 500 m^2/ gram of charcoal.

During air sampling using charcoal tubes, air is drawn through a tube and airborne chemicals are *adsorbed* onto the surface of the charcoal, where they are bound by van der Waals forces. *Adsorption* differs from *absorption*, which is a process where a chemical is taken up (dissolving) by a gas or liquid into solution. Materials used in adsorption processes are termed sorbents. Charcoal is among a group of sorbents that are used for air sampling. Many small nonpolar organics are readily adsorbed on the surface of the charcoal, where they form a monolayer that can easily be removed upon desorption. The coverage of the surface is dependent on the concentration (or pressure) of the adsorbate (the chemical being adsorbed onto the surface). This has been described by the Langmuir equation at a specified temperature (each temperature is called an isotherm)—sometimes called the Langmuir isotherm:

$$\theta = \frac{\alpha P}{1 + (\alpha P)}$$

P is the gas pressure or concentration of the adsorbate, α is the Langmuir adsorption constant, and θ is the percent of surface covered. The constant α decreases as the temperature increases and is dependent on the specific binding energy of the adsorbate (different adsorbates have different binding energies).

It is reasonable to consider the bed of charcoal as a type of chromatographic column. Once an adsorbate A is adsorbed, it may be displaced by another different adsorbate B that has a higher binding energy. When this happens the original adsorbate A will be pushed off the charcoal. For the desorption process, charcoal is often treated with carbon disulfide that displaces and dissolves the adsorbate into solution, rendering it available for analysis by gas chromatography with flame ionization detection or mass spectrometric detection. Carbon disulfide is used because it is nonpolar, extracts nonpolar analytes very effectively, and is so volatile that it does not interfere with gas chromatographic analyses.

Activated charcoal is often used in filters to remove nonpolar airborne contaminants in *ductless* laboratory hoods. These are hoods that filter laboratory air but subsequently recycle the filtered air directly back into the lab. See Section 7.2.3. However, remember that if a chemical with a stronger binding energy comes in contact with the charcoal, the bound adsorbates would be released. This means that if you are working in a ductless hood with a charcoal filter, the chemical will be released into your laboratory and the potential opportunity for exposure would be significant. Currently, some new ductless hoods are being marketed for chemistry labs. If properly used and maintained, these may see application in some limited circumstances, but we believe that ducted chemical hoods as described in Sections 7.1.4 and 7.2.3 should be the standard in most chemistry labs.

Just before sampling, the ends of the tube are snapped off and the end of the tube nearest the smaller 50-mg section of charcoal is attached with tubing to an air sampling pump. Air is drawn at a specific rate over a defined period, such as 200 mL/min for 4 hours. Afterwards, the tube is capped and returned to the laboratory for analysis. A solvent such as carbon disulfide is added to each section of charcoal and an aliquot of each desorbed sample is analyzed using gas chromatography with flame ionization detection. Results are reported in µg/L or ppm.

The results are often compared against established exposure limits, such as OSHA's PELs, ACGIH's TLVs, or NIOSH's RELs (see Section 6.2.2). In Incident 6.3.1.1 the NIOSH recommended ceiling value for formaldehyde was 1 ppm for a 30-minute sampling period and the samples exceeded this level. As a result, additional efforts were made to reduce exposure by improving the performance of the hood. A resampling after corrective measures demonstrated that these changes were effective.

Biological Monitoring

Biological monitoring (aka biomonitoring) is another method for assessing chemical exposure.[6-8] This is not a new technique; it has been around for a long time but in recent years it has become an important tool for assessing exposure to environmental chemicals. This approach involves measuring parent chemicals, their metabolites, adducts, or other biomarkers of exposure in human specimens such as blood, serum, plasma, urine, or other human tissues or fluids.

Biomonitoring provides an assessment of overall integrated measure of exposure and absorption of a chemical in one's environment (laboratory, workplace, or home), in one's foods, or in one's water, and can confirm exposure to a chemical. When measuring potential exposure to chemicals in the environment, it can be difficult to provide a good estimate as to what exposure really occurred because there are variables with the extent of exposure and the absorption of the dose. For example, even though a chemical was present in the environment one does not know how much a person breathed, absorbed through his/her skin, or ingested in his/her food or water. Because biomonitoring is a body measurement, it is also possible to link body concentrations to observed health effects. (See *Special Topic 6.3.1.1* Assessing Exposures to Mercury and Its Compounds in Laboratory Incidents).

SPECIAL TOPIC *6.3.1.1*

ASSESSING EXPOSURES TO MERCURY AND ITS COMPOUNDS IN LABORATORY INCIDENTS

Most laboratory operations are safe and exposure to chemicals can be safely managed. However, there are incidents in which laboratory workers have been exposed to toxic chemicals and measures of exposure have played a role in understanding the causes of the incidents. While there is a broad variety of laboratory incidents involving many chemicals, one chemical is responsible for more incidents than any other—mercury and its compounds. This is due to the toxicity and volatility of mercury and its compounds.

In Incident 6.2.2.1 a widespread contamination by mercury was discovered in a group of chemical laboratories.[9] The incident was recognized when a chemistry graduate student showed up with signs of mercury poisoning. Measurement of urinary mercury levels showed that the student had 1500 µg Hg/L—an abnormally high concentration. Further investigation of other students, professors, technicians, and university employees showed abnormal levels of urinary mercury ranging from 28 to 2500 µg Hg/L. As the investigation continued, airborne concentrations of mercury clearly showed that laboratories were contaminated with mercury—14 of 16 laboratory rooms had measurable mercury airborne concentrations ranging from 2 to 1000 µg/m^3. Clearly, exposure assessment played a key role in this incident in recognizing the incident as mercury poisoning and in helping locate the source of the mercury.

In 1997 a well-known chemist died of dimethylmercury poisoning from a single spill that penetrated one of her protective gloves and was absorbed into her body.[10] (See Incident 7.2.2.1.) Some five months after the exposure, the professor developed signs of poisoning and her whole blood concentration was 4000 µg Hg/L; a blood concentration indicative of toxicity was known to be 50 µg Hg/L. Further investigation of a 15-cm sample of the patient's hair showed that the mercury exposure had occurred at the time of the spill. Hair grows slowly and the hair length is proportional to time so that it can be used to detect and estimate time of exposure to heavy metals. Measurements of the permeability of dimethylmercury through disposable latex gloves (the kind she used) demonstrated that it penetrated these virtually immediately (within 15 seconds or less). This latter discovery resulted in the recommendation to use a highly resistant laminated glove when using dimethylmercury. Although these measurements were unable to help in the recovery of the patient, they did help in providing a clearer understanding of the cause of the incident.

Biomonitoring has been used for monitoring compliance with rules in the use of sports performance enhancing chemicals and has been used for many years to detect the use of illicit drugs. It is used for measuring blood alcohol levels and enforcing laws against driving intoxicated. In more recent years biomonitoring has been used in making public health decisions involving occupational and environmental chemical exposures. For example, biomonitoring clearly demonstrated that the removal

NHANES II blood lead measurements found a
substantial decline in blood lead levels, 10 times
more than predicated by environmental modeling

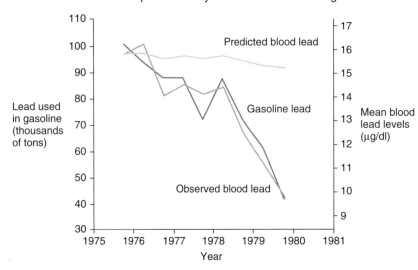

FIGURE 6.3.1.4   Lead Levels in Gasoline and Blood Lead Levels in the U.S. Population from 1975 to 1981. Data collected from the CDC's National Health and Nutrition Examination Survey II (NHANES II) show a strong correlation between blood lead levels and the decrease in lead in gasoline. Data from environmental modeling predicting blood lead levels do not correlate as well as actual measurements or with decreasing gasoline values. (From the Centers for Disease Control and Prevention (CDC), National Center for Environmental Health, Division of Laboratory Sciences.)

of lead from gasoline resulted in a correlated and definitive decline in the blood lead concentrations among the U.S. population[7] (see Figure 6.3.1.4).

Biomonitoring methods are available for chemicals commonly used in laboratories. For example, a common laboratory solvent, toluene, can be measured directly in blood, and its metabolite, hippuric acid, can be measured in urine. Collection of urine samples is often favored because it is not an invasive procedure and does not require a needle stick. Urine is treated with acid and extracted with an organic solvent and the organic layer containing the hippuric acid is separated and dried. The hippuric acid is converted to a derivative that is suitable for gas chromatography with flame ionization detection or mass spectrometric detection, or alternatively the underivatized hippuric acid is measured by liquid chromatography with ultraviolet detection or mass spectrometric detection.

The American Conference of Governmental Industrial Hygienists (ACGIH) has developed *biological exposure indices* (BEIs), which are concentrations of chemicals or their metabolites in biological samples resulting from exposure to an 8-hour shift at the chemical's TLV. As described above, the BEI for toluene is based on the concentration of hippuric acid in urine, which allows for an interpretation of the results of biological monitoring where toluene is being used.

Summary

Measurements that assess the extent of exposure can be important tools in a laboratory safety program. Air sampling has long served as an aid to occupational safety and health professionals who seek to ensure that chemical exposures are minimized. As biomonitoring continues with improved methods and builds a substantial mass of data to aid in interpretation, it will become a very significant technique for assessing exposure to chemicals. The utility of these techniques are illustrated in Special Topic 6.3.1.1 about measuring mercury exposures.

RAMP

- *Recognize* lab hazards that present risks of exposures.
- *Assess* the risk level from various hazards by using air sampling and biomonitoring when possible to do so.
- *Minimize* all exposures to laboratory chemicals using PPE, laboratory ventilation, and other means of managing exposures.
- *Prepare* for emergencies involving laboratory chemical exposures.

References

1. R. Dove. BrainyQuote; available at http://www.brainyquote.com/quotes/authors/r/rita_dove.html (accessed December 23, 2008).

2. R. Patnode. NIOSH Health Hazard Evaluation Report, HETA 82–368-1308, Appalachian Laboratory For Occupational Safety and Health, Morgantown, West Virginia, U.S. Department of Health and Human Services, Public Health Service, Centers for Disease Control, National Institute for Occupational Safety and Health, 1982.

3. OSHA's Laboratory Standard. Available at http://www.osha.gov/pls/oshaweb/owadisp.show_document?p_table=STANDARDS&p_id=10106 (accessed January 2, 2008).

4. National Institute for Occupational Safety and Health. *NIOSH Manual of Analytical Methods*, 5th edition; available at http://www.cdc.gov/niosh/nmam/ (accessed December 23, 2008).

5. SKC, Inc. Available at http://www.skcinc.com/ (accessed December 28, 2008).

6. K. Sexton, L. L. Needham, and J. L. Pirkle. Human biomonitoring of environmental chemicals. *American Scientist* **92**:38–45 (2004).

7. Special Biomonitoring Issue. *Journal of Chemical Health and Safety* **15**(6):5–29 (2008).

8. Centers for Disease Control and Prevention. National Biomonitoring Program; available at http://www.cdc.gov/biomonitoring/ (accessed December 26, 2008).

9. K. D. Rose, E. W. Simpson, and D. Weed. Contamination by mercury in chemical laboratories. *Journal of the American College Health Association* **20**:197–199 (1972).

10. M. B. Blayney, J. S. Winn, and D. W. Nierenberg. Handling dimethylmercury. *Chemical & Engineering News* **75**(19): (May 12, 1997).

QUESTIONS

1. Which environment is *not* a situation where monitoring the atmosphere is required or recommended?

   (a) Confined spaces
   (b) Where radioactive materials are used
   (c) Ordinary chemistry laboratories
   (d) Laboratories where there is reason to believe that exposure to substances is above the "action level"

2. Which kind of sampling system uses a pump to draw air over a sorbent material?

   (a) Active
   (b) Passive
   (c) Dynamic
   (d) Constant-pressure

3. Charcoal tubes usually have two sections of charcoal since

   (a) Different analytes require difference kinds of charcoal in order to be sufficiently adsorbed
   (b) The second section is present to collect breakthrough or overloading in the first section
   (c) Each section of charcoal will be used for different methods of analysis
   (d) The second tube is a protective measure to prevent the escape of gases

4. What is a typical sampling time for an active tube procedure?

   (a) 4 minutes
   (b) 200 minutes
   (c) 4 hours
   (d) 200 hours

5. Biomonitoring is the process of measuring the

   (a) Concentration of microbiological species (viruses and bacteria) in the air
   (b) Concentration of microbiological species (viruses and bacteria) in the water
   (c) Chemicals in human tissue or fluids
   (d) Death rate of scientists working in laboratories

6. When conducting biomonitoring procedures it is

   (a) Easy to determine if a chemical arose from exposure in a lab environment versus in a nonlab environment
   (b) Usually impossible to link the results with observed health effects
   (c) Usually difficult to directly link the results to exposure levels
   (d) Usually possible to link the results to the route of exposure

7. In the process of biomonitoring it sometimes is desirable to derivatize an analyte so that

   (a) It is easier to measure by a method such a UV or mass spectroscopy detection
   (b) The chemist doing the sampling does not get exposed directly to the analyte
   (c) There is less chance of contaminating an LC column
   (d) All of the above

6.3.2

WORKING OR VISITING IN A NEW LABORATORY

Preview   This section presents questions and ideas to consider when starting work in a new laboratory or visiting a new laboratory.

The whole secret of the study of nature lies in learning how to use one's eyes.

George Sand, *Nouvelles Lettres d'un Voyageur*[1]

INCIDENT 6.3.2.1   NITRIC ACID SPLASH ON FACE[2]

A chemist was using a dropper to transfer a few milliliters of concentrated nitric acid to each of a series of test tubes. She was wearing a lab coat and gloves but wore her own prescription glasses rather than chemical splash goggles. As she was carrying out this operation, the acid bottle broke, splashing nitric acid on her face. She immediately rinsed her face using the nearby eyewash while one of her colleagues called 911. After washing her face and eyes for 15 minutes, she took a shower to wash her hair that also been splashed with nitric acid. She changed clothes and the emergency personnel took her to the hospital for treatment.

What lessons can be learned from this incident?

Prelude

When you first enter into a new lab, either as a visitor or as a new student or as a new employee who will be working there, you will be walking into a room with many potential hazards. It would be nice to think that various instruments and pieces of lab equipment are working properly, safety equipment is available, that any ongoing experiments are being conducted safely, and that other people already in the lab are knowledgeable about the chemistry being done and have good safety habits. However, a "safety-conscious" person should take a moment to examine the surroundings, a new unexplored environment. We can better learn why and how to conduct this examination from the habits of emergency responders.

Emergency Scenes—Conducting the "Scene Survey"

When an emergency responder, such as a firefighter or an Emergency Medical Technician (EMT), is called to the scene of some emergency, the first step before even entering the scene is to stand back for a moment and conduct a quick "scene survey." At the scene of a fire, firefighters assess what is on fire, the nature of the fire (sometimes indicated by the color of the smoke), the probability of occupants still inside the building (based on time of day, cars in the driveway, reports from bystanders, etc.), additional dangers nearby (such as a propane tank), weather conditions (such as wind direction and velocity), and the availability of resources (such as the closest fire hydrant). All of these clues give a better picture of both the potential hazards and the best way to attack the fire. Ignoring these clues likely puts rescuers and others at higher risk.

Similarly, an EMT coming upon the scene of a multicar accident needs to do a "scene survey." This can involve a quick estimation of the number of vehicles involved, the number of patients, fire hazards, leaking fluids, downed power lines, unusual contents in vehicles, vehicle stability, and unusual circumstances (such as a wild animal that may have caused the accident!) Upon closer inspection, the

Laboratory Safety for Chemistry Students, by Robert H. Hill, Jr. and David C. Finster
Copyright © 2010 John Wiley & Sons, Inc.

EMT must determine the possible need for the use of extrication equipment before a patient can be accessed and treated. In fact, some accidents can be spread out over hundreds of feet. Even the effort to make sure that all patients are identified is not straightforward. To further complicate the EMT's task, the flow of traffic must be kept away from the scene to protect the caregivers and victims.

For both fire and accident scenes, does the initial scene survey suggest that you have enough resources and personnel on hand (or arriving), or do you need more?

Special incidents require even more thought. Firefighters wonder: "Is this a fire in a residence, or in a manufacturing facility with many other potential hazards?" EMTs consider: "I am about to treat a gunshot wound; is the shooter still here?" Special circumstances require extra factors in the scene survey.

In many situations, it is easy to be drawn into one aspect of an emergency scene much like a "moth to a flame" while other important factors go unnoticed. Missing "the whole scene" can have tragic results if some unnoticed hazard causes harm to a rescuer. This removes a rescuer from the event and adds another patient.

When you are about to begin work in a new lab, or even when you are visiting a laboratory, the moment you walk into the new territory (or "foreign environment") you potentially encounter all of the hazards that the environment might present and you may not be familiar with all of these hazards.

New Labs — Conducting the "Scene Survey"

Students walking into college classrooms see pretty much the same "scene" for both small and large classrooms. There are seats (fixed or movable), tables (fixed or movable), or chairs with armrests, whiteboards or chalkboards, perhaps bulletin boards on a wall, a projector screen, and audiovisual equipment that may include projectors and computers. Some science classrooms might have a chemical hood. However, the hazards are minimal. No one conducts a "scene survey" (except perhaps with regard to who the rest of the students are!) when walking into a classroom.

What are the hazards in a lab? Well, perhaps many. Academic labs for introductory classes likely have only limited numbers of chemicals and equipment, placed there for a particular lab. Bunsen burners aren't hazards until the gas stopcock is opened (assuming that it was *closed properly* by the last user!) and/or matches or strikers set the gas on fire. Electrical outlets are usually abundant, but no more hazardous than home outlets (except when near ubiquitous sinks). Most hazards in introductory labs are readily identified. Introductory organic labs are not much more hazardous, although the likelihood of encountering flammable solvents is much higher. So, your first few years of college chemistry classes likely do not provide the opportunity to have you conduct a "scene survey" since the labs you enter are not filled with unexpected hazards. It's easy to get complacent about walking into labs.

Let's walk into an advanced lab or a research lab. There are pretty clear rules about what chemicals (and quantities of chemicals) should be stored in a cabinet or stockroom and what should not be stored in hoods or on open benches. But, correctly or incorrectly, in research labs there are often many *more* chemicals and many *more kinds of chemicals* sitting in fume hoods or even on open benches in some labs. Are there bottles of flammable liquids or strong acids or bases "out on the bench"? There are likely more pieces of equipment and instrumentation. What are the safety hazards with these devices? Have they been modified in any fashion, and, if so, were they modified safely? Are there any high-voltage sources in the lab? Do any of the chemicals represent special hazards such as concentrated acids or bases, strong oxidants or reductants, peroxide-forming solvents, or shock-sensitive compounds? Are there cryogens in the lab? What compressed gases are present? Are the tanks secure, and are the correct regulators attached? Are the chemical hoods working? What emergency equipment is available, such as fire extinguishers (which ones: ABC, CB, A, D?), emergency eyewash stations, and safety showers? Is there a clear path to exit from every part of the laboratory? Where are the exits? Is there a phone in the lab, and do you need to dial "9" first to get an outside line? What personal protective equipment (PPE) are other lab occupants using? (If you see someone with a full face shield or wearing arm-length gloves you can be sure that something "unusual" is being done.) Without knowing the answers to these questions, you put yourself at additional risk by not knowing the full range of hazards that may be present.

Of course, the answers to many of the questions posed in the paragraph above will not be obvious with a quick "scene survey." This tells you that there may be many safety hazards present that are *not* obvious. This is the reason to be continually alert in the lab, seeking out more information all of the time.

And, are you wearing the appropriate eye protection the moment you walk into the lab, regardless of what chemicals you expect, or don't expect to find? Are you wearing inappropriate footwear like sandals or shoes that do not cover your feet adequately?

Finally, what kind of lab did you just enter as you read through the list of questions above? Perhaps you were thinking about a chemistry lab. In fact, if you were "walking into a chemistry lab" you probably did have your safety goggles on because the use of safety goggles is required in virtually all chemistry labs. But, maybe it was a biochemistry, molecular biology, or genetics lab in the biology department. Or maybe it was a geology or physics lab. *The chemicals and other hazards don't know what kind of lab they are in*. After studying much of the content of this book, you will hopefully become attentive to the range of hazards in all *chemistry* labs and appropriately anticipate them. It's a bit harder to remember that all *science labs present hazards* and some of these are likely to be hazards specific to a type of laboratory that is different from your experiences.

Finally, there's one more hazard: other lab occupants. The level of safety training and awareness varies considerably from lab to lab, and building to building. While you may be thinking of safety all of the time, your peers (in college and beyond) might not be so inclined. Many injuries and exposures occur due to *someone else's* mistake. The safety culture in a given lab is often "handed down" from experienced (although perhaps not wise) occupants to new occupants. "We've always done it this way and no one has ever gotten hurt." The tragedies of the in-flight disintegration of the Space Shuttles *Challenger* (1986) [3] and *Columbia* (2003) [4] both represent episodes where known safety hazards were ignored because they had not caused problems "before."

Working in New Labs: RAMP

When walking into a new lab, or even visiting a lab for a few minutes, stop at the door and do a quick scene survey. What you don't know about the laboratory can hurt you. When starting to work in a new research laboratory, take time to learn carefully about the experiments and equipment. Don't necessarily rely on the current occupants to have assessed all safety hazards thoughtfully.

It's pretty easy to apply RAMP (*recognize* hazards, *assess* risks, *minimize* hazards, and *prepare* for emergencies) as long as you do at least a cursory "scene survey."

- *Recognize* the variety of hazards for individual labs.
- *Assess* the level of risk presented by the laboratory hazards by making your own judgments, rather than trusting the past experiences of others.
- *Minimize* the risk by using appropriate personal protective equipment and following procedures carefully when performing new experiments and working with new equipment.
- *Prepare* for an emergency by learning what safety equipment is available in the lab and where it is located.

References

1. G. SAND. *Bartlett's Familiar Quotations*, 17th edition (JUSTIN KAPLAN, ed.), Little, Brown, and Co., Boston, 2002, p. 461, line 15.
2. University of Delaware, Office of Campus and Public Safety, Department of Environmental, Health, and Safety. Laboratory Incidents at the University of Delaware: Researcher Splashes Nitric Acid on Face; available at http://www.udel.edu/ehs/nitricacidsplash.pdf (accessed September 5, 2009).
3. Report of the Presidential Commission on the Space Shuttle *Challenger* Accident, 1986. Available at http://history.nasa.gov/rogersrep/genindex.htm (accessed September 28, 2009).
4. *Columbia* Accident Investigation Board Report, 2003. Available at http://caib.nasa.gov/news/report/default.html (accessed September 28, 2009).

QUESTIONS

1. Why do emergency rescuers perform a "scene survey" before taking any actions at the scene?

    (a) OSHA regulations require that they do so.

    (b) They want to determine if someone else has already taken care of the emergency.

    (c) They want to document the scene with digital photographs.

    (d) They want to determine if there are circumstances that may compromise the safety of patients, victims, and safety personnel.

2. How are advanced or research laboratories different from laboratories for introductory classes?

    (a) They have more chemicals in the laboratory.

    (b) They have more kinds of chemicals in the laboratory.

    (c) They have more instruments and kinds of equipment in the laboratory.

    (d) All of the above.

3. Which kind of nonchemistry lab may contain many hazardous substances?

    (a) Molecular genetics laboratories

    (b) Physics laboratories

    (c) Geology laboratories

    (d) All of the above

4. It is sometimes true that the current occupants of a lab

    (a) Won't be aware of some lab hazards

    (b) Will be aware of some lab hazards but generally dismiss the hazard since experience has shown that the hazard does not cause safety problems

    (c) Will be aware of some lab hazards

    (d) All of the above

6.3.3

SAFETY PLANNING FOR NEW EXPERIMENTS

Preview   This section discusses how planning can help you prepare to conduct an experiment safely.

Plans are nothing; planning is everything.

Dwight D. Eisenhower[1]

INCIDENT 6.3.3.1   LARGE-SCALE REACTION OUT OF CONTROL[2]

A researcher began a new experiment that she had not previously attempted. This was a large-scale reaction involving the addition of diaminopropane (DP) to potassium hydride (KH) in a 2-liter, three-neck flask. As she mixed 100 mL of DP to 150 g of KH under a nitrogen atmosphere, the mixture began to froth and foam—a common occurrence with hydride reactions. She placed stoppers on the flask but the pressure began to build and blew off the stoppers, splashing her with the reaction mixture.

What lessons can be learned from this incident?

Planning and Preparation for Safety Cannot Be Afterthoughts

As Eisenhower and other generals have observed, planning is a critical part of preparing for battle, but once the battle starts, plans are not worth much. The same is true of laboratory research: planning for safety is essential because once you begin the experiment it's often too late to stop what is going to happen. Good preparation for experiments requires careful consideration of its potential hazards. This can be summarized in the RAMP acronym which has permeated this book:

- *Recognize* the hazards in your new experiment, focusing on the highest hazards.
- *Assess* the opportunities for exposure, and redesign the experiment if the exposure opportunities are too risky.
- *Minimize* the hazards by using appropriate techniques and equipment and ensuring that at-risk behavior is controlled.
- *Prepare* for emergencies by identifying needed safety equipment and procedures—practice using this equipment and procedures.

We have made efforts in this book to prepare you to carry out work in the laboratory and to conduct experiments safely. Nevertheless, we can only deal in basic principles and cannot foresee all future experiments. Because every experiment is different, its hazards require careful consideration to minimize opportunities for adverse events or incidents. Your safety and the safety of others depend on you consistently and repeatedly applying what you have learned. This is a continuous process that will not end when you finish this book or receive your degree, since if you continue to work in laboratories you will encounter new hazards that require you to learn about them. Often the hardest part of the task is recognizing hazards *before* you encounter them.

So let's think about how you can plan and prepare for an experiment that you have not previously done. This may be a research effort doing something new or may be carrying out an experiment that was previously reported by others. This process can be a formal one involving a written plan or a

Laboratory Safety for Chemistry Students, by Robert H. Hill, Jr. and David C. Finster
Copyright © 2010 John Wiley & Sons, Inc.

checklist, or it can be an informal one in which you think about what you are doing rather than writing or sketching out a plan. You may want to try the formal at first to get the feel of this; whatever you do, however, requires that you think carefully and strategically about the experiment by "walking through" each step of the process.

Planning a New Experiment

Planning for a new experiment can be carried out in a systematic manner, using the RAMP principles. We have provided a checklist in Table 6.3.3.1 to aid you in your efforts to plan for experiments.

Before you even begin thinking about using the checklist, you already know several things that you need to do to minimize hazards in your laboratory work. You already know that you need to be thinking about ways to contain the hazards of your experiment or reaction, ways to prevent exposure to your skin and lungs, and ways to reduce risks. For example, minimizing quantities is always a strategy that you should consider in your experiments.

Think about the potential hazards that you may encounter by *recognizing* these in this new experiment. What chemicals will be used and where can you find information about their hazardous properties and the hazardous properties of the products of this experiment? Keep in mind that there are also physical hazards and hazards from equipment that you might use. Consider devising a flow chart of each step of the experiment and identifying potential hazards. Recognizing hazards is one of the most difficult tasks in safety and you must learn this skill early. Lastly, remember that not all hazards are created equal; consider *all* hazards but focus on those hazards that carry the greatest risk.

Assess how you might be exposed to these hazards. There are a limited number of exposure pathways and you need to focus on how to prevent these exposures. Keep in mind that if your assessment shows that there is a very good chance of being exposed to a hazard, then you need to rethink how you are doing this and find a better, safer way to do this experimental work. Incident 6.3.3.1 illustrates this point. If the researcher had not done this before, she should have used a much smaller scale reaction to gain valuable experience. Always consider doing experiments on a small scale initially. Always be more cautious with large scale reactions since their dynamics are often different than smaller scale reactions.

Think about how you can *minimize* these hazards to prevent exposure. Chemistry has developed many techniques for controlling exposures using various kinds of safety equipment or methods that have been designed to minimize exposures. It is important that you identify and actively use these methods in the correct manner. Think about the fact that most incidents happen because of at-risk behavior—people doing things that they should not be doing, including not minimizing exposure opportunities.

Think about what kinds of emergencies you may encounter in carrying out this experiment and *prepare* for them. You will find that the kinds of emergencies (fires, explosions, chemical exposure) that might happen are common to many reactions so you can take steps to be prepared for these. As in any other job or task at home, if you work in a laboratory for any significant amount of time, you will experience an emergency. Prepare and practice for those emergencies.

Thinking is the key here—critical thinking. Planning requires that you think about what you are going to do. You must not only think about the chemistry that you want to accomplish but you need to always think about how to prevent incidents. While carrying out experiments is almost certainly a solo performance, preparation for an experiment is not. It requires not only that you investigate various information resources but that you contact persons with experience, especially those people who you know have some experience in carrying out new experiments such as your instructor, advisor, mentor, guest researchers, graduate students, or health and safety professionals.

Finally, we encourage the application of the principles of green chemistry as described in Sections 1.1.2, 1.2.2, and 1.3.4. "Green" reactions are inherently safer reactions.

6.3.3 SAFETY PLANNING FOR NEW EXPERIMENTS

TABLE 6.3.3.1   Checklist for Safety Planning of New Experiments

Recognize chemical hazards

- Locate information about known hazards of each reactant, solvent, and predicted products (see Section 3.2.2)
- For reactive chemicals, check *Bretherick's Handbook of Reactive Chemical Hazards*[3]
- Ask an instructor, advisor, or graduate student about experience with these chemicals
- Identify hazard classes of these chemicals
- If hazard information is unavailable, examine hazards of structurally similar chemicals, but remember that structural differences can markedly change hazards

Recognize physical hazards

- Identify non-STP conditions: high or low temperatures and/or pressures
- Identify gases being used
- Identify electrical, radiation, and sharps hazards
- Examine area for housekeeping

Recognize hazards of equipment being used

- Ensure that you receive training in the use and operation of this equipment
- Ensure equipment is appropriate for this operation
- Find information about equipment hazards from manufacturer's guide
- Ensure all safety guards are in place and operational
- If this equipment is powered by electricity, ensure it is UL® certified
- If this is specialty equipment constructed locally, identify all areas of potential contact points with open power sources
- Identify equipment parts that present potential person–equipment hazardous interactions—moving parts, pinch points, doors, lids, heated/cooled parts
- Identify any potential sources of X-ray or gamma radiation

Assess potential for exposure to these hazards

- Identify potential opportunities for skin exposure to hazards
- Identify potential opportunities for inhalation exposure to hazards

Minimize exposures to hazards

- Acquire and use the minimum quantities of all hazardous chemicals
- Eliminate ingestion hazards by prohibiting eating, drinking, or taking medicines in the lab
- Eliminate eye hazards by wearing chemical splash goggles
- Carry out new reactions on a small scale before carrying out a large-scale reaction
- Identify equipment, methods, and practices that will minimize exposures to hazards
- Identify personal protective equipment to prevent skin exposure to hazards
- Identify safety equipment that will contain vapors, gases, aerosols, or particulates generated by the experiment
- Identify methods to prevent person–equipment hazardous interactions

Prepare for emergencies

- Identify procedures to be followed if exposure to a hazardous event occurs
- Identify, locate, and procure safety equipment for spills, fires, and medical emergencies
- Learn how to operate safety equipment
- Identify and learn appropriate first aid procedures
- Consider meeting with emergency response personnel/on-site medical personnel to discuss potential emergency situations

References

1. Dwight D. Eisenhower. BrainyQuote; available at http://www. brainyquote.com/quotes/authors/d/dwight_d_eisenhower.html (accessed December 30, 2008).
2. American Industrial Hygiene Association. Laboratory Health and Safety Committee: Laboratory Incidents—Explosions. Mixing diaminopropane and potassium hydride; available at http://www2. umdnj.edu/eohssweb/aiha/accidents/explosion.htm#Diaminopropane (accessed December 30, 2008).
3. *Bretherick's Handbook of Reactive Chemical Hazards*, 7th edition, Vols 1 and 2 (P. Urben, ed.), Elsevier, New York, 2006.

QUESTIONS

1. General safety principles taught in this book

    (a) Are applicable for "routine" experiments but are often not relevant when conducting specialized experiments

    (b) Are sufficiently detailed that their use in all circumstances will prevent lab accidents

    (c) Might not provide enough detail to alert chemists to all of the kinds of hazards that can be encountered

    (d) Are good for undergraduate experiments but much less applicable in the chemical industry

2. When planning a new experiment, the list of items to consider is

    (a) Relatively short

    (b) Quite long, and it is best to always write out a detailed "assessment of risk" for every experiment

    (c) Quite long, but most items can usually be ignored

    (d) Quite long, and careful review of all hazards and possible risks is wise

3. When recognizing hazards, it is best to consider

    (a) All hazards equally

    (b) The overall risk level of the least hazardous steps since they are the most likely to occur

    (c) The overall risk level of the most hazardous steps since they are the most likely to occur

    (d) The overall risk level of the most hazardous steps since they are the most likely to present the greatest danger

4. One way to reduce risk in some experiments is to run the reaction

    (a) At a higher temperature, which will likely complete the reaction more quickly, leaving less time for mistakes to occur

    (b) On a small scale since any adverse incident will then also be "on a small scale"

    (c) On a large scale so there is more solvent to absorb any unexpected heat release

    (d) In a closed system so that any possible explosion is better contained

5. Preparing for emergencies is

    (a) Required by OSHA

    (b) Prudent so that in the event of an accident and subsequent lawsuit, you can demonstrate that the incident wasn't your fault

    (c) Prudent since being able to respond quickly and properly can minimize injuries to personnel and damage to property

    (d) All of the above

CHAPTER 7

MINIMIZING, CONTROLLING, AND MANAGING HAZARDS

ONCE YOU have learned to recognize hazards and to assess the risks of exposure, the next step in the process represents another principle of safety: *minimizing, controlling, and managing your exposure to hazards*. This chapter presents a wide range of topics that will help you learn about personal protective equipment and other strategies and equipment to minimize the risks of working in a laboratory. A well-educated chemist can easily spend a lifetime in the lab without suffering any harm—if the right knowledge is applied the right way at the right time.

INTRODUCTORY

7.1.1 Managing Risk—Making Decisions About Safety   Presents an introduction to the process of risk management and how individual scientists or regulatory agencies decide what is "safe."

7.1.2 Laboratory Eye Protection   Describes the eye protection needed when working in the laboratory, and how you can protect your eyes against chemicals in introductory labs.

7.1.3 Protecting Your Skin—Clothes, Gloves, and Tools   Describes ways to protect your skin from exposure to hazardous chemicals in introductory labs.

7.1.4 Chemical Hoods in Introductory Laboratories   Discusses the use and operation of laboratory hoods as a key part of your arsenal of safety equipment to protect you from exposure and to keep chemical aerosols, vapors, fumes, particulates, and odors under control.

INTERMEDIATE

7.2.1 More About Eye and Face Protection   Discusses various kinds of eye protection needed in advanced labs.

7.2.2 Protecting Your Skin in Advanced Laboratories   Describes ways to protect your skin from exposure to hazardous chemicals in advanced labs.

7.2.3 Containment and Ventilation in Advanced Laboratories   Continues the discussions in Section 7.1.4 with additional features of laboratory hoods, other means by which lab air is kept safe to breathe, and personal protective equipment used to avoid breathing contaminated air.

ADVANCED

7.3.1 Safety Measures for Common Laboratory Operations Touches briefly on safety measures that can prevent incidents during various common laboratory operations.

7.3.2 Radiation Safety Provides an overview of the principles of radiation safety in the laboratory.

7.3.3 Laser Safety Discusses the basics of lasers and safety practices and procedures and other safeguards when using lasers in the laboratory.

7.3.4 Biological Safety Cabinets Describes biological safety cabinets and methods for containment used in handling infectious biological microorganisms in the laboratory.

7.3.5 Protective Clothing and Respirators Discusses personal protective measures that might be employed in advanced chemical laboratory operations.

7.3.6 Safety in the Research Laboratory Presents factors to consider when you are working in a research laboratory and performing new, non-course-related experiments.

7.3.7 Process Safety for Chemical Operations Provides a brief overview of process safety and its importance in chemical safety and hazard recognition.

7.1.1

MANAGING RISK—MAKING DECISIONS ABOUT SAFETY

Preview This section presents an introduction to the process of risk management and how individual scientists or regulatory agencies decide what is "safe."

Take calculated risks. That is quite different from being rash.

George S. Patton, U.S. Army general (1885–1945)[1]

INCIDENT 7.1.1.1   NITRIC ACID AND ETHANOL EXPLOSION[2]

A student was preparing a solution reported in a published paper using nitric acid and ethanol. He mixed the two chemicals and did not notice anything unusual, such as a rise in temperature. He capped the bottle and went home. Later that night the bottle exploded, sending glass shards across the lab more than 30 feet away. Another bottle more than 5 feet away containing oil was broken by the blast. Since no one was present when the explosion occurred there were no injuries.

What lessons can be learned from this incident?

Prelude

The risk assessment process was presented in Section 6.1.1. It involves an understanding of the variety of hazards that are associated with any particular chemical or chemical experiment and an estimation of the likelihood for exposure to the hazards. The level of hazard can be quantified using data such as permissible exposure limits (PELs) or NFPA fire hazard ratings, for example. Even with these "hard" data, there is considerable uncertainty in these values since, especially for toxicological data, the methods of hazard assessment are fraught with statistical uncertainties, human variations in response, and animal-to-human extrapolations. Assessing the "likelihood" of exposure is highly subjective and situation dependent. Together, most risk assessments in labs are fairly crude but a general assessment of a chemical procedure as "relatively safe" or "relatively dangerous" is possible and reasonable.

Scientists are constantly estimating risks, either thoughtfully or casually, and then making decisions about how to proceed with particular experiments with appropriate cautions. This section describes the methodologies used to decide how much risk to accept. This is the process of *risk management*.

Risk management can be conducted either formally or informally. We can view formal risk management as some well-defined process by which the overall level of risk is coupled with the willingness to accept a certain level of risk, or mandate a particular level of safety, in some regulatory process. The Food and Drug Administration (FDA) weighs risk levels of food additives or the use of over-the-counter or prescription medications with the probable positive and negative effects and then decides on a "reasonable level of safety" when setting regulations. Less formally, but using essentially the same process, a chemist working in a lab on a new experiment assesses the risk, considers the benefits, and makes decisions on how to proceed with regard to the scale of the experiment, the equipment and chemicals to be used, and various safety procedures to be followed. We discuss both of these situations, formal and informal risk management, but first we consider the question: "What is safe?"

Laboratory Safety for Chemistry Students, by Robert H. Hill, Jr. and David C. Finster
Copyright © 2010 John Wiley & Sons, Inc.

Acceptable Risk: Deciding What Is Safe

People often come to different conclusions regarding whether a given level of risk is either acceptable or unacceptable. Otherwise stated, we differ on deciding "what is safe." This applies to a wide variety of situations ranging from a "safe" level of alcohol consumption to a "safe" speed to drive on a highway to the "safe" way to invest money in the stock market. What is "safe" depends on both an objective analysis of risks and perceived level of risk (as discussed in Section 6.1.1) and a value judgment of what level of risk is acceptable. Two people might equally enjoy the thrill of skydiving, but might differ on their willingness to take the risk associated with the activity. American society affords considerable (but not total) freedom in letting individuals make personal decisions about safety.

American governance has evolved in a fashion that has led to an enormous number of laws and regulations at local, state, and federals levels. These laws and regulations are designed to protect the "public" from irresponsible behavior that endangers others. These kinds of risks are viewed as nonvoluntary, requiring legal protections, while taking personal risks, such as skydiving or snow boarding, are regarded as voluntary risks that need no regulation. As a community we decide on upper limits to travel on roads by establishing speed limits. OSHA sets many limits on occupational hazards to protect workers. These workplace hazards are regarded as nonvoluntary and we don't want to unwillingly expose workers to hazards without protections. The FDA decides on the safety of our food and drugs. These decisions are often controversial and frequently have financial implications for individuals, businesses, and insurance companies. Liberals and conservatives and Democrats and Republicans (and other minor parties) bring different philosophies and value judgments to these decisions about both who should be deciding what is safe and the level of safety that is acceptable. It's a messy, complicated process. But, eventually, someone or some agency has to decide on the "safe level."

With this backdrop on understanding the complexity involved in answering "What is safe?" let's consider the formal and informal risk management processes.

Formal Risk Assessment: Laws and Regulations

We'll limit our discussion to some examples of federal regulations related to health and safety since we are all subject to these laws and some of them bear directly on how we function in labs.

Health and safety laws can be divided into three categories that reflect different philosophies and operational assumptions.[3]

- "Zero-risk" laws are based on the goal that "perfect" safety is required and that nothing should be "weighed" in some cost–benefit or risk–benefit analysis.
- "Balancing laws" are based on weighing some level of acceptable risk against the benefits associated with some activity.
- "Technology-based" laws recognize that the zero or reasonable risk cannot be attained due to the limits of some technology. The regulated level of exposure is based on the lowest feasible level that some technological solution can produce.

"Zero-based" laws are uncommon. A good historical example is the Delaney Clause (1958) of the Federal Food, Drug and Cosmetic Act (1938), which states that the FDA "shall not approve for use in food any chemical additive found to induce cancer in man, or, after tests, found to induce cancer in animals." This is a very "extreme" condition since it completely bans the use of any (natural *or* synthetic) chemical that is believed to cause cancer at *any* dose in *any* species. In 1988 the Environmental Protection Agency changed its policy in interpreting the Delaney Clause to include the criterion of "*de minimus*," which refers to a risk level "too small to be concerned with." That is, minimal or negligible risks are acceptable.

The "absolute" nature of a zero-risk law is inconsistent with the practice of epidemiology and risk assessment since we can now easily detect the presence of many chemicals (in food, water, or air) at levels far below which any risk of harm can be measured. In the extreme, this philosophy would simply prohibit many activities in labs and in industry. For example, since there is always a slight chance of falling off a ladder, a zero-based OSHA law about ladders would prohibit their use. Similarly, since there is a chance of a power failure or fan belt breaking, which would shut down the operation of chemical hoods, OSHA would be required to ban such devices in labs. Zero-risk laws usually arise from some unreasonable fear that "simplistically" solves the problem of risk by preventing use entirely.

The irrational and unscientific position of zero-based laws naturally leads to "balancing laws," which are very common. The Safe Water Drinking Act (SWDA, 1974), Toxic Substances Control Act (TSCA, 1976), and the Federal Insecticide, Fungicide, and Rodenticide Act (FIFRA, 1947) all represent regulations that seek to determine reasonably safe exposures of various substances. As you can imagine, the official determination of "safe" can be highly controversial. (This is discussed more below.)

There are some environmental circumstances where the ability to prevent pollution or reduce pollution is limited by technology. This leads to "technology-based" laws. An excellent example of this is the current EPA regulation regarding radon gas to 4 pCi/L in homes (see *Special Topic 5.3.8.1 Radon—A Significant Public Health Concern*). At this level, the predicted chance of lung cancer is 23,000 per million (which is well above the accepted rate of 1 per million for carcinogens) but since current, reasonable mitigation technology *cannot* reduce radon below 4 pCi/L, this level is set as the "action level." The EPA does not say this is a "safe level" but only that higher levels present greater risk (see *Special Topic 7.1.1.1 The Precautionary Principle*). Therefore, to *minimize* risk, the EPA established this "action level." Some parts of the Clean Air Act (1963) and Clean Water Act (1974) have technology-based regulations.

SPECIAL TOPIC *7.1.1.1*

THE PRECAUTIONARY PRINCIPLE

Let's assume that someone wants to sell a chemical product with some presumed benefit to the consumer and some regulatory agency is entrusted with protecting the public with regard to this particular product. Should the regulatory agency require that the producer *prove that it is safe* or should the producer require that the regulatory agency *prove that it is dangerous* before being allowed to ban the product?

It is sometimes relatively easy to establish that a product is dangerous: some simple testing of the product with animals or humans may show some adverse effects that warrant some regulatory action. However, proving that something is safe is not so easy. If a chemical at a concentration of 0.001 M shows no deleterious effects in tests, we can state that there is no evidence of a hazard. But, the limits of the test are important since "only the conditions of the test were tested." Can we extrapolate animal tests to humans? Not always. Was the dose oral or dermal? What "effects" were examined? Simple "illness?" Was blood chemistry tested? Were tissue samples taken? What was the time frame of the test? Days? Years? Thus, we cannot prove the chemical is completely safe, we can only establish an absence of effects within the examined parameters. And what about 0.0001 M or less? Proving something is safe, in a universal sense, is not possible. At best, we can say "it is safe as far as we know."

Recently, the Precautionary Principle has become a guiding philosophy for some groups that examine environmental and health issues. It states[4,5] the following:

> When an activity raises threats of harm to human health or the environment, precautionary measures should be taken even if some cause and effect relationships are not fully established scientifically. In this context the proponent of an activity, rather than the public, should bear the burden of proof.

While not asking for "perfect safety," the Precautionary Principle puts the burden of proof on the proponents of an activity to establish that it is reasonably safe, rather than putting the burden of proof on opponents to show that the activity is dangerous. Many environmental problems can be viewed from these opposing views. Global warming presents a useful example. Those who stand to benefit from the continued combustion of fossil fuels, which raise global atmospheric carbon dioxide levels, claim that the burden of proof that global temperature is rising (over decades) has not been definitively established or definitively linked to carbon dioxide levels. Furthermore, the projections of future environmental effects are based on complicated atmospheric models that contain many uncertainties and assumptions. Thus, they would want the burden of *proof of an adverse outcome* to be firmly established before taking action to reduce carbon dioxide output.

Opponents of this position, in part using the Precautionary Principle, claim that the threat of global catastrophe is serious and that the burden of proof of a "safe" future is upon the proponents of fossil fuel combustion. And, in the absence of such proof, we should suspend and/or reduce carbon dioxide generation given the level of threat.

Furthermore, by the time we establish "further proof of harm" it will be too late to meaningfully reduce carbon dioxide levels and future environmental problems will be unavoidable.

What position is taken on matters such as these can reflect underlying philosophical positions of each side (as well as vested interests in the benefits of continued use of fossil fuels).

There are hundreds of case studies in environmental and health care debates where there is some, but not definitive, evidence of an adverse effect of the use of a chemical or drug. Those using the Precautionary Principle argue that some evidence of hazards should prevent an activity until it is proved to be reasonably safe. Opponents argue that the activity should continue until proof of the hazard is definitive. In some instances, such as global warming, the concern is that by the time the "proof" is definitive too much damage will have been done, leading to an irreversible outcome.

Risk–Benefit and Cost–Benefit: Making Informal Decisions in the Lab

What factors are weighed when deciding what level of risk is acceptable? Let's consider the concepts of risk–benefit and cost–benefit as they relate to laboratory research. Risk–benefit means that we consider how much risk we are willing to accept to achieve a certain benefit. What are we willing to risk to successfully accomplish something? Cost–benefit refers to the cost in effort and money to achieve a certain benefit, such as reducing a risk.

Laboratory research has been responsible for many of the discoveries that have improved our lives today. It will likely reap the same benefits in the future. However, there are risks in laboratory research: carrying out experiments that have not been done before, making new compounds that have not been made before, and handling hazardous agents to explore disease processes so that treatments and cures can be found. Scientists choose to take these risks because of the benefits that they may personally receive and that our society may collectively receive from their work. Prudent scientists do all that is feasible to minimize exposures to hazards in the laboratory but know that there will always be risk. They use personal protective equipment, such as eye protection and skin protection, and other safety equipment that removes inhalation exposures. However, scientists know that explosions may occur, exposures may happen, and injuries may result, but nevertheless they are willing to seek a balance between risk and benefit. (In fact, *we know that, statistically, explosions, exposures, and/or injuries will happen*. The fact that they *can* be prevented doesn't guarantee that they *will* be prevented. This heightens importance of the "P" in RAMP: *prepare* for emergencies!)

Scientists, like other workers, have learned that they can do only so much to protect themselves but at some point the cost in time and in dollars can become too much for the sometimes small additional reduction in risk. Thus, a research chemist will likely scale up a reaction rather than do lots of smaller, safer reactions to acquire the amount of product needed because the cost of the additional work to carry out many smaller, but safer reactions becomes too costly. Similarly, sometimes under the pressure of timelines or competition (yes, there is competition in academia too), safety measures may be bypassed to save time and beat the competition. Consider the adage, "time equals money." In this way the laboratory researcher is carrying out a cost–benefit analysis and arriving at the conclusion that the extra cost of safety is not worth the benefit. While there can be some compromise between the cost of reducing risks and the resulting benefits, it is important that the chemist understands that reducing the "cost" in this way increases the risks of adverse incidents and puts everyone in the lab at greater risk.

Some scientists may be too quick to accept a certain level of risk. The field of chemical health and safety has greatly matured in the past few decades and the best current practice is to take the necessary steps to reduce risks to be as low as possible. The reality is that not all scientists judge risk–benefit and cost–benefit the same way. These are sometimes personal decisions, but they are also institutional decisions affected by faculty (in colleges and universities) and managers (in industry). It is probable that someday you may face a dilemma if your personal decision is not the same as the institutional decision. Resolving this difference is not easy. Hopefully, the information and philosophy of this book will help in that decision. See *Special Topic 7.1.1.2* A Case Study in Risk Management—The Tragedy at Bhopal, India.

SPECIAL TOPIC *7.1.1.2*

A CASE STUDY IN RISK MANAGEMENT—THE TRAGEDY AT BHOPAL, INDIA

In December 1984 in Bhopal, India, there was an industrial incident that killed 3800 people in one night, may have caused an additional 15,000 deaths, and injured as many as 500,000 people with 100,000 of those injuries being permanent. This is considered to be the worst industrial incident in history. What happened, and how did it happen?

The Union Carbide Company had a manufacturing facility in Bhopal, a city with a population of 800,000, that synthesized pesticides using methylisocyanate (MIC). MIC is a clear liquid with a sharp odor. The boiling point is 39 °C and it has a vapor pressure of 348 mm Hg at 20 °C. The odor threshold is 5 ppm. The IDLH is 3 ppm and the PEL/TLV is 0.02 ppm. This places it in Hazard Class 1 (see Table 6.2.2.2). MIC is used as an intermediate in the synthesis of carbamate pesticides. The plant at Bhopal was manufacturing Sevin®.

At 1 A.M. on December 3, 1984 a safety valve failed and released 40 tons of MIC. Within hours, thousands of people had died and the city was littered with corpses of humans and thousands of animals and birds. Local hospitals were overwhelmed, in part because personnel did not know what the toxic agent was.

What series of mistakes occurred?[6,7]

- Although there was no need to store MIC on site since it could be used immediately after its synthesis, it was being stored in two containers that held 40 and 15 tons. The plant had been shut down and was in the process of being dismantled, which is why the MIC was being stored.

- With a BP of 39 °C, it is necessary to refrigerate the liquid when being stored. One report indicates that the electricity to the refrigeration units had been turned off to save money and another indicates that the refrigerator coolant had been drained and relocated to another part of the plant.

- A faulty valve allowed 1 ton of water to mix with the MIC. A backup system to prevent water from entering even if the valve fails was not in place. The addition of water causes a highly exothermic reaction and at high temperature some of the MIC will degrade to hydrogen cyanide (HCN). (Reports at hospitals later indicated symptoms of HCN poisoning and some patients responded favorably to treatment with sodium thiosulfate, which is an effective treatment for HCN poisoning.)

- A scrubber system to neutralize any escaping gas with a NaOH solution had been turned off.

- A system to ignite any escaping gas using a burning flare had been turned off to save fuel.

- A water-jet system that is used to capture/dissolve any escaping gas had insufficient pressure to reach the necessary height to capture the gas.

- An alarm system was turned off soon after the release "to avoid causing panic."

- The plant did not have many of the safety equipment and procedures as a similar plant in West Virginia since India does not have the same health and safety regulations as the United States, and even though the local government was aware of many concerns about plant safety it was reluctant to place safety and pollution burdens on the plant since this might lead to the plant being closed and subsequent economic loss for the community.

Where did the process of risk management fail in this episode? Let's use RAMP.

- Since there was a sister plant in West Virginia operating safely, it seems probable that the chemical engineers and managers in Bhopal could have, and surely should have, *recognized* the high level of hazard associated with manufacturing, storing, and using MIC. (Perhaps they did, in fact, recognize the hazards but lack of regulations, and local culture and financial pressures, made it easy or necessary to ignore or accept the hazards.)

- As indicated in the bulleted items above, it is likely that the Bhopal personnel could have *assessed* the probabilities associated with the various hazards and determined that the risk level was very high, particularly in light of the plant being located in a heavily populated city.

- The several bullets above indicate that the risk management process failed to *minimize* or *manage* the risks. Indeed, extremely hazardous processes usually do have several "layers of protection" and Bhopal personnel took active steps to override many safety systems. *They took many known risks.* Most disasters are the result of multiple failures to take known safety precautions. Often we are protected by multiple layers of safety, but when these layers are continually removed, we are put in jeopardy. As each layer is removed and nothing happens (but the risk increases), the operators become more comfortable in time, then seek to remove more safety measures until soon many safety measures have been deliberately removed and the risk of an event and exposure is greatly increased.

- The failure of the local hospitals to know the nature of the disaster indicates a lack of *preparation* since emergency services should be informed of the nature of potential disasters. Furthermore, the alarm that was a form of "preparation" was inactivated shortly after it sounded. The hospital systems were not prepared for such an emergency and did not know what was causing the overwhelming cases of poisoning entering their hospitals. It is doubtful that they knew what chemicals were being used in the plant. The site for this plant was selected because the health and safety laws were weak and oversight was practically nonexistent. Thus, preparation for this kind of emergency was unlikely and was probably not even considered a possibility.

As with almost all incidents, it is easy to identify the cause(s) of the event and the procedures that could have easily prevented this from occurring. It is probably inaccurate to characterize the Bhopal disaster as an "accident" since this implies a certain statistical randomness to the event. In this incident, people took overt actions that caused the event to occur, removed safety systems that could have minimized the effects of the initial mistake, and failed to adequately prepare for the disaster.

Risk Management in the Lab: RAMP

Laboratory incidents can be small or large, and have either fairly trivial outcomes or disastrous, even fatal, outcomes. In almost all cases, incidents happen because someone took a *known* risk. If risk assessment and risk management procedures are thoughtfully addressed, virtually all lab incidents are preventable.

All of the steps in RAMP are part of the risk assessment and management process.

- *Recognize* the hazards associated with chemicals and procedures. Consider *all* of the hazards of chemicals and *all* of the things that can go wrong, not just the obvious and likely event.

- *Assess* the probabilities of the hazardous event occurring.

- *Minimize* the hazards by eliminating them and/or taking steps to reduce the probability of the event occurring. Do not take unnecessary, known risks.

- *Prepare* for emergencies that might result if *anything* goes wrong since probabilities can rarely be reduced to zero. Have an action plan in mind for spills and exposures.

References

1. GEORGE PATTON. The Quotations Page; available at http://www.quotationspage.com/quote/2905.html (accessed September 20, 2009).
2. Hong Kong University of Science and Technology, Health, Safety, and Environmental Office. An explosive nitric acid–ethanol mixture; available at http://www.ab.ust.hk/sepo/tips/ls/ls005.htm (accessed September 4, 2007).
3. *Chemical Risk: A Primer*. American Chemical Society information pamphlet, Department of Governmental Relations and Science Policy, American Chemical Society, 1155 Sixteenth Street, N.W., Washington, DC, 1996.
4. Science and Environmental Health Network. Precautionary Principle; available at http://www.sehn.org/wing.html (accessed September 20, 2009).
5. C. RAFFENSBERGER and J. TICKNER (editor). *Protecting Public Health and the Environment: Implementing the Precautionary Principle*. Island Press, Washington, DC, 1999.
6. E. BROUGHTON. The Bhopal Disaster and Its Aftermath: A Review. *Environmental Health*. Biomed Central; available at http://www.ehjournal.net/content/4/1/6#IDAJLL1K (accessed September 20, 2009).
7. K. BHAGWATI. *Managing Safety*, Wiley-VCH, Weinheim, 2006, pp. 23–26.

QUESTIONS

1. The process of risk assessment is

   (a) Highly quantified, with low uncertainty, with regard to toxicology data
   (b) Highly quantified, with low uncertainty, with regard to exposure measurements
   (c) Fraught with uncertainty in many situations
   (d) Both (a) and (b)

2. Risk management

   (a) Can involve quantitative calculations of how much risk to accept

(b) Can involve crude estimates of how much risk to accept

(c) Should follow some process of risk assessment

(d) All of the above

3. In general, people tend to

(a) Accept about the same level of risk

(b) Differ widely with regard to the level of risk they are willing to accept

(c) Assess risk in the same fashion

(d) None of the above

4. American governance has generally

(a) Avoided the complicated task of risk assessment and management

(b) Avoided establishing laws about risk management since not everyone agrees how to do this

(c) Set laws based on a single philosophy about risk management

(d) Established different kinds of laws related to risk management that are situation dependent

5. "Zero-based" laws are

(a) Fairly uncommon

(b) Based on the notion that we should balance risks and benefits

(c) Based on the notion that we should balance costs and benefits

(d) Based on the concept of a *de minimus* criterion for assessing risk level

6. The SWDA, TSCA, and FIFRA are examples of

(a) Zero-based laws

(b) Balancing laws

(c) Technology-based laws

(d) A combination of (a) and (b)

7. Technology-based laws can change over time since

(a) "Action levels" will change as better toxicology risks are identified

(b) The degree to which a particular technology can minimize exposure might improve over time

(c) The goal is to always be as close to a "zero-based" law as possible

(d) The balance between risk and benefit will change depending on what political party is in legislative control

8. When scientists are working in labs, the process of risk management most closely resembles

(a) A "zero-based" approach to taking risks in labs

(b) A "balancing" approach to taking risks in labs

(c) A combination of (a) and (b)

(d) A "technology-based" approach to taking risks in labs

9. With regard to risk assessment and risk management in a laboratory, scientists

(a) Make judgments based on their education

(b) Make judgments based on the local safety and management culture of an organization

(c) Tend to all generally perform these tasks in the same way

(d) Both (a) and (b)

10. Incidents in the laboratory can be virtually eliminated if

(a) Scientists take only known risks

(b) Scientists eliminate all possible hazards in a laboratory while still performing necessary experiments

(c) Risk assessment and risk management processes are carefully considered in all activities in the laboratory

(d) Scientists take only unknown risks

7.1.2

LABORATORY EYE PROTECTION

Preview   This section describes the eye protection needed when working in the laboratory, and how you can protect your eyes against chemicals in introductory labs.

Young cat, if you keep your eyes open enough, oh, the stuff you would learn! The most wonderful stuff!

Dr. Seuss[1]

INCIDENT 7.1.2.1   ACID SPLASH IN EYES[2]

Chromic acid was being used by a student to clean glassware, when he accidentally splashed the acid into his eyes. He began screaming, and a nearby student ran into the laboratory to find the student with the acid on his face and in his eyes. His goggles were hanging around his neck, instead of over his eyes. He was wearing hard contact lenses. Together they used an eyewash to flush his eyes for 15 minutes, and they were able to remove the contact lenses during the washing. The student was taken to a nearby emergency room for further treatment. His eyesight was not permanently damaged.

What lessons can be learned from this incident?

The Eyes Have It! And You Should Want to Keep Them

There are probably more stories about lab incidents that caused, or nearly caused, injury to eyes than any other category of incident. Not surprisingly, splashes and explosions of chemicals can occur in any lab at any time. Because of this, perhaps the single most important safety rule in every chemistry lab is *wear appropriate eye protection at all times*. This section discusses exactly why this is so important, and how and why this rule sometimes gets violated. Like the cat in Dr. Seuss' tales, keeping your eyes open (and working!) in chemistry labs is important!

Some students resist wearing appropriate splash goggles in the lab because (1) it's an uncommon behavior outside the lab and it's easy *not* to wear them even when you are at risk, (2) goggles impede one's vision to some degree, and (3) goggles are uncomfortable to wear. Even though all of the factors can be overcome fairly easily, it takes time when first taking a chemistry course to get used to wearing appropriate goggles.

The truth is that incidents are *likely* to happen, eventually, in chemistry labs that can cause mild or serious damage to the eyes and it's necessary to be prepared for this possibility. Ask any chemist or chemistry teacher if they have ever experienced such an event and you will surely get a handful of stories based on personal experience. You never know when you might save your eyes from a splash from something you are doing or something *your neighbor* is doing. Incidents, sometimes called "accidents" (even though we avoid this term since it might imply some lack of preventability), are never planned! You cannot know when you might save your sight by simply wearing eye protection.

Almost all solvents and chemical solutions can damage the eyes. The stronger the concentration for acids, bases, oxidizing agents, reducing agents, and other solutes, the greater the risk for damage. Strong acids are very hazardous, but strong bases are even worse. Many substances can cause permanent eye damage and a loss of vision.

Laboratory Safety for Chemistry Students, by Robert H. Hill, Jr. and David C. Finster
Copyright © 2010 John Wiley & Sons, Inc.

Types of Eye Protection

There are different types of eye protection that depend on the hazards and their severity in the laboratory where you might be. Let's examine these types, why different kinds of eye protection are available, and what is required in chemistry labs.

There are several models of "impact only" eye protection. These are sometimes used as "visitor's glasses" and their name alone suggests that they are not appropriate for anything other than a casual stroll through a manufacturing facility where the wearer is not "doing anything" other than observing. Other forms of "impact only" eye protection are shown in Figure 7.1.2.1 and their inadequacy against splashes is apparent from the explosion of dye shown on the mannequins. The exposure of the eyes is very significant since liquids can easily "get around" the sides of the glasses, even when the exposure is directly from the front.

Safety glasses, with or without side shields (Figure 7.1.2.2), provide some protection against flying shrapnel in some nonchemistry labs or manufacturing facilities where shrapnel is the only hazard. Since safety glasses do not provide adequate protection against splashes, they are not appropriate for chemistry labs.

Because splashes are common incidents in chemistry labs, it is necessary to wear eye protection that completely protects the eyes against splashes. The only satisfactory eye protection is the chemical splash goggle. Figure 7.1.2.1 shows "impact only" goggles, which, although better than safety glasses, do not provide complete protection against splashes. The best, and only acceptable, forms of eye protection for chemistry labs are the "chemical splash goggles," which, when worn properly, provide maximum protection against both impact and splash exposures (see Figure 7.1.2.3). The absence of dye in and around the eyes shows the excellent protection for the eyes in the event of some splash or spray of chemicals in the face.

Safety Glasses With Vented Side Shields (Impact Only)

Safety Glasses With Nonvented Side Shields (Impact Only)

Visorgogs® (Impact Only)

Impact Safety Goggles (*Impact Only*)

Chemical Splash Safety Goggles (Impact and Splash Protection)

FIGURE 7.1.2.1   Degrees of Eye Protection from Simulated Splashes Using Varied Types of Eye Protection. The mannequin faces on the right show how effective each type of eye protection is against a chemical splash. Only the chemical splash goggles provide adequate protection and are the only acceptable form of eye protection in chemistry laboratories. (Courtesy of Linda Stroud, Ph.D., Science & Safety Consulting Services. Copyright © 2008 Science & Safety Consulting Services, Inc., Raleigh, NC.)

FIGURE 7.1.2.2   Various Safety Glasses. Safety glasses with side shields can provide some protection against flying shrapnel but are ineffective against chemical splashes. Only glasses that are ANSI/ASSE Z87.1 are acceptable.

FIGURE 7.1.2.3   Chemical Splash Goggles. These goggles provide excellent splash protection and are relatively comfortable to wear, if sized properly.

The Best Features of Chemical Splash Goggles

Chemical splash goggles must also protect against shrapnel in any explosion that sends chunks of glass or metal toward the face. Two organizations, the American National Standards Institute (ANSI) and the American Society of Safety Engineers (ASSE), developed a standard for eye protection—*Occupational and Educational Personal Eye and Face Protection Devices, ANSI/ASSE Z87.1—2003*.[3] This standard describes standard specifications for various types of eye protection in various occupational settings. Only goggles marked with *ASSE Z87.1* are acceptable. This designation indicates adequate thickness to reasonably protect against shrapnel.

Goggles must make a tight seal around your eyes to ensure protection. The seal around your eyes protects your eyes from chemical splashes or sprays. The flange or edging that fits around your face needs to be soft and pliable and fit snugly against your face to prevent liquid from seeping into your eyes in the event of a splash and some models may lack this feature. There are many stories of splash incidents where poorly fitted goggles still allowed liquids to "leak" around the edges and expose the eyes because the seal was not complete. Since every face is a little different, you may find that you have to try different goggles to find the ones that are best for you. Some goggles may meet the ANSI/ASSE standard but they may not be suitable if they do not make a tight fit around your eyes.

A comfortable fit is important because you may not wear protective eyewear if it is not comfortable. Most colleges and universities probably have recommended goggles that are available in bookstores. If there are multiple options, it's best to try on several styles to see what works best for you. You can also find information on the Web about eye and face protection.[4,5] Your college or university will likely recommend what goggles to wear and these will be available in the campus bookstore.

Chemical splash goggles can be either vented or unvented. Vented goggles should be indirectly vented to prevent splashes from getting behind the goggles. Vented goggles are recommended since unvented goggles can "fog up" and impede vision. Even vented goggles can do this, too, depending on the wearer and the local conditions of humidity. Commercial antifogging solutions are available that can be sprayed on the inside of the goggle lens to minimize "fogging."

Other Types of Laboratory Eye Protection

This section is devoted to the argument that chemical splash goggles are the best and only acceptable form of eye protection in chemistry labs. It is not likely that more eye protection would be necessary in introductory general chemistry labs or introductory organic chemistry labs. Advanced labs may present other hazards and require different kinds of protection. This is discussed in Section 7.2.1 with a description of face shields and lab shields. Also, eyes must be protected against excessive visible and ultraviolet (UV) radiation. Protection against laser beams is discussed in Section 7.3.3 and protection against other sources of visible and UV light is also presented in Section 7.2.1.

Contact Lenses in the Laboratory—A Matter of Choice

After contact lenses became popular in the 1970s, the National Institute for Occupational Safety and Health (NIOSH) recommended contacts *not* be worn in the workplace. The main concern was that chemicals would get trapped behind the contact lens and attempts to wash the eyes with water would be much less effective. However, in 2005 after considering available scientific evidence, NIOSH recommended workers be permitted to wear contact lenses, as long as there were no regulations or other medical/safety recommendations against wearing contact lenses.[5] *Wearing contact lenses does not reduce the requirement to wear eye protection—you should still wear chemical splash goggles in the laboratory over your eyes.* Contact lenses should still be removed as quickly as possible in the event of any contamination in the eyes.

Some safety professionals recommend that students who are wearing contact lenses place some mark on the side of the safety goggles (such as a red dot or sticker) to indicate the presence of contact lenses. This can remind anyone assisting the victim in some exposure to the eye(s) that the contacts are present and should be removed. However, others must know that this special designation, a sticker or red dot, means that you are wearing contacts.

"Can I Ever Not Wear My Goggles in the Lab?"

The need for eye protection is dictated by the presence of potential eye-damaging chemicals. In the rare instance that your lab has no chemicals, eye protection is not required. This includes an "experiment" or lab exercise reading MSDSs or using computers for computational chemistry. Your instructor will announce such circumstances and if the lab is deemed absent of chemicals.

Some advanced labs might present rare circumstances where chemical splash goggles are not required and either only safety glasses are required or no eye protection is necessary. However, it is *your* responsibility to protect your eyes and your vision. Developing the habit of wearing chemical safety goggles is a good and prudent practice. This is discussed more in Section 7.2.1.

Sometimes you may have finished the "chemical" activity in a lab and simply be working on calculations, writing notes in your lab notebook. Since other students may still be working with chemicals, you must keep your goggles on while in the lab to prevent unintended exposure.

Finally, if you store your goggles in a lab drawer or locker in the lab, you should put on your goggles as soon as you walk into the lab. This assumes that the goggles are needed only in this lab, and not elsewhere in other labs. It is wise to minimize the time you are not wearing your goggles, particularly at the end of an experiment when other students may still be working with chemicals nearby. In this circumstance, remove your goggles, store them, and exit the lab quickly.

Working in Labs and Protecting Your Eyes: RAMP

The rule is: *always wear chemical splash goggles in labs where chemicals are used.* As noted above, laboratories *without* splash hazards are "the exception, not the rule." Any lab involved in the use of liquids or potentially hazardous or infectious substances requires chemical splash goggles. Chemicals are used in chemistry, biology, physics, and geology laboratories, and in art departments and theater scene shops. *The chemicals don't know what lab they are in*!

Sections 2.1.1, 2.1.3, and 2.1.4 present more information about what to do in the event of chemical exposure to your eyes.

- *Recognize* the different kinds of hazards that may cause injury to eyes. *Don't* assume that non-chemistry labs *don't* require eye protection.

- *Assess* the level of risk based on the possibility of exposure in the laboratory, and the relative danger posed by various chemicals.

- *Minimize* the risk using appropriate eye protection, including chemical splash goggles, face shields, and other levels of protection.

- *Prepare* for emergencies by knowing the location of eyewashes and the proper response procedures for responding to emergencies involving eye exposures.

References

1. TED GEISEL (Dr. SEUSS). *I Can Read with My Eyes Shut*, Random House, New York, 1978.
2. FARIBA MOJTABAI and JAMES A. KAUFMAN (editors). Learning by Accident, Vol. Laboratory Safety Institute, Natick, MA, 1997; modeled after incident #90, p. 15.
3. American National Standards Institute/American Society of Safety Engineers. *Occupational and Educational Personal Eye and Face Protection Devices*, *ANSI/ASSE Z87.1—2003*, ANSI, Washington, DC.
4. Occupational Safety and Health Administration. Eye and Face Protection eTool; available at http://www.osha.gov/SLTC/etools/eyeandface/ppe/selection.html (accessed April 11, 2008).
5. National Institute for Occupational Safety and Health. *Current Intelligence Bulletin 59: Contact Lens Use in a Chemical Enivironment*, NIOSH Publication No. 2005-139, NIOSH, Cincinnati, OH, 2005; available at http://www.cdc.gov/niosh/docs/2005-139/ (accessed April 11, 2008).

QUESTIONS

1. Eye protection should be worn

   (a) At all times in chemistry laboratories

   (b) At all times in chemistry laboratories unless explicitly instructed that it is not necessary

   (c) Only when hazardous chemicals are being used in laboratories

   (d) Only when the probability of splashing chemicals is reasonably high

2. What eye protection is the only kind acceptable in chemistry laboratories?

   (a) Safety glasses with side shields

   (b) Impact safety goggles

   (c) Safety glasses with vented side shields

   (d) Chemical splash goggles

3. The ANSI Z87.1 standard ensures that eyewear protects against

   (a) Splashes of corrosive chemicals

   (b) The impact of shrapnel

   (c) Splashes of organic solvents

   (d) Fogging

4. Wearing contact lenses in chemistry labs

   (a) Is allowed only when organic solvents are *not* in use

   (b) Is allowed as long as chemical splash goggles are also worn

   (c) Is not allowed

   (d) Is allowed only when chemicals with corrosive vapors are *not* in use

5. It is OK to take off eye protection in chemistry laboratories

   (a) Once you are done using chemicals, even if those around you are still using chemicals

   (b) When you are writing notes in a lab notebook and not actively using chemicals

   (c) When working at a chemical hood with the safety shield down

   (d) If, at the end of the lab, you need to store your goggles in a drawer or locker in the lab and quickly exit the lab

6. Appropriate eye protection should be worn in

   (a) Chemistry labs

   (b) Biology labs where chemicals are used

   (c) Geology labs where chemicals are used

   (d) All of the above

7.1.3

PROTECTING YOUR SKIN—CLOTHES, GLOVES, AND TOOLS

Preview   This section describes ways to protect your skin from exposure to hazardous chemicals in introductory labs.

Beauty's but skin deep.

John Davies of Hereford[1]

INCIDENT 7.1.3.1   NaOH LAB PREPARATION[2]

Many years ago a freshman college student got a part-time job working as a laboratory assistant. One of his first tasks was to prepare a large volume of concentrated NaOH solution. He wore safety glasses, but took no other precautions. Chemical hoods did not exist in these labs. A 2-L beaker was nearly filled with water and the appropriate amount of solid NaOH was added. The solution became very hot and fumed considerably. After cooling, the solution was dispensed. Later in the day he went home and left his clothes to be washed. The next day his mother asked "What did you do yesterday? I washed your shirt and pants and they came out shredded to pieces."

INCIDENT 7.1.3.2   ACID SPLASH ON LATEX GLOVES[3]

A doctoral chemist was working with concentrated sulfuric acid. She was wearing latex gloves and some of the acid splashed onto the gloves. The acid burned a hole in the glove. Although she flushed her hand in the sink for 15 minutes to remove the acid, she received a second-degree burn at the site of the splash.

What lessons can be learned from these incidents?

Protecting Your Body from Chemical Exposures—Barriers Are Your Defense

A *barrier* is a wall, fence, or some other structure built to bar passage. In the movie *King Kong* (Universal Studios, 2005) the inhabitants of the island built a giant wall to protect themselves against the giant ape. When it comes to chemical exposures, you need to be thinking in a similar vein—what you can use as a barrier to protect your body from exposure.

You only come with one body and one skin and it is up to you to protect what is yours. Chemical exposures to our skin can have very serious consequences and as you work in the laboratory, you need to consider that it is not like sitting in a classroom. The clothes that you wear to classes may not be appropriate for the lab since shorts, skirts, sandals, and tops that expose a bare midriff all make it easy for a chemical splash to come in contact with your skin.

This section will discuss appropriate lab clothing and also "extra" protective measures such as gloves and common lab tools.

Appropriate Clothing in the Lab

The guiding principle in this section is: *avoid having lab chemicals come in contact with your skin*. Each college and university will establish its own guidelines for what is required. We present here the general rationale that leads to the specific rules you should follow.

First, any barrier (clothing) that covers the skin provides some limited ability for chemicals to come in contact with skin. Thus, long sleeves instead of short sleeves for shirts are better. Long pants instead of shorts or skirts are better. (Your college may have rules that prohibit shorts and skirts.) Some labs require or recommend lab coats or aprons. These provide additional protection but would likely not protect the legs below the knees if shorts or skirts are worn. While clothing provides some minimal barrier, it's obvious that acid-soaked clothing will be in contact with the skin, so clothing must also be removed quickly in the event of some accident involving liquids. Use an emergency shower for rapid washing of spills on your body. This is discussed more in Chapter 2 about emergency response.

Because splashes and spills almost invariably head toward the floor, it is important to always wear shoes that completely cover the feet. Open-toed shoes, sandals, or any other footwear that does not cover the foot should not be worn in the lab.

Deciding on the correct clothing is not so hard, but doing so early in the morning when your lab doesn't start until 2 P.M. can make it hard to remember this easy task. Warm weather also encourages clothing that is not suitable in the lab. Many students have been sent out of the lab, usually due to poor planning, to get more appropriate clothing. Think ahead.

Guarding Those Hands — One of Your Most Valued Tools

Perhaps the most common incidents when working in the laboratory involve cuts or punctures, especially to the hands, from handling glassware (pipets, beakers, flasks, vials, etc.) or cutting tools (scissors, knives, scalpels, razor blades, etc.). The first step toward prevention of these cuts is to be aware that these can occur and the second is to dispose of used or broken glassware properly. Do not put used or broken glass into a normal trash container where custodial staff may be cut or become contaminated. Most labs have a broken glass container and you should seek this out as the proper place to dispose of used glassware (see Figure 7.1.3.1).

Your hands are the most likely part of your body to be exposed to chemicals and you should take precautions to protect your hands. Working carefully in the lab is the first step in this process but

FIGURE 7.1.3.1   Broken Glass Container. Broken laboratory glass, if uncontaminated by chemical residues, should be placed in a container such as this.

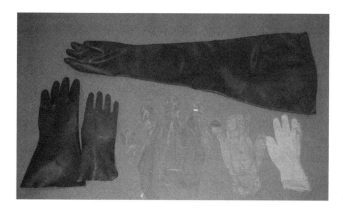

FIGURE 7.1.3.2 Various Types of Protective Gloves Used in Laboratories. Many styles of gloves, made with many different kinds of materials, are available. It is important to select the right glove to protect against the particular hazard. Degree of protection and dexterity vary considerably. There is no "universal" glove.

there are some additional steps that will minimize the likelihood of getting chemicals on your hands. Frequent hand washing is a good practice in the laboratory, especially after using gloves.

Wearing protective gloves in the lab is an obvious way to minimize contact with chemicals in the event of a spill or splash. Figure 7.1.3.2 shows a variety of gloves for various purposes in laboratory work. The main features of the gloves are the size, thickness, cuff length, and the material from which they are made.

One of the most important things to know about gloves is that no one glove material protects against all chemicals and any given material protects only against some chemicals. An online search of "glove protection chart" will provide lists of chemicals and the gloves that are appropriate for their use. You are *not* likely to encounter over 90% of these chemicals in introductory classes. Mostly you will encounter salts and other solids, which will not likely cause rapid damage to the skin, or solutions of acids and bases, which may cause damage depending on the nature of the reagent and its concentration. For solid materials, any glove will provide adequate protection.

What about solutions and solvents? In introductory classes you are likely to have available disposable nitrile gloves. Latex gloves used to be the "standard" glove in labs but increasing numbers of people have had allergic responses to latex in recent years and these gloves are largely being replaced in labs and in hospitals (see *Special Topic 7.1.3.1* Latex Allergies). Nitrile gloves have fairly good protective properties for dilute solutions of acids and bases and various salt solutions. They are good for some organic solvents, but not for others. Thus, nitrile gloves are a reasonable choice in introductory chemistry labs. The best gloves for organic labs and other advanced courses are discussed in Section 7.2.2.

SPECIAL TOPIC *7.1.3.1*

LATEX ALLERGIES

Latex products are derived from a liquid that comes from rubber trees (natural rubber). Latex is widely used in today's modern world—elastic bands in clothing and undergarments, latex or rubber gloves, balloons, baby bottle nipples, rubber bands, toys, condoms, bandages and tapes, and disposable diapers and other products.

When the human immunodeficiency virus (HIV)–acquired immunodeficiency syndrome (AIDS) connection was finally recognized years ago, the health care industry sought ways to minimize and prevent exposures to blood that might contain HIV. The most significant measure recommended for handling blood was the use of protective gloves. Since latex had been used for surgeries, the health community started using latex rubber gloves in much larger volumes. A number of health care workers developed allergies to latex—actually to the protein of natural rubber. These allergies led to the use of nonlatex gloves in the health care setting.

The symptoms of these allergies can range from mild skin conditions to life-threatening conditions (this is very dependent on the person). Latex glove allergies often produce bumps, sores, redness, cracking, or raised areas on the hands. Using nonlatex products such as nitrile gloves can prevent this kind of allergic reaction. For those who do have latex allergies, avoidance of all contact with latex is essential for well-being.

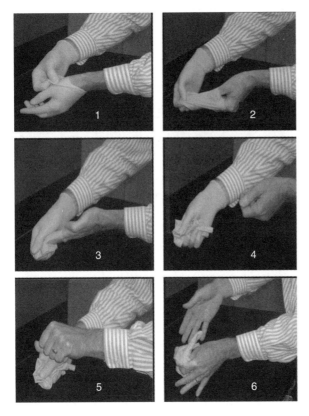

FIGURE 7.1.3.3   How to Remove Lab Gloves to Minimize Hand Exposure. When removing the first glove, turn it inside-out (so that any contamination becomes wrapped up inside the glove), and hold this inverted glove in the palm of the other hand. When removing the second glove, turn this inside-out as well and pull it over the first glove. When done, the exposed, nested gloves will have only the inside surface of the second glove exposed to the hands.

Most disposable gloves are 4–5 mil thick (1 mil = 0.001 inch) This provides reasonable dexterity and modest protection against a puncture. (Some operations in advanced labs or in stockroom work may require thicker gloves that provide more protection.) Usually you will have size selection of small, medium, or large. Selecting the right size is important since gloves that are too small will be hard to put on without tearing and gloves that are too large will reduce the dexterity of your hands and fingers. Gloves that are "too thick" and/or "too loose" can become a safety hazard since handling glassware or small objects becomes more difficult.

As the name suggests, these gloves are disposable. When you are through using them, take them off by "reverse peeling" them starting at the wrist so that they turn inside-out. This will place any contamination on the (new) "inside" of the glove so that no chemicals are transferred to your bare hand in the process of removing and throwing them away (see Figure 7.1.3.3).

When Should You Wear Gloves?

Wearing gloves may reduce your dexterity and you may find that after a time your hands sweat and the gloves become uncomfortable. Plus, disposable gloves cost money! The best policy is to wear gloves when there is a reason to do so, but not otherwise. Unlike safety goggles, which should be worn 100% of the time, the risk of exposing your hands to chemicals that cause damage is not "constant" and wearing gloves 100% of the time may, in fact, make you *less safe* rather than *safer* for some operations. And, make sure to select the right glove for the task since not all gloves protect against all chemicals. This is discussed more in Section 7.2.2.

It is important to always consider the hazards of the chemicals you are using. Handling "common salts" or using solutions of innocuous salts or low concentrations of acids and bases is not very hazardous. Most solids encountered in introductory chemistry (but *not all*) don't pose much of a risk for skin exposure. Nevertheless, contacts with solid chemicals should be avoided or minimized. Furthermore, some solids encountered in advanced labs may pose potential for contact dermatitis. Contact dermatitis is the most common occupational medical condition—this can be caused by allowing skin to come in contact with some solids.

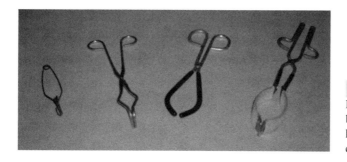

FIGURE 7.1.3.4 Tongs and Forceps Used in Laboratories. Use the right tool for the job. The beaker tongs with the rubber coatings are good for hot beakers but should not be used for really hot crucibles since this will melt the rubber.

Many solutions used in introductory classes don't pose much risk for skin exposure. Before starting any lab, your instructor will alert you about the hazards of the chemicals to be used in a particular experiment. (If not, ask!) Chemicals that should be handled with gloves on in a precautionary fashion will likely be identified. The policies of your college or university will end up dictating if you wear gloves most of the time or rarely.

As you encounter more chemicals, and more kinds of chemicals, in advanced courses the use of gloves will likely increase. This topic is discussed more in Section 7.2.2.

While this book is largely about chemistry laboratories, many students will also be working in biology labs. Any time you are working in an environment where bloodborne pathogens (BBPs) are present, *gloves are mandatory*. Emergency responders and others working in medical fields wear gloves all of the time. This topic is discussed further in Section 4.3.3 and Chapter 2.

Hot Materials: Gloves or Tools?

In introductory labs you are likely to encounter hot beakers, flasks, or test tubes. Handling these with your bare hands is obviously dangerous. While there are special gloves with excellent insulation that are available in most research labs, these are quite expensive and not generally available in introductory labs. And, the thin gloves discussed won't provide any thermal protection.

To handle hot objects, it is best to use beaker tongs and forceps (see Figure 7.1.3.4). These are not interchangeable tools. Beaker tongs are appropriate for small-to-medium size beakers and usually have a rubber coating on the tips to provide good "grip." But, they are not good for the small necks of Erlenmeyer flasks or the small diameters of test tubes. For these pieces of glassware, you should use forceps that are able to safely hold objects of a smaller diameter.

Skin Protection in Introductory Courses: RAMP

There are chemicals used in introductory chemistry labs that can cause damage to the skin and appropriate precautions should be taken to prevent skin contact. Some procedures involve the use of hot glassware, such as beakers of boiling water, and appropriate tools should be used to handle hot objects.

- *Recognize* lab situations where your hands, face, and arms are potentially exposed to chemicals or hot objects.
- *Assess* the level of risk for exposure based on lab procedures.
- *Minimize* the risk for exposure by using appropriate gloves (probably nitrile gloves) and/or tools to handle hot objects.
- *Prepare* for an unexpected exposure by knowing the correct emergency response procedure.

References

1. *Bartlett's Familiar Quotations*, 17th edition (J. KAPLAN, ed.), Little, Brown, and Co., Boston, 2002, p. 231, line 20.
2. R. H. HILL. Personal account of an incident.
3. American Industrial Hygiene Association. Laboratory Health and Safety Committee. Laboratory Incidents. Wrong Gloves Lead to an Acid Burn; available at http://www2.umdnj.edu/eohssweb/aiha/accidents/chemicalexposure.htm#Lead (accessed September 20, 2009).

QUESTIONS

1. Which form of protection is commonly used and/or available in chemistry laboratories to protect skin?

    (a) Disposable and nondisposable gloves
    (b) Clothes
    (c) Laboratory aprons and coats
    (d) All of the above

2. Which parts of the body are at-risk for exposure to chemicals?

    (a) The face
    (b) Hands, arms, legs, and feet
    (c) Midsection
    (d) All of the above

3. Which statement about laboratory gloves is *true*?

    (a) Most gloves protect against most chemicals.
    (b) Only nondisposable gloves protect against most chemicals.
    (c) No glove material protects against all chemicals.
    (d) Latex gloves are the best choice for most people working in laboratories.

4. When using gloves in the lab you should

    (a) Use whatever glove is available since that is likely appropriate for the chemicals being used
    (b) Use the thickest glove that is available since that will provide the best protection

    (c) Learn the proper technique for removing the gloves to avoid skin contamination
    (d) Make sure you identify the right-hand and left-hand versions of disposable gloves

5. What reason(s) might cause you not to wear gloves during a certain procedure?

    (a) Wearing gloves sometimes become uncomfortable and might reduce dexterity.
    (b) Disposable gloves cost money and shouldn't be used unless some reasonable risk presents itself.
    (c) Some laboratory procedures present no significant risk of skin exposure to the hands.
    (d) All of the above.

6. When handling hot objects in the laboratory,

    (a) Disposable gloves generally provide adequate protection
    (b) It is best to select the correct tool to pick up a hot object to make sure that a secure grip is established
    (c) It is always best to let a hot object cool to room temperature before handling it
    (d) Beaker tongs can almost always conveniently be used to handle any hot object, regardless of size and shape

7.1.4

CHEMICAL HOODS IN INTRODUCTORY LABORATORIES

Preview   This section discusses the use and operation of laboratory hoods as a key part of your arsenal of safety equipment to protect you from exposure and to keep chemical aerosols, vapors, fumes, particulates, and odors under control.

Emperor Qin Shihuang succeeded in his effort to have the walls joined together to fend off the invasions from the Huns in the north after the unification of China.

From *History of the Great Wall of China*[1]

INCIDENT 7.1.4.1   SOLVENT FIRE DURING RECRYSTALLIZATION[2]

A student needed to purify a chemical using the process of recrystallization. He needed to use a very flammable solvent but he discovered that the laboratory hood that he intended to use was cluttered with no room for his experiment unless he first cleaned it up. So he decided to carry out the recrystallization on the benchtop. As the solvent boiled he added the crystals of the compound that he wanted to purify. Suddenly the vapors ignited and a fire ensued. Luckily the fire was extinguished and the student only received minor burns.

What lessons can be learned from this incident?

Don't Let Chemicals Take Your Breath Away!

Many liquids that you use in chemistry labs have significant vapor pressures. And, some solids sublime sufficiently that they produce the gaseous form of the compound. The very fact that you can smell any chemical indicates that some of the substance is in the gas phase, and most chemicals have discernible odors (see *Chemical Connection 7.1.4.1* "If I Can Smell It, Am I in Danger?"). The only useful, sweeping generality about the safety of breathing chemicals is that many are not safe to breathe. The occupationally acceptable levels are threshold limit values (TLVs) and permissible exposure levels (PELs) that are described in Section 6.2.2. But in the vast majority of laboratories, there are no systems in place to measure or monitor the levels of contaminants in the air. Since we know that many chemicals are not safe to breathe, our starting assumption will always be: it is best not to breathe vapors associated with lab chemicals. So, just as the Great Wall of China kept out invaders, we wish to keep noxious gases out of the lab air that we are breathing, and we can do this using chemical laboratory hoods.

CHEMICAL CONNECTION 7.1.4.1

"IF I CAN SMELL IT, AM I IN DANGER?"

There is no connection between the toxicology of a chemical and its odor. However, we can ask the statistical question: "If I can smell this chemical, am I breathing a harmful concentration?"

Laboratory Safety for Chemistry Students, by Robert H. Hill, Jr. and David C. Finster
Copyright © 2010 John Wiley & Sons, Inc.

Odor thresholds are concentrations that 50% of a test population can detect.[3-5] However, individuals display a very wide range of sensitivity to odor: about 15% of a given population cannot detect odors that exceed the threshold by a factor of 4. The "width" of the odor threshold can be as high as two orders of magnitude for many substances. (Thus, if a threshold is 10 ppm, some people can smell the chemical at 0.1 ppm and some cannot smell it at 1000 ppm.)

Another factor that influences odor detection is "odor fatigue," which is the phenomenon of decreased sensitivity to odor after some exposure time. This has important implications for laboratorians who may become desensitized to the odor of a harmful chemical, perhaps unknowingly exposing themselves to inhalation hazards for extended periods of times. They may think that the chemical is no longer present when, in fact, they are simply unable to smell it anymore.

How do odor thresholds compare with "dangerous" levels for inhaling chemicals? You may not yet have learned about OELs (occupational exposure limits) and IDLH (immediately dangerous to life and health) values, but these are concentrations in air that are just what the names imply: the limits that are allowed in occupational settings and levels that present "immediate" hazards. (These are discussed more in Section 6.2.2.) We can use these values to compare odor thresholds to "safe" levels to breathe.

The 3M 2009 Respirator Selection Guide[6] has a lengthy list of chemicals, many of which have data for odor thresholds, OELs, and IDLH values. Using the first 35 (of about 150) chemicals with all three values listed, we see the ranges of these three values in Table 7.1.4.1. The ranges of the data are considerable. As expected, IDLH values are higher than OEL values.

TABLE 7.1.4.1   Odor Thresholds Compared with OEL and IDLH Values for Various Chemicals

	Minimum	Maximum
Odor threshold, ppm	0.001	100,000
OEL, ppm	0.05	5,000
IDLH, ppm	6	50,000

Table 7.1.4.1 gives some clue to the answer to the question: "Can odor be used to detect the harmful level of a chemical?" The general answer is "yes," but with many cautions. Most of the time the median odor threshold is above (ratio > 1) or considerably above (ratio > 100) the OEL and IDLH values, as shown in Table 7.1.4.2. However, for 6% (IDLH) and 26% (OEL) of the time the chemical is harmful *below* the odor threshold. Also, taking into consideration the enormous range of individual variation in odor thresholds, it's hard to know how a single individual should interpret this data.

TABLE 7.1.4.2   Percentages of Ratios for IDLH/Odor Threshold and OEL/Odor Threshold

	Percentage where ratio > 1	Percentage where ratio > 100
IDLH/odor threshold	94%	77%
OEL/odor threshold	74%	28%

So, while it may be likely that you can "smell it before it harms you," that surely is not *always* the case. The best practice in labs is not to be working in an environment where chemical odors are present. If you do smell chemicals, the ventilation and containment are not adequate. This requires the use of chemical hoods for many procedures and requires that chemicals be stored in well-ventilated areas or cabinets.

Federal regulations do not require the use of chemical hoods in labs, but they do require that the air in labs is safe to breathe. Since the only really effective way to meet this "performance standard" is through the use of chemical hoods, these devices are very common in modern laboratories.

This section is an introduction to the design and use of the "chemical hood." Like any sophisticated tool, it is more complicated than it first seems. This section will review the basic information that allows you to use hoods safely in most common lab operations. Section 7.2.3 should be considered "required reading" for chemists working in research labs since it continues the discussion of the design and use of hoods and other kinds of ventilation systems.

We will use the phrase "chemical hood." Historically, the phrase "fume hood" was very common and is still in use today but it is really a misnomer. Technically, fumes are small solid particles that are formed when solid is vaporized and which then recondense into a solid form. But chemists often use the term to include gases and vapors. Of course, a chemical hood will also contain and remove aerosols, dusts, and other particulates that are airborne.

How Does "Design Follow Purpose" for a Chemical Hood?

Since many chemical procedures potentially expose chemists to vapors and gases that may be harmful, using a chemical hood allows a means by which the escaping gases are both contained and subsequently removed in a fashion that eliminates the possibility of inhalation. Thus, the chemical hood is a "box" that should allow for air flow in only one direction—away from the person standing at the hood. The exhausted gases are drawn out at the back of the hood. There is usually a large transparent shield (or "sash") that slides vertically in the front, which allows a chemist's hands and arm to reach inside to perform manipulations but also protects the face and neck. Figure 7.1.4.1 shows a picture of a modern laboratory hood. The vertical panel may have a "stop" about 18 inches up so that ordinarily the shield *cannot* go high enough to expose the face and neck. It is possible to override this stop in situations where greater access is needed to the interior of the hood. Some hoods are designed with horizontally sliding panels, which provide advantages for some procedures (see Figure 7.1.4.2).

The chemical hood is outfitted with a sink, water, electricity, natural gas, and house vacuum since many lab operations may be performed in the hood that need these utilities. Some hoods will have "racks" of interconnecting horizontal and vertical bars that make it easy to clamp glassware. The exhaust fan should be located outside of the building so the exhaust duct is under negative pressure with respect to the laboratory and building. This means that the air at higher pressure in the lab and building is drawn or pushed into the negative or lower pressured hood and air duct.

Modern chemical hoods might also have sensors that detect the flow rate of air through the hood and/or alarm systems that indicate inadequate air flow. Some hoods have exhaust fans that operate at slower speed when the sashes are closed and higher speeds when these are up or open. Some hoods also have infrared motion detectors so that when someone is working at the hood the fan is in "high"

FIGURE 7.1.4.1 Laboratory Hood with a Vertical Sash. Sashes should always be closed except when arms and hands need to be accessing something inside the chemical hood. This both saves energy (for many modern hoods) and provides a shield against explosions and fires.

FIGURE 7.1.4.2   Laboratory Hood with Horizontal Sashes. These hoods are less frequently used but can offer similar protection as the vertical sash as long as they are not removed and the openings are kept to minimum size during operations and use.

mode but when no one is there it is in "low" mode. Chemical hoods should always exhaust the air to the outside of a building and this requires that "make-up" air be supplied to the lab. Since in most parts of the country this air needs to be routinely heated, cooled, and/or dehumidified, this is an extremely costly feature of lab ventilation. Using different exhaust fan speeds, based on the real-time need to protect lab workers only when necessary, can save considerable energy and money since hoods are often on all of the time.

Depending on when the lab in which you are working was constructed, the sophistication and features of the chemical hoods may vary considerably. We describe here general features of good, modern hoods. If you are working in older labs, the hoods may not have all the features described here. You need to know the design, function, and (perhaps) limitations of the chemical hood that you are using in order to work safely.

How Do You Use a Chemical Hood Safely?

Chemical hoods should be used whenever chemical operations are performed that will generate aerosols, vapors, and gases that may be harmful to health. In some labs they are used 100% of the time as a general precaution even when 100% of the procedures don't need to be in a hood.

Before using a chemical hood, the most important step is to check to make sure that the exhaust fan is on. This might be the default condition for some hoods that either cannot be turned off (easily) or are routinely left on. For hoods with "on/off" switches, though, an "on" position doesn't guarantee the fans are working! Belts on motors can break, electrical problems may cause a circuit breaker to activate, or other problems might occur. An easy way to check for air flow is to hold a strip of tissue paper near the bottom of the half-closed sash. Airflow will make the paper deflect toward the inside of the hood. Lightly tapping a chalkboard eraser near the back of the hood also works since the chalk dust will quickly get exhausted (if the hood is on!). Take caution, though, since these tests indicate that the fan is *on*, but don't necessarily indicate if airflow is *adequate*. We discuss more about this below.

Before placing any chemical inside the chemical hood, you may need to "build" your apparatus. A common example of this would be the construction of a distillation or reflux apparatus by connecting the appropriate pieces of glassware. It is at this point that the sash "stop" may need to be temporarily overridden in order to get adequate access to the hood. Relowering the sash will reengage the stop. Once chemicals are introduced to the chemical hood, the sash should never be raised more than half-way or above the stop. This still allows for reasonably good access to the equipment inside.

A reasonable face velocity for airflow at the front of the hood is about 100 feet/minute (±20%). You should know that this airflow velocity alone is *not* a defining criterion for adequate hood performance. Containment of airborne contaminants is the essential and primary function of the hood. Hood effectiveness is its ability to capture and contain airborne chemicals within its boundaries. Several factors are interrelated in achieving this containment. In addition to face velocity, other equally important factors that contribute to containing airborne chemicals include what is in the hood, how you carry out operations, and the design of the hood. Thus, the mild "draft" of the face velocity ideally minimizes any aerosols or gases from escaping from the hood into the lab but the three other factors also have a marked influence on a hood's performance.

- First, the interior of the hood needs to be sufficiently uncluttered so that airflow is not impeded from the front to the exhaust areas at the bottom rear of the hood. Excess equipment inside causes erratic airflow patterns that might allow some gases to escape through the front of the hood even though the overall airflow appears to be "inward."

- Second, it is fairly easy to disturb the air at the face of the hood in ways that compromise the 100-feet/minute flow rate—one of several factors that influence hood performance. A very modest walking speed of 3 miles/hour is about 260 feet/minute, so just *walking* by the hood can draw gases back out of the hood. Rapid arm movements also disrupt the airflow. Why not just increase the flow rate at the face to 200–300 feet/minute to minimize air from escaping through the face of the hood? Such high velocities sometimes cause considerable turbulence in the hood and may actually increase the likelihood of contaminated air escaping through the face of the hood.

- Third, since it is impossible to avoid some turbulent air patterns inside the hood, any apparatus inside the hood should be placed on 2-inch blocks to allow better airflow, and all work should be at least 6 inches inside from the front of the hood. This will help minimize the chance of vapors escaping through the face of the hood.

Whenever your hands and arms don't need to be inside the hood, the sash should be closed completely. This optimizes the containment and removal of the air and gases inside. The sash provides a useful shield that will better contain an explosion or fire. And a closed sash may reduce the need for make-up air, depending on the design of the hood.

Since chemical hoods are so well ventilated, it sometimes becomes a pretty common practice to store chemicals inside hoods. This is a very bad practice since the volume of storage usually accumulates to inhibit good airflow in the hood, particularly when some apparatus is then constructed in this crowded space. This compromises the function of the hood and the ability to use the hood safely. Furthermore, as Incident 7.1.4.1 describes, a crowded hood makes it difficult to use the hood all of the time and this inevitably leads to experiments being conducted on open benches instead of inside hoods.

Finally, hoods should be tested once a year with devices that can measure the airflow to determine if the airflow is adequate. While you may not get involved with this testing, you should be able to locate an inspection sticker that indicates the date and airflow rate at the most recent inspection. A sticker from the inspection should indicate the height of the sash that produces 100 fpm. This, of course, still does not guarantee that it is working properly "today." If you have any doubts about the airflow, you should contact someone who can test the airflow rate. *Working at a chemical hood with a faulty fan can be extremely dangerous since the protection that you think you have is absent*!

The Larger Picture of Lab Ventilation: RAMP

This section is only a very brief introduction to the design and function of a chemical hood that should allow you to use these devices safely in common lab operations. The issue of overall lab ventilation is much larger and involves a consideration of lab design, the quality of all of the air in a lab, and other kinds of local containment and exhaust systems besides the "standard" chemical hood. Section 7.2.3 will present a more thorough analysis of these issues.

- *Recognize* when the use of particular chemicals and chemical procedures may produce airborne contaminants in a lab.

- *Assess* the level of risk based on the possibility of exposure and the relative danger posed by various chemicals.

- *Minimize* the risk by working in a chemical hood and following procedures that prevent contaminated air from escaping through the face of the hood.

- *Prepare* for emergencies by anticipating accidental escape of contaminated air and planning in advance what to do if that happens.

References

1. TravelChinaGuide.com. History of the Great Wall; available at http://www.travelchinaguide.com/china_great_wall/(accessed October 23, 2008).
2. R. H. HILL. Personal account of an incident.
3. http://www.trustcrm.com/ectny/respiratory_advisor/oshafiles/appendixe.html.
4. J. E. AMOORE and E. HAUTALA. Odor as an aid to chemical safety: odor thresholds compared with threshold limit values and volatilities for 214 industrial chemicals in air and water dilution. *Journal of Applied Toxicology* **3**(*6*): 272–290 (1983).

5. G. LEONARDOS, D. KENDALL, and N. BARNARD. Odor threshold determinations of 53 odorant chemicals. *Journal of Air Pollution Control Association* **19**:91–95 (1969).
6. 3M Occupational Health and Environmental Safety. 2009 Respirator Selection Guide; available at http://solutions.3m.com/wps/portal/3M/en_US/Health/Safety/Resources/Media-Library/?PC_7_RJH9U5230 GOF802N6GVOSC2CG5_nid=2K1CVJZK74be64R007W6Q4gl (accessed July 13, 2009).

QUESTIONS

1. What is *true* about breathing chemicals in a laboratory?
   - (a) Many liquids have large-enough vapor pressures to produce gases.
   - (b) If you can smell a chemical, then you are inhaling the chemical.
   - (c) Most chemicals used in laboratories are probably safe to breathe.
   - (d) Both (a) and (b).

2. Chemical hoods are
   - (a) Required in all laboratories that use liquids and solutions
   - (b) Required in all laboratories that use any chemical that is harmful to breath
   - (c) Required only in industrial chemistry laboratories
   - (d) Not required per se in laboratories but are used as a means of achieving a safe environment for breathing

3. Which of the following are good features of a chemical hood?
   - (a) A sash that can be partially raised for access to the interior
   - (b) A sash that can be removed when necessary to make interior access easier
   - (c) Access to utilities such as water, vacuum, and natural gas
   - (d) An on/off switch
     - (a) I, III, and IV
     - (b) II, III, and IV
     - (c) III and IV only
     - (d) I and IV only

4. The best chemical hoods
   - (a) Are on at "full volume air flow" at all times to ensure maximum safety
   - (b) Adjust the rate of air flow to keep the laboratory environment safe while minimizing energy use
   - (c) Have on/off switches so that air flow can be turned off when the hood is not being used

   - (d) Automatically shut down if the air pressure in the hood gets too high

5. Before using a chemical hood, one should
   - (a) Momentarily turn it off, and then on again
   - (b) Test for airflow
   - (c) Lower the sash completely
   - (d) Check to make sure all of the utilities, such as water and vacuum, work

6. Tests for airflow
   - (a) Should be conducted annually
   - (b) Should be conducted with the sash completely lowered
   - (c) a and b
   - (d) Can only be conducted when the hood is *completely* empty

7. Airflow face velocity should be
   - (a) About 100 ft/s, but values above this are always better
   - (b) About 100 ft/s, and values either much higher or much lower are unsafe
   - (c) About 100 ft/min, but values above this are always better
   - (d) About 100 ft/min, and values either much higher or much lower are unsafe

8. Equipment placed inside a chemical hood
   - (a) Should be placed as close to the front of the hood as possible for easy access
   - (b) Should be placed at least 6 inches inside the front of the hood
   - (c) Should always be placed as low as possible inside the hood
   - (d) Both (a) and (c)

9. The sash of a chemical hood should generally be left
   - (a) Open so that experiments can be monitored more carefully

(b) Open so that the maximum amount of air can flow through the hood

(c) Closed except when the chemist needs access to the interior of the hood

(d) Closed except when the hood is off

10. Chemical hoods should

(a) Not be used for storage since this limits the ability to use them for experiments

(b) Not be used for storage since this inhibits good airflow

(c) Be used for storage when there is no other place to store laboratory chemicals

(d) Both (a) and (b)

7.2.1

MORE ABOUT EYE AND FACE PROTECTION

Preview   This section describes various kinds of eye protection needed in advanced labs.

The sight of you is good for sore eyes.

Jonathan Swift, *Polite Conversation*[1]

INCIDENT 7.2.1.1   "BUMPED" SOLUTION[2]

A student was removing and disassembling a reflux condenser when the solution "bumped" and splashed the solution onto her face and chest. She was wearing chemical splash goggles but a poor seal of the goggles with her face still allowed some solution to enter her eyes. She was ushered to a safety shower and eventually examined at an emergency room. There was no permanent damage to her eyes.

INCIDENT 7.2.1.2   AZIDE EXPLOSION[3]

A postdoctoral researcher was purifying an organic azide compound using vacuum distillation when an explosion occurred. The ceramic top of the heating mantle fractured and fragments were embedded in his face. Eye protection was being worn.

What lessons can be learned from these incidents?

The Basics: A Review

Section 7.1.2 has already introduced the rule: *always wear chemical splash goggles in labs where chemical are used*. This section will consider situations where perhaps *less* eye protection is needed and situations where *more* eye protection is smart. But first let's review the central argument for chemical splash goggles.

Most labs have bottles of solvents or solutions of acids, bases, oxidizing agents, and reducing agents used in chemical reactions. Thus, the "default condition" is the possibility of a splash hazard always being present, and chemical splash goggles are almost always required. The only time not to mandate the use of chemical splash goggles is in labs where the risk of splashes is zero. Even then, you may be putting yourself at risk if you forget to put them on when you return to the lab where chemicals are being used.

Safety Glasses and Visitor's Glasses

What kinds of labs might not require safety goggles? Some labs in chemistry departments might be those devoted entirely to instrumentation or computational modeling. It might be possible, though unlikely, that some labs would require *no* eye protection if the hazard assessment indicates no liquids or solutions and there is also no chance for any kind of high-pressure or low-pressure explosion that would produce shrapnel. Such labs are uncommon. However, wearing eye protection is so simple and your eyes are so valuable that you may want to consider wearing eye protection anyway.

Laboratory Safety for Chemistry Students, by Robert H. Hill, Jr. and David C. Finster
Copyright © 2010 John Wiley & Sons, Inc.

FIGURE 7.2.1.1   Face Shield. A face shield can be used to get even more and better protection for the face and neck. Chemical splash goggles must still be worn under the face shield.

Some labs that have zero risk from splashes might still pose risks for explosions. In these labs, safety glasses or "visitor's glasses" with an ASSE Z87.1 marking and side shields on them are likely the best choice for eye protection. Visitor's glasses sometimes fit easily over prescription eyeglasses. It is also possible to get safety glasses with prescription lenses that protect against shrapnel. Even for explosions, though, safety goggles are much more protective than safety glasses or visitor's glasses.

When Do You Want More than Chemical Splash Goggles?

Protecting one's eyes is an extremely high priority at all times since eyesight is important and the eye's surface is vulnerable to attack by many agents. Furthermore, splashes and shrapnel are not good for exposed skin either! Laboratory workers usually have most of their skin covered to some degree with clothing, and only hands and the face are exposed. Other sections of this book deal with protecting hands using gloves. Here we consider protecting the face and neck area.

The skin is a reasonable barrier against many chemicals, but strong acids, strong bases, good oxidants, and good reductants react with skin in a deleterious fashion. Assessing the likelihood of exposure is part of the process of risk assessment, and there are some experiments and manipulations in the lab that increase the risk to a level where the prudent person wants more protection than chemical safety goggles provide. The next choice is the addition of a face shield, shown in Figure 7.2.1.1. Face shields come in varying lengths and varying degrees of "wrap-around." As you can guess, bigger shields, in both directions, are better than smaller shields. Wearing safety goggles under the face shield is still a requirement.

Moving beyond personal protective equipment, chemists routinely use safety shields (Figure 7.2.1.2) for some reactions. While it may seem odd to run a reaction with the *expectation* of an explosion, it is simply true that some reactions are more prone to explosions than others even under controlled circumstances. Furthermore, even experienced chemists have had reactions go out of control unexpectedly. Sometimes such incidents are later understood and sometimes they are not. Safety-conscious chemists develop a good sense of how hazardous a reaction might be. A basic safety principle is to overprotect rather than to underestimate a hazard.

All chemical hoods have either horizontal or vertical safety shields (sometimes called sashes or doors) and these shields are present both to provide control of airflow and another layer of protection in the event of liquid or shrapnel explosions. The default position for the chemical hood shield should

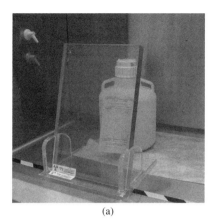
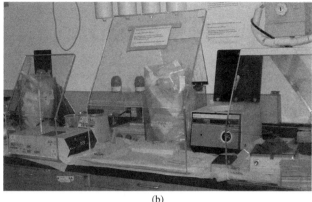

(a)   (b)

FIGURE 7.2.1.2   Safety Shields. Parts (a) and (b) show different styles of portable safety shields. These are designed to withstand impacts without shattering or easily falling over.

always be "closed" except when actually accessing equipment and chemicals in the hood. While using shields might be ergonomically cumbersome at times, the extra protection they offer is worth the small effort.

Eye Protection for Ultraviolet and Visible Radiation

In addition to splashes and shrapnel, eyes are sensitive to ultraviolet (UV) and visible radiation. While "normal" levels of visible light are not harmful, exposure to either visible or UV lasers presents a significant hazard. Lasers are discussed separately in Section 7.3.3. Ultraviolet radiation hazards are discussed in Section 5.3.7 and special safety glasses and goggles are available to protect your eyes against this radiation.

Occasionally, laboratory workers use torches that emit intense light for sealing vials or making custom laboratory glassware. You should seek advice from your instructor or a safety professional on the selection of the proper safety glasses to prevent exposure to the intense radiation when using these torches.

Working in Labs and Protecting Your Eyes and Face: RAMP

The best rule is: *always wear chemical splash goggles in labs where chemicals are used*. Deviations from this rule should be carefully reviewed. Advanced labs may present special hazards and the use of face shields and other shields should be considered depending on the assessment of risk of exposure.

- *Recognize* the different hazards that may cause eye injuries.
- *Assess* the level of risk based on the possibility of exposure and the nature of the operation and the relative hazard of chemicals being used.
- *Minimize* the risk using appropriate eye protection, including the consideration of a face shield and other protection to supplement protection by chemical splash goggles.
- *Prepare* for emergencies by knowing locations of eyewashes and emergency showers, and proper procedures for responding to emergencies involving eye exposures.

References

1. *Bartlett's Familiar Quotations*, 17th edition (J. KAPLAN, ed.), Little, Brown, and Co., Boston, 2002, p. 299, #7.
2. American Industrial Hygiene Association. Laboratory Health and Safety Committee. Laboratory Safety Incidents. Chemical Splash in the Eyes in Spite of Goggles; available at http://www2.umdnj.edu/eohssweb/aiha/accidents/chemicalexposure.htm#Spite (accessed September 22, 2009).
3. American Industrial Hygiene Association. Laboratory Health and Safety Committee. Laboratory Safety Incidents. Phenyl Azide Compound Erupts During a Vacuum Distillation; available at http://www2.umdnj.edu/eohssweb/aiha/accidents/explosion.htm#Phenyl%20Azide (accessed September 22, 2009).

QUESTIONS

1. Safety glasses are a reasonable alternative to splash goggles

   (a) In labs where there is no splash hazard
   (b) In labs where the only hazard is from shrapnel from explosions
   (c) If the glasses also have side shields
   (d) All of the above

2. Face shields should be worn

   (a) When the level of risk to the face and neck is higher than normal
   (b) For eye protection when splash goggles are inconvenient or unavailable
   (c) To protect facial skin from UV damage
   (d) All of the above

3. Most splash goggles

   (a) Also protect against UV radiation
   (b) Will filter out most of the damaging visible light from lasers
   (c) Both (a) and (b)
   (d) None of the above

4. Safety shields are used when

   (a) No other eye or body protection is needed
   (b) There is some higher probability of an explosion
   (c) No other eye protection is available
   (d) It is ergonomically more convenient than using splash goggles or face shields

7.2.2

PROTECTING YOUR SKIN IN ADVANCED LABORATORIES

Preview   This section describes ways to protect your skin from exposure to hazardous chemicals in advanced labs.

> *She was doing routine manipulations. She was a careful experimentalist operating in new labs in a state-of-the-art facility. She had no idea she was in peril. None of the chemists here would have felt in peril.*
>
> Russell Hughes, a colleague of Karen Wetterhahn (Incident 7.2.2.1) [1]

INCIDENT 7.2.2.1   FATAL EXPOSURE THROUGH INAPPROPRIATE GLOVES[1-4]

In August 1996 a chemistry professor was using a pipet to transfer some dimethylmercury into an NMR tube. Dimethylmercury was known to be toxic. She was wearing gloves and a lab coat, working in a fume hood, and had even cooled the liquid to minimize the vapor pressure. She inadvertently spilled a few drops on the latex gloves she was wearing. Tests later showed that dimethylmercury penetrates latex gloves in about 15 seconds. (One MSDS recommended latex gloves, one recommended neoprene gloves, and one recommended "chemically impervious gloves.") The professor removed the gloves and washed her hands. Five months later she started to develop symptoms of organomercury poisoning and despite aggressive treatment she went into a coma one month later and died in June 1997. (This episode is discussed further in *Special Topic 6.3.1.1*)

What lessons can be learned from this incident?

Advanced Labs: More Chemicals, More Hazards

Let's recap the basic set of guidelines described in Section 7.1.3 about how to best protect your skin in introductory labs:

- Use gloves when appropriate to do so. These will probably be nitrile gloves.
- Wear clothing that covers most of your body, within reason. Shorts, skirts, sandals, and tops that expose bare midriffs are unwise and perhaps forbidden. Long sleeves are better than short sleeves.
- Wear closed-toe shoes that don't expose the feet: no sandals, flip-flops, or other open shoes.
- Use tools such as beaker tongs and forceps to handle hot glassware.

These guidelines offer adequate protection in introductory courses where most solutions are aqueous and organic solvents and solutions are uncommon. However, as you proceed in the chemistry curriculum you will certainly encounter many more chemicals, and many more categories of chemicals, that present significant hazards to the skin. You are more likely to encounter *very* hot and *very* cold materials. As the hazards increase in number and severity, so should the level of protection.

This section expands the general topics of appropriate clothing, using appropriate gloves, and using appropriate tools in advanced (postintroductory) labs in order to protect the skin.

Laboratory Safety for Chemistry Students, by Robert H. Hill, Jr. and David C. Finster
Copyright © 2010 John Wiley & Sons, Inc.

Clothing

The chance of spills and splashes in advanced labs is no more or less than in introductory labs, but the chemicals to which you may be exposed are likely to be more hazardous. It is more likely that you will be using organic solvents, stronger acids and bases, and stronger oxidants and reductants. In all cases, more protection is better than less protection, and many students elect to wear lab coats as an additional layer of protection against spills, splashes, and explosions.

While many solids don't pose a significant hazard if in brief contact with the skin, experience has shown that continued exposure to some solid chemicals poses a risk of irritation, contact dermatitis, or other detrimental corrosive effects. These are more likely to be encountered in advanced labs. Some examples are methyl ethyl ketone, formaldehyde, NaOH pellets, elemental Na, and phenol. (Contact dermatitis is one of the 10 leading occupational diseases in the United States.)

Gloves: Many Options and No Perfect Glove

Section 7.1.3 concludes that the need for gloves in *introductory* labs is probably "occasional" and that the most likely gloves to be used are disposable nitrile gloves. These gloves are still the best "general use" glove with regard to the permeability of various solvents and solutions, but the wide variety of organic solvents potentially used in organic and advanced labs requires a much more careful review of glove choice. In fact, the rule of thumb should not be "use nitrile" but "pick your gloves carefully." Incident 7.2.2.1 is a dramatic and tragic example of unknowingly using the wrong gloves.

In most instances it is best to use disposable gloves. These are usually 4–5 mil thick, which provides good dexterity although only modest protection since the thickness of a glove (1 mil = 0.001 inch) is an important factor in the ability of the glove to protect the skin (see *Chemical Connection 7.2.2.1 Testing Protective Gloves for Chemical Permeability*). The possibility of contamination is obviously reduced since these disposable gloves are intended for one-time use. When you are through using them, take them off by "reverse peeling" them, starting at the wrist so that they turn inside-out. This will place any contamination on the (new) "inside" of the glove so that no chemicals are transferred to your bare hand in the process of removing and throwing them away (see Figure 7.1.3.3).

CHEMICAL CONNECTION *7.2.2.1*

TESTING PROTECTIVE GLOVES FOR CHEMICAL PERMEABILITY

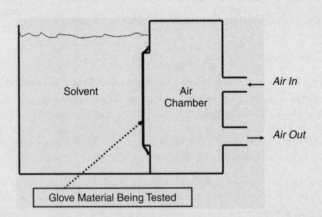

FIGURE 7.2.2.1 Permeation Test Cell. This diagram illustrates the construction of an apparatus to test the ability of glove materials to withstand permeation by a solvent.

As shown in Figure 7.2.2.1 the glove material is attached as an interface between two cells with the exterior surface of the glove material facing the liquid to be tested. The interior side of the glove faces a cell with either air or dry nitrogen. (Sometimes this cell is filled with water or other solution.) The air or solution is tested periodically for the presence of the liquid being tested, usually by gas chromatography.

Gloves are tested for permeation and degradation. Permeation is the ability of a chemical to penetrate the material through small pores. Degradation is a destructive change in the glove material. These two features are not necessarily related to each other. Obviously, if a liquid reacts with the glove material in a destructive fashion, it is not a useful glove material for that chemical.

Two characteristics of the glove material related to permeation are tested: breakthrough time and rate of transfer. Breakthrough time is the time for the first appearance of the material in the cell on the right. This is obviously important since it represents how long the gloves "last" until the skin is exposed to the chemical. However, the rate of transfer through the material is also important. A glove material that has a very short breakthrough time but a very slow rate of transfer may be judged as "good" for that chemical since the overall exposure will be very small. A long breakthrough time with a high rate of transfer might be judged "poor" since once the material fails the exposure is significant.

Glove rating guides typically list ratings such as "excellent," "good," and "poor" or other descriptive words, or number or color systems that indicate similar ratings. It is important to understand the rating system and also make sure you know exactly what glove is being rated with respect to the material and the thickness. Obviously, a 4-mil glove is less protective than a 28-mil glove.

There are some general guidelines about what kinds of gloves are most resistant to various categories of chemicals but these guidelines should serve as a signpost of where to start to look for an appropriate glove rather than a definitive criteria list. There are too many exceptions to the rules.

Latex gloves are made of natural rubber. They are very flexible and thin and have excellent resistance against aqueous solutions, including acids and bases of modest concentration. They are not generally resistant to organic compounds. Their use over the years has decreased significantly since latex-based allergies are now more common.

Nitrile gloves are made of a synthetic rubber polymer (see *Chemical Connection 7.2.2.2* Polymers for Protective Gloves). They are available in thin versions that rival the tactile properties of latex, but they usually don't exhibit the allergic side effects of latex, and they are much more difficult to tear or puncture. You may recognize them as they often come in blue, purple, or green colors. Nitrile gloves are a good general purpose glove for many laboratory uses (as long as you remember that there is no universal glove and that you need to investigate to identify the best choice of gloves in all situations). Nitrile gloves protect well against aliphatic hydrocarbon solvents but can be degraded by ketones, aromatic hydrocarbons, esters, and aldehydes. For more hazardous chemicals, gloves should be thicker and longer.

CHEMICAL CONNECTION 7.2.2.2

POLYMERS FOR PROTECTIVE GLOVES

Protective gloves used in laboratories are made of synthetic rubber. Rubber was discovered in the late 1700s and it was found to be a very useful product—it could be stretched, squeezed, or mashed and came back to its original state. It was waterproof and airtight and didn't wear out easily. It was found that rubber was made of a natural polymer resulting from the joining of thousands of isoprene molecules ($CH_2=CHCCH_3=CH_2$). As the use of rubber increased, there were shortages that spurred chemists to look for synthetic alternatives. During World War I, the first synthetic rubbers were developed. Over the years, we have found increasing uses for these products in our many everyday activities.

Synthetic substitutes for rubber are polymers that are made by joining together small molecules called monomers into long chains of millions of these units. Nitrile gloves are made of acrylonitrile–butadiene (AB) rubber, a polymer of acrylonitrile ($CH_2=CH-CN$) and butadiene ($CH_2=CH-CH=CH_2$). Acrylonitrile adds chemical resistance to the polymer (but too much can make the rubber less flexible) and butadiene adds elasticity. A typical AB rubber might be made of 4 butadiene molecules with 1 acrylonitrile molecule as one unit (the monomer), joined together as thousands of repeating units into a very long chain. During the polymerization process these long chains are tied to each other by "crosslinks" to make the basic AB rubber polymer that is molded into today's resilient protective gloves. The process also involves a catalyst, an emulsifier, an activator, and some other additives. Some other examples are neoprene, a polymer of 2-chloro-1,3-butadiene, and butyl, a polymer of isobutylene (or 2-methylpropene).

The gloves that probably come closest to providing protection against a very wide range of chemicals (much broader than general purpose nitrile gloves) are laminates—they are made by combining layers of different kinds of polymers. The best known is Silver Shield® made by North Safety and is reported to be resistant to more than 280 chemicals. Laminates can offer good protection against many laboratory chemicals but provide less dexterity, fit more loosely around the tips of the fingers, and do not fit snugly as do latex or general purpose nitrile gloves. To prevent mechanical damage to these gloves, some have recommended the use of laminates under a second pair of different gloves. It is best to use the smallest possible size of laminate so that you get the best dexterity.

"Heavy Duty" Gloves: Thicker Is Better

Some gloves are particularly designed for handling specific types of chemicals in industrial situations, and these gloves have very good properties against certain classes of chemicals. They are made of other synthetic polymers, such as butyl, neoprene, fluoropolymers, polyvinyl chloride, and polyvinyl alcohol. However, many of these gloves are thicker and lack the dexterity of latex or general purpose nitrile gloves. Nevertheless, they may offer better protection than latex or general purpose nitrile for specific chemicals, and they should be considered, particularly if working with extremely or highly toxic chemicals. These gloves are also often much longer and can cover the arm up to the elbow or even higher.

Gloves for Hot and Cold Items

Apart from exposure to chemicals, it is important to protect the skin from extreme temperatures. As you know, there are many times when hot objects and hot glassware are encountered in chemistry labs. Sometimes it is easier to use beaker tongs or forceps but in other instances using hands and fingers provides better control of the object. Gloves are available that provide thermal protection (and the gloves described above for *chemical* protection are *not* good at *thermal* protection.) Often, these will be mittens instead of gloves, but these are the "tool of choice" for some operations.

In advanced labs you are also more likely to encounter very cold materials, usually in the form of chunks of dry ice or liquid nitrogen. The potential for very rapid frostbite is considerable and these cold materials must always be handled with care. Gloves or mittens specifically designed for use with cold substances are available and should always be used when handling dry ice or using liquid nitrogen or liquid helium where the chance for a spill is possible.

Online Resources for Gloves

There are many online resources that can provide information about the compatibility of gloves and various chemicals. Some of these are commercial and some are academic web sites. Table 7.2.2.1 provides a sampling of web sites, with annotations. As you know, web sites change frequently so doing your own search may be the wisest course of action. In particular, if you have a particular glove that you intend to wear for a particular chemical, you should try to access the web site of *that* manufacturer for *that* glove to make sure you have exactly the right information regarding degradation and permeation.

RAMP

- *Recognize* lab situation where hands, face, and arms are potentially exposed to chemicals or hot objects.
- *Assess* the level of risk for exposure to particular chemicals based on lab procedures.
- *Minimize* the risk for exposure by using appropriate gloves.
- *Prepare* for an unexpected exposure by knowing the correct emergency response procedure.

TABLE 7.2.2.1   Online Resources About Chemically Resistant Gloves

---

http://www.allsafetyproducts.biz/site/323655/page/74172
- This chart is sorted by categories of chemicals. Glove types are indicated; no thickness information is given.

http://safety.nmsu.edu/programs/chem_safety/hazcom_PPE-resistance_guide.htm
- This chart has an alphabetical list of chemicals. Glove types are indicated; no thickness information is given.

http://safety.nmsu.edu/programs/chem_safety/hazcom_PPE_glove_guide.htm
- General categories and characteristics of glove materials are provided.

http://www.bestglove.com/site/chemrest/default.aspx
- Allows users to search by either specific chemical or specific gloves. Extensive information is provided.

http://www.des.umd.edu/ls/gloves.html
- Alphabetical listing of chemicals. Several glove materials are listed, with thickness indicated.

http://www.saftgard.com/anonymous/SolvaGard1.pdf
- Information on nitrile gloves for many chemicals is provided.

http://www.mapaglove.com/ChemicalSearch.cfm?id=1
- This is a searchable database.

http://www.northsafety.com/ClientFormsImages/NorthSafety/CorpSite/E8D15F2E-1F59-454F-B8F0-147FA2B9D81D.pdf
- This is a PDF file with many gloves and chemicals listed.

http://www.ansellpro.com/download/Ansell_8thEditionChemicalResistanceGuide.pdf
http://www.ansellpro.com/specware/
- This chart has an alphabetical list of chemicals. Glove types are indicated; relatively thick gloves were tested.

References

1. K. ENDICOTT. The trembling edge of science. *Dartmouth Alumni Magazine* 22–31 (April 1998).
2. M. B. BLAYNEY, J. S. WINN, and D. W. NIERENBERY. Handling dimethylmercury. *Chemical Engineering News* 7 (May 12, 1997).
3. T. Y. TORIBARA, T. W. CLARKSON, and D. W. NIERENBERG. More on working with dimethylmercury. *Chemical & Engineering News* 6 (June 16, 1997).
4. J. LONG. Mercury poisoning fatal to chemist. *Chemical Engineering News* 11–12 (June 16, 1997).

QUESTIONS

1. A difference between "introductory" and "advanced" laboratories, with regard to safety, is
   - (a) The chance of spills and splashes is less since students are more skilled
   - (b) The need for protecting the skin is less since experiments use less hazardous chemicals
   - (c) Lab experiments are monitored more closely and, therefore, are safer
   - (d) None of the above

2. Which chemical is generally considered "safe" with regard to dermal exposure?
   - (a) Formaldehyde
   - (b) NaOH pellets
   - (c) Phenol
   - (d) None of the above

3. In organic and other advanced labs,
   - (a) It is still true that nitrile gloves protect adequately for almost all chemicals
   - (b) Latex gloves are preferred over nitrile gloves
   - (c) Glove material must be selected carefully based on the chemicals in use
   - (d) "Double-gloving" is the preferred technique

4. What category of glove material provides the most protection against the widest range of chemicals?
   - (a) Synthetic polymers
   - (b) Natural polymers
   - (c) Laminates
   - (d) Polyvinyl chloride and polyvinyl alcohol

5. "Heavy duty" gloves generally
   - (a) Are thicker
   - (b) Are longer
   - (c) Provide poorer dexterity
   - (d) All of the above

6. Thermally insulating gloves should be used when
   - (a) Nitrile or butyl gloves are not available
   - (b) Handling hot or cold objects
   - (c) A high degree of dexterity is needed
   - (d) "Double-gloving" is not possible

7. Selecting the correct glove to protect against a particular chemical is
   - (a) Fairly difficult but will always be listed in an MSDS

(b) Difficult and therefore it's best to use double-gloved nitrile and latex gloves

(c) Easy since there are only three main kinds of gloves and each is known to protect against particular categories of chemicals

(d) Easy for commonly used chemicals since there are several databases of glove characteristics freely available on the Internet

7.2.3

CONTAINMENT AND VENTILATION IN ADVANCED LABORATORIES

Preview This section follows Section 7.1.4 and discusses additional features of chemical hoods, other means by which lab air is kept safe to breathe, and personal protective equipment used to avoid breathing contaminated air.

Fresh air impoverishes the doctor.

Danish proverb[1]

INCIDENT 7.2.3.1   OSMIUM TETROXIDE EXPOSURE[2]

A researcher using osmium tetroxide in a chemical hood began feeling ill and believed he was being exposed. He left the laboratory but after he began feeling better, he returned to the lab. He began feeling ill again after he returned to work in the hood and he stopped again, closed the hood, and left the lab. Facilities maintenance was called to check the hood for its performance, but the osmium tetroxide was still present in the hood and a safety policy required that hazards must be removed before maintenance is allowed to respond. After the osmium tetroxide was removed, the maintenance staff found that the hood was not working properly and fixed the problem. While the researcher had experienced limited exposure neither he nor the maintenance worker were injured in this incident. [Osmium tetroxide is an extremely hazardous solid with a vapor pressure of 7 torr at 20 deg C. Its high toxicity is reflected by LC_{LO} (rat) = 40 ppm (4 h) and OSHA PEL = 0.0002 ppm.] (See *Chemical Connection 7.2.3.1* Sublimation Hazards.)

What lessons can be learned from this incident?

Prelude

Section 7.1.4 provided an introduction to the design and use of modern chemical hoods. Knowing basic procedures for using hoods correctly is an essential skill for any chemist working in a lab. However, there is more to know about hoods and lab ventilation in general that is important when working in advanced and research labs.

Most of the discussion below assumes that you are working in a relatively modern laboratory. By this we mean that the lab itself is supplied with HVAC (heating, ventilation, air conditioning) and that there is at least some kind of chemical hood available. Some older labs may have chemical hoods but their design and function may be well below the standards that are used today. Nonetheless, the available hoods are the ones that you have to use (or choose *not* to use!) in any lab, so it is smart to understand completely what quality of hood you are using and to use it wisely.

The Lab Room Itself

It is unlikely that you will ever find yourself working in a lab without "conditioned" air or some means by which the overall atmosphere in the lab is constantly being refreshed. The rate at which, on average, the entire volume of air in the lab is replaced is called the "turnover rate" or "air exchange rate."

Laboratory Safety for Chemistry Students, by Robert H. Hill, Jr. and David C. Finster
Copyright © 2010 John Wiley & Sons, Inc.

Even without the presence of chemical hoods (which necessarily remove air from the lab) there will be HVAC inlets and exhaust vents in the lab. There is no OSHA standard on the turnover rate in labs, although a value of six is considered a reasonable rate. At this rate, *assuming complete mixing*, about 98% of the air is replaced in 1 hour. Since it is reasonable to expect that some chemical vapors may occasionally be present in a lab (although prudent use of a chemical hood should virtually eliminate this), a reasonable turnover rate should maintain a healthy atmosphere.

CHEMICAL CONNECTION 7.2.3.1

SUBLIMATION HAZARDS

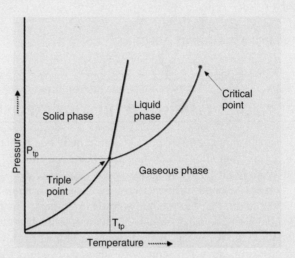

FIGURE 7.2.3.1   Phase Diagram. Sublimation of a substance can occur at particular pressure–temperature conditions below the triple point, in the lower left of the diagram where the compound can move from the solid phase to the gas phase without going through the liquid phase. Although it is more common to experience a liquid vaporizing than a solid subliming, anytime you can smell a solid, you are experiencing the result of sublimation.

Sublimation is the transition process in which a solid chemical goes directly into the gaseous state without going through the liquid state. If you examine a phase diagram that shows the three states of a chemical—solid, liquid, and vapor at varied pressures and temperatures—you will see lines that separate these phases and a point where they all intersect called the triple point (see Figure 7.2.3.1). At the triple point all three phases of the chemical are present in equilibrium. At temperatures below the triple point only the solid and gas phases can exist and the line connecting these two regions are pressure–temperature combinations where the substance will sublime (solid to gas) or deposit (gas to solid).

Ordinarily, we expect to be exposed to inhalation dangers from gases and liquids since we commonly experience odors from liquids and we know that liquids have vapor pressures. Usually, we are less concerned about vapors from solids even though the very fact that we can smell many solids readily demonstrates that there are molecules in the gas phase. However, substances that readily sublime can create significant inhalation hazards.

Some common chemicals that sublime are iodine and carbon dioxide ("dry ice"). If one examines a bottle of iodine in a chemical stockroom you are likely to see the crystals of I_2 that have deposited on the underside of the bottle cap. These arise from the solid having sublimed and then redeposited. Since iodine is reasonably toxic by inhalation, working with the solid can be hazardous. Dry ice is commonly used in many chemistry labs as a cooling agent. Small pieces of dry ice "disappear" as they sublime (without melting). Similarly, when using a carbon dioxide fire extinguisher, the solid white material rapidly sublimes after having been discharged from the fire extinguisher. In large enough quantities, carbon dioxide can become a simple asphyxiant. Other chemicals that are less hazardous but readily sublime are naphthalene (used as moth balls) and dichlorobenzene (used as a toilet deodorant).

Incident 7.2.3.1 involved potential exposure to osmium tetroxide (OsO_4). OsO_4 is hazardous due to its high toxicity and the fact that it easily sublimes at room temperature and pressure. It has little or no warning properties since you can't smell it at concentrations that can seriously harm you. (See also *Chemical Connection* 7.1.4.1 "If I Can Smell It, Am I in Danger?")

The main problem with this analysis is the assumption of "complete mixing." The ability of the air to mix in the room will be dictated by the positions of the inlet and exhaust vents and whether doors are opened or closed. (Most labs do not have windows that can open.) In a fast-flowing river there can be very "quiet pools" of water at the edges depending on the outline of the bank of the river and the placement of rocks or other obstructions. These disturbances are sometimes called "eddies." Similarly, even in a high-turnover room, there are can "pockets" of air that go virtually undisturbed. Little can be done about this, so it is good to know, for example, that at some weighing station in the corner of a lab, there may be little air turnover.

In order to keep any airborne lab contaminants *inside* labs, the HVAC systems of lab-containing buildings are ideally designed to keep the pressure in the labs slightly lower than the pressure in hallways outside the labs. This is commonly referred to as a "negative" lab pressure (relative to the hallway) and this is easily detected by noting if lab doors (that open *into* the hallway) are *slightly* hard to open due to the pressure difference. Obviously, the reverse situation (higher pressure in the lab than in the hallway) represents a serious imbalance in the building's HVAC design or operation since this could result in chemical vapors or aerosols entering the hallway and beyond. Propped-open doors or open windows will compromise the goal of negative lab pressure.

Most HVAC systems in *nonlaboratory* buildings enjoy the advantage of being able to recirculate the interior air of a building, which saves tremendously in the costs to heat, cool, and dehumidify the air. Since lab air must always be considered "potentially contaminated," it is not recirculated and the exhaust vents usually lead to roof-mounted vents and new, fresh air must be continuously drawn from outside the building as make-up air. Make-up air is the air that replaces the air that was exhausted by ventilation. With this in mind, it is energy efficient to minimize the turnover rate and the use of chemical hoods. Since safety is a priority, most buildings are designed with an inherent high cost for HVAC.

There are known instances of the exhaust and intake ports for lab air being placed inappropriately close to each other on the roof of a building, or where the intake is downwind of the exhaust. This leads to contaminated air being drawn back into the building (called entrainment). These shocking situations are usually discovered when chemical odors "appear" in labs unrelated to the source of the odor and suggest that the make-up air is not completely "fresh." There is no easy fix to this engineering blunder but it is good to be aware of this situation.

Various Kinds of Chemical Hoods

There are many variations to the main theme of the chemical hoods.[3,4] A good basic design as shown in Figure 7.2.3.2 often includes the following:

- A movable front face shield (or sash)—vertical or horizontal variations.
- A workspace inside—ideally with an "indented" floor to capture spills.
- Small cup sinks in the rear of the cabinet.
- On the front of the hood, switches for lights and perhaps to turn off the hood.
- A rear wall with adjustable ventilation slots (baffles) allowing exhaust at low, middle, and high points in the hood.
- Side panels that may be fitted with vacuum, air, gas, and water.
- A smooth air foil across the front face of the hood floor that allows air to be drawn in from outside the hood and across the floor of the hood to capture vapors heavier than air.
- Air foils on the two sides that provide for the smooth flow of air as it is drawn into the hood.
- Gauges on the hood that measure the air pressure in the exhaust duct to indicate that it is working properly. Other hoods may have gauges that measure airflow (velocity) into the hood face.
- A front opening that can be varied in size by moving the sash to a desired position; air is drawn inward through this opening and upward into the exhaust to capture vapors lighter than air as well as aerosols. Figure 7.2.3.3 shows the flow of air into and out of the hood.

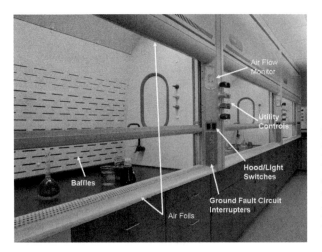

FIGURE 7.2.3.2 Features of a Laboratory Hood. It is important to understand the function of the various parts of a chemical hood to use a hood effectively. If the chemical hood is used without understanding the dynamics of airflow, you may think you are protected from exposure to the interior atmosphere of the chemical hood when you are not. (Courtesy of Labconco, Inc., Kansas City, MO).

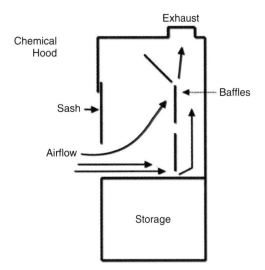

FIGURE 7.2.3.3 Airflow in a Laboratory Hood. This idealized situation can be compromised by inappropriate materials in the hood, poor airflow, or rapid arm movement that can allow the interior atmosphere to flow back into the laboratory.

- The motor that exhausts the hood should be located outside of the building—these are often found on the roof of the building housing laboratories.

- An exhaust system with sufficient negative air pressure to capture airborne chemicals. Many have recommended that the hood should provide an average flow rate of 100 ft/min (ranging from 80 to 120 ft/min in some places). Nevertheless, you should know that this airflow velocity alone is *not* the defining criterion for adequate hood performance—there are other factors such as what is in the hood, how you carry out operations, and the design of the hood that contribute to containing airborne chemicals. Testing hood performance is a good prudent practice that should be performed on an annual basis or more often if the performance of the hood is thought to be compromised.

Some hoods are on all the time, but by various means the fan rate may be adjusted to minimize airflow when the hood is not in use and to maximize airflow when the hood is being used and/or the sash is open. The desire to use the lowest flow rate that still maintains a safe working condition is driven by the desire to be energy efficient.

Some easy ways to check for airflow are described in Section 7.1.4, but it is prudent to perform regular measurements to ensure proper fan operation. Anemometers are devices that measure wind velocity and a common form of this for lab use is a hot-wire anemometer, where the resistance of an exposed hot wire is measured. As air flows over the wire, it is cooled so one can measure flow rate by measuring resistance. The simplicity of this procedure is greatly compromised by the need for repeated measurements at many locations on the open face of the hood. Thus, accurate measurements are not quick and easy but with a good protocol, an average face velocity can be measured. An alternative

FIGURE 7.2.3.4 Vaneometer[TM] Used for Measuring Airflow in Laboratory Hood. These inexpensive devices can give a reasonable indication of airflow under a particular set of circumstances. (Courtesy of Dwyer Instruments, Inc., Michigan City, IN.)

method to measure face velocity of hoods is with the swing vane velometer or Vaneometer[TM] (see Figure 7.2.3.4). This is a small hand-held meter that is relatively inexpensive and can measure air velocity from 25 to 400 ft/min.

The sashes on chemical hoods can be made of several different kinds of transparent material but none of them is truly explosion-proof. The sash should always be lowered whenever internal access is not being made. This optimizes the capture and removal of gases, offers maximum protection to anyone standing in front of the hood, and will provide some protection in the event of a fire or explosion.

Hoods with horizontally sliding sashes can be particularly helpful in some operations where "hands-on" control is required. The panels are usually narrow enough to reach around, which provides complete vertical protection for the body (see Figure 7.1.4.2).

Since chemical hoods are so well ventilated, it can become a common practice to store chemicals inside the hoods. This is a very bad practice since the volume of storage usually accumulates to inhibit good airflow in the hood, particularly when some apparatus is then constructed in this crowded space. This compromises the function of the hood and the ability to use the hood safely. Furthermore, as Incident 7.1.4.1 describes, a crowded hood makes it difficult to use the hood all of the time and this inevitably leads to experiments being conducted on open benches instead of inside hoods.

Specialized Chemical Hoods

There are two special types of chemical hoods that we will mention only briefly—perchloric acid hoods and radioactive materials hoods.[4,5]

Operations using perchloric acid for digestions or heated dissolutions should be performed in "perchloric acid hoods" that are specifically constructed for this purpose. Perchloric acid, being extremely reactive, can produce explosive perchlorates with metals that are touch- or shock-sensitive. Thus, the interior lining of these hoods must be constructed of nonreactive materials (such as stainless steel) with sealed seams to facilitate decontamination. These hoods are normally fitted with equipment for "washing down" the interior of the hood exhaust system after perchloric acid use to prevent the buildup of explosive perchlorate residues. The entire exhaust system must be independent of other exhaust systems connected to other chemical hoods. These hoods must be clearly labeled as "perchloric acid work only." Anyone using perchloric acid should seek special training in these operations.

Similarly, scientists working with radioisotopes (which is common in chemistry and biology labs where tracers are used) should work in chemical hoods designed for radioisotope work and allow for relatively easy decontamination. These hoods are fitted with HEPA and charcoal filters at the hood–exhaust

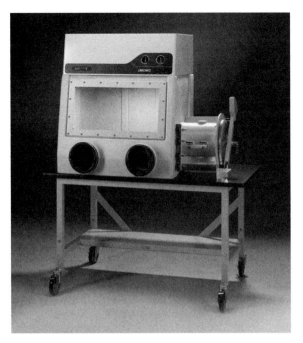

FIGURE 7.2.3.5   Glovebox. Gloveboxes are usually used in chemistry laboratories more for the protection of chemicals than lab personnel. Chemicals that are air- and/or water-sensitive can be used in these closed environments. This picture does not show a tank of nitrogen or argon that would usually be used for the interior atmosphere or various pumps and recirculating systems. The antechamber on the right is used to transfer materials in and out of the glovebox without allowing the laboratory atmosphere to enter the interior of the glovebox. (Courtesy of Labconco, Inc., Kansas City, MO.)

duct interface to prevent contamination of air ducts and environmental release of radioactive materials. These hoods are typically stainless steel to facilitate cleaning.

Gloveboxes

Finally, we briefly mention the "ultimate" chemical containment system, which is the glovebox. Many variations of gloveboxes exist, ranging in cost from hundreds of dollars to tens of thousands of dollars. The basic design is that the atmosphere inside the glovebox is entirely contained and manipulations inside the box are performed by using gloves through ports in the front of the box (see Figure 7.2.3.5).

Gloveboxes are most often used for handling oxygen- and/or water-sensitive compounds in some inert atmosphere of nitrogen or argon. They usually have recirculating systems that constantly purge the interior atmosphere of water and/or oxygen. Access to the interior for equipment and chemical is usually provided through some secondary chamber that can be evacuated. The weak link in the use of gloveboxes is the glove and it is essential to establish a program to maintain the integrity of the glovebox gloves.[6]

Some gloveboxes are used to contain highly toxic chemicals, radioactive materials, or highly pathogenic microbiological organisms.[7,8] These gloveboxes use normal air but it is often filtered going in and as it is exhausted. Using any glovebox can be a difficult and tedious procedure. Any use of a glovebox requires special training by an experienced mentor.

Other Containment Systems

There are a handful of other kinds of containment systems that one might encounter in chemistry labs. These are generally considered to be significantly inferior to chemical hoods but in certain circumstances may see limited application.

Some labs have "local" exhaust systems where some exhaust port is located at or near a workspace, providing some removal of gases. These can be "overhead snorkels" that are connected to an exhaust system with flexible tubing that can be placed over a workspace (see Figure 7.2.3.6). Examples of uses for these types of local exhaust systems include gas chromatographs, mass spectrometers, and atomic absorption spectrophotometers. In some cases, such as exhausting a vacuum pump, local exhaust ventilation can be very effective and very useful. Since these systems cannot guarantee 100% removal of the gases,

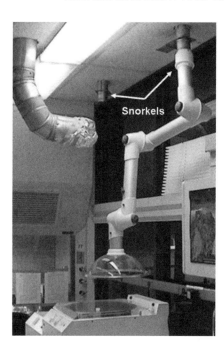

FIGURE 7.2.3.6 Laboratory Ventilation Snorkel. These snorkels can be used in limited situations when "spot" ventilation is required but they do not effectively remove all air contaminants from a particular location.

and often may be quite less effective than that depending on the particular arrangement of equipment, overhead snorkels must be used very carefully and only in situations where 100% exhaust is not required.

Other devices that mimic, but do not replicate, a chemical hood are "boxes" that roughly enclose a workspace on several sides that have an exhaust port at the rear or on the top. Sometimes these are clear plastic boxes placed over balances to provide some removal of vapors or aerosols during weighing (see Figure 7.2.3.7). Use of these systems should also be carefully scrutinized since they may not provide 100% capture of gases and odors.

Another category of hoods are stand-alone, self-contained systems, known as ductless hoods. These hoods should not be used for volatile chemicals. They are not connected to an external exhaust but instead have an internal fan that captures contaminants, sends them through a filter, and then returns the filtered air to the lab—if the contaminant is not adsorbed onto the filter it is passed back into the air in the room. The effectiveness of these hoods depends on the nature of the filter and the frequency with which it is changed. While these might be appropriately used in limited situations, they should not generally be considered an appropriate substitute for a chemical hood and should never be used for volatile, flammable, or toxic chemicals. Ductless hoods have many limitations and their use must be carefully evaluated for each and every operation. In some institutions, ductless hoods are not permitted or recommended.

Closely related to chemical hoods are biological safety cabinets (BSCs).[6-9] These resemble chemical hoods and stand-alone hoods but also have a HEPA (high efficiency particulate air) filter as part of the exhaust system that filters 99.97% of the particles greater than 0.3 microns in diameter. The filter captures microbiological and chemical aerosols but not vapors or gases. The filtered air is exhausted outside the building, inside the room, or recycled back through the cabinet. There are many variations on this style of cabinet designated as Class I, IIA, IIB1, IIB2, IIB3, and III. The filters in BSCs do not filter at the "chemical/molecular level" and BSCs should not be used as "chemical hoods." Section 7.3.4 discusses BSCs in more detail and references[4-8] are web sites and books with more information about BSCs.

RAMP

- *Recognize* when the use of particular chemicals and chemical procedures may produce airborne contaminants in a lab. Chemical hoods should not be used for storing chemicals.

FIGURE 7.2.3.7   Laboratory Balance in Plastic Box. Some boxes can be constructed to allow equipment to be used inside.

- *Assess* the level of risk based on the possibility of exposure and the relative danger posed by various chemicals.

- *Minimize* the risk by working in a chemical hood and following procedures that prevent contaminated air from escaping through the face of the hood. Usually, devices other than chemical hoods and PPE are not appropriate protection.

- *Prepare* for emergencies by anticipating accidental escape of contaminated air and planning in advance what to do if that happens.

References

1. The Quote Garden. Available at http://www.quotegarden.com/health.html (accessed September 22, 2009).

2. University of Delaware, Office of Campus and Public Safety, Environmental Health and Safety. Exposure to Osmium Tetroxide; available at http://www.udel.edu/ehs/osmiumtetroxide.html (accessed October 23, 2009).

3. A. KEITH FURR (editor). *CRC Handbook of Laboratory Safety*, 5th edition, CRC Press, Boca Raton, FL, 2000, pp. 137–169.

4. L. DIBERARDINIS. Laboratory Chemical Hoods. In: *Handbook of Chemical Health and Safety* (R. J. ALAIMO, ed.), ACS/Oxford University Press, Washington, DC, 2001, pp. 299–306.

5. N. W. COUCH, J. J. NICHOLSON, and S. R. PRAKASH. Lab Design/Radiosynthesis Lab Design. In: *Handbook of Chemical Health and Safety* (R. J. ALAIMO, ed.), ACS/Oxford University Press, Washington, DC, 2001, pp. 496–501.

6. M. E. COURNOYER, J. M. CASTRO, M. B. LEE, C. M. LAWTON, Y. H. PARK, R. LEE, and S. SCHREIBER. Elements of a glovebox glove integrity program. *Journal of Chemical Health and Safety* **16**(*1*):4–10 (2009).

7. D. EAGLESON. Isolation Technology. In: *Handbook of Chemical Health and Safety* (R. J. ALAIMO, ed.), ACS/Oxford University Press, Washington, DC, 2001, pp. 487–495.

8. Centers for Disease Control and Prevention/National Institutes of Health. *Primary Containment for Biohazards: Selection, Installation, and Use of Biological Safety Cabinets*, 2nd edition, Public Health Service, U.S. Department of Health and Human Services, September 2000; available at http://www.cdc.gov/od/ohs/biosfty/bsc/bsc.htm (accessed April 30, 2009).

9. R. W. HACKNEY, Jr. Biological Safety Cabinets. In: *Handbook of Chemical Health and Safety* (R. J. ALAIMO, ed.), ACS/Oxford University Press, Washington, DC, 2001, pp. 307–313.

QUESTIONS

1. What is a reasonable "turnover rate" for air in a chemistry laboratory?

   (a) One "room volume" per hour
   (b) Three "room volumes" per hour
   (c) Six "room volumes" per hour
   (d) Six "room volumes" per 8-hour day

2. In a laboratory, it is safe to assume that, with a proper turnover rate for room air,

   (a) All of the air in the lab gets exchanged frequently enough

(b) Some of air will be exchanging frequently enough, but not all areas of the laboratory undergo adequate air mixing

(c) All of the air in the lab gets exchanged frequently enough as long as no windows are open

(d) All of the air will get exchanged enough as long as the hoods are working at full capacity

3. A properly "balanced" building, with regard to air pressures, is one where the laboratories are at a

(a) "Negative" pressure compared to hallways outside the lab, thus preventing airflow into the lab

(b) "Negative" pressure compared to hallways outside the lab, thus preventing airflow out of the lab

(c) "Positive" pressure compared to hallways outside the lab, thus preventing airflow into the lab

(d) "Positive" pressure compared to hallways outside the lab, thus preventing airflow out of the lab

4. Air that is exhausted from a chemical hood

(a) Is recirculated into the building

(b) Is recirculated, but only back into the laboratories

(c) Is recirculated, but only back into the laboratories after passing through a chemical filter

(d) Is not recirculated back into any part of the building

5. If the "make-up" air that comes into a laboratory "smells like chemicals" this means that

(a) The hoods need to be set to a higher flow rate

(b) Laboratory windows should be opened to allow more fresh air into the laboratory

(c) The intake ducts, usually on the roof of the building, are located "downwind" of the hood exhausts

(d) The exhaust duct work should be disconnected from the intake ductwork

6. Which of the following are good features of a chemical hood?

(a) A movable front shield or sash

(b) A fixed front shield or sash

(c) Adjustable panels (baffles) at the back of the hood

(d) Fixed panels (baffles) at the back of the hood

(e) An exhaust motor located just above the hood

(f) An exhaust motor located outside the building

(g) I, IV, V

(h) II, III, V

(i) I, III, VI

(j) II, IV, VI

7. An anemometer measures

(a) The difference in air pressure at two locations

(b) The velocity of airflow

(c) The direction of airflow

(d) Nonstatic lab pressure

8. Vertical sashes should be closed except when

(a) Measuring the airflow of a hood

(b) Access to equipment inside the hood is necessary

(c) There is some chemical reaction occurring inside the hood

(d) One expects an explosion

9. Horizontal sashes are useful in situations where

(a) A vertical sash would be inconveniently heavy to raise and lower

(b) It is necessary to keep some part of the shield "open" to maximize airflow

(c) It is desirable to have access to some equipment while still protecting the face and body

(d) All of the above

10. What two categories of materials require special engineering for chemical hoods?

(a) Radioactive substances and perchloric acid

(b) Radioactive substances and bloodborne pathogens

(c) Toxic gases and radioactive gases

(d) Toxic gases and bloodborne pathogens

11. Gloveboxes

(a) Are used when it is necessary to completely isolate lab air from the air inside the glovebox

(b) Sometimes have argon or nitrogen atmospheres so that air- and water-sensitive compounds can be handled

(c) Use gloves that can develop leaks over time and should be checked periodically for integrity

(d) All of the above

12. Overhead "snorkels"

(a) Are a cheaper but "almost-as-adequate" substitute for a chemical hood

(b) Are used for "local ventilation" where 100% capture is not essential

(c) Are commonly used over analytical balances that are used to weigh toxic gases

(d) All of the above

13. Biological safety cabinets

(a) Are a common substitute for chemical hoods

(b) Cannot be used in chemistry laboratories according to federal regulations

(c) Filter the air before it is exhausted

(d) Are always exhausted outside the building

7.3.1

SAFETY MEASURES FOR COMMON LABORATORY OPERATIONS

Preview This section touches briefly on safety measures that can prevent incidents during various common laboratory operations.

Our responsibility is to do what we can, learn what we can, improve the solutions, and pass them on.

Richard P. Feynman, American physicist and Nobel Laureate[1]

INCIDENT 7.3.1.1   EXPLOSION USING NONEXPLOSION-PROOF REFRIGERATOR[2]

A laboratory worker had prepared some samples that contained a flammable solvent and he put them in the refrigerator before leaving that evening. During the night the power in the building went off for several hours, but was finally restored some time during the early morning hours. Moments later there was an explosion, followed by a fire that destroyed and damaged a large part of the building. The refrigerator was not explosion-proof, was not suited for storage of flammables, and was located in an office area.

What lessons can be learned from this incident?

Learning Laboratory Operations and Safety Precautions from Others

There are a myriad of different kinds of operations in the laboratory, many of which are very specialized. When you are carrying out laboratory operations using predesigned experiments as part of an academic course, the safety risks have likely been addressed and pointed out by your instructors. However, when you begin to carry out a research project, it is likely that you will need to employ techniques that you have not used previously. While it is possible to find written instructions for some operations, such as those described in Incident 7.3.1.1, most are best learned from more experienced chemists or scientists who pass these along to others. Learning techniques and procedures from others is an essential part of your educational process. We stress that you must find and consult experienced personnel in your laboratory such as your research mentor, instructor, professor, or graduate students, to learn as much as possible about specific techniques. Ask them for help. Not everyone's experience is the same, so you may have to search to find someone who has the experience you need.

There are many specialized techniques and often they are learned on-the-fly as the need arises for you to carry out a given reaction. Nevertheless, there are also many common laboratory operations that can be discussed in a general way with regard to standard safety precautions. Some of these are presented below. Several books provide valuable practical guidance and these are listed later in this section under "Other Sources of Information." You should look for information that describes the operation you need to carry out.

Laboratory Safety for Chemistry Students, by Robert H. Hill, Jr. and David C. Finster
Copyright © 2010 John Wiley & Sons, Inc.

Reducing Scale Reduces Risk

When carrying out a new reaction, remember that reducing the scale reduces the risk. Consider reducing the scale of a normal reaction by about a factor of 10; this is sometimes called "miniscale" and uses about a gram of material. Reducing the scale to around 25–100 mg is called microscale chemistry. Using microscale laboratory equipment is a good way to learn about the reaction and its success as well as learning about handling the chemicals involved in the process and reducing associated risks. (See below, "Other Sources of Information.") Usually the largest obstacle to reducing the scale is the need for reduced scale equipment and your laboratory may not have this available. Nevertheless, the use of miniscale or microscale techniques can reduce your risks and also reduce waste because you will need less starting materials and generate less waste.

Common Laboratory Operations and Equipment—Safety Considerations

The following discussion focuses on safety considerations for several common laboratory operations and devices. This is not intended to be comprehensive, but rather seeks to cover those things that many working in chemistry will encounter.

Working Alone   You should not work alone in a laboratory. Use the buddy system. If you are separated by a wall or out of direct observation, then you must devise a system to periodically check your buddy. Your buddy should know what you are doing, particularly if this involves hazardous operations. Someone should always be in the near vicinity so that if something happens, an explosion or incident, your buddy can have an opportunity to get you assistance. Explosions, fires, splashes, spills, or contact with chemical or physical hazards could result in life-threatening incidents that require immediate action. If you are alone you are putting yourself at high and unnecessary risk.

Glassware, Needles, Scalpels, and Other "Sharps"   Perhaps the most common incidents in laboratories involve cuts and punctures from broken glassware, needles, or cutting tools. "Sharps" is a generic term often used in the medical field to describe needles or other items that can puncture skin. This term has also been adopted in many chemical labs to refer to needles, broken glass, and other lab materials that can puncture skin. It is important that you dispose of them in containers specifically designed for that purpose; do not put sharps in the normal trash can. Custodial personnel have been injured by inappropriate disposal of these materials. Furthermore, should you be working with biological organisms, these sharps could become a source of infection.

Weighing   Since the process of weighing a solid or liquid involves the transfer of the substance from a container to some holding device on a balance, this can easily result in contamination of a surface near the vicinity of the transfer. Surface contamination presents an opportunity for exposure. This can be significant if the chemical being weighed is very toxic and can be absorbed through the skin. Prudent measures include covering the area with plastic-backed absorbent paper and decontaminating the area after each weighing, especially if the contamination is visible. Plastic-backed absorbent paper is frequently used in many laboratory operations to prevent the spread of contamination and prevent the formation of aerosols that might be produced from liquid spills or splashes. Appropriate gloves should be worn to prevent skin contact (see Sections 7.1.3 and 7.2.2).

Some solids may develop static charges, making them very difficult to handle without "blowing" the solid around. Antistatic weighing boats, antistatic brushes and guns, and other procedures for weighing are available to prevent and minimize this problem.[3,4] The development of static charge can be particularly prevalent inside glove boxes with dry, inert atmospheres.

Mercury-Containing Thermometers   As indicated in incidents elsewhere in this book (Sections 4.2.1 and 6.2.2), exposure to vapors from mercury that has accumulated in the flooring of laboratories over extended periods of time can become significant. While there has been a concerted effort in many places to eliminate mercury thermometers from laboratories where possible and replace them with thermometers that do not contain mercury, it is still possible to find mercury-containing thermometers. If you break a thermometer or find spilled mercury, the spilled mercury needs to be cleaned up immediately. Mercury travels easily and gets into small cracks in the floor where it might not be readily visible. Special techniques are required to clean up mercury spills and only trained personnel should do this. Make sure, also, that you do not inadvertently walk on the mercury, contaminate your shoes, and extend the spill wherever else you may walk. Notify someone in charge immediately.

Heating   This is an essential part of laboratory work. Wherever possible, you should use steam heating for your operations since this is much safer than hot plates. However, steam heating is limited to a maximum temperature of 100 °C. Some heating devices can be potential ignition sources and it is important that you recognize this and prevent contact of flammable vapors with these devices. If you are using a hot plate to heat flammables, be sure that it is spark-proof by looking for a label that indicates this, but also remember that some flammables can ignite just on contact with hot surfaces. Particularly, be wary of old hot plates unless you can determine that they are spark-proof since many older models are not spark-proof.

Heating mantles contain a heating element and are covered with a fire-resistant material such as fiberglass. They are used with variable transformers that control the voltage and thus control the heat of the mantle. Mantles are perhaps one of the safest ways to provide heating in your experiments.

Oil baths used for heating vessels should utilize oils that do not smoke, such as silicone oil. They are heated with heating elements that can be bought commercially or sometimes they are heated with a laboratory-made heating element that is controlled by a variable transformer. There are several potential hazards from these: burns from touching hot surfaces or splatters of hot oil, potential for fires if the flash point of the oil is exceeded, and potential for electrical shock from unguarded terminals (especially with lab-made oil baths).

Heat guns (the scientific version of a high-powered "blow drier") have a motor that draws in air and passes it over a heating coil that easily becomes red-hot. These are potential sources of ignition and they should never be used around flammables.

Ovens are often used to dry glassware or some chemicals such as drying agents. You should not put any glassware that has been washed with a flammable solvent, like acetone, into a drying oven until the solvent has completely evaporated and the "air" inside the glassware has been flushed with air or nitrogen. You should never use ovens to treat or dry a volatile or flammable chemical. Besides the danger of explosion, you could be exposed to toxic vapors since there is typically no exhaust on ovens and once the door is opened the vapors will escape into the lab. You should avoid the use of mercury thermometers in ovens (since a mercury spill inside a hot oven will release mercury vapor) and use a substitute temperature-measuring device.

Microwave Ovens   Microwave ovens have become a popular tool in many laboratories, but while they can be very useful their misuse can have serious consequences. Explosions, burns, and microwave and chemical exposures have occurred from inappropriate use of microwave ovens. Below is a list of safety measurements for microwave ovens in the laboratory.[5-7]

- Do not heat flammable liquids in a microwave oven. Domestic ovens are not spark-proof, and laboratory ovens may or may not be spark-proof. Look for labels indicating "spark-proof."
- Do not heat radioactive or other hazardous chemicals in microwave ovens since any incident or release will contaminate the oven.
- Do not override or modify safety interlocks or switches or other mechanical or electrical systems.
- Do not put wires or tubing between the door and its sealing gasket; always ensure the door is properly sealed when closed.
- Never heat food or drink in microwave ovens being used in laboratory operations.
- Do not put metal objects such as aluminum foil or magnetic stirring bars in the oven.
- Do not heat sealed or loosely sealed containers. Screw cap containers with loose caps have exploded. Use laboratory wipes to cover the containers during heating.
- Do not overheat solutions and remember that solutions and containers can be very hot. Use protective face/eye equipment and protective gloves. Superheated water can "explode" when the container is moved, causing severe burns.

- Use only laboratory grade microwave ovens for heating chemicals. These ovens have built-in safety features such as a venting/exhaust system, an interlock that prevents operation if the fan or venting system fails, a sensor to detect organics in the air, a temperature probe with feedback system to prevent overheating, a pressure release mechanism, and a rotator and stirrer.

Ultrasonic Cleaners   Ultrasonic cleaners are used in many laboratories for cleaning small parts. While the high-frequency noise (16–100 kHz) can be very irritating, it has not been reported to produce permanent damage through airborne transmission. However, it has been reported to cause fatigue, headaches, nausea, and tinnitus (ringing in the ears).[8] You should not come in direct body contact with solutions undergoing ultrasonic cleaning since this could damage tissues; don't put your fingers in these vibrating solutions. You may want to consider acoustically insulating the ultrasonic cleaner using foam or other sound-absorbing materials.

Chromatographic Techniques   Chromatographic techniques play a key role in the purification of chemicals in many laboratories in varied forms, such as column chromatography and thin layer chromatography (TLC). These techniques make use of adsorbents, such as silica gel and alumina, and are often in powdered forms that can cause irritation to skin, eyes, and the respiratory tract. Silica gel is amorphous silica and does not exhibit the adverse effects of its hazardous relative crystalline silica that can cause silicosis and lung cancer upon prolonged exposure. Column and TLC separations present opportunities for exposure to solvents and are best carried out in a chemical hood.

Distillation and Refluxing   Distillation and reflux apparatuses are often water-cooled using tubing from a lab water supply to the condenser and another length of tubing to return the water from the condenser to a cup sink. Because there can be fluctuations in water pressure, it is essential that tubing be secured at the water source and on the nipples of the condenser with wire or clamps. It is also essential that the tubing going into the cup sink be secured since pressure could force the tube out, producing a flood. There have been many incidents of unsecured tubing coming loose under water pressure fluctuations and this results in several problems: flooding (particularly for reactions running overnight), failure to cool the vapors in the reaction so they are released into the environment, and potential hazardous conditions from contact with electrical equipment. Commercially purchased water-cooling pumps must have tubing secured with clamps. Devices are also available that monitor water flow and turn off the electrical supply to a heating device if the flow is not adequate. A common error is having the water pressure too high in condenser systems. Usually, a very low flow rate is adequate for cooling.

Recrystallization   Recrystallization often involves the use of flammable solvents that are heated and evaporated to produce a concentrated solution of the substance to be crystallized. Care must be taken to ensure that ignition sources are eliminated from the area of operation since many fires have resulted from vapors evaporating from these solutions being ignited by an ignition source, such as a non-spark-proof hot plate. Use steam heating wherever possible for these operations.

Extraction   Extractions are a common laboratory operation. These usually involve the use of separatory funnels but may also involve capped tubes. Shaking two immiscible solvents often generates a slight pressure but it easily generates aerosols. Exposures can occur in several ways. When using the separatory funnel you will shake the mixture and turn the tapered spout into the air, and then open the stopcock at the bottom to release the air pressure. This can generate and release aerosols. It is best accomplished so that the spout is opened within a chemical hood to capture and remove the aerosols. Another potential exposure scenario is if the stopcock, stopper, or cap leaks, or if the stopcock slips out of place. Wearing protective gloves can help prevent exposures to your hands. Capped tubes will also build up pressure and produce aerosols, so you should work in a chemical hood when opening them.

Stirring   Stirring a reaction mixture is important not only to ensure effective mixing of reagents but to ensure that there are no "hot spots" that could occur and "bump" the reaction mixture out of the containment during heating. (Bumping can also be minimized by the use of "boiling chips" or other small materials that provide nucleation sites which make small bubble formation easier.) There are many kinds of stirrers, including magnetic stirrers (magnetically connected to a Teflon®- or glass-coated magnetic stirbar) and motor driven stirrers with a shaft that goes to the bottom of the flask. If motor driven or electric stirrers are used it is important to ensure that the motors in these devices are spark-proof; this is essential if you are using a flammable mixture. Also remember that vigorous mixing can produce aerosols so you should consider using a hood.

Centrifuging   Separating layers of liquids or separating solids from solutions by centrifuging is a frequent and important method in a chemist's toolbox. There are a wide variety of centrifuges from simple benchtop models with low RPM (revolutions per minute) Values of 0–500 up to very fancy and expensive floor models with RPM values up to 30,000. These various models each have different levels of safety features. It is important to understand the particular features of the centrifuge you are using either by studying the instruction manual or seeking help from an experienced user.

It is important to minimize opportunities for breakage of tubes by ensuring that the centrifuge is balanced, since an unbalanced centrifuge can lead to broken tubes. If you use a centrifuge to any extent, you will likely encounter a broken tube. This creates two problems: aerosol formation and a spill that needs to be cleaned up. It is sometimes possible to recognize that a tube has broken from the noise in the centrifuge. If you suspect this is true, for centrifuges with closed tops you should allow time (30–60 minutes) for the aerosols to settle before opening the top of the centrifuge.

If you are handling particularly hazardous materials, you may want to find a centrifuge that can be fitted with capped chambers for the tubes, so if a tube does break the contents are contained within the capped chamber. This is very important with potentially infectious materials since aerosols from infectious materials can spread illness. Over the years many clinical chemists and medical assistants have contracted infectious diseases from centrifuge operations where broken tubes produced aerosols.

Vacuum Pumps, House Vacuum, and Water Aspirators   Many operations require the use of vacuum or some reduced pressure. Oil-based vacuum pumps are common in many labs. All vacuum pumps should be fitted with guards to prevent contact with the moving belt and wheel, and pumps without such guards should not be used or protection should be built around them. Either the vacuum pump should be exhausted through local exhaust ventilation or the incoming air should be filtered with a charcoal or similar filter that removes the vapors before they enter the pump. This latter filtering method should also be used for the house vacuum since this prevents vapors from being exhausted into the engineering space where the pump is located. Alternatively, it is common to also use a "trap" between the operation and the pump or house vacuum. These traps, often side-armed flasks, are designed to capture liquids or gases/vapors that should not enter the pump. Sometimes they are cooled to effect better capture. Pumps can also be very noisy and are sometimes enclosed in sound-reducing enclosures.

Refrigerator/Freezer Usage   Flammable chemicals or samples containing flammables should only be stored in refrigerators or freezers that are explosion-proof or specifically designed for storage of flammable liquids (see Figure 7.3.1.1). In the "laboratory" refrigerators, the internal light and thermostat switches have been eliminated or moved outside. Ordinary refrigerators sold for home use are not safe for the storage of flammable chemicals. There have been many devastating incidents from ignition with subsequent explosion and fire from flammables stored in inappropriate refrigerators. Imagine how this can happen. When the power goes off, the inside and its contents begin to warm. If left for long enough the flammable containers build up pressure and tops "pop" off, potentially allowing the flammable vapors to reach explosive limits inside the refrigerator. When the power comes on and switches spark the result is an explosion. Laboratories and whole buildings have been destroyed or seriously damaged in these incidents (see *Chemical Connection 7.3.1.1* Volume of Flammable Liquid Needed to Reach an Explosive Concentration in a Refrigerator).

It is critically important that food should never be stored in a refrigerator that is intended for chemical storage nor should lab chemicals be stored in refrigerators intended for food storage. Anyone who has ever opened a chemical refrigerator, with its unavoidable "chemical smell," would not likely wish to store a bagged lunch in there!

FIGURE 7.3.1.1   Explosion-Proof Refrigerator for Laboratories. These refrigerators are usually hard-wired into electrical supplies (so that they cannot accidentally become unplugged) and they have spark-proof interiors (internal switches and lights have been removed.)

CHEMICAL CONNECTION *7.3.1.1*

VOLUME OF FLAMMABLE CHEMICAL NEEDED TO REACH AN EXPLOSIVE CONCENTRATION IN A REFRIGERATOR

To understand why you should not store flammables in a refrigerator that has not been specially designed for this purpose, we will calculate the volume needed to cause an explosion. For these calculations:

- Dimensions of the internal space of a typical refrigerator = 22 in. × 28 in. × 42 in.
- Solvent to be used as the model is acetone, with a lower explosive limit (LEL) = 2.5%.
- Assume the temperature in the malfunctioning refrigerator reaches 30 °C and 1 atm.
- Assume that the volume of liquid will become completed vaporized.

Step #1:.   Internal volume of refrigerator in liters

$$V = (22 \text{ in.} \times 28 \text{ in.} \times 42 \text{ in.})(2.54 \text{ cm/in.})^3 (1 \text{ L}/1000 \text{ cm}^3) = 420 \text{ L}.$$

Step #2:.   Volume of acetone when LEL is reached = (420 L)(0.025) = 11 L.

Step #3:. $n = PV/RT = (1 \text{ atm})(11 \text{ L})/(0.082 \text{ L-atm/K} - \text{mol})(303 \text{ K}) = 0.43 \text{ mol}.$

Step #4:.   Volume of acetone that will explode if vaporized and exposed to an ignition source

$$V_{\text{liquid}} = (0.43 \text{ mol})(58 \text{ g/mol})/(0.79 \text{ g/mL}) = 32 \text{ mL}$$

Thus, only 32 mL of acetone that is vaporized within this refrigerator will produce the LEL and can result in an explosion if an ignition source is present—and remember this is probably not the only flammable chemical in the refrigerator.

Other Sources of Information

As noted earlier, consulting with mentors and other persons who have experience with particular experiments and instruments is often the best way to learn about safety protocols and details. We list here a few printed resources that also contain useful information. Searching for information online may or may not produce reliable information and it is important to carefully assess the quality of the information.

- R.J. Alaimo (editor). *Handbook of Chemical Health and Safety*, Oxford University Press, New York, 2001.
- A. Ault. *Techniques and Experiments for Organic Chemistry*, 6th edition, University Science Books, Sausalito, CA, 1998.
- R.J. Errington. *Advanced Practical Inorganic and Metalorganic Chemistry*, Blackie Academic and Professional, London, 1997.
- J.W. Lehman. *Microscale Operational Organic Chemistry: A Problem-Solving Approach to the Laboratory Course*, Prentice Hall, Englewood Cliffs, NJ, 2004.
- McMaster University. Microscale Laboratory Techniques, 1997/1998; available at http://www.chemistry.mcmaster.ca/~chem206/labmanual/microscale/complete.html (accessed January 17, 2009).
- National Research Council. *Prudent Practices in the Laboratory: Handling and Disposal of Chemicals*, National Academy Press, Washington, DC, 1995; available at http://books.nap.edu/catalog.php?record_id=4911#toc (accessed January 16, 2009).

References

1. R. P. FEYNMAN. BrainyQuote; available at http://www.brainyquote.com/quotes/authors/r/richard_p_feynman.html (accessed January 16, 2009).
2. R. H. HILL Personal account of an incident.
3. ALDRICH-SIGMA. Antistatic devices; available at http://www.sigmaaldrich.com/labware/labware-products.html?TablePage=19810758 (accessed January 27, 2009).
4. Staticmaster® Brushes and Ionizing Units. Available at http://www.2spi.com/catalog/photo/statmaster.shtml (accessed January 27, 2009). Zerostat® antistatic gun. Available at http://www.dedicatedaudio.com/inc/sdetail/6193 (accessed January 27, 2009).
5. H. M. KINGSTON, P. J. WALTER, W. G. ENGLEHART, and P. J. PARSONS. Chapter 16: Microwave Laboratory Safety; available at http://www.sampleprep.duq.edu/dir/mwavechap16/mwave.htm (accessed January 23, 2009).
6. B. L. FOSTER. Safety Issues Related to Microwave-Enhanced Chemistry. Online Chemistry Course (OLCC), Paper 2.b. 2004; available at http://science.widener.edu/svb/olcc_safety/papers/foster_2.pdf (accessed January 23, 2009).
7. Occupational Health and Safety Unit, University of Queensland. Safe Use of Microwave Ovens in Laboratories; available at http://www.uq.edu.au/ohs/pdfs/microwaveguideline.pdf (accessed January 23, 2009).
8. National Research Council. *Prudent Practices in the Laboratory: Handling and Disposal of Chemicals*, National Academy of Sciences, Washington, DC, 1995, pp. 118–119.

QUESTIONS

1. What is an effective safety measure when running a new reaction?

   (a) Keep the temperature low.

   (b) Run the reaction on a small scale.

   (c) Run the reaction for just a short time.

   (d) Avoid using any catalysts.

2. When using the "buddy system" to avoid working alone, you should

   (a) Be in direct visual contact with each other, if possible

   (b) Make sure that your buddy knows the hazards that are associated with the experiment you are performing

   (c) Make periodic contact with your buddy if you are not in the same room

   (d) All of the above

3. Under what circumstance can static charge cause difficulty in transferring a solid during weighing?

   (a) High humidity

   (b) A solid that easily sublimes

   (c) Wearing nitrile gloves

   (d) Working in a glovebox

4. Mercury spills

   (a) Are not much of a concern since elemental mercury has such a low vapor pressure

   (b) Are not much of a concern since mercury is primarily toxic by ingestion

   (c) Must be cleaned up using special techniques

   (d) Can effectively be swept up with a small broom and dustpan

5. What is the main hazard when working with hot plates?

   (a) Burns

   (b) Electrical shorts

   (c) Igniting flammable vapors

   (d) Vaporizing ordinarily nonvolatile liquids

6. Domestic microwave ovens used for chemical reactions in laboratories

   (a) Are nearly as safe as laboratory-grade ovens

   (b) Can safely be used as long as the safety-interlock is over ridden

   (c) Should be wiped down inside before using them for heating food

   (d) None of the above

7. Ultrasonic cleaners

   (a) Work using frequencies well beyond what can be heard by humans

   (b) Can also be used as a heat source for chemical reactions

(c) Use solution into which you can place your fingers since the ultrasonic waves cause no tissue damage

(d) Are sometimes acoustically insulated from the surroundings

8. It is safe to inhale powdered silica gel and powdered alumina used in chromatography since

(a) They have been treated to make them safe

(b) The particles are too small to cause respiratory irritation

(c) The particles are too large to cause respiratory irritation

(d) None of the above

9. What is a common problem in setting up distillation and reflux apparatuses?

(a) Failure to secure the cooling lines to the condenser

(b) Failure to secure the cooling line in the sink

(c) Running the cooling water at too high a pressure

(d) All of the above

10. What is a common hazard when using a separatory funnel?

(a) The release of aerosols when venting the funnel

(b) Heat buildup in the funnel

(c) Pressure buildup in the funnel

(d) Both (a) and (c)

11. The main hazard associated with using centrifuges is

(a) Broken tubes

(b) Aerosol formation from spinning the sample too rapidly

(c) Unbalanced samples leading to excessive vibration and rotor destruction

(d) Spilling samples since centrifuge tubes have round bottoms

12. A common hazard associated with vacuum pumps in the lab is

(a) Poor atmosphere quality due to the pump oil that escapes

(b) Exposed belts and moving parts

(c) Flammability from leaking pump oil

(d) None of the above

13. Refrigerators used in laboratories

(a) Are designed the same as those used in domestic kitchens

(b) Can also store food and drinks as long as the laboratory chemicals are on a different shelf

(c) Have detectors indicating flammable vapor concentrations

(d) None of the above

7.3.2

RADIATION SAFETY

Preview This section provides an overview of the principles of radiation safety in the laboratory.

In the Radiation Laboratory we count it a privilege to do everything we can to assist our medical colleagues in the application of new tools to the problem of human suffering.

Ernest Lawrence[1]

INCIDENT 7.3.2.1   EXPOSURE TO AIRBORNE PLUTONIUM[2]

A researcher and a colleague were working in a facility that handled plutonium within gloveboxes. One researcher was replacing a valve that connected two gloveboxes when a monitor that detects the presence of airborne radiation alarmed. Both scientists immediately left the laboratory. An emergency response team arrived and surveyed each employee for radioactive contamination. The researcher who was replacing the valve had contamination on his face, head, and chest. Further investigation showed that nasal swabs of this employee had readings of 5000–7000 disintegrations per second. The other employee had nasal swab readings of 60–90 disintegrations per second. The exposure of the employee with the highest reading was estimated to be twice the background level that will be delivered over the next fifty years or about 650 mrem/year. This employee was treated using chelation therapy to help remove heavy metals (i.e., plutonium).

What lessons can be learned from this incident?

Laboratory Safety for Ionizing Radiation — The Radiation Safety Program

This section provides an overview of radiation safety in the laboratory. If you use sources of ionizing radiation your institution will be subject to regulation by the Nuclear Regulatory Commission (NRC) or a designated state agency. It requires that your institution obtain a license for handling these materials and that a Radiation Safety Program (RSP) must be established for the protection of individuals using radiation as well as colleagues and the general public.[3] You must comply with the specific procedures, practices, and policies established under your institution's RSP. The RSP will apply to you if you are using confined, sealed sources or unconfined, open sources. The RSP is developed and managed by the Radiation Safety Officer (RSO) or a Radiation Safety Committee (RSC). Most institutions have an RSO. To acquire and use radioisotopes you must have approval by the RSO and RSC. If you do not comply with their requirements, you may be exposing yourself and others to unacceptable levels of radiation. Furthermore, you may lose your privilege of using radioisotopes. Since you will be subject to the specific RSP of your institution, we will address the basic principles of radiation safety that will be used in your RSP.

Seeing Those Invisible Rays

You learned some of the basics about ionizing radiation in Section 5.3.8, including the common types of radiation: α particles, β particles, γ rays, X rays, and neutrons.[4,5] The basic difference between chemical safety and radiation safety is that the latter is sometimes more difficult because you can't

Laboratory Safety for Chemistry Students, by Robert H. Hill, Jr. and David C. Finster
Copyright © 2010 John Wiley & Sons, Inc.

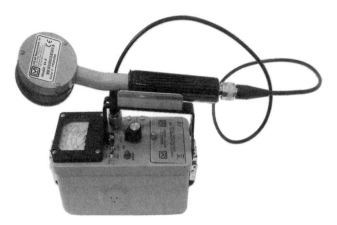

FIGURE 7.3.2.1 Ludlum Geiger–Müller Counter. These portable meters are routinely used to monitor an area for the presence of radioactive materials during and after handling these materials in laboratory operations. (Courtesy of Ludlum Measurements, Inc.)

see, smell, taste, or feel radiation if you are being exposed. Nevertheless, even though you personally cannot detect this ionizing radiation, it is capable of doing harm to you if you are exposed. Radiation causes harm by releasing energy as it interacts with human tissue so if you are using ionizing radiation you need to be able to detect it so that you can be sure your protection measures are working. To be able to determine the hazards of these materials requires that you use methods that will help you "see" these invisible rays.

There are several kinds of radiation detectors and we will briefly examine only the most common ones. There are two principal types of detectors that directly detect radiation: detectors having ionization chambers filled with gas and scintillation detectors. *Passive* detectors measure radiation exposure indirectly; they have matrices that react and are modified by radiation in such a way that past exposure can be estimated. These latter detectors are the radiation badges worn by people working in areas where radiation is used.

It is important to know what kinds of radiation a given detector senses. Otherwise, there may be radiation present that goes undetected.

Geiger counter probes may be round flat types called a "pancake" or they may be tubular in design. The counter has a scale that can be adjusted for the appropriate range of disintegrations, and a meter that "jumps" or "ticks" as the disintegrations are detected (counted); some counters may have digital displays. Figure 7.3.2.1 shows a typical Geiger counter. This detector is very useful in finding point sources of ionizing radiation. Geiger counters are often used by laboratory researchers to conduct rapid "real-time" scans to determine if radiation can be overtly detected as the result of contamination. Requirements for more exact measurements are made with other kinds of detectors such as liquid scintillation counters (see below). Most Geiger counters will not detect α or low-energy β particles because they cannot penetrate the normal film that covers the probe. However, these particles can be detected using a special kind of Geiger counter probe.

Gas flow proportional counters use a gas ionization detector that produces ionizations as the radiation penetrates the chamber. These are not portable detectors, but they can be used to measure some radiation energies and are able to distinguish between α and β radiation.

The second type of active detector is the scintillation detector that works on the basis of scintillation materials (scintillators) interacting with radiation to emit light that is proportional to the radiation energy. These detectors are useful for measuring radiation from β and γ emitters and neutrons. However, the detection efficiency depends on the proper selection of a scintillator that is sensitive to a particular kind of radiation. Liquid scintillation counters are very commonplace where radiation is used. They are usually not portable and require collection of a sample for analysis. The samples are mixed with a special "cocktail" that consists of an organic solvent and an additive that fluoresces. When radiation energy is emitted, its energy is transferred to the fluorescent additive that emits light, and this in turn is detected by a photomultiplier tube. It is a very sensitive method of measuring radiation and is often utilized for detecting common β emitters in wipe tests of areas used for handling radioisotopes, such as in laboratory tagging experiments with tritium or phosphorus-32.

Radiation badges are passive devices. Their detection matrix is modified when it is struck by radiation; these devices are called dosimeters and they are often an important part of any radiation safety program. The radiation film badge is used by people regularly working in an area where radiation sources exist.[6] Radiation interacts with film and when developed allows for an estimation of the radiation received by the badge—this assumes that you wear the badge during radiation-handling operations. For most common laboratory use, these badges are used for some period (1–3 months) and then sent off for reading. So if you were exposed during that time you would not know this until some time after the fact. Nevertheless, this provides an assessment of one's overall exposure during that time period.

A second type of "badge" detector is a thermoluminescence dosimeter that traps energy from radiation in a thermoluminescent material that releases this energy when it is heated. These detectors have largely replaced film badges because they are very sensitive and have a linear response over a wide range of doses as well as dose rates.

As you can see from the discussion above, there are several ways to detect radiation but each has its purposes and limitations. If you use or are in an area where ionizing radiation is being used, then your institution's RSO will determine what detectors will be employed.

ALARA—Minimizing the Dose

The philosophy of radiation safety is similar to chemical safety: minimize exposures. Because of the differences in properties, radiation protection procedures are different from those for chemical hazards. Nevertheless, as in the case of chemicals, the "dose makes the poison," so it is important to limit the exposure (dose) of radiation.

As with chemical toxicity, there is no question that high doses of radiation cause biological damage. However, as discussed in Section 5.3.8, there are two models for the effects of radiation at low doses and we don't know with certainty if there is a "low dose" (or threshold) below which no damage is done. Therefore, it becomes difficult to set "acceptable" radiation doses, particularly in the context of a natural background of radiation to which we are all necessarily exposed. In medicine, we have accepted the notion that the use of radioisotopes in the body and the use of X rays are warranted when the overall known benefits outweigh probable or certain negative side effects.

In laboratories, the use of radiation and radioisotopes is necessary for many experimental procedures but since the means to eliminate and reduce exposure are available, the prevailing rule has become to keep the dose "as low as reasonably achievable," aka *ALARA*. This principle helps decide which procedures and practices are to be followed to keep the dose to the ALARA level. ALARA does not mean that no exposure is allowed, but rather that the user of radioisotopes has looked at each step in the handling process and determined how to minimize exposures for a given step or procedure. Furthermore, the NRC has established maximum dose limits for radiation and these are strictly enforced; once someone reaches these limits that person may no longer work with radiation. For the whole body, the maximum yearly dose limit is 5 rem while isolated organ or tissue exposures of 50 rem are allowed and eye lenses are allowed a maximum of 15 rem/year (see Section 5.3.8). The dose limits are set higher for organs because it is believed that these organs can absorb this much radiation over an annual period without causing a health effect on that organ. For example, it is believed that eye lenses can receive up to 15 rem of radiation on an annual basis without producing radiation-induced cataracts.

Protection from Sealed Sources of Radiation in Laboratory Instrumentation

In the laboratory you may be exposed to instruments that generate radiation (such as X-ray diffraction instruments) or may have sealed sources that contain hazardous levels of radioactive materials (such as γ irradiators or analytical detectors). You may also be exposed to unsealed, open sources of radiation. Both types require special training and education to learn proper procedures and handling precautions about these radiation hazards. We discuss first how to protect yourself from sources of radiation in laboratory instrumentation.

When you are working with radiation and/or radioactive sources, you will use three methods to minimize your exposure: time, distance, and shielding. You can reduce the dose from a radiation source by limiting the time that you are exposed to the radiation. You can minimize the dose by increasing the distance between you and the source. This is very effective with most α and β sources since they don't travel far in air. Radiation emissions follow the inverse square law, which states that the dose rate falls by the square of the distance. This means that if you double the distance between you and the source that the intensity of the radiation will fall by a factor of 4.

The use of shielding is effective in decreasing exposure to sources of radiation. Often sealed sources are heavily shielded so that you may work around them without being exposed. Lead shielding is often used in these instruments since most instrumentation utilizes γ or X-ray radiation.

Before using any instrument that utilizes intense sources of radiation, it is essential that you receive training in the safe use of this equipment. There will be written safety instructions that you should read and follow to prevent exposure to the radiation source. Always have a survey meter (Geiger counter) available to scan the area for leaks, and observe and follow warning lights. When using instruments that generate radiation, do not leave these instruments unattended when they are in the operational mode. Never tamper with the instrument safety interlocks or try to bypass the shutter mechanisms, and repairs should only be carried out by trained technical staff or repair/maintenance technicians from the instrument manufacturers. Do not uncover a sealed source and allow you or others to be exposed to this source. X-ray diffraction instrumentation offers potential for very hazardous exposures if you are exposed to the primary beam, so you must ensure that you are never exposed to an open X-ray source that is operational. Large irradiators, used to inactivate biological materials, also have very high hazard sources of γ radiation, but if you always follow proper instructions and do not bypass safety measures you will be protected.

Protection from Open Sources of Radiation in the Laboratory

Some chemists might not encounter any radioactive isotopes for their entire careers. Other chemists and some laboratories use them daily, most often as tracers in chemical and biochemical reactions. As noted above, the use of radioisotopes is highly regulated.

"Open sources" of radiation refer to solids or solutions that contain radioisotopes. As with any other chemical, the chance for spillage or contamination is always present. Combining the principles of using time, distance, and shielding with other basic laboratory safety practices can effectively prevent exposures from radioisotopes.

When using open sources of radiation, you should frequently monitor the work area for contamination as well as monitoring all personnel after they finish work but before they exit the work area. If contamination is found, you must take steps to decontaminate and remonitor afterwards to ensure that all contamination has been removed. (See Incident 5.3.8.1 as an example of not following this rule.) If you are using radioisotopes in operations that could generate open aerosols, you should be working in a chemical hood. If this involves volatile isotopes, such as $^{131}I_2$ or volatile organic compounds containing ^{14}C, you should use a chemical hood that has been fitted with a filter to remove these volatiles from the hood exhaust.

You should use portable shielding when you openly handle radioisotopes. The type of shielding that you should use depends on the type of radiation. Shielding is usually not used for α emitters since these particles can only travel a few centimeters in air and they do not penetrate the skin. When using β emitters, you should use acrylic see-through shielding; this kind of shielding is commonly used in many laboratories. When using sources that emit γ rays or X rays, you must use a very dense material such as lead. These shields can be lead "bricks" or lead aprons. You probably recall that when you get an X ray the technician covers the part of your body that is not being X rayed with a lead apron. Lastly, although unlikely, some laboratorians could be exposed to neutron-emitting materials and they must be shielded with water or paraffin wax blocks. (See *Chemical Connection 7.3.2.1* Radiation Shielding.)

Ingestion of radioactive materials is, of course, prevented by the basic laboratory rules of no eating, drinking, or mouth-pipetting. Good housekeeping is an essential practice in radiation work areas, perhaps more so than "normally" since radioactive contamination and spreading the contamination is more insidious than with nonradioactive substances. You should consider using plastic or metal trays to contain unintentional spills.

CHEMICAL CONNECTION *7.3.2.1*

RADIATION SHIELDING

Radiation shielding is an important tool in preventing and minimizing exposure. However, selection of shielding is not always straightforward. Charged particles such as alpha radiation (helium nuclei) or beta (electrons) radiation are more easily shielded because their charges react with the materials they encounter. Alpha particles (most at 4–6 MeV and some higher) have a very high linear energy transfer (LET) rate, the ability for radiation to cause ionization and transfer energy to other matter. However, due to their high mass, α particles are easily shielded by just a thin sheet of paper. Intact skin will not likely be penetrated by α particles. The greatest risk of α-particle emitters is from internal ingestion or inhalation (see *Special Topic 5.3.8.1* Radon—A Significant Public Health Concern).

Beta particles (2 MeV and lower) have generally medium LET rates and their ability to penetrate a material depends on their energy. You should use low-Z (atomic number) materials such as plastics for β emitters. Do not use high-Z nuclei materials for β emitters (such as lead) because thin lead shielding of β emitters can produce "bremsstrahlung" X rays as unintended by-products, and a β emitter now becomes an X-ray emitter. Low-energy β particles can be stopped by a thin film; however, high-energy β particles require much thicker plastic shielding made of Plexiglass® (a polymethylmethacrylate polymer) or polyvinyl chloride plastic.[8] A common recommendation is to use a 3-cm Plexiglass shield to protect against beta radiation. Nevertheless, it is always prudent to test the shielding with a Geiger counter to ensure that it is effective.

Gamma rays are more difficult to shield than charged particles. Because γ rays are low LET, they tend to easily pass through many materials until they become low energy and then they are more effectively absorbed. This makes γ radiation more hazardous than α or β radiation. Gamma radiation of moderate or high energy can produce secondary electrons and X rays upon impact with shielding, and this secondary shower can move through several inches of lead shielding before it is stopped. Gamma radiation is best shielded by high-density, high-thickness materials with a high Z, such as lead. The amount of lead depends on the energy of the gamma radiation.[8]

Neutrons can be generated in the laboratory by some sources used in physics or engineering laboratories. Shielding for neutrons is very different from γ radiation. Materials that have efficient moderating properties and large absorption cross sections are best for neutron protection, and materials with a lot of hydrogen are good for this purpose. Water is often used as a shielding material for neutrons, but other materials high in hydrogen content include paraffin wax or polyethylene blocks. However, if you are potentially exposed to neutrons, you should consult with your RSO because this can be a complicated decision that requires expert knowledge. Selecting the wrong shielding can be very hazardous.

While unrelated to lab safety per se, it is also interesting to note that spent nuclear fuel rods are usually stored under large pools of water. The water serves both to thermally cool the hot fuel rods and also to effectively absorb radiation. The blue glow seen in these pools is due to *Cherenkov radiation*, which is emitted when a charged particle passes through an insulating material at a speed greater than the speed of light in that medium.

Use personal protective equipment to protect yourself against radiation exposures. As in all laboratory operations, you should wear safety goggles in the event of a splash. The most common kinds of incident with radioisotopes are spills and splashes, which can lead to contamination of surfaces as well as individuals who are carrying out the procedures. It is really important that you do as much as possible to prevent exposure to your skin. You should wear long-sleeve shirts, protective gloves with long cuffs, a full length laboratory coat, and pants.

Radioactive materials must be locked in a secured container or a secured storage area. You must not leave radioactive materials unsecured in an unattended laboratory at any time. This means if you are using radioisotopes actively you will need to keep the laboratory locked if you leave. Any visitors to the laboratory entering areas where radiation is stored must be escorted. Your RSP may require added efforts to secure radioactive materials.

Lastly, you should prepare for radiation emergencies or incidents, such as spills, splashes, or overt exposures.[7] This requires preplanning and acquiring the materials and equipment necessary to handle these situations. Your RSO will be very helpful in this regard.

RAMP

- *Recognize* that radiation can be hazardous (and that it is different from chemical hazards) and recognize the potential for exposures to external and internal radiation sources.

- *Assess* the overall risk level based on the type of radiation and probability for exposure.

- *Minimize* exposure by using the ALARA principle and using time, distance, and shielding combined with other personal protective methods. Follow procedures and protocols established by the RSP and RSO.

- *Prepare* for emergencies involving radiation, especially spills and splashes, and contact your RSO to learn more about emergency procedures.

References

1. ERNEST LAWRENCE. BrainyQuote; available at http://www. brainyquote.com/quotes/quotes/e/ernestlawr352111.html (accessed December 4, 2008).
2. Los Alamos National Laboratory. Employee Receives Radioactive Dose in Lab Incident; available at http://www.lanl.gov/news// index.php/fuseaction/home.story/story_id/1602/view/print (accessed April 20, 2009).
3. P. E. HAMRICK. Ionizing Radiation: Radiation Safety Program Elements. In: Handbook of Chemical Health and Safety (R. J. ALAIMO, ed) Oxford University Press, New York, 2001, Chapter 62, pp. 428–435.
4. D. D. SPRAU, P. E. HAMRICK, and F. L. VAN SWEARINGEN. Ionizing Radiation: Fundamentals. In: *Handbook of Chemical Health and Safety* (R. J. ALAIMO, ed.), Oxford University Press, New York, 2001, Chapter 63, pp. 436–446.
5. H. J. ELSTON. The Chemical Hygiene Officer's radiation protection primer. *Journal of Chemical Health and Safety* 15 (1): 14–19 (2008).
6. B. WILSON. Radiation Emergency Response, Decontamination, PPE. In: *Handbook of Chemical Health and Safety* (R. J. ALAIMO, ed.), Oxford University Press, New York, 2001, pp. 446–452.
7. M. P. CARMINE. Dosimetry. In: *Handbook of Chemical Health and Safety* (R. J. ALAIMO, ed.), Oxford University Press, New York, 2001, pp. 452–458.
8. University of Guelph, Environmental Health and Safety. A Guide to Nuclear Radiation Shielding (UG-RSOG-025-2004); available at http://www.uoguelph.ca/ehs/uploads/2008/11/guide-to-shielding.pdf (accessed December 16, 2008).

QUESTIONS

1. What organization licenses the use of sources of ionizing radiation and radioisotopes?

   (a) Nuclear Regulatory Commission
   (b) Atomic Energy Commission
   (c) A state agency
   (d) Either (a) or (c)

2. What is the main difference between chemical safety and radiation safety?

   (a) Radiation is generally more dangerous than chemicals.
   (b) Radiation is generally less dangerous than chemicals.
   (c) It is more difficult to detect exposure to radiation using the senses.
   (d) Regulations require radiation safety officers to oversee the use of radiation and radioactive isotopes but regulations do not require chemical safety officers to oversee the use of chemicals.

3. Geiger–Müller counters

   (a) Are passive detectors
   (b) Are active detectors
   (c) Come in both active and passive configurations
   (d) Measure only X and γ radiation

4. Scintillation detectors

   (a) Are usually not portable
   (b) Measure α and β radiation
   (c) Use fluorescence as a means of detecting radiation
   (d) Both (a) and (c)

5. Radiation badges

   (a) Are active detectors
   (b) Are usually worn for a few months to detect accumulated dose
   (c) Are dangerous to handle
   (d) Usually have internal calendars that identify when radiation was detected

6. The ALARA rule states that radiation doses in laboratories be

   (a) Zero
   (b) Less than 5 rem/year
   (c) Minimized as much as possible
   (d) No more than one-half of the natural background level

7. The NRC allows annual exposures of

   (a) 5 rem for the whole body
   (b) 50 rem for isolated organ and tissue
   (c) 15 rem for eyes
   (d) All of the above

8. What methods(s) can be used to protect yourself from radiation?

   (a) Limit the time of exposure.
   (b) Increase the distance from the source of exposure.
   (c) Use shielding.
   (d) All of the above.

9. Which forms of radiation are most effectively reduced by distance?

   (a) Alpha and beta
   (b) Beta and gamma

(c) Alpha and gamma

(d) Alpha, beta, and gamma

10. What material is often used for radiation shielding?

(a) Tin

(b) Gold

(c) Lead

(d) Plutonium

11. When using intense sources of radiation,

(a) Always have a survey meter at hand to check for leaks

(b) Override interlocks and safety devices only when safe to do so

(c) Uncover a sealed source only when you know the radiation level will be low

(d) Allow for exposure to the beam, such as an X-ray beam for a diffraction instrument, for no more than 1 second per day

12. When using open sources of radiation,

(a) Monitor the area once a week for contamination

(b) Use acrylic shielding for X rays

(c) Work in a chemical hood if there is a chance the radioactive source may become volatile

(d) Use a lead apron when working with α emitters

13. When using radioisotopes, good housekeeping and use of personal protective equipment are

(a) Generally less important than when working with non-radioactive chemicals

(b) Generally more important than when working with non-radioactive chemicals

(c) Crucially important and all procedures must be approved by the RSO

(d) Not important

7.3.3

LASER SAFETY

Preview This section discusses the basics of lasers and safety practices and procedures and other safeguards when using lasers in the laboratory.

Hokey religions and ancient weapons are no match for a good blaster at your side, kid.

Han Solo (Harrison Ford), *Star Wars*[1]

INCIDENT 7.3.3.1   EYE DAMAGE FROM A PULSED LASER[2]

An undergraduate student was working with an experienced researcher using two different kinds of Class IV lasers. One laser was a "particle-generating" laser that produced suspended particles in a chamber, and the other laser was an LIBS (laser-induced breakdown spectroscopy) laser that vaporized suspended particles. The student bent down to look at the suspended particles and experienced a flash. A red-brown substance was seen in her eye and later medical evaluation confirmed that the student had a retinal traumatic hole burned in her eye due to a pulsed laser. Neither the researcher nor the student wore eye protection. No one immediately reported the incident.

INCIDENT 7.3.3.2   SHOCK FROM LASER POWER SUPPLY[3]

A researcher was working on a laser with a 17,000-volt power supply when he noticed some condensation on one of the contacts of the power supply. He decided to wipe the moisture away with a tissue but touched the anode and received a sudden shock. He staggered into the hall and collapsed into the arms of a passerby after uttering "I've been shocked." His heart had stopped and cardiopulmonary resuscitation (CPR) was started. Emergency personnel arrived and used a defibrillator to restore his normal heartbeat. He was able to explain what happened. Safety interlocks had been bypassed allowing him to be exposed to this high voltage.

What lessons can be learned from these incidents?

Laboratory Lasers

The word "laser" is an acronym derived from "light amplification by the stimulated emission of radiation." Lasers are monochromatic beams of light that are in phase and moving in the same direction. Usually a single wavelength is emitted from a source, although a few sources emit multiple wavelengths. Lasers have become an integral part of our everyday world from scanning product bar codes for instantaneous pricing at most local retail stores in the United States to their use in compact disk (CD) and digital video disk (DVD) players, in laser printers, and in laser light shows.

Lasers are also used in our laboratories and, depending on their use, they can present significant hazards for those working in laboratories that may require careful attention to safety measures. The principal hazard concern for lasers is the potential for serious eye damage from exposure to the laser beam. These are essentially burns (thermal), although some photochemical injuries can result. It is also possible to receive skin burns from high powered lasers. Other hazards also exist from the potential for electrical shock from power supplies. The most powerful lasers are generated with very high voltage

Laboratory Safety for Chemistry Students, by Robert H. Hill, Jr. and David C. Finster
Copyright © 2010 John Wiley & Sons, Inc.

power supplies. If the products of laser interactions are toxic, ventilation may be needed. Some lasers can be powerful enough to start a fire upon contact with combustible or flammable materials.

The highest dose (exposure) that can be received safely by the eye from a laser beam is called the maximum permissible exposure (MPE) measured in watts/cm^2 or joules/cm^2. At the MPE a laser has virtually no probability of causing damage if an eye exposure occurs. MPE is dependent on several factors: laser energy, laser wavelength, incident time upon the target, and the light-source spatial distribution (also called spatial coherence).

Classes of Lasers

Over the past few years a better understanding of lasers has resulted in an evolution of classification systems for lasers. Before 2002 the older system of classification used Roman numerals with the most hazardous class being Class IV (see *Special Topic 7.3.3.1* Old Laser Classification). As experience with lasers grew, the classification system was updated to include new defining specifications. Beginning in 2002 a newly revised classification system was phased in and was fully implemented in 2007. It is based on International Electrotechnical Commission (IEC) Standard 60825-1/ANSI Z136.1—2007 that separates lasers into four classes. Class 1 is the least hazardous, Class 4 is the most hazardous, and there are new subclasses.[4] The classification of lasers is dependent on the dose of radiation that can be received from a laser. A brief description of these laser classes is presented in Table 7.3.3.2.

SPECIAL TOPIC *7.3.3.1*

OLD LASER CLASSIFICATION

While the classifications in Table 7.3.3.1 are now obsolete, it is possible that you may encounter these labels and it is good to be generally aware of their meaning when working with lasers.

TABLE 7.3.3.1   Old Laser Classes

Class	Criteria
I	Safe; eye damage not possible due to low output power or laser is enclosed
II	Blinking reflex prevents eye damage; output ≤ 1 mW and only visible light lasers
IIa	Low power Class II lasers that with continuous viewing >1000 seconds can cause retina burn
IIIa	Output ≤ 5 mW; beam power density ≤ 2.4 mW/cm^2; dangerous with optical instruments that change beam diameter or power density
IIIb	Direct or specular reflection viewing of beam can cause damage; power 5–500 mW; high-powered lasers are a fire hazard and can cause skin burns
IV	Output power >500 mW; can cause severe, permanent eye damage; diffuse reflections can be hazardous to eye or skin

Eye Protection from Lasers

The most important safety measure that you must remember for *all* lasers is never stare directly into the beam. You must always avoid eye exposure to lasers. Eye injuries, including permanent eye damage, are the most common serious result from laser exposure. If you are working with Class 3B or Class 4 lasers, you must make a conscious effort to wear eye protection. All persons working in this area should also be wearing laser eye protection.

While "wearing eye protection" in the form of special safety goggles that protect the eyes from the laser radiation sounds easy, just as in the situation with gloves, there are no "universal goggles" that protect the eye from all wavelengths of light. (Indeed, such "universal protection" would be an opaque goggle in the visible region, as well as UV and IR regions that would render the user sightless in the

TABLE 7.3.3.2   IEC Laser Classes

Class	Maximum permissible exposure (MPE)	Maximum power	Safety precautions
1	Cannot be exceeded		None
1M	Cannot normally be exceeded unless focusing or optics narrow the beam		Do not use with magnifying lens
2	Blinking reflex provides protection, even when using optical instruments	≤0.001 W (1 mW)	
2M	Blinking reflex provides protection	≤0.001 W (1 mW)	Do not use with magnifying lens
3R	Small risk of damage if MPE is exceeded	≤0.005 W (5 mW)	Do not view beam directly
3B	Direct viewing unsafe; diffuse reflections are safe	≤0.5 W for 315–1400 nm (near infrared), and ≤30 mJ for 400–700 nm pulsed lasers	Protective eyewear, key switch, safety interlock
4	Skins burns, permanent eye damage, fires upon contact with flammables	>0.5 W	Protective eyewear, key switch, safety interlock

lab!) Protection from lasers has to be addressed with specific wavelength hazards in mind, in addition to concerns about the intensity of the laser light. So, eye protection can be challenging for many reasons.

Damage to the retina is most likely from Class 3B and Class 4 lasers in the region of 400 to 1400 nm. Damage to the cornea is caused by lasers with radiation in the infrared and far-infrared region, 1400 nm to 1 mm.

The lenses used in laser safety goggles are rated for their optical density, OD:

$$OD_\lambda = -\log_{10}(I/I_0)$$

where OD is optical density at the specific wavelength λ, I is intensity after filtering, and I_0 is initial intensity from the light source. OD is the same as the absorbance, A, when considering how a solution absorbs light in visible spectroscopy.

The greater the OD, the more effective the lens for that wavelength of light; however, this turns out be a trade-off since the vision through these lenses can also be reduced as the optical density increases. OD values of 2–7 are common, with the higher numbers representing many orders of magnitude of reduction in transmitted light at the filtering wavelength. The filtering material is also characterized by the "percent visible light transmitted" (%VLT), ranging from 10% to 90%.

Laser glasses or goggles will have information printed on them that identifies the OD in various wavelength ranges. You should only wear eye protection that is specifically designed for the laser light you are using. Since lasers can produce either CW (continuous wave) or pulsed light and this light can vary significantly in intensity, it is best to check with the manufacturer's manual for the laser that you are using to determine the best eye protection in terms of wavelength and OD.

The lens selection effective for specific wavelengths often results in colored lenses, sometimes very dark colors, and the result can be a reduction in the ability to see indicator lights, perhaps carry out delicate hand operations, or the failure to see tripping hazards. Nevertheless, it is important to learn to work with these safety goggles, since many incidents have occurred because an "experienced" researcher decided that it was okay to do this work without safety goggles. Rockwell Laser Industries provides a database of laser-related accidents.[5] When sorting the database by "scientific" settings, 67 incidents are listed and in 45 (67%) of these incidents no eye protection was worn.

We also note that laser safety glasses and laser goggles are not chemical splash goggles. Persons using lasers and wearing laser eye protection may or may not be handling solutions that pose a splash hazard but if such hazards exist in labs where lasers are used, chemists need to be mindful of what level of splash protection is afforded by the laser eye protection that they are using. If there is potential for a splash, a face shield should be worn.

Safety Measures for Handling Lasers

Before becoming involved in laser work, learn all you can about the particular laser and setup you will be using. Consult with your advisor or other trained and experienced persons who have used the particular laser system that you will be using. Remember, though, that even experienced users can become lax in handling lasers since long-term experience with any potential hazard (including lasers) can affect judgment about the real risks of the hazard. You must make every effort to avoid direct or reflected exposure to laser light. OSHA requires that laboratories using nonenclosed lasers have appropriate signage and labels. Be sure to look for these signs in labs where lasers might be used.

Lasers should be built or enclosed in protective housing. Class 3B and Class 4 lasers should have key switches and safety interlocks to prevent opening without shutting down the beam. You should never bypass, disable, or circumvent safety interlocks. Lasers are assembled on horizontal optical benches to reduce the opportunities for beams to stray in nonhorizontal directions that might result in eye contact. However, this is not a foolproof solution since inadvertent bumping of a mirror might result in a beam being projected in an unintended direction. The more complex a setup the more likely this is to happen. Never allow your eyes within the horizontal plane of the optical bench. Minimize your chances of exposure by wearing laser safety goggles.

Never shine a laser on a reflective surface. You must seek to prevent using objects that might reflect a beam inadvertently into the eyes. You should not wear jewelry or watches that could result in inadvertent scattering of a beam. Tools should have "matted" (rough, nonreflective) surfaces to prevent mirror-like ("specular") reflections.

Many lasers have high-voltage power supplies, some ranging into thousands of volts, so electrical safety is a critical and substantial consideration in working with lasers. See Incident 7.3.3.2 as an example of a near-fatal electrocution. Make sure your advisor or an experienced person "walks you through" the electrical components of the system. Good grounding of all electrical equipment including laser power supplies is a requirement for safety. Additionally, many lasers use water cooling and leaks combined with these power sources make for potential hazard situations. Installing the optical bench above the floor (at least 10 inches) is an important safety measure in case of flooding.

Many larger institutions have a Laser Safety Officer (LSO) who has been assigned the duty of reviewing the safety measures for using lasers in that institution. If your institution has an LSO, you should establish a working relationship with this person and learn as much as possible about safety measures. The LSO can assist you in the proper selection of laser safety glasses or goggles.

OSHA does not have specific standards regulating the use of lasers, but rather cites the OSHA General Duty Clause (see Section 3.3.1) that covers all workplace hazards and also cites the personal protective equipment regulations as being applicable (29 CFR 1910.132, 1910.133).[6] OSHA has issued the publication *Guidelines for Laser Safety and Hazard Assessment* for guidance.[7] Additionally, the American National Standards Institute (ANSI) has issued standards for the safe use of lasers (Z136.1—2000), including a standard especially for educational institutions (Z136.5—2000). The International Electrotechnical Commission established a revised standard for lasers (IEC 60825-1) in 2007.[3]

You may also encounter another kind of laser in your classroom in the form of the laser pointer. Learn more about the hazards of these smaller lasers in *Special Topic 7.3.3.2* Laser Pointers.

SPECIAL TOPIC *7.3.3.2*

LASER POINTERS

Laser pointers are commonly used in many presentations today and they are likely used in your educational institution. Most laser pointers are Class 2 lasers and their power is very low (<1 mW) but there are also some Class 3R pointers ranging up to 5 mW in power. Battery-operated red laser-diode pointers are the most common and least expensive lasers emitting at 630–680 nm. Lasers also come in other colors including green, yellow-orange, and blue. Our eyes respond differently to varied wavelengths and lasers in the green region (520–570 nm) appear to be bright due to greater sensitivity of the eye in low-light situations.

Laser pointers should be labeled with power and wavelength. Unlabeled laser pointers may not be safe and caution should be exercised in purchasing or using these pointers. Class 2 lasers should be labeled with the word "caution." More powerful laser pointers may be labeled with "danger." You should only use Class 2 or 3R laser pointers that have less than 5-mW output and operate in the red region (630–680 nm).

There is very little hazard from Class 2 lasers or even Class 3R pointers. However, you should know that there are no restrictions on the sale of laser pointers in the United States and more powerful laser pointers exceeding the MPE and having power in excess of 5 mW may be found as imports from other countries. The Food and Drug Administration (FDA) has issued a warning that laser pointers should not be considered toys and should not be used by children without supervision. Short exposure to these more powerful laser pointers can result in temporary flash blindness.

There have been no reports of eye injuries from Class 2 lasers but there is one report of a child staring into a 5-mW laser beam at close range for 10 seconds and suffering a blind spot for 3 months.[8]

When using or handling laser pointers, it is prudent to follow these recommendations:

- Never look directly into the beam of a laser pointer.
- Never aim a laser pointer at a person.
- Never aim a laser pointer at a reflective surface.
- Never view a laser pointer through a magnifying lens.
- Never allow a child to use a laser pointer without supervision.

Laser Safety: RAMP

When working with lasers, the main issues to identify are the wavelength and power of a laser since these two factors will guide the necessity for and selection of safety equipment and procedures. Since eye damage is the most likely hazard when working with lasers there is likely little immediate first aid procedures to use if an exposure occurs. Labs should have a plan to transport someone to an ophthalmologist or emergency room.

- *Recognize* and understand the hazards of lasers you will be using.
- *Assess* the risks posed by lasers in the laboratory.
- *Minimize* the risks of using lasers by actively taking safety measures to prevent direct and indirect beam exposure to the eye and by wearing appropriate eye protection, learning about the hazards of lasers and their electrical components, and obeying signs and labels pertaining to laser use.
- *Prepare* for emergencies with lasers by having a response plan for eye exposure.

References

1. HAN SOLO, played by HARRISON FORD, *Star Wars*, George Lucas, Director, Lucasfilm, 1977; available at http://www.imdb.com/title/tt0076759/quotes (accessed January 4, 2009).
2. Public Affairs Office. Findings, recommendations of laser investigation team presented at briefing, November 18, 2004; available at http://www.lanl.gov/news/index.php/fuseaction/home.story/story_id/5896 (accessed January 4, 2009).
3. American Industrial Hygiene Association's Laboratory Safety and Health Committee Incident Reports. Electrical Shock from Laser Power Supply; available at http://www2.umdnj.edu/eohssweb/aiha/accidents/electrical.htm#Power (accessed January 5, 2009).
4. International Electrotechnical Commission. IEC 60825-1 (2007) Table of Contents (only); available at http://webstore.iec.ch/webstore/webstore.nsf/Standards/IEC%2060825-1?openDocument (accessed January 23, 2009). American National Standards Institute. ANSI Z136.1 Table of Contents; available at http://webstore.ansi.org/RecordDetail.aspx?sku=ANSI+Z136.1-2007 (accessed February 2, 2009).
5. Rockwell Laser Industries. Laser Accident Database; available at at http://www.rli.com/resources/accident.aspx (accessed January 27, 2009).
6. OSHA. Laser Hazards Standards; available at http://www.osha.gov/SLTC/laserhazards/standards.html (accessed January 23, 2009).
7. OSHA. *Guidelines for Laser Safety and Assessment —STD 01-05-001 —Pub 8-1.7*; available at http://www.osha.gov/pls/oshaweb/owadisp.show_document?p_table=DIRECTIVES&p_id=1705 (accessed January 23, 2009).
8. M. A. MAINSTER, B. E. STUCK, and J. BROWN, Jr. Assessment of alleged retinal laser injuries. *Archives of Ophthalmology* **122**: 1210–1217 (2004).

QUESTIONS

1. MPE stands for

   (a) Multiple pulse exposure
   (b) Minimum permissible exposure
   (c) Maximum permissible exposure
   (d) Maximum photon emission

2. The old and new classification systems for lasers use, respectively,

   (a) Roman numerals and Arabic numbers
   (b) Arabic numbers and Roman numerals
   (c) Roman numerals (for both)
   (d) Arabic numbers (for both)

3. Eye damage is most likely from beam exposure to laser class

   (a) 1 and 2
   (b) 3 and 4
   (c) 3B and 4
   (d) Only 4

4. Which features influence the choice of eye protection against laser light?

   (a) Wavelength and power
   (b) Wavelength and duration (pulse or CW)
   (c) Power and duration (pulse or CW)
   (d) Wavelength, power, and duration (pulse or CW)

5. The majority of eye safety incidents in laser labs are caused by

   (a) Not wearing any eye protection
   (b) Wearing the wrong eye protection
   (c) Using the laser beam inappropriately
   (d) None of the above

6. When using lasers in an environment where these is a splash hazard, the best eye protection is

   (a) Chemical splash goggles with lenses that protect from exposure to laser light
   (b) Glasses that protect from exposure to laser light plus a face shield
   (c) Alternating between splash goggles and laser glasses, as necessary
   (d) Either (a) or (b)

7. Incidents and eye damage from using lasers can arise from

   (a) Overriding key switches and safety interlocks
   (b) Stray, unexpected reflections of the laser beam
   (c) Failing to minimize the presence of shiny surfaces
   (d) All of the above

8. Lasers often have

   (a) Low-voltage power supplies that are nonetheless dangerous due to the large current
   (b) High-voltage power supplies
   (c) Cooling mechanisms that use flowing water
   (d) Both (b) and (c)

7.3.4

BIOLOGICAL SAFETY CABINETS

Preview This section describes biological safety cabinets and methods for containment used in handling infectious biological microorganisms in the laboratory.

Infectious diseases introduced with Europeans, like smallpox and measles, spread from one Indian tribe to another, far in advance of Europeans themselves, and killed an estimated 95% of the New World's Indian population.

Jared Diamond[1]

INCIDENT 7.3.4.1   MENINGITIS IN THE LAB[2]

A microbiologist working in a laboratory handled specimens from patients with infectious diseases. He developed symptoms including malaise, fever, and myalgia and went to a local emergency room. He was given antibiotics and discharged, but he returned the next day with tachycardia and was hypotensive. He died three hours later. Investigation revealed that he had handled and cultured samples from a patient who had *Neisseria meningitides*, the leading cause of bacterial meningitis. Analyses of specimens from this microbiologist revealed that he had become infected from handling the infectious samples. It was learned that he did this work on an open bench and did not use a biological safety cabinet, eye protection, or a mask in his work on these samples.

What lessons can be learned from this incident?

Containment—An Essential in Handling Infectious Agents

The hazards of microbiological infectious agents are different from most chemicals in that it is possible to contract disease from exposure to just a few microorganisms. With the overlap between biochemistry and molecular biology, it is possible that you could encounter some of these microorganisms in an advanced chemistry or biology laboratory.

Biosafety is the term that describes measures to safely handle microbiological agents in the laboratory. The emphasis in biosafety is the same as in chemical safety and laboratory safety: prevent exposure. When handling microbiological agents, the biosafety community makes every effort at "containment" so that exposure does not occur. Aerosols are produced by many laboratory operations and these aerosols can provide significant opportunities for exposure if not contained.

Fomites are the other primary concern in microbiological laboratories. These are inanimate objects, substances, or surfaces that can serve as a point of exposure to an infectious organism from one person to another. A fomite might become contaminated with an infectious agent that has a significant half-life on an environmental surface, and if a person contacts that fomite the infectious organisms can be transferred to him/her at the point of contact. If it is your naked hand, you might unknowingly touch your mouth or nose and transfer this organism to a place where it can be ingested or inhaled, and this could lead to an infection. This should help you understand the importance of hand washing not only in laboratories using microbiological materials, but also when you use the restroom.

Many organisms do not live for long in the environment, but there are some that do and these can present opportunities for exposure. Routine surface decontamination is an integral part of all processes

Laboratory Safety for Chemistry Students, by Robert H. Hill, Jr. and David C. Finster
Copyright © 2010 John Wiley & Sons, Inc.

(a)

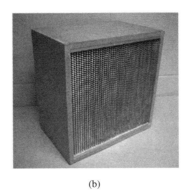

(b)

FIGURE 7.3.4.1 Components of a HEPA Filter. (a) Borosilicate glass woven filter with aluminum separators (small folds) and (b) HEPA filter mounted in a wooden box. (From Figure 1 in CDC/NIH, *Primary Containment for Biohazards: Selection, Installation, and use of Biological Safety Cabinets*, 3rd edition, 2007; available at http://www.cdc.gov/od/ohs/biosfty/primary_containment_for_biohazards.pdf.)

involving the handling of infectious materials. Dilute bleach (1:100 dilution of bleach) or alcohol (ethanol or isopropanol) are routinely used to wipe down surfaces (see *Chemical Connection 7.3.4.1 Decontamination Using Chemicals*).

Just as the chemical hood is the primary device for controlling and minimizing exposure to chemicals, the biological safety cabinet (BSC) is the primary device used to prevent exposure to microbiological agents. There are several types of BSCs and the selection of the appropriate BSC is dependent on what microbiological agent is being handled, what procedures are to be used, and if volatile chemicals will be used. We will briefly describe each below. For more detailed information about BSCs, you should consult an excellent reference that describes these devices including their varied purposes and uses, how they are installed and work, and how they can be tested and evaluated.[3]

HEPA Filters Trap Microorganisms

High efficiency particulate air (HEPA) filters were developed by the military in the 1940s and 1950s. They were designed to remove 99.97% of 0.3-μm particles from air and they also remove larger and smaller particles with better efficiency. Viruses and bacteria are removed by HEPA filtration and these filters are a critical part of the design of BSCs.

HEPA (pronounced "hep-uh") filters are made of borosilicate glass fibers woven into a single sheet that has been treated with a binder that makes it water-repellent (see Figure 7.3.4.1). The HEPA sheet is folded into pleats, which increases the area of the filter. The pleats are divided by aluminum separators that prevent the filter from collapsing during operation. The filter is placed in a frame made of metal, wood, or plastic (see Figure 7.3.4.1). HEPA filters should always be handled with care since rough treatment or dropping could result in small tears and holes that would damage the efficiency of the filter. To check for this kind of damage, HEPA filters must be tested once they are put in place. Because HEPA filters remove particulates from the air, it is important to minimize large-scale production of dusts or other aerosols that will be trapped and significantly reduce the life of a filter.

Once a HEPA filter has been used in a biological safety cabinet it has been potentially contaminated with microorganisms and special care must be taken when it is removed so that the personnel carrying out this operation are not exposed to these contaminants. There are recommended procedures for bagging out "dirty" HEPA filters.

HEPA filters can be used in two ways with a BSC. All BSCs except Class I (see below) have HEPA filters that remove particles from incoming air to provide "clean" air that protects the microbiological materials being handled within the BSC. A second HEPA is used to remove microbiological materials from the exhaust.

BSC Classes

BSCs are specially designed for handling microbiological agents or infectious materials and are classified according to design and setup. A comparison of the various classes of BSCs is presented in Table 7.3.4.1.

7.3.4 BIOLOGICAL SAFETY CABINETS

TABLE 7.3.4.1 Biological Safety Cabinet Classifications[3]

Class	Protects[a]	HEPA Filter[b]	Recirculated[c]	Face velocity (ft/min)	Ducted[d]	Chemicals[e]	Notes[f]
I	P, E	Out	No	75	No	NV only	
I	P, E	Out	No	75	Yes, H	NV, V	1
II, A1	P, Pr, E	In, Out	Yes, 70%	75	No	NV only	2
II, A1	P, Pr, E	In, Out	Yes, 70%	75	Yes, C	NV only	2
II, A2	P, Pr, E	In, Out	Yes, 70%	100	No	NV only	3
II, A2	P, Pr, E	In, Out	Yes, 70%	100	Yes, C	NV, V	3
II, B1	P, Pr, E	In, Out	Yes, 30%	100	Yes, H	NV, V	
II, B2	P, Pr, E	In, Out	No	100	Yes, H	NV, V	
III	P, Pr, E	In, Out	No	N/A	Yes, H	NV, V	4

[a]Protects: P = personnel; Pr = product; E = environment.
[b]HEPA filter: In = filters incoming air; Out = filters exhaust air.
[c]Recirculated: 70% recirculated (30% exhausted); or 30% recirculated (70% exhausted).
[d]Ducted: H = hard ducted; C = canopy unit.
[e]Chemicals: NV = nonvolatile chemicals; V = volatile chemicals. Keep chemicals to minimal (termed "minute" and "small") amounts; the concentrations must be far below lower explosive limit.
[f]Notes: 1. May require special duct to outside, in-line charcoal filter, explosion-proof motor and spark-proof electrical components. 2. Air pressure in the internal plenum is positive with respect to the room. 3. Air pressure in the internal plenum is negative with respect to the room. 4. N/A = not applicable. This is a glovebox.

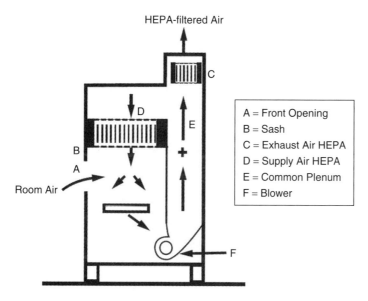

A = Front Opening
B = Sash
C = Exhaust Air HEPA
D = Supply Air HEPA
E = Common Plenum
F = Blower

FIGURE 7.3.4.2 Schematic of Class II, Type A1 Biological Safety Cabinet. Air enters the BSC in the front opening, where it is drawn through a grill in the cabinet floor by a blower (F) and into the common plenum (E). Part of this air is filtered through the supply HEPA filter (D) to provide clean air to the working surface, and the other portion of the air is exhausted through the exhaust HEPA filter (C). (From Figure 3 in CDC/NIH, *Primary Containment for Biohazards: Selection, Installation, and Use of Biological Safety Cabinets*, 3rd edition, CDC/NIH, Washington, DC, 2007; available at http://www.cdc.gov/od/ohs/biosfty/primary_containment_for_biohazards.pdf.)

We will discuss BSCs in general here, but each class of BSC has individual differences. The discussion below only briefly describes the variations in BSCs; you should consult other sources for a more complete explanation of varied types of BSCs.[3]

In general, BSCs are designed so that air drawn into the front of the hood is drawn downward by an internal exhaust motor that pulls the air through a HEPA filter (see Figure 7.3.4.2). A portion of this air is recirculated and is blown down from the top of the hood to provide a protective curtain of air to the product. A portion of the air is exhausted as it passes through a HEPA filter. The exhausted air usually is returned back into the laboratory room air. In rare instances, a BSC will exhaust air to the outside of the building though ductwork.

When handling microbiological agents, the purpose is usually to identify the organisms, purify the organisms, or "grow" them to make a larger quantity of material. It is very possible that normal air can contain microscopic organisms and if the operations to identify, purify, or propagate the microorganisms of interest are to be effective, it is important to protect these "products" from contamination. The design

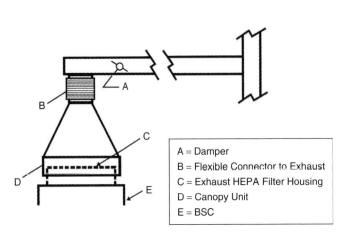

A = Damper
B = Flexible Connector to Exhaust
C = Exhaust HEPA Filter Housing
D = Canopy Unit
E = BSC

FIGURE 7.3.4.3   Canopy Unit Exhaust for Class II Type A1 BSC. The canopy unit (sometimes called the thimble) is designed to fit over the exhaust of the BSC but not tightly. There is a gap of about 1 inch between the canopy and the BSC exhaust so that air is also exhausted from the room. Canopy units are often used in small lab rooms where some exhaust ventilation is needed to remove chemical contaminants that might be in the air. (From Figure 4 in CDC/NIH, *Primary Containment for Biohazards: Selection, Installation, and Use of Biological Safety Cabinets*, 3rd edition, CDC/NIH, Washington, DC, 2007; available at http://www.cdc.gov/od/ohs/biosfty/primary_ containment_for_biohazards.pdf.)

of most BSCs incorporates incoming HEPA filters that remove these unwanted airborne organisms from normal air. It allows for protection of the personnel using the BSC. And it allows for protection of the environment from microbiological contamination as the exhaust air is HEPA filtered as it is exhausted. See the "Protects" column in Table 7.3.4.1.

Generally, the face of the BSC is set at a fixed opening and the face velocity of air entering the cabinet is 75 or 100 ft/min, depending on the class. Class III BSCs are gloveboxes and do not have an opening. The motors that operate the fans inside the BSCs are not spark-proof. This means that flammable or combustible chemicals cannot safely be used in BSCs due to the possibility of fire or explosion. And volatile chemicals cannot be used if the BSC is exhausted back into the laboratory room.

BSCs are designed to handle microbiological agents and are not designed to handle chemicals. It is possible to use small amounts of nonvolatile chemicals. You will see that some classes allow minimal amounts of volatile chemicals and this is only when they are ducted. The CDC/NIH BSC book[3] calls these minimal amounts "minute" or "small" but it is not clear what these terms mean. However, the use of flammable or volatile chemicals in BSCs may present a very real hazard from fire or explosion if concentrations are not carefully controlled. Concentrations of flammable chemicals must never be allowed to approach the lower explosion limit. Procedures involving other than "minute" or "small" amounts of flammable or volatile chemicals should be carried out in a chemical hood, not a BSC.

Some BSCs are exhausted through canopy units (also called thimble units). Figure 7.3.4.3 shows a canopy unit. This unit is not directly connected to the BSC but rather is placed in close proximity to the exhaust. There is about an inch of space between the BSC exhaust pipe and the inside lower lip of the canopy unit. The canopy unit serves two purposes: it removes any volatile chemical vapors that might be exhausted from the BSC or in the laboratory and it provides a source of ventilation for the laboratory room especially when the BSC is off, thus preserving the directional airflow into the laboratory.

Using BSCs

The proper use of BSCs requires training by an experienced user of BSCs or a formal training course. We provide minimal details about the use of BSCs to give you some understanding of the process, but this is not a substitute for specific BSC training. Using a BSC requires a number of specific operations that require prudent precautions.

Most BSCs are fitted with a gauge that measures the pressure drop across the HEPA filter. There is no set value for this—it depends on the particular BSC—but when you use a BSC you should note the position of the gauge and watch for changes relative to this position. You should check this each time you use the BSC and monitor the gauge to look for abnormal increases or decreases in pressure. Increased pressure drop could mean that the filter is becoming full or is clogged, and it needs to be replaced. A sudden decrease in pressure could mean that the filter has broken down and there is a hole that has resulted in a pressure loss. This requires attention since it could mean that the BSC is not protecting you.

BSCs make use of laminar flow—the smooth flow of air in layers down and across the front of the BSC. Due to the nature of the laminar flow in a BSC, it is important to minimize disruptions to this airflow. Getting everything you need beforehand and placing it in the BSC is a good practice. Clean materials must be kept separated from "dirty" materials. Minimizing arm movements within the BSC is important during operations. You will need to move your arms straight in and out, rather than moving them side to side since this latter movement causes disturbances in the airflow.

The biosafety level (BSL) will determine the kind of personal protective equipment needed.[4] Class I and Class II BSCs can be used at BSL 1 to BSL 3. (See Section 4.3.3 for more information about BSLs.) Specially designed Class II BSCs can be used at BSL 4 and utilize air-supplied protective suits. Class III BSCs can be used for BSL 4 work without suits.

The grille across the front floor of the BSC must not be blocked by equipment or supplies, or by the operator. You must raise your arms above the grill so that air sweeps around your arms rather than allowing your arms to lie upon the grille. As in the chemical hood, you must work as far as possible inside the hood, but at least 4 inches further in from the front grille. Equipment and materials inside the BSC can disturb the airflow and it is important to keep these to the minimum of what is necessary for a particular operation only. Equipment that can generate aerosols should be as far back in the BSC as possible so that these aerosols are captured by the hood. If aerosols are near the front of the cabinet they could escape. If a vacuum system is used, it must be HEPA filtered. It is important to keep your clean materials at least a foot from aerosol-producing operations.

Surfaces should be regularly decontaminated with an alcohol or bleach. Decontamination is especially important if the BSC has been shut off and is now being turned on for use. You will need to allow the BSC a few minutes to reach operational status before using it. After the daily use of the BSC, all surfaces should be decontaminated.

Plastic-backed absorbent paper should be used on the working surface behind the grille for operations, but care must be taken not to cover the grille. This will contain spills and minimize the production of aerosols should drops of liquid hit the surface. It also reduces cleanup time.

A discard pan inside the BSC is often used for waste materials. Some BSC users put their waste inside biohazard bags positioned inside the cabinet rather than in a discard pan. Decontamination of waste or contaminated glassware can be by chemical treatment or by steam heating in an autoclave. If chemical decontamination is used, the disinfection solution should be added to the discard pan at the beginning of the operation. For autoclaving contaminated materials should be placed in the discard pan or into biohazard bags that can be transferred to an autoclave. The biohazard bags or discard pans must have enough water in them to generate steam that sterilizes the contents upon autoclaving. Bags should be taped closed and will need to be placed into a pan or tray to trap any leakage.

Maintaining and Certifying BSCs

BSCs are certified by testing upon installation and they must be recertified at least once a year or after repairs or maintenance. BSCs should have a sticker indicating the date of the last certification. BSCs generally require more maintenance than chemical hoods because they have motors, moving parts, and filters. Carrying out maintenance is not a trivial matter since a BSC must first be sterilized to protect maintenance workers. This is usually accomplished by using formaldehyde, hydrogen peroxide, or chlorine dioxide gases or vapors within the hood. (See *Chemical Connection 7.3.4.1* Decontamination Using Chemicals.)

CHEMICAL CONNECTION *7.3.4.1*

DECONTAMINATION USING CHEMICALS

The use of microbiological agents can create an opportunity for contamination of surfaces, equipment, and materials that could be sources of exposure. Microbiologists frequently use chemicals to decontaminate these surfaces, equipment, and materials.

Surface decontamination of BSCs can be carried out using 70–85% ethanol or isopropanol.[6] These alcohols are effective against most microbiological organisms except for spores or nonlipid viruses. Concentrations of these alcohols below 50% are not effective. It is important to limit the use of these alcohols in BSCs due to their flammability and the fact that components of the BSCs are not spark-proof.

Bleach is often used to decontaminate surfaces. Ordinary household bleach, containing about 5% sodium hypochlorite, is diluted 1:100 to produce a solution of about 500 ppm. This solution is effective against all microbiological organisms, including spores. However, it is corrosive to some surfaces. If it is used in a BSC, it must be followed by a rinse with sterile water to ensure that the stainless steel surfaces do not corrode. Sometimes a detergent is combined with the dilute bleach solution to make a very effective disinfectant.

Ethylene oxide (ETO) is a gas and it is widely used as a chemical sterilant in medical facilities and hospitals. Special ETO decontamination equipment is available commercially. It is effective for microbiological organisms, but its effectiveness depends on temperature, concentration, exposure time, and humidity. Often it can penetrate packaging. Materials treated with ETO must sit 24 hours to allow excess ETO to dissipate.

Formaldehyde has frequently been used in the past for decontamination of BSCs as well as entire laboratories. Because of its toxic properties and regulations associated with it use, the use of formaldehyde has declined. Generally, formaldehyde is generated by heating paraformaldehyde (e.g., in an electric fry pan) within the BSC or within the laboratory. The decontamination is carried out for at least 8 hours. The BSC or laboratory must be allowed to ventilate or air out so that the formaldehyde is dissipated.

Hydrogen peroxide vapor (HPV) is used for decontamination of BSCs. It is effective against a wide range of bacteria, viruses, and fungi. Equipment such as computers, complex instruments, and electronic parts are *not* adversely affected by HPV and there is no residue from its use. Generation of HPV requires special equipment and careful placement of units to ensure adequate decontamination. However, it is an environmentally friendly disinfectant since its products are oxygen and water.

Chlorine dioxide is used for laboratory decontamination. It was also widely used in decontamination of facilities contaminated with anthrax. It comes in the form of wipes, foams, and burstable sachets, as well as the gaseous form. It is effective against a wide range of bacteria, viruses, and fungi.

The selection of a decontamination agent is dependent on several factors—and as you might guess cost is one of these. For routine microbiological work in a BSC, alcohols are frequently used to avoid the corrosive nature of bleach. However, some organisms may only be killed by bleach and so it may be utilized. Bleach is also frequently used to decontaminate ordinary laboratory countertops, since it is cheap and can be made quickly and easily. Chlorine dioxide is more expensive, but its broad effectiveness and the introduction of wipes, foams, and sachets have made it more accessible.

For sterilizing small pieces of equipment, ethylene oxide is widely used—but it requires the purchase of special equipment. HPV also requires special equipment, but this can be used for decontaminating entire laboratories and BSCs. Formaldehyde is still used by many because it does not require special equipment and is relatively cheap—nevertheless, regulations may limit the use of this technique.

NSF/ANSI Standard 49–2008[5] sets specifications for the design, performance, and testing of Class II BSCs. BSCs meeting these requirements are labeled with the NSF logo. BSCs should be tested at least annually by a professional certified in BSC testing. There are courses available that provide appropriate training and certification. It is prudent to allow only a certified professional to test your BSC.

Testing of BSCs involves evaluation of the inflow velocity of air, the downflow velocity of air within the cabinet, and airflow smoke patterns, a HEPA filter leak test, a cabinet leak test, and an electrical connection integrity test (must meet Underwriter's Laboratory requirements), as well as other tests for lighting, vibration, and noise.

RAMP

- *Recognize* that microbiological organisms can present significant hazards.
- *Assess* the potential for exposure to these organisms.
- *Minimize* exposure to microbiological agents by using biological safety cabinets.
- *Prepare* for emergencies that might occur from handling microbiological agents.

References

1. JARED DIAMOND. BrainyQuote; available at http://www.brainyquote. com/quotes/authors/j/jared_diamond.html (accessed January 9, 2009).
2. Centers for Disease Control and Prevention. Laboratory-acquired meningococcal disease—United States, 2000. *Morbidity and Mortality Weekly Report* 51 (7): 141–144 (February 22, 2002); available at http://www.cdc.gov/mmwr/preview/mmwrhtml/mm5107a1.htm (accessed January 13, 2009).
3. CDC/NIH. *Primary Containment for Biohazards: Selection, Installation, and Use of Biological Safety Cabinets*, 3rd edition, CDC/NIH, Washington, DC, 2007; available at: http://www.cdc.gov/od/ohs/ (accessed January 9, 2009).
4. CDC/NIH. *Biosafety in Microbiological and Biomedical Laboratories*, 5th edition, CDC/NIH, Washington, DC, 2008; available at http://www.cdc.gov/od/ohs/ (accessed January 13, 2009).
5. NSF International (NSF). *American National Standards 8*. Institute (ANSI). NSF/ANSI Standard 49–2004. Class II (laminar flow) biosafety cabinetry. NSF/ANSI, Ann Arbor, MI, 2008.
6. University of North Carolina. Chapter 15. Decontamination and Disposal, August 2000; available at http://ehs.unc.edu/manuals/bsm/BSM15.pdf (accessed January 14, 2009).

QUESTIONS

1. What is the most important laboratory method by which exposure to microbiological agents is prevented?
   (a) Wearing masks
   (b) Wearing gloves
   (c) Containment
   (d) Incineration of waste

2. Fomites are
   (a) Inanimate objects or surfaces
   (b) Small microbiological organisms
   (c) Agents used to decontaminate surfaces
   (d) Filters used in BSCs

3. HEPA stands for
   (a) High energy particle accelerator
   (b) High efficiency particulate air
   (c) High exhaust product access
   (d) High efficiency plastic article

4. HEPA filters are
   (a) Used to clean the air entering a BSC
   (b) Used to clean the exhaust from a BSC
   (c) Commonly found in face masks
   (d) (a) and/or (b)

5. BSCs are designed to protect personnel and
   (a) The lab environment
   (b) The contents of the BSC
   (c) The lab environment, but not always the contents of the BSC
   (d) Either (a) or (c)

6. In BSCs, the air is circulated so that
   (a) Some of the air returns to the BSC
   (b) Some of the air is exhausted to the room
   (c) Some of the air is exhausted from the laboratory
   (d) (a) and either (b) or (c)

7. BSCs can typically
   (a) Handle most reagents that are used in chemical hoods
   (b) Handle only very small amounts of volatile chemicals
   (c) Be used as a substitute for a chemical hood
   (d) Handle flammable reagents as long as the LEL is exceeded

8. Face velocities for BSCs should be
   (a) Well below 75 ft/min to avoid turbulence in the cabinet
   (b) Well below 75 ft/min to avoid back-pressures from the HEPA filter
   (c) 75 or 100 ft/min
   (d) Always above 100 ft/min so that no air can escape back into the lab

9. A canopy unit in a BSC
   (a) Helps provide laboratory exhaust when the BSC is off
   (b) Is placed over, but not directly connected to, the BSC
   (c) Removes chemical vapors from the BSC
   (d) All of the above

10. Which class of BSC is a glovebox?
    (a) I
    (b) II
    (c) III
    (d) II and III

11. The gauge that measures the pressure drop across the HEPA filter
    (a) Can be used to detect a hole in the filter
    (b) Can be used to detect a clogged filter
    (c) Both (a) and (b)
    (d) Should be disabled when using chemicals in the BSC

12. Laminar flow describes the
    (a) Smooth flow of air in layers down and across the front of the BSC
    (b) Splitting of the exhaust air between recirculated and nonrecirculated paths
    (c) Smooth flow of air across the laminated surface of the BSC
    (d) Flow of air through the HEPA filter

13. Good practice inside BSCs involves
    (a) Keeping all materials and equipment at least 4 inches from the front, back, and sides of the unit
    (b) Keeping your arms on the grille when working
    (c) Generating aerosols, if necessary to do so, near the back of the unit
    (d) Never using a vacuum system inside a BSC

14. Good practice when using BSCs involves
    (a) Decontamination at least once a week

(b) Using plastic-backed absorbent paper that drapes across the grille

(c) Collecting and discarding, or decontaminating, waste carefully

(d) All of the above

15. BSCs should be

(a) Recertified annually

(b) Recertified after repair or maintenance

(c) Sterilized after, but not before, maintenance

(d) Both (a) and (b)

7.3.5

PROTECTIVE CLOTHING AND RESPIRATORS

Preview   The section discusses personal protective measures that might be employed in advanced chemical laboratory operations.

Experience teaches only the teachable.

Aldous Huxley, novelist[1]

INCIDENT 7.3.5.1   *t*-BUTYLLITHIUM FIRE AND FATALITY[2]

A research chemist was carrying out a procedure using *t*-butyllithium in pentane. She was taught a technique for handling this pyrophoric chemical using a special bottle with a septum that could be pressurized with an inert, dry gas. This allowed a sample to be withdrawn with a syringe. As she attempted to withdraw an aliquot from the bottle, the syringe's plunger popped out of the barrel, and the *t*-butyllithium and pentane splashed on her, igniting immediately. She did not have on a lab coat. The nitrile gloves and synthetic sweater she was wearing caught fire. Although there was an emergency shower nearby, she ran away from it. A colleague in the same laboratory smothered the flames using his laboratory coat and emergency personnel were called. She received second and third degree burns over 40% of her body and died a few weeks later.

INCIDENT 7.3.5.2   MISUSE OF A RESPIRATOR[3]

A research chemist transferred a solvent, methyl *t*-butyl ether, from three 4-L bottles to a large carboy in a chemical hood, an operation that required about 2 minutes. He decided to use a full face mask air-purifying respirator for added protection during this operation. That evening he began to experience a scratchy throat and cough that developed into a respiratory infection. His doctor suspected chemical exposure as the cause. Investigators found that the chemical hood was working properly and that earlier assessments showed no exposure occurring during these operations. Investigators found that the researcher had no recent fit-testing and training in the use of respirators. He had been using this respirator intermittently over the past 6 or 8 years. The respirator filter cartridge selected and used was not effective for solvents. It had been stored in the open air for years next to stored chemicals and among them was a sensitizer. It was surmised that the respirator and cartridge over the years could have absorbed chemicals and fungi due to improper storage. The researcher was diagnosed with allergic tracheobronchitis which was believed to be related to repeated chemical exposures in his laboratory.

What lessons can be learned from these incidents?

Protective Clothing for the Laboratory

Laboratory coats are made to protect you and your clothes against contamination, splashes, and potential exposures. Always put on your laboratory coat when you enter the laboratory, handle hazardous materials, or are involved in hazardous laboratory operations. In the event of a spill or splash, it is easier to remove a lab coat than have to remove your clothing. Keeping them buttoned and the sleeves down provides you with a good degree of protection. You should never wear lab coats in public places or

---

eating places. This may be common practice in hospitals among medical staff, but it is a poor practice to wear potentially contaminated clothing into a public forum. They should be laundered at work and not at home.

Laboratory coats provide a protective layer of clothing between you and the hazard. If you are using flammable, pyrophoric, or other reactive chemicals, you should use fire-resistant lab coats. While lab coats made of 100% cotton or polyester/cotton blends offer some protection against splashes, spills, or contamination, these coats have little or no fire resistance. Laboratory coats made of Nomex®, Indura®, or Excel ® are principally made of 100% cotton combined with fire retardant threads and are fire resistant. (Alas, however, they don't come in white.) Washing does not remove their fire resistance. The importance of wearing a fire-resistant lab coat for some laboratory operations is illustrated by Incident 7.3.5.1.

When working in the laboratory you should also consider the potential flammability of your own personal clothing. Many synthetic materials are flammable and once ignited burn rapidly, increasing your chances of serious burns. If you are using flammables, pyrophorics, or highly reactive materials, you may want to select cotton clothing over synthetic materials.

Lab suppliers offer a variety of lab coats and suits with some chemical-resistant properties. These products are made of synthetic polymers, like polyethylene or polypropylene, and some are coated with a material that helps them repel chemicals. These lab coats come under a variety of trademark names, such as Tyvek®, Proshield®, NexGen™, Posiwear®, Qorpak®, and Suprel®. Descriptions of their protective value vary and may include terms like chemically and biologically inert; nontoxic liquids, sprays, or aerosols; light splashes; liquid-proof barrier; water-based chemicals; fluid repellent; and chemical resistant. While these coats and suits provide protection against some chemicals and small splashes, it is clear from the descriptions that their chemical-resistant properties are limited and these may be permeable to many chemicals. Consider these synthetic lab coats as you would protective gloves; there is no universal protective lab coat that protects you against all chemicals. You must seek out the best information about these based on your use and the need to minimize exposures.

Laboratory aprons are often underutilized for protection. They can prevent chemical exposure to the front of your body and should be considered for use over your lab coat. Rubber aprons have been used for many years and are thick enough to offer considerable protection against spills and splashes of many chemicals. Lab aprons made of the same kind of material as chemical-resistant lab coats are also available.

Respirators — The Last Resort

Respirators, a form of personal protective equipment (PPE), can be used to avoid breathing contaminated air. Respirators are more commonly used in industrial and factory environments where contaminated atmospheres may be present or airborne contaminants cannot be contained by engineering controls. However, the use of respirators in laboratories should be exceedingly rare since their use implies uncontrolled hazardous substances in the laboratory atmosphere. Almost all laboratory chemical operations involving release of volatiles or aerosols can be managed by the appropriate use of a laboratory chemical hood, glovebox, local exhaust ventilation, or other means. There should be few instances where respirators are needed. *Respirators are the last possible resort as protective devices* when chemical hoods or other containment devices cannot be used. An exception to this is during the cleanup of chemical spill, an emergency operation most often carried out by experienced personnel.

Before you wear a respirator you must ensure that you (1) have selected the correct respirator, (2) are medically fit to wear a respirator, (3) are properly fitted for a respirator, and (4) have been trained in wearing the respirator.

There are three categories of respirator PPE: simple masks, systems with cartridges, and self-contained breathing apparatus.

Simple disposable masks (Figure 7.3.5.1) are those that are often worn by health-care personnel, particularly in "emergency" environments such as ambulances and hospital emergency rooms. These masks are designated with letters (N = not resistant to oil, P = resistant to oil) and numbers that indicate the percent efficiency of the filter. Common values are 95 (95%), 99 (99%), and 100 (= 99.97%). Most

FIGURE 7.3.5.1   Disposable Mask Respirator. These disposable masks are easy to use but do not provide a tight seal against the face. Hospital personnel often wear these kinds of masks to protect the patient; they are not used to protect health care personnel (such as in surgery). These masks should rarely be considered for use in a laboratory. (Copyright 3M, St. Paul, Minnesota.)

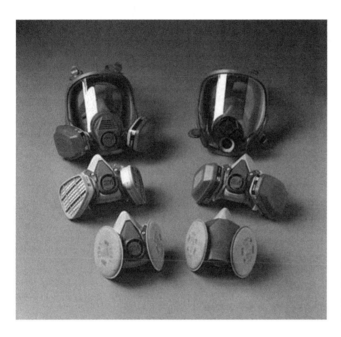

FIGURE 7.3.5.2   Cartridge Respirators. As with disposable masks, these respirators should rarely be considered for use in a laboratory. (Copyright 3M, St. Paul, Minnesota.)

health-care masks are N95 masks. Since these devices are particulate filters they can protect against aerosols and dusts, but not gases. They should generally not be considered for use in labs. They do not provide a "tight seal" against the face and leakage around the edges is common.

Cartridge respirators (Figure 7.3.5.2) must provide a tight seal around the mouth and nose. This forces all inhaled air through a cartridge filled with absorbent material. "Half-face" models cover the mouth and nose and "full face" models cover the entire face. There are several kinds of cartridges and each is designed for specific uses. OSHA has a color code[4] indicating the chemicals effectively filtered by each cartridge (see Table 7.3.5.1). While these devices can be purchased in hardware stores for home use, occupational settings require an elaborate OSHA-mandated Respiratory Protection Plan with fit testing procedures to ensure a proper face seal. Thus, only trained, authorized personnel should use these devices. While available cartridges can filter several kinds of gases, these are not recommended for lab use since their use infers a laboratory has a hazardous and poor-quality atmosphere. They are more commonly used in many industrial settings.

Finally, the self-contained breathing apparatus (SCBA) is the device most commonly used by firefighters (Figure 7.3.5.3) or during HAZWOPER (hazardous waste operations and emergency responders) operations. SCBA provides, as the name implies, a supply of clean air from air tanks carried on the

TABLE 7.3.5.1   Color Coding for Selected Respirator Cartridges and Canisters[4]

Chemicals	Cartridge/canister color code
Acid gases	White
Hydrocyanic acid (hydrogen cyanide)	White with 1/2-inch green stripe completely around cartridge near bottom
Chlorine	White with 1/2-inch yellow stripe completely around cartridge near bottom
Organic vapors	Black
Ammonia	Green
Carbon monoxide	Blue
Acid gases and organic vapors	Yellow
Acid gases, organic vapors, ammonia	Brown
Radioactive materials, except tritium or noble gases	Purple (magenta)
Particulates—P100	Purple
Particulates—P95, P99, R95, R99, R100	Orange
Particulates free of oil—N95, N99, N100	Teal

FIGURE 7.3.5.3   Self-Contained Breathing Apparatus (SCBA) Being Used by Firefighters Outside a Residential Fire. Chemists would not wear SCBA. These devices are used only by emergency personnel or someone doing routine work in a confined space. The arrow points out author David Finster, who is also a volunteer firefighter.

back of the responder. Considerable training is required to use an SCBA. Only in specialized industrial settings are they likely to be used. Few chemists would ever use an SCBA during their career as scientists.

Before Using a Respirator

Respirators *require* medical evaluation, proper fit-testing, and training.[4] You should never just pick up or select to use a respirator unless you have been properly fitted and trained. One of the common failures of respirators is a poor fit that allows the user to have a false sense of security when in fact the respirator is not providing protection. Selection of the use of a respirator should be a last resort, and it is best to seek out professional assistance in selecting respirators. See Incident 7.3.5.2 about failure to be properly fitted and trained in respirator use.

Respirators would likely be needed during emergency operations or in preparation for emergency operations. For example, if an operation called for the use of hydrogen cyanide gas, a researcher might want to have the proper respirators on hand in case of an unexpected release. Self-contained breathing apparatus (SCBA) are required for oxygen-deficient atmospheres, such as might be encountered in confined spaces. *Any* respirator use requires the wearer or wearers involved in the emergency preparedness effort to have a medical evaluation, proper fit-testing, and training to be able to effectively carry out emergency operations.

RAMP

- *Recognize* lab circumstances requiring additional protective clothing or the use of a respirator.
- *Assess* needs for additional PPE or for reconsideration in performing experiments. Select the *correct* additional clothing and/or the *correct* respirator to protect against the specific hazard.
- *Minimize* exposures to chemicals by wearing and using *appropriate* PPE and ensuring safety equipment is providing the necessary protection.
- *Prepare* for emergencies by thinking about "worst case" scenarios and knowing the proper use and location of emergency equipment.

References

1. A. HUXLEY. BrainyQuote; available at http://www.brainyquote.com/quotes/authors/a/aldous_huxley.html (accessed January 29, 2009).
2. A. ANDERSON. Research assistant burned in chemical fire. *Daily Bruin*, January 1, 2009; available at http://www.dailybruin.ucla.edu/news/2009/jan/01/research-assistant-burned-chemical-fire/ (accessed January 16, 2009). Additional details reported on Division of Chemical Health and Safety List Serve [DCHAS-L@LIST.UVM.EDU] on Wednesday, January 7, 2009, 7:29P.M. by DEBBIE M. DECKER. C. MCGOUGH. Assistant dies of fire injuries. *Daily Bruin*, January 21, 2009; available at http://dailybruin.ucla.edu/news/2009/jan/21/ assistant-dies-fire-injuries/ (accessed January 23, 2009). J. N. KEMSLEY. Learning from UCLA: Details of the experiment that led to a researcher's death prompt evaluations of academic safety practices. *Chemical & Engineering News* **87**(*31*):29–31, 33–34 (August 3, 2009).
3. R. HILL. Incident related by a colleague.
4. Occupational Safety and Health Adminstration. OSHA Bulletin, General Respiratory Protection Guidance for Employers and Workers; available at http://www.osha.gov/dts/shib/respiratory_protection.pdf (accessed April 30, 2009).

QUESTIONS

1. Laboratory coats should be worn
   (a) When you wish to have an extra layer of protection in the laboratory
   (b) Only when conducting experiments that present unusual hazards
   (c) All day long, even when moving to public places in a building
   (d) Both (a) and (c)

2. Most laboratory coats are
   (a) Flame-resistant
   (b) Chemical-resistant
   (c) Make of 100% cotton
   (d) None of the above

3. Respirators are
   (a) Commonly used in laboratories since they are easy to use and require no training
   (b) Commonly used in laboratories but require training and fitting
   (c) Not commonly used in laboratories
   (d) Used only in laboratories when the HVAC system and chemical hoods are not working

4. Most simple, disposable masks
   (a) Work effectively against most chemicals
   (b) Protect against aerosols and particulates but not against gases
   (c) Provide a reasonably tight seal around the face if worn properly
   (d) Have labels on them indicating how many times they can be used before being discarded

5. Cartridge respirators
   (a) Provide a tight seal around the mouth and nose
   (b) Sometimes involve eye protection
   (c) Can only be used in an occupational setting by authorized individuals
   (d) All of the above

7.3.6

SAFETY IN THE RESEARCH LABORATORY

Preview This section presents factors to consider when you are working in a research laboratory and performing new, noncourse-related experiments.

If we knew what it was we were doing, it would not be called research, would it?

Albert Einstein[1]

INCIDENT 7.3.6.1   RUNAWAY RESEARCH REACTION[2]

A senior student was working on a research project involving a reaction between two reactants using sodium hydride. He and his professor set up the research experiment using a three-neck flask fitted with a condenser in a chemical hood. To the flask was added one of the reactants in a dry solvent and sodium hydride. While cooling the flask with an ice bath, the second reactant was added slowly to the flask but there were no indications of a reaction. They continued to add the second reactant until it was all added without an apparent indication of a reaction. After allowing the reaction mixture to warm, they observed the reaction for an additional hour and, not seeing signs of a reaction, they decided to leave for the evening. Before leaving the building they checked the reaction a last time. Now the reaction was clearly taking place and a foaming green solution was bubbling out of the top of the condenser and running on the floor of the hood. They began cooling the reaction to slow it down and after about an hour got it under control without further incident.

What lessons can be learned from this incident?

Undergraduate Research

Research, the lifeblood of science, leads us to new discoveries, new understandings, and new technologies that can improve our lives. Nevertheless, research is a blade that cuts both ways: it can be exciting, thrilling, and rewarding, but it can also be frustrating and filled with potholes, wrong turns, dead ends, and even danger. Research is doing something that has not been done before, and when we do this we don't know what the outcome will be. Based on what we know now, we can predict, plan, and prepare for what we think will happen, but we can't know what will happen until we do it.

Many undergraduate science majors take on academic-year or summertime research projects at their home institutions, at other colleges or universities, or at some chemical industry location in the form of an internship. These are valuable experiences since they allow students to experience both the joys and frustrations of cutting-edge research. Working in a research lab is often a very different experience from performing "academic" lab experiments that are associated with courses in a chemistry curriculum. Many new hazards can be encountered and the level of supervision can vary considerably in comparison to "lab courses," where experiments have been pretested, hazards are well known, and supervision is constant.

Considerations When Performing Research

First, when starting to perform some experiment that is new to you, you should make sure that some experienced person is available for consultation. In many situations, you will first run some experiments under the watchful eye of a graduate student, faculty member, or other mentor who can guide you carefully with regard to the particular techniques that you are encountering. In fact, chemical and safety information is often transmitted rather informally when working side-by-side, particularly when running experiments for which there may be little or no formal written instructions. Feeling "ignorant" at first is normal and expected, and it is best to adopt an attitude of asking lots of questions to make sure you understand what you are doing and why you are doing it. It is probable that you will be working with chemicals and equipment that you have not used before. *When in doubt, ask*.

Hopefully, the lab in which you are working has already established a strong tradition of safety and you will be advised about the hazards of various procedures and chemicals. However, this is not always true, and the culture of some labs may be less than ideal in this regard. Perhaps some hazardous procedures are conducted with such frequency (and without mishap) that safety measures are by-passed to some degree since "it always works OK." While we can attribute this to "human nature," this realization does not reduce a lurking hazard that might lead to a mishap, particularly in the hands of a novice researcher. So, perhaps more so than in course-related lab experiments, you must take on an extra measure of personal responsibility for your own safety. Asking questions, and even challenging procedures that seem unsafe, is your responsibility.

Most of the hazards that you will encounter in a research lab are discussed in this book: toxic compounds, flammable solvents, high-voltage sources, strong oxidizing or reducing agents, and so on. So, the general training in safety that this book provides will be an excellent background in staying safe in a research lab. However, since some of what you might do really is "doing something that no one else has done before," you may encounter specific hazards that no one knows about! Obviously, the synthesis of a new compound necessarily produces a substance of unknown toxicity. It is best to assume that any new compound is somewhat toxic and handle it with that hazard in mind. Or, perhaps some new reaction is highly exothermic in an unexpected way or produces an unexpected gas. It is important to anticipate "worst case scenarios" at all times and to make reasonable preparations for events that can pose exposures or other hazards. It is wise to review procedures and hazards with more experienced researchers when designing such experiments.

Safety Steps in Research

When preparing for research, learn as much as you can about the hazards of the chemicals that you will be using, and the potential products that you might expect *before you begin*. If you are using new equipment, learn about potential hazards and consider what might go wrong. Seek the proper safety equipment to protect yourself and others against potential incidents. Minimize potential for exposure and prepare yourself for emergencies that you can imagine might happen.

When working in a new lab, perhaps at a new university or industrial research lab, it is important to acquaint yourself with the safety equipment that is available. Take time to find the location of fire extinguishers, safety showers and eyewashes, spill kits, and emergency exits. Locate natural gas shut-off valves and telephones. Know the address of the building in which you are working if you have to call 911.

It is also important to think carefully about handling wastes. You likely already know that waste disposal is an important consideration with regard to safety, environmental impact, and cost. As discussed elsewhere in this book (Sections 8.1.1 and 8.3.3), "down the sink" is almost never the appropriate waste disposal procedure. Check with your research mentor about how to safely handle chemical wastes that arise from research experiments.

While few traditional "academic" lab experiments last more than a few hours, it is not uncommon to perform a reaction in a research lab that may have to run overnight or even for days. Leaving any active chemical reaction unattended poses additional risks and you should discuss this situation with your mentor. Any unattended reaction should have a sign clearly posted nearby that indicates (to a trained chemist) what reaction is occurring, when it was started, when it is due to finish, the main chemicals involved, and contact numbers in the event of question or emergency. Some of this information, but probably not all of it, would be useful to emergency personnel.

A perennial issue that arises when working in research labs is whether or not you can, or should, work alone in the lab. Unlike course-related lab experiments that are usually designed to start and end at convenient times, "real" experiments sometimes demand your presence at times other than 8 A.M. to

5 P.M. Many labs have strict rules about "working times" and some labs are more flexible. As a rule, it is not safe to be alone in a laboratory, and you should seek a nearby partner to keep each other safe.

Obviously, running a dangerous experiment when you are alone should not be allowed and certainly shouldn't be attempted. You will need to learn the local rules about working "alone." If they allow some version of being "alone" then you need to take responsibility for your own safety and decide if even "following the rules" does not provide enough of a safety factor in your work. Much of this can be situation-dependent. Entering data in a computer or taking an IR spectrum might be acceptable in some circumstances but performing almost any chemical reaction is probably not.

RAMP

Throughout this book we have placed an emphasis on RAMP. When you move into the research arena, this is especially where you will need to practice these principles.

- *Recognize* the hazards of chemicals, equipment, and procedures that are new to you.
- *Assess* the risk of exposure, particularly taking into account that you are unfamiliar with new procedures and equipment.
- *Minimize* the risk by anticipating "what can go wrong" and "worst case scenarios" and take appropriate precautions, including PPE.
- *Prepare* for emergencies by anticipating what they might be and how to respond to them.

References

1. A. EINSTEIN. BrainyQuote; available at http://www.brainyquote.com/quotes/quotes/a/alberteins148837.html (accessed March 9, 2009).

2. R. HILL. Personal account of an incident.

QUESTIONS

1. How does performing a research experiment differ from "course-related" chemical experiments?

    (a) The outcome may not be known.
    (b) The level of real-time supervision may be much less.
    (c) Unexpected hazards may lurk.
    (d) All of the above.

2. When visiting a new lab during a summer research internship, it is safe to assume that

    (a) No one will ask you to perform some experiment that has risks associated with it
    (b) Safety is, and always has been, a top priority in the laboratory
    (c) Most safety information will appear in written instructions for an experiment
    (d) None of the above is necessarily true

3. When faced with a new experiment that is similar, but not identical, to many other experiments previously accomplished in the laboratory, you can assume

    (a) That no new hazards will be present
    (b) That someone else will inform you of potential hazards
    (c) That it is your responsibility to investigate the potential hazards before conducting the experiment
    (d) None of the above

4. When synthesizing a new compound, you should assume that

    (a) It has a toxicity similar to related compounds
    (b) It may be toxic and should be handled carefully
    (c) Wearing nitrile gloves will prevent skin exposure
    (d) If you can't smell it, then you are not breathing it

5. When conducting an experiment overnight,

    (a) You should never leave the laboratory while the reaction is running
    (b) You should consult with a mentor about what extra safety precautions should be taken for an unmonitored experiment
    (c) You should leave an ABC fire extinguisher near the reaction
    (d) You should leave the laboratory lights on as a signal to others that a reaction is being run

6. As a general rule,

    (a) You can work alone in a laboratory as long as you feel safe
    (b) You should never work alone in a laboratory
    (c) You can work alone in a lab as long as there is someone else in the building
    (d) You can work alone in the lab as long as you have a cell phone with you and can call someone in an emergency

PROCESS SAFETY FOR CHEMICAL OPERATIONS

Preview   This section provides a brief overview of process safety and its importance in chemical safety and hazard recognition.

Your own safety is at stake when your neighbor's wall is ablaze.

Horace, Roman poet[1]

INCIDENT 7.3.7.1   EXPLOSION FROM SCALE-UP OF REACTION[2]

Chemists had prepared an amide using a published procedure without incident. The reaction involved adding magnesium nitride (Mg_3N_2) to an aryl ester in methanol at 0 deg C followed by sealing the tube, mixing for an hour at room temperature, and finally heating to 80 deg C for several hours. The researchers then scaled up the reaction to produce more of the desired compound. The reaction was performed in a thick-walled borosilicate tube in an ice bath. An explosion occurred within a minute after removing the tube from the ice bath. Inquiry to the authors of the original paper revealed that they stirred the mixture in a room temperature water bath for 1 hour to control an exothermic reaction forming ammonia from Mg_3N_2.

What lessons can be learned from this incident?

What Is Process Safety and Why Is It Important?

Process safety is a term that describes developing an understanding of the hazards of a chemical process and developing methods to safely manage and minimize the risks of the process. Its purpose is to prevent incidents from unexpected or unplanned deviations from the ordinary chemical process. Process safety is widely used in the chemical industry to produce "inherently" safer chemical processes and products and to identify appropriate safeguards to manage the risk from those hazards which cannot be eliminated.[3-5] Many graduating chemistry students go on to work in industrial chemical operations, where they will encounter process safety. However, as a student you should know that process safety really begins with the chemistry developed in the laboratory. It is upon this laboratory work that industry builds its products and its business. Thus, as a student, you should have some understanding about process safety and how work in the laboratory can impact industrial chemical safety. Learning the fundamentals of process safety can help you minimize and avoid incidents in your laboratory work and in future work in the chemical industry. Remember that "process" incidents can occur at any scale of operations, including in the laboratory as illustrated by Incident 7.3.7.1.

There have been catastrophic events that were the result of the failure to properly manage the safety of chemical operations. These events not only were disastrous to the companies but were devastating to the surrounding community that suffered the loss of friends and neighbors, property damage, and economic losses from the negative impact of these incidents. Additionally, these incidents not only damaged the companies publicly and financially, but they reinforced the public's negative attitude toward the chemical industry in general. A direct result of these incidents is an emphasis on process in the chemical industry.

Perhaps the best known example occurred in Bhopal, India, in 1984. A highly toxic gas, methyl isocyanate (MIC), was released from a chemical plant into the surrounding community. The MIC cloud immediately killed more than 3800 people and 10,000 more died within a few weeks and another 15,000 to 20,000 suffered premature death from the MIC exposure over the next 20 years.[6] There are also estimates that thousands of others died from diseases related to MIC exposure. Investigations found that water had entered into the MIC tank, causing a large pressure buildup and a relief valve opened, releasing the MIC gas. The root causes and underlying factors contributing to the incident reinforced the need for process safety in the chemical industry.[7,8] (This episode is described further in *Special Topic 7.1.1.2* A Case Study in Risk Management—The Tragedy at Bhopal, India.)

In the United States the U.S. Chemical Safety and Hazard Investigation Board (CSB) investigates chemical incidents. A web site provides a list of current investigations and a list of completed investigations.[9] These case studies provide excellent analyses of the incidents that are often due to failures of safety systems and the lack of strong process safety. The CSB web site also includes videos illustrating incidents and these can be very useful in understanding how incidents have occurred in the past.

Process Safety—A Brief Review of Some Basic Concepts

Process safety involves many of the topics that we have covered in this book—flammability, explosions, toxicology, reactive chemicals, and runaway reactions[3-5] (see Section 5.3.10 on runaway reactions). It also involves many of the basic concepts in chemistry including kinetics and thermodynamics.

Developing process safety for an operation involves the experience, knowledge, and expertise of many people; it is most often a team effort among chemists, chemical engineers, design engineers, safety specialists, industrial engineers, employees, and managers. This team examines each step of a process using *process hazard analysis* (PHA) to identify hazards and potential areas that could lead to hazardous conditions. They ask the question "What can go wrong?" with this process. This hazard identification step is essential because you can *never* manage the risk from a hazard that you have failed to identify. Since many processes that fail involve runaway reactions, there is an emphasis on what might happen if a reaction reaches this uncontrollable condition. Safeguards are considered that might be put in place to prevent, mitigate, or minimize the consequences of these events.

Chemists can play an early role in process safety when they develop the original synthetic process that is used for a product made and sold in large quantities. As the synthetic process is developed in the laboratory it can have a dramatic effect on the safety of an industrial process adopted to produce this product.

The chemist should consider the following strategies of process safety.[3-5,10] These will sound familiar to you, and they are similar to the goals of green chemistry (see Sections 1.1.2, 1.2.2, and 1.3.4).

- *Minimize* the amounts of hazardous chemicals to be used in the process.
- *Substitute* or replace hazardous materials with less hazardous materials.
- *Moderate* the conditions of the reaction process by finding ways to carry out reactions under lower pressure and temperatures.
- *Simplify* the process by seeking to remove or minimize the steps and complexity of the process.

Reactions that are best for scale-up to pilot plant and industrial scale would ideally possess the following traits:[4]

- Result in high yields with little or no by-products
- Occur very quickly when reactants come in contact with each other
- Utilize a single-phase system with low viscosity
- Operate at ambient temperature and pressure
- Are not strongly exothermic
- Are relatively insensitive to small changes in operating conditions

While these traits should be sought in developing reactions, you should always keep in mind that a chemical may have a less hazardous property in one area but may be more hazardous in another area. For example, dichloromethane is a common solvent that is chosen for use because it is nonflammable. However, dichloromethane has significant toxic properties, including the potential to cause cancer and cardiac effects.[11] So it is not unusual that there is actually a trade-off in selecting less hazardous materials.

Process Safety Management (PSM) — A Systems Approach

In the late 1980s and early 1990s catastrophic events using highly hazardous chemicals in the chemical industry evoked strong public reactions that led for calls to improve safety.[12] OSHA initiated steps to develop a process safety management regulation utilizing a systems safety approach. Although industry had in the past strongly opposed regulation, they welcomed and supported efforts to implement a standard for process safety management.[13] Strong cooperation between government, industry, and labor resulted in the standard 29 CFR 1910.119 Process Safety Management of Highly Hazardous Chemicals (PSM).

This is a performance standard that allows the operator of a manufacturing plant to determine specifically how the standard is met. As long as the approach includes certain elements, the industry is compliant with the standard. While there are several elements, the most important aspect of the standard is its safety systems approach utilizing a team approach to effectively accomplish PHA of the entire operation—this approach seeks to identify the potential hazards in each step of the process and then establishes procedures to prevent or minimize serious incidents. In a sense PHA seeks to answer the questions "What can go wrong?" and "What can we do to prevent this?" The standard requires investigation of each incident to systematically find and remedy the root causes.

While the chemical industry has embraced PSM and most companies vigorously follow PSM, there still continue to be failures by some companies to use PSM and these failures have resulted in major chemical incidents. PSM will continue to be an important tool in preventing catastrophic events in the future. PSM can and does save lives every day, and this safety systems approach will be an important part of anyone's job that entails large scale operations.

Analysis of a Major Incident — Failure to Use PSM

To illustrate the value of PSM, we offer a brief description of a serious industrial incident, the "Synthron Explosion" at Morganton, North Carolina, on January 31, 2006,[14] including its causes and root causes, with pointers to show where PSM could have been more effectively used. The following brief summary was derived from the incident investigation report by the U.S. Chemical Safety and Hazard Investigation Board (CSB). You may want to read the entire 17-page report.

Brief Description of Incident   This facility manufactured powder coatings and paint additives by a process using free-radical polymerization of acrylic monomers in flammable solvents. A customer requested more of an additive than the facility would normally make in a single batch, so managers decided to scale up the operation to produce the required amount for the order. The managers and most operators in this plant had been with the plant less than a year, and none of them had previous experience in manufacturing polymers.

The 1500-gal reaction vessel was charged with additional monomer, however, the energy released by this scaled up reaction was double the normal amount and the cooling capacity of the reactor was exceeded, resulting in a runaway reaction. The reaction vessel had an access port called a manway that was supposed to be fastened with 18 clamps to maintain pressure in the system; however, only four clamps were used to seal this port. The pressure in the reactor increased until the gasket between the manway and the reaction tank gave way, releasing solvent vapor into the open plant. The vapor cloud encountered an ignition source resulting in a catastrophic explosion. One severely burned worker died a few days after the incident, 14 other workers (the total workforce was 17) were injured, the facility was destroyed, and other off-site structures were damaged.

Summary of Causes of the Incident   The CSB found many factors that contributed to the incident:

- Lack of hazard recognition

- Inadequate process safety information
- Ineffective control of process changes
- Lack of automatic safeguards to control the reaction
- Improper bolting of the manway (access port)
- Poor operator training
- Inadequate emergency drills
- Inadequate corporate oversight of process safety

Lessons Learned   A safety system must be in place to:

1. Identify and control reactive hazards—ask "What can go wrong?"
2. Examine the safety consequences of changes in normal operations
3. Maintain equipment capabilities for reactive chemistry operations
4. Train personnel in hazard recognition and process safety
5. Plan and prepare for emergencies
6. Ensure that corporate management establishes and maintains a program to manage reactive hazards using accepted industry good practices, including PHA and PSM

Learning More

The CSB web site is an excellent source for reading thorough reports about major chemical accidents. It is often the case that the list of "what went wrong" is, in retrospect, a list of completely preventable errors and usually there are several errors that act in combination to lead to catastrophic events. All of this can be applied to much smaller, academic lab projects, however. Applying the basic questions of PHA to all laboratory experiments will prevent accidents and incidents almost 100% of the time. Reading through the various incidents that serve as introductions to most sections of this book invariably leads to the observations that "that could have easily been prevented if" basic safety rules had been followed.

RAMP

We conclude, as usual, with RAMP as a guiding "checklist" for performing experiments in the lab.

- *Recognize* hazards using process hazard analysis. This is essential in industrial process operations.
- *Assess* the risks of the hazards, including asking "What can go wrong?"
- *Manage* the risks involved in chemical processing operations through the use of the safety systems approach. This is essential for the safety of plant operations.
- *Prepare* for emergencies, an essential part of normal operations in the chemical process industry.

References

1. HORACE. BrainyQuote; available at http://www.brainyquote.com/quotes/quotes/h/horace138534.html (accessed March 11, 2009).
2. S. CRANE. Chemical safety explosion hazard. *Chemical Engineering News* **87**(*15*):2– (April 13, 2009); available at http://pubs.acs.org/cen/letters/87/8715letters.html (accessed April 16, 2009).
3. R. E. BOLLINGER, D. G. CLARK, A. M. DOWELL, R. M. EWBANK, D. C. HENDERSHOT, W. K. LUTZ, S. I. MESZAROS, D. E. PARK, and E. D. WIXOM. *Inherently Safer Chemical Processes: A Life Cycle Approach*, 2nd edition, John Wiley & Sons, Hoboken, NJ, 2009.
4. D. C. HENDERSHOT. Inherent safety strategies for process chemistry. *Journal of Chemical Health and Safety* **5**(*4*):18–22 (1998).
5. D. C. HENDERSHOT, J. F. LOUVAR, and F. O. KUBIAS. Add chemical process safety to the chemistry curriculum. *Journal of Chemical Health and Safety* **6**(*1*):16–22 (1999).
6. E. BROUGHTON. The Bhopal disaster and its aftermath: a review. *Environmental Health Journal* **4**:6–11 (2005); available at http://www.ehjournal.net/content/4/1/6 (accessed April 16, 2009).
7. A. RICE. Bhophal revisited—the tragedy of lessons ignored. *Asia-Pacific Newsletter on Occupational Safety and Health* **13**:46–47 (2006); available at http://www.ttl.fi/NR/rdonlyres/AF130282-A0AB-4439-8E3C-AFF55CDEF59F/0/AsianPacific_Nwesletter22006.pdf (accessed March 11, 2009).

8. K. BHAGWATI. *Managing Safety*, Wiley-VCH, Weinheim, Germany, 2006, pp. 23–26.
9. U.S. Chemical Safety and Hazard Investigation Board. Available at http://www.chemsafety.gov/ (accessed March 11, 2009).
10. The Institution of Chemical Engineers and The International Process Safety Group. *Inherently Safety Process Design*, The Institution of Chemical Engineers, Rugby, UK, 1995.
11. Occupational Safety and Health Administration. Methylene Chloride (dichloromethane) Standard. 29 CFR 1910.1052; available at http://www.osha.gov/pls/oshaweb/owadisp.show_document?p_table=STANDARDS&p_id=10094 (accessed March 11, 2009).
12. Occupational Safety and Health Administration. Preamble to the Process Safety Management of Highly Hazardous Chemicals; Explo-sives, and Blasting Agents. Section 1.—I. Background; available at http://www.osha.gov/pls/oshaweb/owadisp.show_document?p_table=PREAMBLES&p_id=1039 (accessed April 16, 2009).
13. E. MASON. Elements of process safety management: Part I. *Journal of Chemical Health and Safety* **8**(*4*):22–24 (2001); Part 2, *Journal of Chemical Health and Safety* **8**(*5*):23–26 (2001).
14. U.S. Chemical Safety and Hazard Investigation Board. Final Report No. 2006-04-I-NC, July 31, 2007: Runaway Chemical Reaction and Vapor Cloud Explosion, Synthron, LLC, Morganton, NC, January 31, 2006; available at http://www.csb.gov/completed_investigations/docs/Synthron%20Final%20Report.pdf (accessed March 13, 2009).

QUESTIONS

1. Process safety is a term that describes developing

   (a) An understanding of the hazards of a chemical process
   (b) Methods to safely manage the risks of a process
   (c) Methods to minimize the risks of the process
   (d) All of the above

2. When comparing "academic laboratory processes" and "industrial processes,"

   (a) The difference in scale requires a very different analysis of hazards and risks
   (b) Many of the hazards are the same, except on larger scale in industry
   (c) The level of risk in industrial processes can be much higher due to scale
   (d) Both (b) and (c)

3. For the MIC incident in Bhopal, India, the total estimated human death toll is

   (a) About 3800
   (b) About 10,000
   (c) About 20,000
   (d) Over 35,000

4. What agency in the United States investigates serious chemical incidents?

   (a) EPA
   (b) Congress
   (c) CSB
   (d) None of the above

5. Process safety usually involves

   (a) Chemists and chemical engineers
   (b) Design engineers and safety specialists
   (c) Employees and managers
   (d) All of the above

6. What risk cannot be managed and minimized?

   (a) Fires
   (b) Explosions
   (c) Radiation leaks
   (d) The risk that is not identified

7. What government agency is responsible for 29 CFR 1910.119 Process Safety Management of Highly Hazardous Chemicals?

   (a) CSB
   (b) OSHA
   (c) EPA
   (d) None of the above

8. What questions does PHA address?

   (a) "What can go wrong?" and "What can we do to prevent this?"
   (b) "What can we do to prevent this?" and "Who is responsible for damages?"
   (c) "What can go wrong?" and "Who is responsible for damages?"
   (d) All of the above

9. What circumstances contributed to the "Synthron Explosion"?

   (a) Inexperience by plant personnel with the particular chemistry of the reaction
   (b) Scaling up a reaction without scaling up cooling capacity
   (c) Inadequate number of clamps on a gasket
   (d) All of the above

CHAPTER 8

CHEMICAL MANAGEMENT: INSPECTIONS, STORAGE, WASTES, AND SECURITY

MANAGING CHEMICAL *hazards* is an important and necessary skill for chemists. A chemical analysis or synthesis is one part of the process of doing chemistry. Before an experiment is done, chemicals have to be safely purchased and stored, and afterwards the products of an analysis or reaction need to be handled in a safe and environmentally responsible fashion. While in college other persons will tend to these tasks, but when you are working as a chemist you will need to be much more mindful of the "before" and "after" stages of an experiment. This involves inspecting labs to make sure they present a safe working environment, storing and handling chemicals, using a chemical stockroom, knowing how to handle chemical wastes, and securing your most hazardous chemicals.

INTRODUCTORY

8.1.1 Introduction to Handling Chemical Wastes Describes the basic guidelines for how to handle chemical wastes that are generated in academic laboratories.

INTERMEDIATE

8.2.1 Storing Flammable and Corrosive Liquids Introduces the storage of flammable and corrosive chemicals in the laboratory.

ADVANCED

8.3.1 Doing Your Own Laboratory Safety Inspection Describes how you can conduct an inspection of your own laboratory and provides a simple checklist to assist you.

8.3.2 Managing Chemicals in Your Laboratory Provides guidance for students who may be assigned responsibility for ordering, receiving, and tracking chemicals.

8.3.3 Chemical Inventories and Storage Describes some basics of chemical storage including inventory, cabinetry for chemical storage, and how to group chemicals for safe storage.

8.3.4   Handling Hazardous Laboratory Waste   Describes the management of hazardous chemical waste from the laboratory.

8.3.5   Chemical Security   Discusses chemical security and measures for preventing theft of high hazard chemicals and other laboratory materials that might be used for untoward purposes if procured by persons with malevolent purposes.

8.1.1

INTRODUCTION TO HANDLING CHEMICAL WASTES

Preview This section describes the basic guidelines for how to handle chemical wastes that are generated in academic laboratories.

> *We have met the enemy and he is us!*
>
> Pogo (a comic strip character, 1948–1975) by Walt Kelly[1]
> [Originally used in his Pogo Comic Strip on Earth Day, 1971]

INCIDENT 8.1.1.1   THE STINKING BUILDING[2]

An analytical chemist working on the third floor of a laboratory building was using ethanedithiol in a reaction and had completed his work. He decided that he would pour the waste down the sink, which was connected to all of the other laboratory sinks. The stinking odor from the ethanedithiol smelled like the mercaptan odorant added to natural gas. The occupants of the building thought there was a gas leak and the evacuation notice was given for the entire building. The fire and police departments arrived, while the many building occupants huddled outside. Soon thereafter the local news media arrived. After investigation the source of the odor was discovered. The news media reported the inappropriate disposal of a hazardous chemical.

What lessons can be learned from this incident?

Prelude

You may be reading this section fairly early in your college experience, and this may seem odd since we have located this section in Chapter 8 of the book. Logically, it is near the end of the book since so many other sections are about how to function safely *during* experiments while handling waste is usually at the *end* of an experiment. However, since you will likely be generating some kind of laboratory waste during your first lab experiment this is a topic that needs to be addressed early on in your lab experience.

Americans of all ages have grown up in a culture that largely disposes of unwanted substances by either putting them in a trash can or pouring them down a sink. And for most of us, we don't much concern ourselves about what happens to our waste after that. "Out of sight, out of mind" is the prevailing mindset. In fact, much of the chemical industry in America treated chemical wastes the same way for decades until the environmental awareness about such practice in the 1960s and 1970s led to federal and state regulations prohibiting these actions. Except for the United States, most of Europe, and a handful of other countries, this practice of disposing of chemical wastes directly into the environment is still common in the 21st century.

Today, however, most students in chemistry labs might quickly and appropriately wonder "if it's safe to put this stuff into the sink or trash can" and that is an appropriate concern. This section will introduce the basic principles of how to process wastes generated in academic labs. Section 8.3.4 will extend this discussion considerably by presenting the guidelines for processing wastes from research labs and industrial labs.

Laboratory Safety for Chemistry Students, by Robert H. Hill, Jr. and David C. Finster
Copyright © 2010 John Wiley & Sons, Inc.

Rule #1: "Think Before You Use the Sink!"

The first and simplest rule about handling laboratory wastes is: *Don't put anything into a trash can or down the sink unless specifically instructed to do so*. While "down the sink" is almost *never* appropriate in research labs, in fact in introductory chemistry classes there are some solutions that can be poured down the sink and the occasional solid product that can be placed in a trash can. Introductory lab experiments are often designed to use and produce innocuous chemicals since (1) the chemical principles that are being taught can be accomplished using these "safer" chemicals, (2) not having to further handle the chemical wastes saves time for stockroom personnel, and (3) this saves money by not having to pay to dispose of chemicals wastes. However, introductory lab courses will also inevitably generate many wastes that *cannot* be poured down a sink or placed in a trash can.

Laboratory instructions routinely include comments about both the various safety hazards associated with a particular experiment and instructions on how to process the chemical wastes. It is very important that these instructions be followed carefully to ensure your safety and to protect the environment. For many decades, and surely since the advent of green chemistry (Sections 1.1.2, 1.2.2, and 1.3.4) chemical educators have tried to teach students about how to handle wastes in academic labs both to "do the right thing" and to prepare them for their futures as scientists functioning in a variety of lab circumstances that may deal with both small and large volumes of wastes. In truth, of course, when one student (of the several hundred thousand students taking general chemistry classes each year in the United States) pours a few milliliters of a solution down the sink that shouldn't be poured down the sink, this is not an environmental catastrophe. (There are, though, other "chemical" reasons not to improperly dispose of solutions down the sink, as we will see below.) Learning the appropriate procedures early on is very important, because it trains you to be conscientious and thoughtful about waste disposal in preparation for a host of situations in your future when disposing of chemical wastes is very, very important.

There are three main reasons why we should be mindful of appropriate waste disposal: environmental protection, safety, and cost. We'll discuss each of these separately, even though they are often linked arguments.

Protecting the Environment

The enterprise of chemistry involves an extremely wide array of chemicals. Some are fairly harmless to the environment, and some are not. (It is probably worth pointing out, too, that "the environment" itself is a collection of chemicals!) In the United States, the Environmental Protection Agency (EPA) determines the laws regarding what chemicals can be disposed of in the air, water, and ground. It is easy to guess that "simple rules" are not easy to construct since there are so many chemicals that pose varying degrees of hazard, at varying concentrations, to the environment. Further complicating this is that we don't even know the toxic effects (on humans and animals) for many chemicals that we are using.

Since processing wastes is inevitably more costly than simply discarding them, there is a tension between the desire to save money (and increase profits in the chemical industry or simply saving money at an academic institution) and the desire not to harm the environment. Should laws be drafted requiring that a substance be *proved to be toxic* before simple disposal is illegal, or to *prove that a substance is safe* before simple disposal is allowed? And what is done in the absence of knowledge about chemical toxicity? In general, the EPA takes the position that something is harmful if it is not known to be safe.

Unlike academic labs, many industrial chemical facilities handle huge volumes of chemicals so while it may be environmentally tolerable to pour 2 mL of Chemical X down the sink it would not be environmentally tolerable to pour 2 million gallons of Chemical X "down the sink."

Disposing of Chemical Wastes Safely

Remember the words of Pogo[1] regarding our environment, "we have seen the enemy and he is us." Being environmentally responsible, we need to take proper care of our laboratory waste. Completely apart from environmental concerns, there are some situations where inappropriate waste disposal is a safety hazard. It could be that Solution X can be poured down the sink alone by Student #1 with little concern for the reactivity of the solution (that is likely already dissolved in water). However, under the sink is the trap, which inevitably would hold some of Solution X (unless very excessive flushing with tap water is used). Then, unknowingly, Student #2 pours Solution Y down the sink with the same understanding that Y doesn't react with water. However, if X and Y react with each other, it is possible that an unsafe reaction could occur in the plumbing, generating toxic fumes or even spewing a hazardous product back into the sink.

Strong acids and strong bases react, and oxidants and reductants react. So, just because a chemical is "safe" by itself in water, doesn't mean it won't react with something else. Some particularly well-known examples of "never down the sink" chemicals are salts of azides (N_3^-) and perchlorates (ClO_4^-) that can form explosive compounds with metal cations. Sulfides (S^{2-}) and cyanides (CN^-) will form poisonous gases in contact with acids. (It is not likely that you'll encounter these anions in introductory courses.)

And it is never wise to put a solid into a sink. Anything that is insoluble in water will end up sitting in the trap, just waiting for a chemical with which it will react.

Disposing of a solid chemical into a common task receptacle is almost never wise, unless specifically told to do so. It is possible, but perhaps not likely, that the chemical will react with something else, but since custodians often reach into wastepaper baskets to simply remove trash—although it is always wiser not to do so—it is possible that someone could be injured or contaminated by the solid.

Apart from environmental and safety concerns, hazardous laboratory waste is heavily regulated by the EPA and improper disposal can lead to significant fines and adverse publicity in the community. (More about this in Section 8.3.4.)

Waste Disposal Costs Money

Chemistry labs inevitably generate wastes that must be disposed of by a qualified waste disposal company. These companies charge very large sums of money for this service and it is not uncommon to pay *more* to have an impure chemical removed than it did to purchase the pure chemical in the first place!

To minimize waste disposal costs, it is imperative that all wastes first be clearly identified. The most expensive waste to dispose of is always the "unknown" waste since it must be treated as if it is extremely hazardous until testing has proved otherwise. And the cost of this kind of testing is high, which drives up the cost of disposal.

Some wastes can be recycled by a waste disposal firm (at some modest profit) and these wastes will be the cheapest to handle (assuming that they are pure or known mixtures). Other wastes can be processed, which costs money to do, and will therefore cost more to dispose of. Some can be incinerated. Finally, many wastes must simply be buried in special containers that are placed in special landfills. This is very expensive, and in the long run it is not a desirable solution.

General Guidelines

There are many other features of the general topic of waste disposal that needn't concern you as a student but are very important in the chemical industry for both manufacturing and research facilities. All of these factors will be discussed more in Section 8.3.4, which is designed to be read by advanced students as they are preparing to leave college.

For most academic courses, a fairly simple set of guidelines will help teach you the basics of waste disposal and save money (which is ultimately your tuition dollars) and time for stockroom personnel.

- The first and most important rule is: Follow instructions! It is easy at the end of a lab experiment to be rushing to finish the experiment, clean and put away equipment, and move on to the next task in your academic life. And it is usually at the end of the experiment when you are likely to take care of accumulated wastes. So, take a little time to be conscientious about this task for reasons of safety, environmental protection, and cost savings.

- Don't pour anything down the sink unless you are instructed to do so. Some solutions can be poured down the sink and some solids can be placed in an ordinary trash can, but this determination will be made by your instructor, not you.

- Most wastes will have specific waste jars available to receive them. Read labels carefully. If you inadvertently dispose of a chemical in the wrong receptacle, tell your instructor! *Not* telling your instructor about this kind of mistake can lead to a serious incident due to an unexpected reaction in the waste jar or some reaction in subsequent steps when the waste is processed.

- Anything that becomes contaminated with a chemical, including paper towels, sponges, or ordinary paper, must be treated separately. Ask your instructor what to do.

- Broken glassware that is contaminated with any chemical must similarly be handled by someone else. Ordinary glassware that is broken and clean should be placed in a special receptacle for broken glass. Handling broken glass can be dangerous and it is best to have a teaching assistant or instructor take care of broken glass.

- Don't mix wastes unless specifically told to do so. There are some situations where this is both safe and no more costly than separated wastes. However, "unmixing" chemicals is generally expensive and sometimes quite difficult.

- All of the steps above assume that you know what the waste is! To be certain about this, all beakers, test tubes, and other containers must always be identified at all times during an experiment with accurate labels. It is easy, at first, to think that you'll remember what is in each of a few beakers as an experiment proceeds, but this is a system that will inevitably fail. Many solutions are clear and therefore indistinguishable.

Following these guidelines and the specific waste instructions associated with each lab experiment is the safe, environmentally sound, and cost-effective procedure to follow.

RAMP

- *Recognize* that chemical waste can be hazardous.
- *Assess* potential exposure paths to chemical waste.
- *Minimize* the hazards of chemical waste through proper handling and disposal.
- *Prepare* for emergencies dealing with hazardous chemical waste.

References

1. WALT, KELLY. *Pogo: We Have Met The Enemy and He Is Us*, Simon and Schuster, New York, 1972; available at http://en.wikipedia.org/wiki/Pogo_(comics) (accessed March 5, 2009).

2. R. HILL. Personal account of an incident.

QUESTIONS

1. Which statement is *true*?

   (a) Up until the 1970s, chemical waste disposal in the United States largely involved direct discharge into the environment.

   (b) In the 21st century in the United States, most chemical waste cannot be disposed of directly into the environment.

   (c) In the 21st century most waste disposal outside the United States and western Europe involve direct discharge into the environment.

   (d) All of the above are true.

2. The best rule for disposal of waste from a chemistry lab is

   (a) Down the sink unless instructed otherwise

   (b) Into trash cans unless instructed otherwise

   (c) Into chemical waste receptacles unless instructed otherwise

   (d) Both (a) and (b)

3. Learning to properly dispose of even very small amounts of chemicals in academic lab experiments is important because

   (a) If hundreds of students all disposed of small amounts, the total would be large and significant

   (b) Sometimes even small amounts of chemicals down the sink can be a hazard

   (c) This practice teaches students the appropriate method of disposal

   (d) All of the above

4. With regard to the disposal of chemical wastes, the EPA assumes that

    (a) All chemicals are hazardous to the environment
    (b) Chemicals are hazardous unless they are known to be nonhazardous
    (c) Chemicals are nonhazardous unless they are known to be hazardous
    (d) Chemicals in small quantities are necessarily not very hazardous, so waste disposal legislation only applies to large quantities of chemicals

5. Apart from environmental concerns, why is it unwise to dispose of chemicals down the sink?

    (a) Solids that are insoluble in water will remain in the sink trap and perhaps react with subsequent chemicals poured down the sink.
    (b) Some chemicals can produce toxic gases when mixed with other chemicals in the sink trap.
    (c) Some chemicals may react unexpectedly with other chemicals in the trap.
    (d) All of the above.

6. Unidentified wastes from chemistry labs

    (a) Are generally the cheapest to dispose of since they are always burned
    (b) Are generally the cheapest to dispose of since they are always buried
    (c) Are generally expensive to dispose of because they are assumed to be dangerous mixtures

    (d) Are generally expensive to dispose of because they must be treated as if they are very hazardous

7. If you inadvertently dispose of a chemical into the incorrect waste container, you should

    (a) Make a note of this in your lab notebook
    (b) Tell your instructor before the end of the laboratory session
    (c) Tell your instructor immediately
    (d) Write an email to your instructor so that there is a legal statement about the event

8. Paper towels that are contaminated with chemicals

    (a) Should simply be put in the trash
    (b) Should be burned in a chemical hood
    (c) Should be placed in the same container that holds waste containing the same chemicals
    (d) None of the above

9. Broken glass that is contaminated with chemicals

    (a) Should be washed thoroughly before disposal in the broken glass container
    (b) Should be handled by your instructor or someone with training about how to handle this situation
    (c) Can be placed in a trash receptacle
    (d) Should be placed in your lab drawer until the chemical self-decontaminates and then it can be placed in the broken glass container

8.2.1

STORING FLAMMABLE AND CORROSIVE LIQUIDS

Preview   This section introduces the storage of flammable and corrosive chemicals in the laboratory.

One is not exposed to danger who, even when in safety, is always on their guard.

Publilius Syrus, Roman writer[1]

INCIDENT 8.2.1.1   DEFECTIVE CHEMICAL STORAGE CABINET[2]

A chemical storage cabinet was "homemade" of thin plywood and particle board shelves and these were held together by a pressed paperboard back panel. The cabinet was used to store organic chemicals. The bottom shelf collapsed but since it only dropped a few inches, there were no broken bottles.

What lessons can be learned from this incident?

Minimizing Risks of Exposure to Corrosives and Flammable Liquids

Probably the two most hazardous classes of chemicals that you will encounter in your early days in undergraduate laboratories are corrosives and flammable liquids. Common inorganic acids include hydrochloric acid, sulfuric acid, nitric acid, and phosphoric acid and the most common bases are sodium hydroxide, potassium hydroxide, and ammonium hydroxide. Organic solvents are the most common flammable liquids and examples of these are acetone, methanol, hexane, toluene, ethyl acetate, and dioxane.

   While you have learned some basics about the hazards of corrosives and flammable liquids, you should know something about the management of these chemicals that require particular chemical storage. Proper chemical storage is important because it allows us to have these chemicals close at hand in case we need them, but we minimize potential for exposure and incidents by keeping these chemicals protected in the proper storage cabinet.

Storing Corrosive Chemicals

Section 5.1.1 describes how corrosives are very hazardous and capable of causing severe damage to skin and other tissues exposed to these materials. However, corrosives include both acids and bases and these two materials are incompatible (see Section 5.2.3), meaning that if they come in contact with each other a violent reaction may occur releasing heat and perhaps even resulting in an explosion. Thus, in storage we want to keep these chemicals separated so that there will not be an incidental bump of bottles that might break the containers and cause these chemicals to mix and react. The first step in storing corrosives is to separate acids from bases. We will also learn about other separations.

   Let's first consider how corrosive storage cabinets should be made. You might think there are some kinds of rules about this, such as regulations, but it turns out there are none for corrosives. Cabinets made of metal are not best for corrosives, since acids react with metals; however, metal cabinets are sometimes painted with a corrosion-resistant epoxy paint that protects the metal from acid attack. Still

Laboratory Safety for Chemistry Students, by Robert H. Hill, Jr. and David C. Finster
Copyright © 2010 John Wiley & Sons, Inc.

it is possible and even likely that painted surfaces can be scratched or scraped over time, exposing bare metal, so it is perhaps best to avoid storing corrosives in metal cabinets.

Most cabinets for corrosives are made of a heavy molded plastic (polyethylene) or wood. Polyethylene cabinets are especially good since they can be made seamless and can contain spills from broken or cracked bottles. Wooden cabinets for corrosives are common. Some institutions make their own wooden cabinets, some have cabinets under hoods or laboratory benches made for chemical storage, and others purchase cabinets from manufacturers and chemical supply houses. Wood cabinets can also be good if they are designed correctly and are made of good strong materials. The bottom should have a lip to contain any spill that might occur. Wood is not always the best choice since some acids can react with wood; however, sealing the wood with something like polyurethane provides a protective barrier to minimize contact between the acid and wood.

Acids are separated from bases with each group in its own cabinet. Organic acids, many of which are flammable, should not be stored with inorganic acids, which are often oxidizers. Thus, do not store acetic acid (an organic acid) with acids, but rather store it in a flammables cabinet since it is combustible. Nitric acid is a strong oxidizing acid and reacts readily with organics such as acetic acid in a violent manner. Concentrated sulfuric acid is a strong dehydrating agent and its reactions with other substances liberate considerable heat. Because they are so reactive, plastic trays are often used under these two acids even in an acid storage cabinet to prevent incidental contact with other acids.

Sodium and potassium hydroxide solutions can be stored with ammonium hydroxide in a separate cabinet for bases. If hydrochloric acid and ammonium hydroxide are stored near each other, they will often become covered with a fine white solid from the reaction of two gases. Can you guess what the gases and solid will be?

Storing Flammable Liquids

When you take organic chemistry laboratory, you will encounter many flammable liquids, especially solvents. A fire is perhaps the most dangerous and most damaging of all potential incidents that might occur in a laboratory. It is important to take all possible steps to reduce and minimize exposures to these flammables. You can minimize exposures by keeping a minimal quantity out in the open laboratory, and keeping the rest of the solvents in a proper flammables storage cabinet.

Flammable liquid storage cabinets are usually made of double-walled steel with tight fitting joints that are welded or riveted. The bottom is designed to catch spills and doors may be self-closing and self-latching (a three-point latch is usually used). Although NFPA 30 allows the use of wooden cabinets for flammables made from 1-inch plywood,[3] many institutions do not allow wooden cabinets. Cabinets must meet performance standards under NFPA codes. Local fire codes may vary in their requirements. Several manufacturers supply flammables storage cabinets that meet code requirements. Figure 8.2.1.1 shows a flammables storage cabinet.

The location of the flammables cabinet is really important for your safety. It should not be located near the exits and must not be an obstacle to escape. Flammables cabinets must not be located near any ignition source. All flammables cabinets must be labeled prominently "Flammable—Keep Fire Away."

Venting of flammables cabinets is not required and for unvented cabinets the bungs (holes designed to connect to vent tubing) are sealed. However, if it is vented to remove chemical odors, the cabinet must be vented from the bottom with make-up air coming in from the top on the opposite of the cabinet. This is because most flammable chemicals that you will encounter are denser than air and tend to collect near the floor or bottom of the cabinet. This configuration for venting makes use of the natural movement of the vapor. For vented cabinets the exhaust and make-up air are connected at the location of the bungs and these bungs are often fitted with flame arrestors (special fine metal screens that prevent the passage of flames).

Limitations on Volumes of Flammables in a Laboratory

For the safety of people in the building, the volume of flammables in a laboratory is limited by the fire code. The higher the fire rating of the laboratory, the larger the volume of flammables allowed.

FIGURE 8.2.1.1 Flammables Storage Cabinet. Flammables liquids and solvents should be stored in the appropriate cabinet. Some cabinets are vented, some are not. This cabinet is not. To vent such a cabinet, the "bung hole" seen at the lower left is connected to a source of "make-up" air and a similar hole at the top of the right side of the cabinet is connected to piping to the exterior of the building. Simply opening these holes (without proper piping) would defeat the fire-rating capacity of the cabinet.

The fire rating depends on the construction of the laboratory, including, for example, wall thickness and the presence of a sprinkler system. Laboratories having flammables ideally have sprinkler systems; however, many do not.

In Section 3.2.1 we describe the recognition of hazards using the GHS Hazard Classification System. The NFPA sets standards under its codes for fire protection and most of these are used by state and federal governments. Because there is a new system coming into play (GHS) and an existing older system (NFPA) there may be some confusion in understanding and applying hazard classifications to current codes. In Table 8.2.1.1 we seek to help you understand the current differences in these two systems.[4]

The NFPA code uses NFPA classifications to identify limits on storage within a laboratory. As can be seen from Table 8.2.1.1 the NFPA and GHS classification systems are not identical. NFPA Classes IA, IB, and IC are similar, but not identical, to GHS Classes HC1, HC2, and HC3. GHS HC3 is defined as flammable, but NFPA classifies a portion of GHS HC3 as "combustible." GHS HC3 compounds with F.P. $\geq 37.8\,°C$ to $\leq 60\,°C$ are considered as combustibles by the NFPA. Similarly, NFPA Class II combustibles are considered to be flammables by GHS. These two systems are not likely to resolve into one system since both are firmly rooted in organizations resistant to change. At this point in time, the NFPA guidelines are the "accepted standard." In time, federal and/or state regulations may choose to prefer the GHS guidelines. In practical terms, either system will provide adequate protection.

You will also note that the flammable hazard classification of a chemical determines the allowable containersize for a chemical. Thus, Class IA or HC 1 are only allowed up to 0.5 L in size for a glass container (most laboratories use glass). There is also a limit as to the amount of flammables allowed outside a flammable storage cabinet at any one time. This is very important for your safety, so when you are finished using a flammable you should promptly return it to the flammables cabinet. This reduces your risk in case there is a fire.

RAMP

- *Recognize* that corrosives and flammables are very hazardous and have special storage requirements.
- *Assess* the risk from handling flammables and corrosives.
- *Minimize* the risk of handling flammables and corrosives by storing them in proper cabinets.
- *Prepare* for emergencies by minimizing the volume of flammables and corrosives out in the open laboratory.

TABLE 8.2.1.1   NFPA 45–2004 Code: Allowable Container Sizes for Flammable and Combustible Liquids/Maximum Quantity Allowed Outside Storage Cabinet per 100 ft^2 Laboratory Space with Sprinkler System for Laboratory Unit Fire Hazard Class C

NFPA class[a]	GHS category[a] (GHS not used by NFPA)	NFPA allowable container sizes	NFPA maximum quantity allowed outside storage cabinet for Fire Hazard Class C[b]
Class IA Flammable (F.P. <22.8 °C, B.P. <37.8 °C)	HC 1 Extremely Flammable (F.P. <23°C; B.P. ≤35 °C)	Glass—0.5 L Metal/approved plastic—4 L Safety can—10 L	7.5 L total Class I combined (includes flammable gases)
Class IB Flammable (F.P. <22.8 °C, B.P. ≥37.8 °C)	HC 2 Highly Flammable (F.P. <23 °C; B.P. >35 °C	Glass—1 L Metal/approved plastic—20 L Safety can—20 L	7.5 L total Class I combined (includes flammable gases)
Class IC Flammable (F.P. ≥22.8 °C, B.P. <37.8 °C)	HC 3 Flammable (F.P. ≥23 °C to ≤60 °C)	Glass—4 L Metal/approved plastic—20 L Safety can—20 L	7.5 L total Class I combined (includes flammable gases)
Class II Combustible (F.P. 38–60 °C)	HC 3 Flammable (F.P. ≥23 °C to ≤60 °C)	Glass—4 L Metal/approved plastic—20 L Safety can—20 L	15 L total for Classes I, II, IIA combined
Class IIIA Combustible (F.P. 60–93 °C)	HC 4 Combustible (F.P. > 60 °C to ≤ 93 °C)	Glass—20 L Metal/approved plastic—20 L Safety can—20 L	15 L total for Classes I, II, IIA combined
Class IIIB Combustible (F.P. >93 °C)		Not specified	

[a]F.P. is flash point; B.P. is boiling point.

[b]Laboratory Unit Fire Hazard Class C is an Instructional Laboratory Unit—above 12th grade and below postcollege graduate level (i.e., undergraduate).

Source: Reprinted with permission from NFPA 45–2004, *Fire Protection for Laboratories Using Chemicals*. Copyright © 2004, National Fire Protection Association, Quincy, MA. This reprinted material is not the complete and official position of the NFPA on the referenced subject, which is represented only by the standard in its entirety.

References

1. Publilius Syrus. BrainyQuote; available at http://www.brainyquote. com/quotes/quotes/p/publiliuss390414.html (accessed March 18, 2009).
2. Princeton University, Environmental Health and Safety. Shelf of Chemical Storage Cabinet Collapses; available; at http://web.princeton.edu/sites/ehs/labsafetymanual/sec11.htm#shelf (accessed March 19, 2009).
3. National Fire Protection Association (NFPA). NFPA 30—Flammable and Combustible Liquids Code—2008 Edition; available at http://www.nfpa.org/aboutthecodes/AboutTheCodes.asp?DocNum=30 (accessed March 18, 2009).
4. NFPA 45–2004. *Fire Protection for Laboratories Using Chemicals*, National Fire Protection Association, Quincy, MA.

QUESTIONS

1. What are the best materials for constructing a cabinet designed to hold corrosives?

   (a) Polyethylene or painted/sealed wood
   (b) Metal or polyethylene
   (c) Polyethylene or unpainted wood
   (d) Metal or unpainted wood

2. What features should a well-constructed corrosive cabinet have?

   (a) Sturdy shelves
   (b) Locks
   (c) Spill-containing shelves and bottom
   (d) Both (a) and (c)

3. Which is *true*?

   (a) Acids and bases can be stored in the same cabinets, but not on the same shelf.
   (b) Acids and bases can be stored together as long as the concentrations are less than 6 M.
   (c) Acids and bases should not be stored together.

(d) Acids and bases can be stored together, but only for periods of less than 1 month.

4. Organic acids, such as acetic acid, should be stored

(a) With inorganic acids
(b) With flammable chemicals
(c) In a separate cabinets designed for organic acids
(d) Any of the above

5. Ammonium hydroxide should be stored with

(a) Strong bases such as NaOH and KOH
(b) Weak acids, since it is a weak base
(c) Other weak bases, but not with strong bases
(d) Nonoxidizing acids

6. Storage cabinets for flammables can be constructed from

(a) Only steel
(b) Steel or thick plywood
(c) Only thick plywood treated with fire-resistant paint
(d) Noncombustible polyethylene

7. Storage cabinets for flammables

(a) Must be vented
(b) Must not be vented
(c) Can be vented or nonvented

(d) Must be vented with flame-arresting openings

8. What is the total quantity of Class I flammables allowed to be stored outside a storage cabinet, per 100 ft^2, for a laboratory unit in Fire Hazard Class C?

(a) 4 L
(b) 7.5 L
(c) 10 L
(d) 15 L

9. What is the allowable glass container size for Class II combustibles for a laboratory unit in Fire Hazard Class C?

(a) 4 L
(b) 10 L
(c) 15 L
(d) 20 L

10. If an organic solvent has a flash point of 19 °C and a boiling point of 45 °C, what is the NFPA class?

(a) IA
(b) IB
(c) IC
(d) II

8.3.1

DOING YOUR OWN LABORATORY SAFETY INSPECTION

Preview   This section describes how you can conduct an inspection of your own laboratory and provides a simple checklist to assist you.

The more important the subject and the closer it cuts to the bone of our hopes and needs, the more we are likely to err in establishing a framework for analysis.

Stephen Jay Gould, American scientist[1]

INCIDENT 8.3.1.1   UNRECOGNIZED HOOD FAILURE[2]

A chemist was working in a hood preparing samples for analysis. He used toluene to extract the samples as part of preparation. As he was working he became aware that he was getting the strong odors of toluene from this procedure, but he could hear the fan motor in the hood. However, as he investigated further he realized that the hood was not drawing air into its exhaust. He stopped his work and called the maintenance department. They checked on the roof of the facility and found that the belt that drove the fan had broken and even though the motor was running the exhaust fan was not working.

What lessons can be learned from this incident?

Maintaining a Safe and Secure Laboratory

As you enter advanced chemistry courses and undergraduate research, you may find yourself working in a laboratory assigned to you or to a group of students. As you work in these advanced labs, you can apply much that you have learned in this book about chemical and other laboratory hazards, assessing and managing the risks of laboratory operations, and preparing for emergencies. These are critical for your survival and well-being during your work in the laboratory. However, when this happens you have been handed the double-edged sword: as you become more independent, you are also expected to become more responsible. In a safety context, this means that you will have responsibility not only for the safety of your experiments and laboratory operations but also for the safety of others who are working around you.

Part of the job of being safe requires that you maintain a safe and secure laboratory. While you can practice all that you have learned to keep your operations safe, you should also take time to "step back" and look at the overall safety of your laboratory by carrying out a safety inspection or survey of your laboratory on a periodic (monthly) basis. We are not talking about inspections that might be conducted by safety professionals within your institution,[3,4] but rather an informal and effective process by you or your colleagues. Your purpose will be to identify hazards that are in your laboratory, but that you have not seen or have not taken time to address. You should conduct these inspections on a periodic basis, depending on how much you work in the lab. Furthermore, you should consider at times switching with someone in another laboratory so that a "new" pair of eyes sees your laboratory and its "hidden" hazards. If your safety inspection reveals hazards that are not being controlled or managed in the proper way, or if it uncovers some deficiencies in equipment or facilities, you should talk with your research advisor or the laboratory principal investigator about finding ways to get these shortcomings corrected.

Laboratory Safety for Chemistry Students, by Robert H. Hill, Jr. and David C. Finster
Copyright © 2010 John Wiley & Sons, Inc.

Humans are creatures of habit. In many labs some conditions and procedures may exist that have existed for years (without incident) but still represent unsafe conditions that pose unnecessary risk. It is often hardest to see unsafe conditions that have become "familiar" to you. Furthermore, when entering a new lab with many preexisting conditions, procedures, and people, it is easy to get trained to function in an environment and manner that had been considered "safe" for years, even though it never was "safe," or that changing rules and norms made "unsafe" today. Stephen Jay Gould's quote reminds us that it is often hardest to clearly and objectively see those things that are familiar, and perhaps important, to us. The thoughtful scientist enters a new lab with new eyes and asks questions. The safety inspection described below provides a framework to review labs that are either new to you, or very familiar to you.

How Do I Conduct An Inspection of My Laboratory?

We have provided a checklist (Table 8.3.1.1) of things to look for, but the key is to keep your mind open to detect uncontrolled hazards. It is easiest to divide your inspection into the broad areas below and then get more detailed with specifics of each of these areas:

- Housekeeping
- Equipment
- Chemical storage and chemical waste
- Facility integrity
- Personal protective equipment
- Safety information
- Procedures

Rather than trying to be comprehensive for each area, it is probably better to focus on one or two areas and just hit the highlights of the others. Change the focus points on the next inspections. Remember that your purpose is to make your lab safer by finding and "fixing" previously unrecognized hazards so that they are managed. Also, remember that the checklist is for you and your lab mates, but as with all lab notes it is prudent to keep this as a reminder of when and what you did on your last few self-inspections.

What Do I Look For?

Housekeeping is a principal cause of incidents (see Section 5.3.6). Labs can become cluttered and it takes initiative to keep them clean and neat.

- Look at exits to ensure they are clear.
- Examine your emergency equipment to make sure it is not blocked or hindered by clutter.
- Check for cables, cords, bottles, cans, boxes, books, and other materials that can be trip hazards.
- Look at benchtops, hoods, sinks, and refrigerators to ensure that they are clean and neat.

Equipment can be hazardous if not properly maintained.

- Look at your chemical hood. Make sure it is working, ensure its vents are not blocked by equipment or bottles, and check the date for its last certification (annual requirement).
- Check your fire extinguisher for its required yearly inspection tag. Make sure your eyewash station is working and free of obstruction. While you will not be able to check your safety shower, at least make sure that it is accessible.
- Examine the spill kit to be sure it is fully stocked.
- Look for unguarded electrical contacts or moving parts on equipment or instrumentation; especially evaluate "homemade" apparatus for safety issues.
- Check for unguarded vacuum pumps with pulleys.

8.3.1 DOING YOUR OWN LABORATORY SAFETY INSPECTION

TABLE 8.3.1.1   Checklist for Inspecting Your Laboratory

Your name_____ Date ___/___/___

Lab Room Number/Building _____

[Check mark assumes that you have made necessary corrections]

Housekeeping

Exits clear/unobstructed	☐	Floors clean, dry	☐
No cables/cords on floor	☐	No bottle, cans, boxes on floor	☐
Benchtops clean/neat	☐	Refrigerators clean/neat	☐
Hoods clean/neat	☐	Sinks clean/neat	☐

Equipment

Chemical hood working	☐	Hood vents unobstructed	☐
Hood doors in place	☐	Hood window clear	☐
Hood certified (___/___/___)	☐	Fire extinguishers unobstructed	☐
Extinguisher certified (___/___/___)	☐	Eyewashes working	☐
Eyewashes unobstructed	☐	Safety shower unobstructed	☐
Shower certified (___/___/___)	☐	Spill kit fully stocked	☐
No open electrical contacts	☐	No open moving parts	☐
No unguarded vacuum pumps	☐	Local exhaust ventilation in-place	☐
Vacuum/pressure devices guarded	☐	Radiation sources guarded	
Refrigerators explosion-proof	☐		

Chemical Storage and Chemical Waste

Chemicals returned to storage	☐	Chemicals in compatible groups	☐
Time-sensitive chemicals removed	☐	No chemicals on high shelves	☐
No heavy objects on high shelves	☐	Gas cylinders secured	☐
Chemical waste properly labeled	☐	Chemicals dated at receipt	☐
Chemicals in inventory	☐		

Facility Integrity

Floors clear/undamaged	☐	Ceiling tiles undamaged	☐
Ceiling tiles—no sign of leaks	☐	No visible signs of plumbing leaks	☐
Lighting functions properly	☐	No burned out lights	☐
Electrical receptacles undamaged	☐	Exit signs in place/working	☐
Door closers operating	☐	Door locks working properly	☐
Chemical storage locks working	☐		

Personal Protective Equipment

Eye protection available/worn	☐	Gloves available	☐
Gloves are properly selected	☐	Lab coats available/worn	☐

Safety Information

Emergency contact information	☐	Hazard signs posted	☐
MSDSs for chemicals being used	☐	MSDSs reviewed	☐
Chemical inventory up to date	☐	CHP available	☐

Procedures

Procedure hazards identified	☐	Procedure hazards minimized	☐
Emergency equipment available	☐	Emergency materials available	☐
Fire extinguisher location known	☐	Eyewash/shower location known	☐
Fire blanket location known	☐	Spill kit location known	☐

- Look at equipment that emits gases, vapors, or aerosols and ensure that they are properly exhausted.
- Examine apparatus under negative or positive pressure to ensure it is guarded or wrapped to protect against implosions or explosions.
- Check for radiation sources, such as lasers, to ensure they are fully guarded.
- Look at your refrigerator and if it is used for storing flammables or combustibles, ensure that it is explosion-proof or safe for flammables.

Chemical Storage and Chemical Waste can be contributing factors for incidents. Often chemicals that are not returned to storage can lead to hazardous situations.

- Check to ensure that your chemicals are properly stored in compatible groups (see Sections 5.2.2 and 5.3.1).
- Look for time-sensitive chemicals and ensure they are disposed of at needed times. (see Sections 5.3.2 and 8.3.3).
- Check on high shelves to ensure that chemicals or heavy objects are not stored there—don't store chemicals higher than 5 feet off the floor.
- Examine gas cylinders to make sure they are strapped or secured.
- Check chemical waste to make sure that it is labeled properly (see Sections 5.1.1 and 5.2.1).

Facility Integrity can contribute to incidents.

- Look for damaged floors or ceiling tiles.
- Check plumbing to ensure there are no obvious leaks.
- Check lighting to ensure burned out fixtures are replaced.
- Check electrical receptacles to ensure they are in good condition.
- Check exit signs and door closers to ensure they are in working order.
- Check locks on doors and secure chemical storage to ensure they are working properly.

Personal Protective Equipment is an essential part of your protection against hazards.

- Ensure that you and others in your lab have and use eye protection.
- Ensure that appropriately selected gloves are available and used in procedures.
- Ensure everyone wears lab coats.

Safety Information is essential to recognize chemical hazards.

- Look for posted emergency contact information and ensure it is up to date.
- Look for safety warning signs and ensure they are up to date and accurate—note that signs are not for you but for others who are not familiar with your lab.
- Check the MSDSs for chemicals that you are using—briefly review them if you have not done so recently.
- Examine your chemical inventory to ensure it is up to date for use by emergency personnel (see Section 5.3.1).
- Look for your lab's Chemical Hygiene Plan.

Procedures are the most likely area where incidents will occur.

- Examine your procedures for hazards and ensure that you have taken steps to minimize these hazards.
- Look for emergency equipment that you might need for these procedures.
- Check for all of your emergency equipment and know its location—fire extinguisher, fire alarm, eyewash, safety shower, fire blanket, and spill kit.

RAMP

- *Recognize* hazards in your lab using the inspection process. Using a formal checklist can help you see previously unrecognized hazards.

- *Assess* the risk of the hazards in your lab recognized during your inspection.

- *Minimize* the risk of the hazards in your lab using control measures and PPE.

- *Prepare* for emergencies for the hazards recognized in your lab during your inspection.

References

1. S. J. GOULD. BrainyQuote; available at http://www.brainyquote.com/quotes/authors/s/stephen_jay_gould.html (accessed May 1, 2009).
2. R. HILL. Account of incident related by a colleague.
3. National Research Council. *Prudent Practices in the Laboratory: Handling and Disposal of Chemicals*, National Academy Press, Washington, DC, 1995, pp. 175–177.
4. B. L. FOSTER. Fundamentals of productive laboratory inspections in academia. *Journal of Chemical Health and Safety* **10**(*1*):28–34 (2003).

QUESTIONS

1. Why is it sometimes hard to detect unsafe conditions in a lab in which you regularly work?

   (a) Unsafe conditions that have not yet caused an incident become accepted as "normal."
   (b) Students are probably not trained to conduct "self-assessments" of laboratory safety.
   (c) Unsafe conditions are usually hard to detect.
   (d) Both (a) and (b).

2. When entering a new lab, what circumstances make it easy to overlook unsafe conditions?

   (a) Unsafe conditions and procedures might have been in place for years without incidents, and therefore considered "safe."
   (b) The habits of current and former occupants often get transferred to new lab workers without much reflection on the safety of those habits.
   (c) It may be uncomfortable for a new person to challenge preexisting conditions that are "normal."
   (d) All of the above.

3. What is a good time period for conducting informal inspections of laboratories?

   (a) Daily
   (b) Weekly
   (c) Monthly
   (d) Annually

4. The best way to conduct inspections is

   (a) To use a checklist and have someone else who does not work in the lab accompany you
   (b) To use a checklist and conduct the inspection alone
   (c) To simply "walk through" the laboratory and see what doesn't "look right"
   (d) To take pictures of all unsafe conditions

8.3.2

MANAGING CHEMICALS IN YOUR LABORATORY

Preview This section provides guidance for students who may be assigned responsibility for ordering, receiving, and tracking chemicals.

Inanimate objects can be classified scientifically into three major categories: those that don't work, those that break down, and those that get lost.

Russell Baker, American journalist[1]

INCIDENT 8.3.2.1   CHEMICAL EXPOSURE AND FIRE FROM AN OLD STORED CHEMICAL[2]

A project in a laboratory required N-methyl-N-nitrosourea. After receipt of the chemical, the project was postponed for another higher priority project and the box containing several bottles of N-methyl-N-nitrosourea was set in a corner of the lab. The box was forgotten until several months later when one of the bottles burst from pressure resulting from decomposition of the chemical. The resulting chemical exposure was so irritating that everyone had to leave the laboratory and the building. Someone returned wearing a self-contained breathing apparatus and moved the box of bottles to the hood. The box caught fire in the hood, but was put out with fire extinguishers and water. No injuries were reported.

INCIDENT 8.3.2.2   CHEMICAL SPILL DURING DELIVERY TO THE LAB[3]

An organization had a group of personnel involved in receiving, delivering, and shipping materials into and out of the institution's laboratories. The personnel delivering the materials were principally high school graduates with little or no science background. Several 4-liter bottles of toluene were received and they were transferred to another box filled with other materials. There was no padding or protection provided for the bottles. During delivery the bottles banged together and one or more of the bottles broke, spilling toluene into the open box. This happened in an atrium area within a building. The delivery person did not know what to do, so he left the box and returned to the shipping department without notifying anyone about the spill. Personnel in the building smelled the strong toluene odor and called the health and safety department. They investigated, found the box with the broken bottles, and cleaned up the spill. Although there were strong odors throughout the building, there were no reported injuries.

What lessons can be learned from these incidents?

Responsibility for Managing Chemicals

Chemicals are used in laboratories everyday, but who is responsible for these chemicals? In the university system it is probably difficult to say since chemicals may be the property of a professor whose grant ordered the chemicals, or the property of the department that bought the chemicals used in teaching laboratories, or the property of the university that owns the entire facility.

At some point in time you may be asked to order chemicals for a laboratory and you may now have some responsibility for these chemicals. This seems like a trivial, perhaps even an unimportant,

Laboratory Safety for Chemistry Students, by Robert H. Hill, Jr. and David C. Finster
Copyright © 2010 John Wiley & Sons, Inc.

task, but nothing could be further from the truth. Managing chemicals is a critical and important function in any laboratory and it takes forethought and careful consideration to do the job safely. You should take this responsibility with the perspective that you can make a difference in keeping your institution safe by doing the right thing.

Before Buying

When you need to purchase chemicals for experiments—and you will, someday—it is important that you take several factors into consideration before you do so.

- *Proper facilities.* Ask if your facility has the capability to handle, manage, and store this chemical. Identify where the chemical will be handled and stored before ordering the chemical.
- *Less hazardous substitute.* Ask if it is possible to substitute a less hazardous chemical before you order.
- *Mini- or microscale use.* Ask if it is possible to carry out the experiments on a mini- or micro scale using only very small amounts of this chemical—thus, you would only need to order a small amount of material.
- *Sharing.* Before you order, you should check to find out if this is available from another laboratory within the laboratory, the department, or university. Inventories are often maintained by the university's health and safety office so that would be a good place to start. If you can locate and borrow the chemical, not only will you save money, but you will save time and reduce existing inventories, thus reducing the risks from potential exposure to this chemical, as well as potential costs of hazardous waste disposal. So sharing or exchanging chemicals is perhaps a winning proposition for all involved; this is yet another good reason to have and maintain an up-to-date chemical inventory.

Other sections in the book (1.1.2, 1.2.2, 1.3.4) discuss the concept and practice of green chemistry. Two of the items above, using less hazardous chemicals and working at a smaller scale, are important steps in green chemistry practice. As pointed out elsewhere in these sections, not only is green chemistry safer, it is also usually less expensive both in the short run and in the long run.

Minimize Chemical Orders

A phrase often heard when discussing the use of chemicals is "cradle to grave" accountability. This means that you are responsible for the safe storage, use, and disposal of a chemical from the time that you receive it until it is consumed in a chemical reaction, recycled, or disposed of by burial or incineration. One of our guiding safety principles is to minimize exposures, and one of the best ways to do this is to simply minimize the amount of chemical that is being ordered. This reduces the risk of handling, storage, and disposal. A particularly good, and free, resource about minimizing the volume of chemicals is the booklet *Less is Better: Guide to Minimizing Waste in Laboratories*,[4] published by the American Chemical Society as part of its green chemistry initiative.

For the past few decades the concept of "just-in-time delivery" has been used throughout U.S. manufacturing facilities in many locations. This process involves ordering supplies just before you need them so that on-site storage is minimized or eliminated. This saves money. This same concept has been used when ordering chemicals, too, since chemical suppliers keep large stocks of materials and can ship them the same day that an order is received. This means that there is less need to take the risk of maintaining large chemical inventories since chemicals can usually be obtained so readily. If you order much more than you need you might also end up with a waste disposal situation that costs far more than the cost of ordering the original chemical. Order *only* what you need.

You may find that you only need a few grams but the supplier only offers the chemical in minimum quantities (e.g., 1-lb bottle) that far exceeds your needs. You should ask if the supplier can provide you with only the amount you need for the same price as their minimum quantity. This may not seem like a good deal, but the short-term "cost inefficiency" may become "cost effective" in the long run when you consider the reduced risk and the cost of disposing of chemical waste. Even if they charge you a little more for less, it will be a better deal. Shop around for other suppliers who can sell you the smallest possible amount of needed chemical.

If you must order a significant amount of a chemical, consider ordering several small bottles rather than one large one. This may cost more, but you will be minimizing risks by handling smaller amounts and you will be maintaining the integrity of the chemical by using only one small bottle at a time since some chemicals may degrade when bottles are opened and exposed to atmospheric moisture. For example, prolonged storage increases the risks of peroxide formation in some compounds such as ethyl ether (see Section 5.3.2).

Receiving Chemicals

The point at which you receive chemicals is the ideal time for several important safety measures:

- Inspect the container to make sure it is intact.
- Receive and review the MSDS for this chemical (required on the first shipment of a chemical to an institution; some companies provide these online).
- Date the container(s); don't damage the original label.
- Label the container with the owner's name.
- Apply your institution's special labels—some universities have color-coded labels to help recognize and sort various types of hazards for storage.
- Enter the chemical into the inventory system (see Section 8.3.3 for information about inventories).

Transport, Transfers, and Movement of Chemicals

Handling chemicals at your institution's shipping/receiving department offers opportunities for exposure to personnel who may not have much education or understanding of chemicals. See Incident 8.3.2.2. If you are expecting the receipt of an especially hazardous material, you should alert this department. You may want to be involved in the transfer of this chemical to your laboratory or advise the receiving personnel about its safe handling and transfer. Perhaps you should tell them that you will open the original box yourself and verify its contents, so it is best for them to call you when it arrives. You may want to consider opening and examining containers of volatile chemicals in a chemical hood in case you encounter a leaking or damaged container. This should make you aware that receiving is a potential area where chemical exposure might occur, and you should take appropriate measures to minimize exposures.

Sometimes others may want to use a chemical that is in another lab within an institution and you should also take precautions when transferring hazardous chemicals within your organization. In fact, some restricted chemicals or biological agents may not be permitted to be transferred without making a formal request to a government agency.

Lastly, you should be cautious about shipping chemicals to other institutions, unless you are sure that you have met the requirements of the U.S. Department of Transportation in regard to proper packaging. There are special training courses that teach techniques in packaging and handling hazardous chemicals during transport operations.

RAMP

- *Recognize* that hazardous chemicals are encountered in ordering, receiving, and transporting operations.
- *Assess* potential ways in which exposure might occur during these operations.
- *Minimize* potential exposures by reducing quantities of hazardous chemicals, finding ways to substitute less hazardous chemicals, or using only very small amounts in experiments.
- *Prepare* for emergencies that might occur during receiving or transporting operations.

References

1. R. BAKER. BrainyQuote; available at http://www.brainyquote.com/quotes/quotes/r/russellbak128254.html (accessed February 25, 2009).
2. N. V. STEERE. *Safety in the Chemical Laboratory, Vol. 2*, Accident Case Histories. *N*-Methyl Nitrosoureas Improperly Stored, Division of Chemical Education, reprinted from *Journal of Chemical Education*, Easton, PA, 1971, p. 121.
3. R. HILL. Personal account of an incident in the late 1990s.
4. American Chemical Society. Task Force on Laboratory Environment, Health, & Safety. *Less Is Better: Guide to Minimizing Waste in Laboratories. 2002*; available at http://membership.acs.org/c/ccs/pubs/less_is_better.pdf (accessed February 26, 2009).

QUESTIONS

1. Who is responsible for the chemicals on a college campus or in an industrial facility?

   (a) Everyone who handles or uses the chemicals
   (b) Only the person who bought the chemicals
   (c) Only the person who uses the chemicals
   (d) The chemical hygiene officer

2. What factors should be considered before purchasing a chemical?

   (a) Whether the chemical can be safely stored and handled on site
   (b) If it is possible to buy a cheaper chemical, even if more hazardous
   (c) If it is possible to buy in larger quantity to get a better "per gram" price
   (d) All of the above

3. When purchasing, for example, 4 L of ethyl ether, it is best to purchase

   (a) A 4-L bottle
   (b) Four 1-L bottles
   (c) A 20-L supply (at reduced cost) and divide this into five 4-L containers
   (d) A 55-gallon drum and remove the quantity of ether "on demand"

4. When receiving a new chemical,

   (a) Quickly put it on a shelf in the proper alphabetical location
   (b) Make sure that the container is intact and that it is, in fact, the correct chemical
   (c) Mark the "expiration date" on the label
   (d) Remove the manufacturer's label and replace this with a local label that is more useful

5. When planning to receive a particularly hazardous chemical at your facility, you should

   (a) Plan to take the day off so that if anything goes wrong you will not be blamed
   (b) Notify appropriate people who may be receiving and transporting the chemical locally to advise them about appropriate handling procedures
   (c) Notify anyone else who may potentially handle the chemical and tell them not to touch anything
   (d) Advise people who might transport the chemical locally to remove all warning labels so that no one gets worried

8.3.3

CHEMICAL INVENTORIES AND STORAGE

Preview   This section describes some basics of chemical storage including inventory, cabinetry for chemical storage, and how to group chemicals for safe storage.

I became almost immediately fascinated by the possibilities of trying out all conceivable reactions with them, some leading to explosions, others to unbearable poisoning of the air in our house, frightening my parents.

Richard Ernst, Nobel Prize, Chemistry, 1991[1]

INCIDENT 8.3.3.1   THE EXPLODING BOTTLE[2]

A chemist had worked in a laboratory at a company for three years. He had "inherited" chemicals from a predecessor. He needed some isopropyl ether for an experiment and found a pint bottle on the shelf. He took it to a sink where he tried to open the bottle. The cap was difficult to open so he held it close to his body to get a better grip on the top, and twisted the cap loose. As the cap came loose, the bottle exploded. The chemist suffered devastating wounds to his abdominal area and lost several fingers in the explosion. Although a nearby colleague reached him quickly and got medical care for him, the chemist died at the hospital of internal injuries.

What lessons can be learned from this incident?

Prelude

Many chemists might spend a career working in research laboratories without giving much thought to stockrooms and chemical inventories. A stockroom might be no more than a place where you visit a counter periodically to retrieve glassware and common chemicals and, hopefully, the person in charge maintains the appropriate supplies in an organized fashion. The purpose of this section is to explain the basic organizational principles that allow for the safe storage of chemicals, both in stockrooms *and* in labs. Furthermore, we'll explain the need for a chemical inventory, in stockrooms *and* labs.

Chemical stockrooms in colleges and universities are likely to be staffed by chemists who are knowledgeable about issues of safety related to chemical storage. Computer-based inventories are common. However, academic labs are sometimes woefully deficient with regard to short- and long-term storage, and local inventories are sometimes absent or not up-to-date, since there can be a constant change-over in personnel, and faculty members who supervise these labs sometimes spend little time in the labs themselves.

General Considerations in Chemical Storage

As a student, you may or may not have visited a chemical stockroom. And even if you have worked in one, you may not have fully considered the challenge of just "storing chemicals" safely. The most obvious feature in a stockroom is likely to be many, many shelves of chemicals stored in alphabetical order since, of course, an alphabetical sorting is the first most logical way to store and find chemicals.

Laboratory Safety for Chemistry Students, by Robert H. Hill, Jr. and David C. Finster
Copyright © 2010 John Wiley & Sons, Inc.

You might have noticed, though, that flammable solvents are stored in special cabinets and that some chemicals are stored in refrigerators. These are clues, but not the full picture, of the complexity of storing chemicals.

Unlike the storage of many other kinds of material (such as books in a library or materials in most warehouses) the storage of chemicals goes far beyond just placing items on shelves for easy retrieval. In fact, managing the storage of chemicals is a continuous process requiring careful separation of incompatible chemicals, reviews of existing inventory on a periodic basis, physical examination for container integrity, tracking of time-sensitive chemicals, and updating the inventory with new chemicals and changes to older chemicals.

Most laboratory chemists use many chemicals and most often these are in relatively small quantities. It is usually only solvents or acids that are used in larger quantities (4-L bottles) since they are used in larger quantities even in research labs. Safely storing chemicals is a challenge since this depends on the number and types of hazards posed by the various chemicals as well as the quantity being stored. Although you may think that the "problems of storing chemicals properly" is a task only for a stockroom manager, the proper storage of chemicals is a critically important part of every chemist's safety efforts because all chemists inevitably store chemicals in labs. In both labs and stockrooms, there are many aspects to chemical storage: where to store chemicals, how long they should be stored before they must be discarded, what environmental conditions need to be maintained, what kind of cabinets or shelving should be used, and what chemicals can be stored together and what chemicals cannot be stored together. This latter aspect refers to chemical compatibility, the most important consideration in chemical storage.

Chemicals need to be stored in approved storage places. You should not store chemicals on the floor (see Incident 5.3.6.1), in laboratory hoods, or in places that clutter the working surfaces for experiments and laboratory operations. Remember part of your job in the laboratory is to minimize opportunities for exposure, so you and others can do your experiments without bumping into, kicking, knocking over, or encountering improperly stored chemicals.

Improper chemical storage has led to many incidents, and some chemical safety experts indicate that perhaps one-quarter of all chemical incidents are related to chemical storage issues.[3] Improper chemical storage has caused or contributed to incidents resulting from:

- Storing incompatibles together
- Failing to track chemicals that form explosive by-products upon extended storage
- Storing chemicals on the floor or other locations where they can easily be encountered by an inattentive person
- Storing chemicals in a cabinet or shelf that itself is incompatible with the chemicals
- Storing larger volumes of flammables than are allowed by codes and regulations
- Events such as fires, earthquakes, or storms that release stored chemicals unexpectedly and further contribute to an already bad situation

While we will provide some general guidelines for chemical storage, there are, perhaps surprisingly, no universal methods that work in all cases. The greater the number of chemicals you have on hand, the greater the possibility that some chemicals are going to be incompatible, not tracked, or improperly stored. This should be viewed as an incentive to keep your chemical inventory to a minimum.

A Good Chemical Inventory — A Continuous Process

A chemical inventory is a list of all the chemicals at a facility or lab. Today, such inventories are almost always located in a computer database. Why do you need an inventory? It will keep you and others safe and secure. Inventories are used for:

- Locating needed chemicals
- Chemical sharing

- Hazard identification
- Classification for storage
- Tracking for time-sensitive chemicals
- Chemical cleanup events
- Planning/space management
- Compliance with codes and regulations (e.g., OSHA's 29 CFR 1910.1450—"The Lab Standard")
- Chemical security
- Information sources for emergency response personnel

In addition to the reasons listed above, in some circumstances, it is *required* that you have a chemical inventory. Most communities in the United States have a Local Emergency Planning Committee (LEPC) that requires that chemical inventories be submitted, mostly for emergency response purposes. Additionally, some insurance agencies and risk management consultants will require chemical inventories.

Chemical inventory systems can be small or large. Keeping track of your own personal store of chemicals is relatively easy to do but you can imagine that chemical storage stockrooms in a college chemistry department are much more challenging situations, particularly with many persons likely having access (with varying degrees of interest and knowledge about the need for a good inventory).

It is both safer and cheaper to keep your chemical inventory to a minimum in a lab. When you consider purchasing chemicals, it is almost always best to minimize the amount that you buy. Buying larger amounts than you need because it is cheaper per unit is a poor practice that, later on, usually drives up the cost of hazardous waste processing if all of a chemical is not used before its shelf life is exceeded. Getting rid of excess hazardous chemical waste often costs 3–10 times the cost of the original purchase.

If there is no existing inventory in the lab in which you are working, you will find that constructing one will be a time-consuming challenge. Nevertheless, it is a good safety practice that will pay back dividends in time and money in the long run. It is likely that, even if your college or university has a central stockroom, your particular laboratory will have its "own" chemicals and will need its own inventory. Having a good inventory of chemicals is especially important to academic institutions where a high turnover of people graduating can generate many "old" chemicals that may have no further use. You don't want to leave unexpected hazards for new students.

If you store chemicals in your laboratory, you need to have a chemical inventory that specifies the chemical name, molecular formula, CAS number, source, container size, date received, and location. Individual labs may also track other information about chemicals. Automated chemical inventory systems may require more information. There are commercial inventory systems available and it is likely that you will not have any choice in this selection.[4–7]

Compatibility Determines Proper Storage

Separating incompatibles during chemical storage is essential. You must learn how to identify incompatibles (see Section 5.2.3 on incompatibles) and then store them away from each other. This is the reason that you should not store chemicals in alphabetical order since this leads to incompatibles being stored near each other. The recommended practice is that you should separate chemicals into compatible groups such as acids, bases, oxidizers, and pyrophorics (see below). However, there are several reasons why the seemingly reasonable practice of separating into chemically compatible groups may also lead to problems:

- Chemicals frequently fall into multiple classes.
  - Nitric acid is an inorganic acid and an oxidizer.
  - Acetic acid is an organic acid and is combustible (not to be stored with oxidizers).
- Identifying or recognizing incompatibles may not be easy.

○ Sources may not identify all incompatible classes.

○ Classification may require advanced chemical knowledge to understand chemical names and potential reactive sites.

• Finding adequate space and storage for all classes requires more space and more management.

The most important thing you should know is that there is no single system that will work for all compounds, and it is going to take some effort to decide how to safely store your chemicals. You must locate information that can be used in determining compatibility. Information about hazard classes is found on most labels and information about incompatibles may be found in MSDSs, *Bretherick's Handbook of Reactive Chemical Hazards*,[8] *Prudent Practices in the Laboratory: Handling and Disposal of Chemicals*,[9] or the *NIOSH Pocket Guide to Chemical Hazards*.[10]

Chemical compatible groupings for storage have been suggested by many, including, for example, the U.S. Coast Guard CHRIS,[11] Texas A&M University,[12] University of Iowa,[13] Massachusetts Institute of Technology,[14] Washington University in St. Louis,[15] and the National Research Council.[16] All of these methods are similar but still have some differences. All are suitable generally, but there may be anomalies with each system for the reasons listed above: some chemicals do not fit in a class, some fit into more than one class, or some fit into a class that has a chemical with which it is incompatible. Each scheme separates chemicals into classes: flammables, oxidizers, corrosives (acids, bases), reactive chemicals, highly toxic, and others. Some first divide chemicals into organic and inorganic compounds. A newly proposed system for separating incompatibles involves first a separation by primary classes (explosives, flammables, corrosives, and caustics), then followed by another separation process into seven classes based on reactivity.[17] There is no system that will work for all compounds in all cases.

Time-Sensitive Chemicals

While many compounds degrade to harmless by-products, some chemicals degrade into more hazardous substances over time and need to be tracked.[18–20] For chemicals that degrade over time, it is necessary to date the label when it was received and when it was first opened. We have previously discussed how some compounds degrade to form dangerous products such as peroxides (see Section 5.3.2). These compounds must be carefully tracked and discarded after a specified period (see Table 5.3.2.2 Peroxidizable Chemicals). Other compounds also need to be tracked by time in storage (see *Special Topic 8.3.3.1* Picric Acid—A Potential Chemical Storage Hazard).

Tracking time-sensitive chemicals is a critical part of a chemical storage plan and it must not be overlooked or neglected. There is an old adage: "When a committee is in charge, no one is in charge." When applied to chemical storage this might become: "If you don't track your chemicals, no one else is going to track your chemicals." You might be at a large institution that has an environmental, health, and safety office that does this, but this is uncommon. Even then, they can only do this with your help.

SPECIAL TOPIC *8.3.3.1*

PICRIC ACID—A POTENTIAL CHEMICAL STORAGE HAZARD

Picric acid, 2,4,6-trinitrophenol, is used in many laboratories, especially for staining in microscopy in biological laboratories. It is sometimes used in organic laboratories for preparing crystalline "picrate" salts. It has also been used for creatinine measurements in clinical and drug screening, and as a preservative for biological specimens.

Picric acid is a close relative to TNT (trinitrotoluene) and is perhaps a more powerful explosive than TNT, especially if it becomes dry. When picric acid is sold to a laboratory, it comes as a "wetted" solid with more than 30% water. The water makes picric acid a "flammable solid," but it is not classified as an explosive unless its water content falls below 30%—when this happens it becomes a Class 1.1.D explosive. In practical terms, this means that picric acid should never be allowed to dry out. Since it is also a strong acid, it should not come in contact with metals or concrete surfaces, where it might form explosive picrates.

It is not uncommon for an "old" bottle of picric acid to be discovered in a dry state—this is a dangerous occurrence and requires attention from emergency personnel and probably the local bomb squad. There are many incidents involving dried out picric acid within university settings. One report described the removal of several bottles of picric acid using a bomb squad robot and as the last bottle of picric acid was placed in the bomb container, it exploded, producing a crater.[18]

If possible you should seek out an alternative for picric acid. If you must use picric acid, order the absolute minimum quantity (solutions are best), date it when you receive it, and monitor it to ensure that it is wet. J. T. Baker indicates that if crystals roll over each other on tilting the bottle, that the picric acid has dried out and it is explosive.[22]

If the moisture content of the picric acid is high, there is little cause for concern. However, if you suspect that a bottle of picric acid is dry, do not touch the container. Do not attempt to screw the cap off a bottle, since the friction of unscrewing the lid could detonate the picric acid. Metal caps are the most dangerous since metal picrate salts may also have formed. Seek assistance from your institution's health and safety office. They will likely call the local bomb squad for safe removal.

Chemical Storage Locations, Cabinets, and Shelving

Now that we have addressed the properties of chemicals that impact storage, let's look at physical requirements. If you have a stock of chemicals in your laboratory, you will likely have chemicals distributed in a number of places as a result of incompatibility. First, we reiterate the policy that chemicals should *not* be stored in hoods (except for very brief times, while "in use"). The temptation to do this is strong since hoods are ventilated and seen as a relatively "safe" place to keep chemicals. The inevitable result of this practice is that hoods become cluttered with bottles of chemicals, which both reduces the effective airflow in the hood and uses space that should be reserved for chemical reactions. (See Sections 7.1.4 and 7.2.3.)

For flammable, volatile, or odorous chemicals, an NFPA-approved flammables storage cabinet should be used.[21] This is typically a yellow metal cabinet fitted with spark-proof ports that can be connected to local exhaust ventilation (see Figure 8.2.1.1). There is no requirement for ventilation but doing so reduces hazards and odors. However, if flammables storage cabinets are vented they must be ducted directly to outdoors.

Any room used to store chemicals should have single-pass exhaust ventilation to keep volatile hazards and odors to a minimum. Laboratories should have no single containers of flammables larger than 4 liters or 1 gallon, and 5-gallon containers of chemicals should be housed in a central location where transfers can safely be made. There are limitations established by fire codes on storage of flammables in various types of facilities, such as a laboratory.[21] A more detailed description of NFPA guidelines for the storage of flammables is presented in Section 8.2.1.

Ideally, chemical stockrooms should not "smell like chemicals." While all stored chemicals should be in leak-proof containers and/or in bottles with tight-fitting caps, it is inevitable that some chemical fumes and odors will escape. Stockrooms should be designed with a constant turnover of air that removes vapors and odors. If you visit, or work in, a chemical stockroom that "smells like chemicals," you are in an environment with inadequate ventilation that is likely hazardous to your health. OSHA requires that the ventilation for these storerooms be capable of exchanging the air six times per hour.

A laboratory can be defined as a single room, a group of rooms, or even an entire floor; it depends on the construction used in the laboratory and its fire rating. The limitations on flammable chemical storage depends on the laboratory construction, the kind of laboratory (teaching lab versus research lab), the fire protection in the laboratory (automatic fire extinguishing system or not, the former allows more storage), and how chemicals are stored (fire-resistant storage cabinets, glass bottles, metal cans, plastic bottles, or safety cans).

Acids should be stored in cabinets made of plastic or wood. Metal should be avoided unless it is coated with a heavy abrasion-proof, acid-resistant paint. Wood cabinets used to store chemicals should be coated with a chemically resistant paint such as epoxy. This is essential for cabinets used for storing oxidizers since they can react with unprotected wood. Reactive materials must be stored in a cool dry place out of direct sunlight. Water-reactive materials should not be stored directly under fire-sprinklers (!) and there may be code requirements for these compounds. Some compounds such as organic peroxides may need to be stored under refrigerated conditions and it is imperative that these only be stored in spark-proof or explosion-proof refrigerators. Conversely, it may increase the hazard if some organic peroxides are stored at too low a temperature. Therefore, it is important to adhere to the recommended storage requirements for specific reactive chemicals.

Other considerations include ensuring that shelves are not overloaded or overcrowded so that chemicals might be pushed out of sight. Shelves that have a front-edge lip can prevent bottles from falling as a result of vibration, tremors, or earthquakes. It is a good practice to secure shelving to walls to prevent them from tipping. A common good practice is to place your chemicals in trays (plastic or metal depending on the chemical) so that spills might be better contained. Chemicals should not be stored above eye level.

When You Are Working in Labs: RAMP

Storing chemicals at first seems like a very tangential matter when one is working on synthesizing new chemicals, analyzing chemicals, or performing other very interesting chemical experiments. However, it is easy to see that improper storage can make a lab a very hazardous environment and easily compromise the safety and success of those working in the lab.

- *Recognize* the need to determine how to best store hazardous chemicals.
- *Assess* potential exposure paths to hazards from chemical storage.
- *Minimize* the chemical hazards through proper chemical storage.
- *Prepare* for emergencies dealing with hazardous materials.

References

1. R. ERNST. BrainyQuote; available at http://www.brainyquote.com/quotes/authors/r/richard_ernst.html (accessed January 31, 2009).
2. N. V. STEERE. Control of Hazards from Peroxides in Ethers. In: *Safety in the Chemical Laboratory*, Vol. 1, Division of Chemical Education, American Chemical Society, Easton, PA 1967, pp. 68–70.
3. F. SIMMONS, D. QUIGLEY, H. WHYTE, J. ROBERTSON, D. FRESHWATER, L. BOADA-CLISTA, and J. C. LAUL. Chemical storage: myths vs. reality. *Journal of Chemical Health and Safety* 15(2):23–30 (2008).
4. J. RAPPAPORT and J. LICHTMAN. Ongoing development of a chemical/biological inventory and safety management solution for Temple University. *Journal of Chemical Health and Safety* 12(5):4–8 (2005).
5. L. M. GIBBS. ChemTracker Consortium—The higher education collaboration for chemical inventory management and regulatory reporting. *Journal of Chemical Health and Safety* 12(5):9–14 (2005).
6. M. E. COURNOYER, M. M. MAESTAS, D. R. PORTERFIELD, and P. SPINK. Chemical inventory management: the key to controlling hazardous materials. *Journal of Chemical Health and Safety* 12(5):15–20 (2005).
7. B. L. FOSTER. The chemical inventory management system in academia. *Journal of Chemical Health and Safety* 12(5):21–25 (2005).
8. P. G. URBEN (editor). *Bretherick's Handbook of Reactive Chemical Hazards*, 7th edition, Elsevier, New York, 2007.
9. National Research Council. Prudent Practices in the Laboratory: Handling and Disposal of Chemicals, Appendix B: Laboratory Chemical Safety Summaries, National Academy Press, Washington, DC, 1995.
10. National Institute of Occupational Safety and Health. *NIOSH Pocket Guide to Chemical Hazards*, September 2005; available at http://www.cdc.gov/niosh/npg/default.html (accessed February 4, 2009).
11. U.S. Coast Guard. Chemical Hazards Response Information System (CHRIS); available at http://www.chrismanual.com/ (accessed February 4, 2009).
12. Texas A&M University. Segregation Based on Hazard Classes; available at http://safety.science.tamu.edu/chemstorage.html (accessed February 4, 2009).
13. University of Iowa, Health Protection Office. Proper Chemical Storage; available at http://www.uiowa.edu/hpo/chemsafety/chemstor.html (accessed February 4, 2009).
14. Massachusetts Institute of Technology. Chemical Storage Table Supplement for Chemical Storage SOG (3–06); available at http://web.mit.edu/environment/pdf/Chemical_Storage_Table.pdf (accessed February 4, 2009).
15. Washington University in St. Louis. Chemical Storage Guidelines; available at http://epsc.wustl.edu/admin/documents/chemical_storage_guidelines.pdf (accessed February 4, 2009).
16. National Research Council. Prudent Practices in the Laboratory: Handling and Disposal of Chemicals, Chapter 4 Management of Chemicals. 4E Storage of Chemicals in Stockrooms and Laboratories, National Academy Press, Washington, DC, 1995.
17. J. J. M. WIENER and C. A. GRICE. Practical segregation of incompatible reagents in the organic chemistry laboratory. *Organic Process Research and Development*, 13(6):1395–1400 X-X (2009). [DOI 10.1021/op900094d].
18. J. BAILEY, D. BLAIR, L. BOADA-CLISTA, D. MARSICK, D. QUIGLEY, F. SIMMONS, and H. WHYTE. Management of time-sensitive chemicals (I): misconceptions leading to incidents. *Journal of Chemical Health and Safety* 11(5):14–17 (2004).
19. J. BAILEY, D. BLAIR, L. BOADA-CLISTA, D. MARSICK, D. QUIGLEY, F. SIMMONS, and H. WHYTE. Management of time-sensitive chemicals (II): their identification, chemistry, and management. *Journal of Chemical Health and Safety* 11(6):17–22 (2004).

20. D. QUIGLEY, F. SIMMONS, D. BLAIR, L. BOADA-CLISTA, D. MARSICK, and H. WHYTE. Management of time-sensitive chemicals (III): stabilization and treatment. *Journal of Chemical Health and Safety* **13**(*1*):24–29 (2006).

21. National Fire Protection Association (NFPA). NFPA 45—Standard on Fire Protection for Laboratories Using Chemicals—2004 Edition; available at http://www.nfpa.org/aboutthecodes/AboutTheCodes.asp?DocNum=45 (accessed March 10, 2009).

22. J.T. Baker Material Safety Data Sheet. Picric Acid, Wet; available at http://www.jtbaker.com/msds/englishhtml/p4556.htm (accessed February 20, 2009).

QUESTIONS

1. Why is storing all chemicals in alphabetical order *not* a good idea?

   (a) It is hard to know how to alphabetize some chemicals since they have multiple names.

   (b) Alphabetical arrangement might place incompatible chemicals next to each other.

   (c) Some chemicals should not be stored on an "open shelf" but instead in a vented cabinet or in a chemical refrigerator.

   (d) Both (b) and (c).

2. How is storing chemicals different from storing books?

   (a) Books do not deteriorate over time to become more hazardous.

   (b) Books are not in containers that might crack or otherwise lose their intergrity.

   (c) Books do not react with each other.

   (d) All of the above.

3. The safest place to store most chemicals in a laboratory is

   (a) On the floor

   (b) In cabinets (for flammable or corrosive chemicals) or on shelves

   (c) In chemical hoods

   (d) In desk drawers, out of sight

4. Maintaining an accurate chemical inventory in a laboratory

   (a) Is required by some Local Emergency Planning Committees

   (b) Is so tedious that safety experts agree it is not worth the effort

   (c) Is almost impossible since the daily use of chemicals makes this impractical

   (d) Is recommended, but not required, by the "Lab Standard"

5. What are some challenges faced by the process of organizing chemicals for storage?

   (a) Chemicals frequently fall into multiple classes.

   (b) Identifying or recognizing incompatibles may not be easy.

   (c) Finding adequate space and storage for all classes requires more space and more management.

   (d) All of the above.

6. The best scheme for storing chemicals is

   (a) The one that works best and safest at your institution

   (b) Found in *Prudent Practices in the Laboratory: Handling and Disposal of Chemicals*

   (c) In the *NIOSH Pocket Guide to Chemical Hazards*

   (d) Located at the OSHA web site

7. Time-sensitive chemicals

   (a) Are rarely used in most laboratories

   (b) Should be marked with the date of receipt and the date when opened

   (c) Should be stored in a chemical vault

   (d) Should always be used immediately upon receipt

8. Which categories of chemicals require special storage considerations?

   (a) Flammables and combustibles

   (b) Corrosives

   (c) Unusually odorous chemicals

   (d) All of the above

9. Shelves that store chemicals should be

   (a) Stabilized so that they cannot easily be tipped over

   (b) Constructed from floor-to-ceiling to save space

   (c) Made of materials that react with chemicals so that leaks can quickly be detected

   (d) All of the above

8.3.4

HANDLING HAZARDOUS LABORATORY WASTE

Preview   This section describes the management of hazardous chemical waste from the laboratory.

The most important environmental issue is one that is rarely mentioned, and that is the lack of a conservation ethic in our culture.

Gaylord Nelson, American politician (1916–2005)[1]

INCIDENT 8.3.4.1   TOLLENS' TEST EXPLOSION[2]

Two students were standing at the window of an unoccupied undergraduate organic chemical stockroom when an explosion occurred. Small pieces of a fragmented bottle hit one student causing minor cuts. A fire broke out just after the explosion and a single sprinkler head activated, spraying the area. The roll down doors automatically closed and a building alarm sounded. The single sprinkler put out the fire. An investigation showed that a bottle of hazardous waste sitting on the bottom of a lab cart had exploded. The fire resulted from cardboard and bottles of ethanol on the floor near the exploded waste container. The hazardous waste bottle was from an undergraduate qualitative experiment to detect aldehydes using the Tollens test, also known as the silver mirror test. One product of this test is known to be explosive fulminating silver that precipitates from the Tollens reagent if the waste is allowed to stand for extended periods. (To better understand this incident, see *Special Topic 8.3.4.1* Precautions for Tollens' Reagent.)

INCIDENT 8.3.4.2   DOWN THE DRAIN[3]

A chemical company began making pesticides in 1980. From that time until 2007 employees poured the waste water from the pesticide production down the floor drain. This untreated hazardous waste went directly into the city sewer system. In 2007 an EPA inspection identified this illegal practice. The supervisor of the operation was sent to prison for 6 months, followed by 6 months home confinement, and a fine of $100,000. The vice president of the company is facing a similar fate.

What lessons can be learned from these incidents?

Prelude

Section 8.1.1 presented an introduction to the issue of handling academic lab wastes in undergraduate, course-related laboratories. Mostly, the instructions in academic labs are to "dispose of the waste in the appropriate container," which necessitates that someone else will appropriately process the waste. Section 8.1.1 also discussed many reasons to be mindful of chemical wastes so as to be environmentally responsible, cost-conscious, and safe.

This section assumes that you may now be working in advanced lab or research labs where the simple instruction "put it in the appropriate waste container" is not helpful since you will handle many more unique chemicals and unique chemical wastes. Furthermore, after graduation you may find yourself in graduate school or the chemical industry where there may be new rules about handling

Laboratory Safety for Chemistry Students, by Robert H. Hill, Jr. and David C. Finster
Copyright © 2010 John Wiley & Sons, Inc.

waste from unique sources. Perhaps you will even work for a small chemical company and *you* will have to figure out what to do with various chemical wastes. With these less-certain situations in mind, let's consider how to handle hazardous wastes in a broader context.

What Happens to Hazardous Waste from the Laboratory?

Well, it just doesn't disappear. Somebody has to do something with it. And this can cost your institution a whole lot of money. It is important that you understand what happens to waste when it leaves your laboratory, and that you take steps to minimize your waste and correctly identify what is in your waste. You should know that what happens to it depends a lot on you! Another reason you should know something about this is that if you go to work as a chemist you will have to handle hazardous waste properly to avoid putting that company at financial risk. We will explain why below.

Most hazardous waste is handled by the environmental, health, and safety (EHS) department in larger universities. It takes special knowledge and handling to do this and the EHS department has professionals who know what to do. However, not all academic or employment situations will provide an EHS department. In smaller institutions someone, perhaps even you, may be designated the job to handle hazardous waste. This assignment may be a challenge since that person will need to learn the laws, regulations, and procedures to properly handle this waste.

The U.S. Environmental Protection Agency (EPA) is the government agency that regulates hazardous waste, including that from laboratories. Under the guidance of the 1976 Resource Conservation and Recovery Act or RCRA (pronounced "reck rah"), the EPA established very strict guidelines for handling hazardous waste that require lots of documentation, including:

- How long you can hold it before you must dispose of it
- Identifying the contents of the hazardous waste
- Packaging it up for transport to its disposal site
- Transporting to the disposal site
- Disposing of the waste
- Obtaining proof of final disposal

You should know that this is an expensive proposition, and what you do when you generate the hazardous waste can have a dramatic effect on the cost for disposal.

SPECIAL TOPIC *8.3.4.1*

PRECAUTIONS FOR TOLLENS' REAGENT

Tollens' reagent is a very mild oxidizing agent used to detect aldehyde groups.[4] Other chemical groups also produce a positive Tollens' test, including some aromatic amines and some α-substituted ketones. This procedure is sometimes called the silver mirror test. The reagent is made from 5% aqueous solution of silver nitrate to which has been added a drop or two of 10% sodium hydroxide to make the solution alkaline. Concentrated ammonium hydroxide (30%) is added until the precipitated silver oxide just dissolves in order to minimize excess ammonia. It must be freshly prepared and should not be stored since the resulting decomposition products are explosive and shock-sensitive.

Tollens' reagent is added to an aldehyde solution and the oxidation produces free silver that coats the side of the flask—thus making a silver mirror.

$$RCHO + 2\ Ag(NH_3)_2OH \rightarrow 2Ag(s) + RCOONH_4 + H_2O + 3\ NH_3$$

Colorless solution $\rightarrow$ Silver mirror (deposited on glass)

This reaction is a dramatic demonstration and is often used to teach students about principles of mild oxidation of aldehydes by silver ions.

While the Tollens reagent is a great teaching tool, precautions are particularly important in the cleanup stages of experiments with Tollens' reagent. Tollens' reagent solutions and waste containing silver ammonium ion have long been known to be unstable upon storage.[5] Tollens described this hazardous property of his reagent in 1882. Thus, these ammonia–silver solutions should be discarded immediately after use. The waste solutions should be acidified with dilute acid before disposal.[6,7] (Disposal is not "down the drain," of course, due to the silver.) They should not be stored in a container to attempt recovery of the silver. Silver nitride, Ag_3N, is believed to be the explosive by-product formed from decomposition of aminesilver hydroxides, $Ag(NH_3)_x OH$. Silver nitride is friction-sensitive even when wet, and when dry is an extremely sensitive explosive. Ammoniacal silver solutions have been known to explode spontaneously in suspensions; see Incident 8.3.4.1.

What Is My Role in Hazardous Waste Disposal?

You have two primary responsibilities: (1) handling hazardous waste properly, including identifying the contents of hazardous waste that you generate, and (2) minimizing hazardous waste.

In your undergraduate laboratories during the first two or three years, the experiments have been designed in advance and this includes steps to deal with hazardous waste. You are told exactly what to do with your waste. It is most often deposited or poured into a container, and the contents of the waste are known and are labeled. In these early labs your choices were very limited. In advanced labs your choices provide more latitude, with responsibility to act more independently.

When you begin to move into more advanced chemistry courses, the structure of laboratory sessions often changes, becoming less prescriptive and more personalized as to your particular assigned laboratory experiment. It is not unusual that there are several different experiments going on simultaneously. It is at this point that your responsibilities for handling hazardous waste change. You need to understand what is required and why.

In the laboratory when it comes to hazardous waste, the really important issue is *identifying* the contents of your hazardous waste. We can hear you groaning now. Why do I need to worry about this stuff, after all it's just waste and I just want to get rid of it. Well, so does your institution. However, if the contents of a hazardous waste container are not labeled, that is, unknown—well, as Tom Hanks said in *Apollo 13*, "Houston, we have a problem." Well, maybe it's not the same magnitude as a marooned astronaut but it is a very real problem.

Hazardous waste cannot be disposed of unless its contents are known or it has been characterized or tested for the presence of various classes of hazardous materials. We don't really want to be put into a situation where we have to do this because it is going to cost us in effort and dollars.

The bottom line is that you need to identify the contents of your hazardous waste on the official form provided by your university. You need to be as accurate as possible, so as you add waste to a container, write it on the label. You should know that if you are not accurate, you may be placing someone who handles your waste at risk since waste is often processed so that it is combined. If something is not labeled correctly, this could result in a reaction when the waste is mixed and a resulting incident.

A little time on your part will save your institution and others enormous amounts of time and money. So think of labeling hazardous waste as a normal and necessary part of your responsibility when working in the laboratory. Just do it!

Also, don't forget that you must always consider the safety side of handling chemical waste. In your prescribed laboratory classes, the instructors told you where to put your waste. In your more independent classes and research, you will have to figure this out yourself. Always consider asking someone who knows.

Perhaps the most important aspect of handling hazardous waste is the possibility of incompatibility. We have already cited incidents where mixing incompatibles in waste containers can lead to adverse events such as injuries, fires, and explosions (see Incidents 1.1.1.1, 3.1.2.1, 5.2.3.1, and 7.1.1.1). Similarly, inappropriate handling of hazardous waste can lead to incidents (see Incidents 8.1.1.1 and 8.3.4.1). You may want to review the section on incompatibles to help ensure that you properly consider what you should and should not mix together (see Section 5.2.3).

Requirement of Compliance with Regulated Hazardous Chemical Waste

While it will not be your responsibility to comply with regulations governing chemical waste, you should have just a brief understanding of the process. If you continue working in the laboratory, chemical waste management can become an issue that impacts you and others.

RCRA requires a significant amount of paperwork that increases if the status of your institution changes from a small waste producer to a large scale producer. Your initial labeling of a waste container is only the start of a lengthy process. The rules for handling hazardous waste depend on how much waste is generated. Under EPA rules[8] there are three classes of hazardous waste generators:

- *Conditionally exempt small quantity generators.* These generate <100 kg/month hazardous waste and <1 kg of acutely toxic waste in a given month; there is no time limit on holding hazardous waste but it cannot exceed 1000 kg.

- *Small quantity generators.* These generate between 100 and 1000 kg/month but cannot exceed 6000 kg on site; there is a 180-day time limit on storage.

- *Large generators.* These generate >1000 kg/month; there is a 90-day limit on storage.

Many states do not recognize the first designation, and even if they did, unless your institution is small, it will most likely fall into one of the two latter categories. Once chemical waste has been generated, picked up from your laboratory, and moved to an accumulation site for preparation for shipping (your institution probably has a waste room somewhere), the clock starts ticking. It is now officially designated as hazardous waste and it must be disposed of within a designated number of days: 180 days for a small quantity generator and 90 days for a large quantity generator.

The EHS department or institution contracts with companies that take care of the actual disposal that involves incineration, treatment, or recovery for fuel. This is an expensive operation, but it is absolutely necessary. These contracting companies will not accept unknown waste. It must be tested and classified as to its hazardous properties. The companies can do this on your site but it is very expensive.

The hazardous waste is usually prepared for shipping and disposal by segregating the waste into comparable groups that are packed into "labpacks." These labpacks consist of an outer container that may be a steel drum or a fiber drum. The inside is filled with containers of waste and an absorbent material such as vermiculite that protects the bottles from breakage and can absorb chemicals in case of a leak. These containers are taken by the contractor to their waste disposal site for ultimate processing and disposal.

Because the EPA regulates hazardous wastes in order to protect the environment (not to prevent human exposures), they define hazardous properties differently than agencies such as OSHA, whose purpose is to protect humans. According to the EPA, hazardous waste is chemical material that has been or will be discarded and it has one or more hazardous characteristics or is on one of four lists (called F, K, P, and U Lists). The characteristics regulated are flammability, corrosivity, reactivity, or toxicity.[9] Toxicity is defined as[1] a liquid with concentrations above those listed on the Toxicity Characteristic Leaching Procedure (TCLP) list; or[2] a solid that when extracted by the TCLP, has a resultant concentration about listed concentrations. This definition has little to do with the human exposure definition of toxicity that you will find in the toxicology sections of this book.

We point out one last important point about hazardous waste that you generate. Your institution is responsible for the ultimate disposal of the hazardous waste so if their contractor does not dispose of the waste properly and it is uncovered years later, your institution (not the contractor) will be held accountable by regulators. Most EHS departments require contractors to provide them with a certificate of ultimate disposal before they pay them.

Minimizing Hazardous Waste — "Less Is Better"

All of this should help you understand why minimizing hazardous chemical waste should be an integral part of all chemical processes. The American Chemical Society publishes a booklet entitled *Less is Better: Guide to Minimizing Waste in Laboratories*[10] that addresses ways to minimize hazardous waste.

It offers the strategies below that have now become more familiar and have been adopted as part of the strategy of green chemistry (see Sections 1.1.2, 1.2.2, 1.3.4).

- Don't order more of a chemical than you need; then you won't have much hazardous waste.
- Substitute less hazardous or nonhazardous chemicals for hazardous ones.
- Use small (mini or micro) scale experiments to minimize the chemicals needed.
- Share chemicals with colleagues.
- Recycle or reuse wherever possible.
- Separate various wastes as much as possible to allow maximum treatment and recovery of chemicals by waste handlers.
- Spread the word about the benefits of waste minimization and other laboratory pollution prevention efforts.
- Consider incorporating, as part of your planned experiments, methods that minimize waste, such as recycling solvents or neutralizing solutions. *Special caution*: Regulations require permits for treatment of waste, and violations carry considerable penalties; ensure that you are not violating these regulations in your steps to minimize waste.[11,12] The EPA and most state regulators allow elementary neutralization (pH adjustment) of hazardous waste without a permit; the EPA does not require permits for recycling.

Hazardous Biological Infectious Waste

Today, there is more and more overlap between chemistry and biology. As a result, some chemistry laboratories generate biological infectious waste if they handle biological cultures or animal or human specimens. These materials also require special handling to ensure that no one will be exposed to an infectious agent.

Biological waste can be taken off-site by a contractor for treatment at their facilities. However, most facilities treat biological waste on site. Materials potentially contaminated with infectious materials are placed in biohazard bags in pans and these pans are treated with high temperature and steam in an autoclave. There are established protocols and procedures for doing this. It is important that no volatile organic solvent be included in this process, since this could lead to exposures or fires.

Biological waste can also be disinfected with chemical agents such as bleach or phenolic disinfecting agents. The most common agent is bleach since it is inexpensive and will likely result in a product that can be disposed of down the sanitary sewer. Other chemical disinfection may result in chemical waste that needs further handing by chemical waste management.

RAMP

- *Recognize* that chemical waste can be hazardous and requires special handling.
- *Assess* the risks of the chemical waste that you might generate, looking for ways to minimize potential for exposures.
- *Minimize* the amounts of hazardous waste you generate in the laboratory.
- *Prepare* for emergencies with hazardous waste in the same way that you would prepare for other chemical hazards.

References

1. G. NELSON. BrainyQuote; available at http://www.brainyquote.com/quotes/quotes/g/gaylordnel215707.html (accessed March 15, 2009).
2. University of California, Riverside. Science Laboratories 1 Building Storeroom Explosion and Fire of May 10, 2005—Final Report; dated August 24, 2005.
3. I. AMATO. Newscripts: Scofflaws' just desserts. *Chemical & Engineering News* **87**(*37*):48 (2009).
4. D. W. MAYO, R. M. PIKE, and P. K. TRUMPER. *Microscale Organic Laboratory: With Multipstep and Multiscale Synthesis*, John Wiley & Sons, Hoboken, NJ, 2000, p. 631.
5. P. J. URBEN (editor). *Bretherick's Handbook of Reactive Chemical Hazards*, 7th edition, Vol. 2 (see fulminating metals, fulminating silver, Tollen's reagent), Elsevier, New York, 2007, pp. 147; 401–402.

6. Comenius—European Cooperation on School Education. Hands-on-Science Project. Chemical Safety Data: Tollens' Reagent; available at http://cartwright.chem.ox.ac.uk/hsci/chemicals/tollens_reagent.html (accessed April 20, 2009).
7. Absolute Astronomy. Tollens' Reagent; available at http://www.absoluteastronomy.com/topics/Tollens'_reagent (accessed April 20, 2009).
8. Hazardous Waste Generators. Available at http://www.epa.gov/waste/hazard/generation/index.htm (accessed August 15, 2009).
9. U.S. Environmental Protection Agency. 40 CFR 262 Subpart K; available at http://ecfr.gpoaccess.gov/cgi/t/text/text-idx?c=ecfr&sid=385ea9ac5d6e0fb83be2351f0836eb09&rgn=div5&view=text&node=40:25.0.1.1.3&idno=40#40:25.0.1.1.3.11 (accessed August xx, 2009).
10. American Chemical Society. *Less is Better*: *Guide to Minimizing Waste in Laboratories*; available at http://membership.acs.org/c/ccs/pubs/less_is_better.pdf (accessed March 20, 2009).
11. U.S. Environmental Protection Agency. 40 CFR 260.10; available at http://www.access.gpo.gov/nara/cfr/waisidx_08/40cfr260_08.html (accessed April 20, 2009).
12. National Research Council. *Prudent Practices in the Laboratory: Handling and Disposal of Chemicals*, National Academy Press. Washington, DC, 1995, p. 208.

QUESTIONS

1. What federal agency oversees the disposal of hazardous waste?
   - (a) OSHA
   - (b) EPA
   - (c) CSB
   - (d) DOE

2. What regulation governs the disposal of hazardous waste?
   - (a) Resource Conservation and Recovery Act
   - (b) The Clean Water Act
   - (c) The Clean Air Act
   - (d) The Hazardous Waste Treatment and Disposal Act

3. The two most important steps in managing hazardous wastes are
   - (a) Identifying the exact components of the waste and minimizing the quantity of waste
   - (b) Identifying the exact components of the waste and finding someone else who can take care of it
   - (c) Making sure that all wastes are in a solid state and minimizing the quantity of waste
   - (d) Making sure that all wastes are in a liquid state and minimizing the quantity of waste

4. How many categories of waste generators are identified by RCRA?
   - (a) One
   - (b) Two
   - (c) Three
   - (d) Five

5. There are limits with regard to the time that hazardous waste can be stored on site for
   - (a) Conditionally exempt small quantity generators
   - (b) Small quantity generators
   - (c) Large generators
   - (d) Both (b) and (c)

6. What name is given to the container that holds hazardous waste?
   - (a) Hazardous waste drum
   - (b) Labpack
   - (c) Hazdrum
   - (d) WasteCon

7. What letters are used by the EPA to identify lists of hazardous characteristics (flammability, corrosivity, reactivity, toxicity) of wastes?
   - (a) F, K, P, and T
   - (b) F, C, R, and T
   - (c) F, K, P, and T
   - (d) F, K, P, and U

8. Biological wastes from chemistry laboratories
   - (a) Can be treated the same as chemical wastes
   - (b) Are treated differently than biological wastes from biology laboratories
   - (c) Are usually treated locally in an autoclave
   - (d) Can be disinfected with many different agents and then discarded down a sanitary sewer

8.3.5

CHEMICAL SECURITY

Preview This section discusses chemical security and measures for preventing theft of high-hazard chemicals and other laboratory materials that might be used for untoward purposes if procured by persons with malevolent purposes.

Distrust and caution are the parents of security.

Benjamin Franklin[1]

INCIDENT 8.3.5.1   RADIOACTIVE IODIDE POISONING[2]

A university safety office was carrying out routine thyroid screening of laboratory investigators when they discovered a female student had abnormally elevated radioiodine levels. The student had not recently used any of this material. Additional investigation showed that a food dish in her apartment had been contaminated with radioactive iodine. The investigation was turned over to law enforcement. Although her boyfriend, a graduate student, did not have authorization to use radioactive iodine, the police investigation found that he obtained radioactive iodine from a laboratory where he worked. He had added the radioactive iodine to his girlfriend's food to poison her. A female roommate also ate some of the poisoned food. Police charged the graduate student with poisoning, assault, and larceny. The state agency responsible for granting the university's license for using radioactive materials continued to monitor the licensee's actions in this matter.

What lessons can be learned from this incident?

Prelude

Academic laboratories are not typically very secure locations. The academic enterprise is, by nature, one that is largely "open" with regard to the exchange of information and this philosophical tone sets a context for laboratory facilities with varying degrees of access. As a matter of convenience, and sometimes necessity, labs may often be left unlocked so that students have access to them, particularly small research labs. Chemical stockrooms are likely to have less "open" access but again as a matter of convenience or necessity they are usually not highly secured areas since some student and faculty access is essential to allow research activities to proceed with few restrictions.

This "openness" of academic laboratories obviously presents tremendous opportunities for unauthorized individuals to gain access to labs to steal chemicals and equipment. Since the terrorist attacks on September 11, 2001 and more generally in the past decade with the increase in illicit drug synthesis (particularly methamphetamines), there has been a heightened awareness nationwide about access to chemicals. Some laws and regulations (described below) have emerged that address this situation. Nonetheless, many academic labs and stockrooms are still relatively unsecure with regard to theft.

This section will describe prudent means by which theft from labs can be prevented, or at least deterred.

What Might Be Stolen?

There are three main categories of chemicals that might be stolen from laboratories: poisons, chemicals for making drugs, and explosives.

Laboratory Safety for Chemistry Students, by Robert H. Hill, Jr. and David C. Finster
Copyright © 2010 John Wiley & Sons, Inc.

Intentional poisonings have been with us since the beginning of civilization. Socrates was forced to take "hemlock" by the government for corrupting the youth of his country. Cleopatra did herself in with the poisonous bite of a snake. The Borgias, a Spanish family that migrated to Italy, became famous for their knack at poisoning. In 1978 a Bulgarian, George Markov, was poisoned by "ricin" through the tip of a specially made umbrella with an injectable tip.[3] In 1995 a cult known as Aum Shinrikyo organized an attack in a Tokyo subway with a chemical warfare agent, sarin.[4] Ukranian leader Victor Yushchenko was poisoned with "dioxin" and exhibited its characteristic "chloracne" mask on his face.[5] While it is possible for people to obtain poisons from many places, some have learned that these materials are sometimes available in purest forms in our laboratories. As these examples illustrate, poisons can be used for a "personal" attack against an individual or to threaten a group or population.

The most commonly synthesized illicit drug today is methamphetamine. While many commonly available materials can supply the chemical precursors and solvents for the preparation of "meth," reagent grade chemicals and solvents are available in laboratories. The availability of chemicals from nonlab sources, however, has made theft of laboratory chemicals relatively rare.

Persons interested in using explosive chemicals usually want to obtain large quantities of chemicals. While academic labs are less likely to be targets for theft of explosive chemicals, there have been cases of chemicals being stolen from universities for mischief or illicit purposes, such as drug production.

The remainder of this section will discuss measures that can, and in some instances must, be taken to prevent the theft of chemicals and equipment from chemistry labs and stockrooms. These comments apply equally to academic and industrial situations. However, for reasons stated earlier, academic labs are far less secure than most industrial labs. Research and development labs and production facilities in the chemical industry can ordinarily and easily be made more secure by limiting access to the area to employees only.

Securing High Hazard Materials in the Laboratory

Laboratory safety and laboratory security are related but differ. Laboratory safety seeks to prevent injury to you or others from laboratory operations. Laboratory security seeks to prevent injury to you and others by restricting access to very hazardous chemical, biological, and radiological materials. Restricted access means that only certain people have access to certain chemicals. These chemicals are secured in order to prevent easy and immediate access by others not having a legitimate need to use them.

There are several elements of a laboratory security plan[6,7] including security of the following.

- **Physical facilities.** Securing the physical facilities is the responsibility of your institution and involves keeping unauthorized persons out of your facilities. In an academic setting, physical security is a challenge since it involves keeping people out of certain areas and as discussed earlier, this is nearly impossible in a practical sense due to the "open" nature of colleges and universities. In industry, it is possible to require that people who enter certain areas have proper identification and even criminal background checks as part of a preemployment review. Industrial situations might also use fences, locked doors, cardkeys, and checkpoints to pass through to enter the work area.

- **Personnel.** Security means being able to trust the people working around you. In academic labs, there is no means by which students can be "screened" with background checks as is the case in industry. As you can see by Incident 8.3.5.1, theft by someone within an organization is a chief consideration of a security program. In fact, it is far more likely that someone within an organization may try to steal very hazardous materials, than a group of terrorists breaking into a place from outside. Thus, there is a lot of emphasis on "internal" security. Even in academic labs, you should take the responsibility to report suspicious activities to authorities or campus police. In industry, visitors to areas using high hazard materials usually are escorted at all times.

- **Materials.** Securing high hazard materials requires taking steps to control and account for the high hazard materials. These high hazard materials are secured in a locked cabinet, drawer, refrigerator,

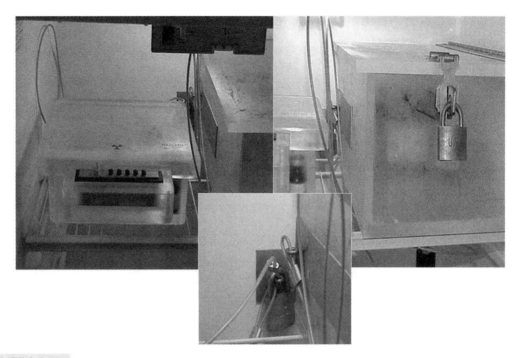

FIGURE 8.3.5.1 Lock Boxes in Refrigerator Secured with Steel Cables. These and other similar systems or containers may be used for securing and protecting especially hazardous materials from theft or misuse.

box, or room when not in use (see Figure 8.3.5.1). This also means that these materials need to be under your control and constant surveillance when being used in an open lab. This has been the practice for radioisotopes for many years. Accounting for these high hazard materials requires taking and maintaining inventories (see Section 8.3.3), including amounts of materials, and ensuring that these materials are accounted for during handling.

- **Information.** Information about your high hazard materials should be available only to those who have a need to know. Here is where there could be conflict with laboratory safety. Laboratory safety often requires that doors be posted to identify hazards within a laboratory, but laboratory security does not want others to know that these materials are in the laboratory. There must be a balance between these issues. People should know that there are hazardous materials in a laboratory, but they should not be given specific information about them. Usually, generic signs about hazards are sufficient for safety purposes without the need to publicly identify specific chemicals. Computer security becomes very important in this regard and inventories should only be available to those who have a need to know such as people working in the laboratory and emergency response personnel.

- **Transfer, shipping, and receiving high hazard materials.** The receipt of high hazard materials should be a formal process requiring signatures indicating transfer to a laboratory. These materials should not just be left on someone's desk, but rather someone should accept the shipment. Sharing chemicals is a common practice, but sharing high hazard materials should become a formal process with accountability on both sides. Sharing with colleagues in outside laboratories also needs to be a formal process. Some organizations adopt the "chain-of-custody" approach requiring everyone who handles these materials to sign and be accountable. The CDC/USDA Select Agent Program established the transport of these materials as a formal process requiring their approval. This has not been established for other high hazard materials, but this system should be emulated to ensure that misuse does not occur.

- **Reporting incident or stolen/lost high hazard materials.** Your institution is likely to have a system for reporting incidents or losses of high hazard materials. This should be a part of your responsibility when handling these materials.

Identifying High Hazard Materials

High hazard materials can fall into three broad classes: chemical, biological, and radiological.

The most dangerous hazardous chemicals would fall into the GHS Hazard Class 1 for toxic substances. (See Section 3.2.1 for an overview of the GHS and Section 6.2.1 for more information about the GHS toxicity ratings.) It is good practice to restrict access to these chemicals. This means that you know where your high hazard chemicals are, how they are being used, and that they are protected from theft and misuse.

The U.S. Department of Homeland Security regulates about 300 chemicals that could be used to make chemical agents or could be used in chemical attacks.[8,9] This program is also known as "Top Screen." While the focus of the regulations is on chemical facilities, the regulations also cover academic institutions, if quantities of these chemicals reach a certain threshold amount. The focus of the regulations is to prevent the release of these chemicals in the environment, theft of these chemicals, or sabotage and contamination by these chemicals.

The most dangerous biological agents are regulated by the U.S. government through the Centers for Disease Control and Prevention, U.S. Department of Agriculture, and U.S. Department of Justice. This means that you cannot possess, use, transfer, or accept these "Select Agents" without a permit to do so. The penalties for violating these regulations can be very severe, including criminal prosecution and jail time. If your laboratory is using these, you will likely know this. But be aware that if you casually encounter these agents, you are still subject to these laws. Security measures are required for handling these materials in the laboratory.

Radioactive materials are regulated by the Nuclear Regulatory Commission (NRC) or a state agency designated to act for the NRC. The use of radioisotopes requires that they be secured or under constant surveillance when in use. Failure to guard these materials could result in revocation of an institution's license to handle them. Thus, security measures are required for handling these materials in the laboratory.

Summary

Since academic labs are generally "open" and not very secure, it is easy as a student not to think about issues of security. You might reasonably expect that this should be more the responsibility of faculty and stockroom personnel. However, many faculty members are probably not very mindful of this since they have spent their lives in academic environments where security is not a matter of great concern. Furthermore, when working in a lab it is easy to get focused on "doing chemistry" and not thinking about security issues. We hope that this textbook will help you constantly to think about "safety issues" though! Hopefully, this section will raise your awareness of this situation and prepare you for future employment circumstances where security may be a matter of greater importance.

RAMP

- *Recognize* that unsecured high hazard chemicals can be stolen and used in illicit and inappropriate ways.
- *Assess* the risks of unsecured high hazard chemicals to determine how they might bring about harm.
- *Minimize* the risks of high hazard chemicals being diverted for inappropriate uses by having and following a security plan that tracks the use and storage location of these materials.
- *Plan* for emergency situations involving inappropriate use of high hazard chemicals including establishing a reporting system to appropriate officials.

References

1. BENJAMIN FRANKLIN. BrainyQuote; available at http://www.brainy quote.com/quotes/quotes/b/benjaminfr122731.html (accessed February 19, 2009).
2. U.S. Nuclear Regulatory Commission. Preliminary Notification of Event or Unusual Occurrence PNO-1-98-052. Intentional Ingestion of Iodine-125 Tainted Food (Brown University), November 16, 1998. U.S. Nuclear Regulatory Commission, Washington, DC, 1998.
3. GEORGE MARKOV. *The Umbrella Assassination*; available at http://www.portfolio.mvm.ed.ac.uk/studentwebs/session2/group12/georgie.htm (accessed February 18, 2009).
4. *Aum Shinrikyo: Once and Future Threat*. Available at http://www.cdc.gov/ncidod/eid/vol5no4/olson.htm (accessed August 20, 2009).
5. *Yushchenko and the Poison Theory*. BBC News; available at http://news.bbc.co.uk/1/hi/health/4041321.stm (accessed February 18, 2009).
6. J. Y. RICHMOND and S. L. NESBY-O'DELL. Laboratory security and emergency response guidance for laboratories working with Select Agents. *Morbidity and Mortality Weekly Report* **51**: RR-19 (December 6, 2002).
7. R. M. SALERNO and J. GAUDIOSO. *Laboratory Biosecurity Handbook*, CRC Press, Boca Raton, FL, 2007.
8. Chemical Facility Anti-Terrorism Standards. Available at http://www.dhs.gov/xprevprot/laws/gc_1166796969417.shtm (accessed February 18, 2009).
9. 6 CFR Part 27. Appendix to Chemical Facility Anti-Terrorism Standards; available at http://www.dhs.gov/xlibrary/assets/chemsec_appendixa-chemicalofinterestlist.pdf (accessed February 18, 2009).

QUESTIONS

1. Most academic laboratories are
   (a) Not very secure
   (b) Fairly secure
   (c) Reasonably secure
   (d) Very secure

2. What categories of compounds are mostly likely to be stolen from chemistry laboratories and stockrooms?
   (a) Poisons, chemicals for making drugs, and acids and bases
   (b) Poisons, chemicals for making drugs, and explosives
   (c) Strong oxidizing agents, poisons, and explosives
   (d) Strong oxidizing agents, chemicals for making drugs, and explosives

3. With regard to physical facilities, how does the academic environment compare with an industrial environment?
   (a) They are about the same
   (b) More security is possible, and likely, in the industrial environment
   (c) More security is possible, but unlikely, in the industrial environment
   (d) More security is possible, but unlikely, in the academic environment

4. What method can be used to maintain control over high hazard materials?
   (a) Lock boxes
   (b) Locked rooms
   (c) Locked drawers
   (d) All of the above

5. How are safety and security sometimes in conflict?
   (a) Laboratory safety requires specific signage about chemicals and hazards and easy access to information, while security requires less, or vague, signage and controlled access to information.
   (b) Laboratory safety requires specific signage about chemicals and hazards and controlled access to information, while security requires less, or vague, signage and easy access to information.
   (c) Laboratory safety requires vague signage about chemicals and hazards and easy access to information, while security requires specific signage and controlled access to information.
   (d) Laboratory safety requires vague signage about chemicals and hazards and controlled access to information, while security requires specific signage and easy access to information.

6. How many chemicals does the U.S. Department of Homeland Security regulate with regard to allowable quantities in labs (without notification)?
   (a) About 30
   (b) About 300
   (c) About 3000
   (d) About 30,000

INDEX

INDEX